Springer Proceedings in Mathematics

Volume 10

For further volumes:
http://www.springer.com/series/8806

Springer Proceedings in Mathematics

The book series will feature volumes of selected contributions from workshops and conferences in all areas of current research activity in mathematics. Besides an overall evaluation, at the hands of the publisher, of the interest, scientific quality, and timeliness of each proposal, every individual contribution will be refereed to standards comparable to those of leading mathematics journals. It is hoped that this series will thus propose to the research community well-edited and authoritative reports on newest developments in the most interesting and promising areas of mathematical research today.

N.S. Narasimha Sastry
Editor

Buildings, Finite Geometries and Groups

Proceedings of a Satellite Conference, International Congress of Mathematicians, Hyderabad, India, 2010

Springer

Editor
N.S. Narasimha Sastry
Indian Statistical Institute
8th Mile, Mysore Road
RVCE Post
Bangalore 560 059
India
nsastry@isibang.ac.in

ISSN 2190-5614 e-ISSN 2190-5622
ISBN 978-1-4614-0708-9 e-ISBN 978-1-4614-0709-6
DOI 10.1007/978-1-4614-0709-6
Springer New York Dordrecht Heidelberg London

Library of Congress Control Number: 2011937764

Printed on acid-free paper

Springer is part of Springer Science+Business Media (www.springer.com)

Preface

This is a collection of invited articles by some active researchers in the interface of the theory of finite groups and their representations, incidence geometries and the theory of buildings. The collection includes eight articles by some of the invited speakers at the Satellite Conference on "Buildings, Finite Geometries and Groups" held at the Indian Statistical Institute, Bangalore Centre, during August 29–31, 2010. The request to the authors of the chapters of this volume was to discuss the latest developments in the topics mentioned above, with emphasis on the basic questions of current interest. This was also the objective of the conference.

I would like to thank all the authors for kindly writing the articles and many colleagues who carefully refereed the chapters in time. I also wish to thank the speakers and the participants at the Satellite Conference.

This conference was one of several such conferences organized at various universities and research institutions in India as a part of the International Congress of Mathematicians held at Hyderabad, India, during August 19–27, 2010. The organizing committee for this Satellite Conference consisted of:

- N.S. Narasimha Sastry (Chairman) – Indian Statistical Institute, Bangalore, India
- Bhaskar Bagchi – Indian Statistical Institute, Bangalore, India
- Navin Singhi – Tata Institute of Fundamental Research, Mumbai, India
- Peter Sin – University of Florida, Gainesville, Florida, USA
- Ravindra Shukla – University of Allahabad, Allahabad, India

The conference was supported by the Department of Science and Technology (DST), India, Indian Statistical Institute (ISI) and the International Mathematical Union (IMU). I thank these organizations for their very generous support. I would also like to thank Springer for kindly publishing this volume. I also thank Dr. Hans J. Koelsch at Springer for his encouragement and helpful suggestions and Nathan Brothers, Elizabeth Loew, Jacob Gallay, Rajiv Monsurate and Harini at Springer for their friendly and efficient cooperation which is instrumental in bringing out this volume in time. Finally, I record my indebtedness to Ms. Asha Lata of the Statistics and Mathematics Unit, Indian Statistical Institute, Bangalore Centre, for her help through all stages of the preparation of this volume.

Bangalore, India N.S. Narasimha Sastry

Talks given in the Conference

F.G. Timmesfeld	Consequence of the theory of abstract root subgroups
Lizhen Ji	Applications of buildings and curve complexes in topology and geometry
Antonio Pasini	Classifying flag-transitive circular extensions of the $F_4(2)$-building
Köhl né Gramlich	On the geometry of global functions fields, the Riemann-Roch theorem, and finiteness properties of S-arithmetic Groups
Navin Singhi	Two problems in Discrete Mathematics
Bhama Srinivasan	Modular representations of finite general linear groups: old and new
Arjeh Cohen	The geometry of extremal elements in a Lie algebra
Arun Ram	Musings towards Elliptic buildings
Roderick Gow	Properties of a 27-dimensional space of symmetric bilinear forms acted on by E_6
T.N. Venkataramana	Generators for Arithmetic Groups
Peter Sin	Oppositeness in Buildings and simple modules for finite groups of Lie type
N.S.N. Sastry	Some equations over finite fields related to simple groups of Suzuki and Ree types
Bertrand Remy	Recent results in the theory of twin building lattices
Radha Kessar	On blocks of finite groups whose defect groups are elementary abelian of order 8
Kay Magaard	Representations of Unitriangular Groups
Murali K. Srinivasan	Symmetric chains, Gelfand-Tsetlin chains, and the Hypercube
Leo Storme	The use of blocking sets in Galois geometries
Bhaskar Bagchi	On characterizing designs by their codes

Participants

Bhaskar Bagchi	Indian Statistical Institute, Bangalore, India (bbagchi@isibang.ac.in)
Sonia Balagopalan	National University of Ireland Maynooth, Ireland (s.balagopalan@gmail.com)
Anirban Bose	Indian Statistical Institute, Delhi, India (anirban.math@gmail.com)
Arjeh M. Cohen	Technische Universiteit, Eindhoven, Netherlands (amc@win.tue.nl)
Shripad Garge	Indian Institute of Technology, Bombay, India (shripad@math.iitb.ac.in)
Roderick Gow	University College Dublin, Ireland (rod.gow@ucd.ie)
Köhl né Gramlich	University of Birmingham, UK (gramlich@mathematik.tu-darmstadt.de)
Gerhard Hiss	Lehrstuhl D fur Mathematik - RWTH Aachen University, Germany (gerhard.hiss@math.rwth-aachen.de)
Lizhen Ji	University of Michigan, USA (lji@umich.edu)
Kay Magaard	University of Birmingham, UK (k.magaard@bham.ac.uk)
Antonio Pasini	University of Siena, Italy (pasini@unisi.it)
S. Kannan	Chennai Mathematical Institute, Chennai, India (kannan@cmi.ac.in)
Radha Kessar	University of Aberdeen, UK (r.kessar@abdn.ac.uk)
Tanusree Khandai	Institute of Mathematical Sciences, Chennai, India (tanusree@imsc.res.in)
Anil Kumar C.P.	Institute of Mathematical Sciences, Chennai, India (anilkcp@imsc.res.in)
Andreas Mars	Technische Universitat Darmstadt, Germany (mars@mathematik.tu-darmstadt.de)
Santosha Kumar Pattanayak	Chennai Mathematical Institute, Chennai, India (santosh@cmi.ac.in)

Arun Ram	University of Melbourne, Australia (aram@unimelb.edu.au)
Bertrand Remy	Institut Camille Jordan, France (remy@math.univ-lyon1.fr)
Binod Kumar Sahoo	National Institute of Science Education and Research, Bhubaneswar, India (bksahoo@niser.ac.in)
Sharad Sane	University of Bombay, Mumbai, India (sharadsane@gmail.com)
N.S. Narasimha Sastry	Indian Statistical Institute, Bangalore, India (nsastry@isibang.ac.in)
Peter Sin	University of Florida, USA (sin@ufl.edu)
R.P. Shukla	University of Allahabad, Allahabad, India (rps@mri.ernet.in)
Anupam Kumar Singh	Indian Institute of Science Education and Research, Pune, India (anupamk18@gmail.com)
Navin Singhi	Tata Institute of Fundamental Research, Mumbai, India (singhi@math.tifr.res.in)
Bhama Srinivasan	University of Illinois at Chicago, USA (srinivas@uic.edu)
Murali K. Srinivasan	Indian Institute of Technology, Mumbai, India (bbagchi@isibang.ac.in)
J.B. Srivastava	Emeritus Fellow, Indian Institute of Technology, Delhi, India (jbsrivas@gmail.com)
Leo Storme	Ghent University, Belgium (ls@cage.ugent.be)
Gernot Stroth	Universtitat Halle, Germany (SGernot@aol.com)
Maneesh Thakur	Indian Statistical Institute, Delhi, India (maneesh@isid.ac.in)
F.G. Timmesfeld	der Justus-Liebig-Universitt Gieen, Germany (Franz.Timmesfeld@math.uni-giessen.de)
T.N. Venkataramana	Tata Institute of Fundamental Research, Mumbai, India (venky@math.tifr.res.in)
D.N. Verma	Tata Institute of Fundamental Research, Mumbai, India (dnverma@math.tifr.res.in)
Markus Wermer	Technische Universitat Darmstadt, Germany (wermer@mathematik.tu-darmstadt.de)
Stefan Witzel	Technische Universitat Darmstadt, Germany (switzel@mathematik.tu-darmstadt.de)
Manoj Kumar Yadav	Harish-Chandra Research Institute, Allahabad, India (myadav@mri.ernet.in)

Contents

On Characterizing Designs by Their Codes

Bhaskar Bagchi

Abstract We observe that the finite regular generalized polygons are characterized by their codes (over any field!): they are the only generalized polygons (with given parameters) which maximise the number of minimum weight words in the dual code. We conjecture that an analogous characterization (over $\mathbb{F}_p$, p the correct prime) holds at least for the point-line designs of a finite dimensional projective space over a prime field. In support of this conjecture, we present a weaker coding theoretic characterization of these design in terms of the notion of "large clubs" introduced here. Along the way, we also prove a combinatorial characterization of the point-line designs of all finite projective spaces: apart from projective planes, these are the only Steiner 2-designs having as many hyperplanes as points.

A similar characterization of the desarguesian projective plane among projective planes of a prime order is not expected, except perhaps rather vacuously. However, we conjecture that the prime order desarguesian planes are characterized by maximising the number of words of the second minimum weight in their dual codes. We state a conjecture, on small linear spaces of prime order, whose validity is shown to imply this conjecture for projective planes.

Subject Classifications: Primary: 51E, 05B Secondary: 94B

Keywords Projective geometry • Dual code • Hamming weight • Designs

---

B. Bagchi (✉)
Theoretical Statistics and Mathematics Division, Indian Statistical Institute,
Bangalore 560 059, Karnataka, India
e-mail: bbagchi@isibang.ac.in

N.S.N. Sastry (ed.), *Buildings, Finite Geometries and Groups*, Springer Proceedings in Mathematics 10, DOI 10.1007/978-1-4614-0709-6_1,

1 Introduction

Use of coding theory in designs has a long history by now. The monograph [1] is an excellent reference. Codes have been used to suggest constructions of designs (c.f. [8] for a nice example). They have also been used in non-existence results. A spectacular example is the proof of non-existence of projective planes of order ten (cf. [10]). Particularly in the presence of a moderately large group of automorphisms, codes can also be a powerful tool for characterization. For example, in [3] coding theory was used to prove that if a finite inversive plane of even order has a point-transitive automorphism group then it is a classical inversive plane. However, despite bright early promises coding theory has not yet been very successful in proving uniqueness results. The purpose of this paper is to suggest that, with a focus on the correct class of problems, this state of affairs can perhaps be improved upon substantially. Unfortunately, the results we present here are not as satisfactory as we had expected. But this is precisely what may attract younger researchers to the very attractive problems that we present here. We particularly wish to draw the attention of the reader to the "$3p - 3$ conjecture" in the last section.

Recall that an *incidence system* (finite throughout this article) is a triple (X, B, I) where X and B are finite sets (whose elements are called points and blocks, respectively) and $I \subseteq X \times B$ is a binary relation (called incidence) between points and blocks. The incidence system (X_0, B_0, I_0) is said to be a *subsystem* of the incidence system (X, B, I) if $X_0 \subseteq X$, $B_0 \subseteq B$ and $I_0 = I \cap (X_0 \times B_0)$. It is called an *induced subsystem* (or subsystem induced on X_0) if B_0 consists of all the blocks in B incident with at least one point in X_0.

It is customary to identify each block β of an incidence system with its shadow (i.e. the set of points incident with β). With this identification, the incidence relation becomes set inclusion $\in$. Although we shall ourselves occasionally fall in line with this custom (so that, for instance, we say that two blocks are disjoint if no point is incident with both), this has certain disadvantages: (a) sometimes blocks come with multiplicities, i.e. the "set" of blocks becomes a multi-set (and codes do not "see" this multiplicity), (b) the notion of subsystems becomes somewhat murky, (c) the obvious duality between points and blocks gets lost and (d) often, in a natural construction of an incidence system, the incidence relation is not set membership.

Given an incidence system $\mathcal{X} = (X, B, I)$ and a field $\mathbb{F}$, the $\mathbb{F}$-ary code $C_{\mathbb{F}}(\mathcal{X})$ of $\mathcal{X}$ is defined to be the linear subspace of $\mathbb{F}^X$ generated by the "indicator" functions χ_β, $\beta \in B$. Here $\chi_\beta(x) = \begin{cases} 1 & \text{if } xI\beta \\ 0 & \text{otherwise} \end{cases}$. The $\mathbb{F}$-linear space $\mathbb{F}^X$ is equipped with a natural "inner product" (non-degenerate symmetric bilinear form) $\langle \cdot, \cdot \rangle$ defined by $\langle w_1, w_2 \rangle = \sum_{x \in X} w_1(x)\, w_2(x)$, for $w_1, w_2 \in \mathbb{F}^X$. The dual $\mathbb{F}$-ary code $C_F^\perp(\mathcal{X})$ is defined to be the orthocomplement of its $\mathbb{F}$-ary code with respect to this inner product:

$$C_{\mathbb{F}}^\perp(\mathcal{X}) = \{w \in \mathbb{F}^X : \langle w, w' \rangle = 0 \;\forall\, w' \in C_{\mathbb{F}}(\mathcal{X})\}.$$

In particular, if $\mathbb{F} = \mathbb{F}_p$ is the finite field of prime order p, we write $C_p(\mathcal{X})$ and $C_p^{\perp}(\mathcal{X})$ for these two codes, and call them the *p-ary code* and the *dual p-ary code* of the incidence system $\mathcal{X}$. Clearly we have $\dim C_p(\mathcal{X}) + \dim C_p^{\perp}(\mathcal{X}) = \#(X)$.

More generally, a code is a linear subspace C of $\mathbb{F}^X$. We define its dual code $C^{\perp}$ exactly as above. For any "word" $w \in C$, its *support* is the point set $\{x \in X : w(x) \neq 0\}$. The Hamming weight of w is the size of its support. The minimum of the weights of the nonzero words of a code is usually called the *minimum weight* of the code.

The following (admittedly somewhat vague) principle is actually well known to experts, even though we are not aware of anybody ever writing it down explicitly.

1.1 The Fundamental Principle of Coding Theory of Designs

The size of the dual p-ary code (for appropriate choice of the prime p) of a design (or incidence system) is a measure of the combinatorial regularity of the design. The larger the dual code, the more regular is the design.

This paper is an attempt to quantize this principle in various special cases. Actually, the original conjecture of Hamada and Sachar (briefly discussed in Sect. 3 below) is perhaps the first such attempt.

2 Generalized Polygons

The *incidence graph* of an incidence system $\mathcal{X} = (X, B, I)$ is the graph with the disjoint union $X \sqcup B$ as vertex set, such that $x \in X$ is adjacent with $\beta \in B$ iff $xI\beta$; and there are no other adjacencies. This incidence system is said to be a generalized polygon if (its incidence graph is connected and) the girth of its incidence graph is double its diameter. If the diameter is n then we talk of a generalized n-gon. For any finite generalized n-gon there are parameters s and t such that each block is incident with $s + 1$ points and each point is incident with $t + 1$ blocks. One talks of an (s, t)-generalized n-gon. It is called *thin* if $s = 1$ and *thick* otherwise. The ordinary polygons (i.e. cyclic graphs with vertices thought of as points, edges thought of as blocks) are the most obvious examples of generalized polygons (hence the name); they are thin. All other examples are called non-trivial. In this section, by "generalized polygon" we shall mean the non-trivial ones.

By a famous theorem of Feit and Higman (cf. [7]), all thick finite generalized polygons have $n = 3, 4, 6$ or 8. The generalized 3-gons (triangles) are just the projective planes. The following result is well known (and easy to prove):

Lemma 2.1. *Let $\mathcal{X}$ be a finite thin (non-trivial) generalized n-gon, say with parameter $(1, t)$. Then $\mathcal{X}$ is the incidence graph of a (t, t) generalized $\frac{n}{2}$-gon. Thus $n = 4, 6, 8$ or 12.*

(In the case $n = 4$, a generalized 2-gon is a trivial system in which each point is incident with each block. This is usually excluded from consideration for instance, in the statement of Feit – Higman theorem quoted above since one implicitly assumes $n \geq 3$ in the definition of a generalized n-gon. Also, in the Lemma above, the case $n = 16$ is ruled out since there is no (t, t) generalized octagon.)

Definition 2.2. Let $\mathcal{X}$ and $\mathcal{Y}$ be generalized n-gons (with the same n). We say that $\mathcal{Y}$ is a sub-generalized polygon of $\mathcal{X}$ if $\mathcal{Y}$ is a subsystem of $\mathcal{X}$. The thick generalized n-gon $\mathcal{X}$ with parameters (s, t) is said to be **regular** if any two points x, y of $\mathcal{X}$ at distance n (in the incidence graph) occur together in some (necessarily unique) sub-generalized n-gon with parameters $(1, t)$. (This definition is due to N.S.N. Sastry).

The following result is actually a reformulation of an old theorem (Theorem 2.8 in [4]).

Theorem 2.3. *(Bagchi and Sastry): Let $\mathcal{X}$ be a thick (s, t) - generalized n-gon, with $n = 2k$ even. (Thus, $n = 4, 6$ or 8). Then, for any prime p, the minimum weight of $C_p^{\perp}(\mathcal{X})$ is at least $2(t^k - 1)/(t - 1)$ and the dual code contains at most $\frac{p-1}{2}(s+1)(t-1)s^k((st)^k - 1)/(st - 1)$ words of this weight. Further, equality holds in the last inequality (i.e. $\mathcal{X}$ maximises the number of minimum weight words in its dual code among all generalized polygons of the same parameter) for some prime p iff $\mathcal{X}$ is regular.*

Thus, if equality holds for some prime then it holds for all primes.

Proof. Usual counts show that the (s, t) generalized $2k$-gon $\mathcal{X}$ has exactly $(s + 1)((st)^k - 1)/(st - 1)$ points, and each point is at distance $n = 2k$ from exactly $s^k t^{k-1}$ points. Therefore, there are $M := (s + 1)((st)^k - 1)s^k t^{k-1}/(st - 1)$ ordered pairs of points at distance $2k$ in $\mathcal{X}$. Similarly, any $(1, t)$ sub-generalized $(2k)$-gon $\mathcal{Y}$ of $\mathcal{X}$ contains $N := 2t^{k-1}(t^k - 1)/(t - 1)$ ordered pairs of points at distance $2k$. Now, it was shown in [4] that the minimum weight of $C_p^{\perp}(\mathcal{X})$ is at least $2(t^k - 1)/(t - 1)$ and the support of a word of this weight must be the point sets of a $(1, t)$ sub-generalized $(2k)$-gon $\mathcal{Y}$ of $\mathcal{X}$. The argument there also shows that each pair of points at distance $2k$ in $\mathcal{X}$ are together in at most one such $\mathcal{Y}$. Therefore, there are at most $M/N = \frac{1}{2}(s+1)(t-1)s^k \cdot ((st)^k - 1)/(st - 1)$ choices for $\mathcal{Y}$ (and hence for the support S of a minimum weight word). Clearly, each such S supports exactly $(p - 1)$ words of $C_p^{\perp}(\mathcal{X})$ (scalar multiples of each other). Hence, the upper bound. Also, if S is the point set of a sub-generalized n-gon with parameter $(1, t)$, then by Lemma 2.1 we can write S as a disjoint union $S^+ \sqcup S^-$ where S^+ and S^- are the points and blocks of the associated (t, t) generalized k-gon. Then, any block of $\mathcal{X}$ is either non-incident with all points in S, or else it is incident with exactly one point from S^+ and exactly one point from S^-. Therefore, the word w_S defined by

$$w_S(x) = \begin{cases} 0 & \text{if} \quad x \notin S \\ 1 & \text{if} \quad x \in S^+ \\ -1 & \text{if} \quad x \in S^- \end{cases}$$

is in $C_p^{\perp}(\mathcal{X})$, and so are all its scalar multiples. Therefore, equality holds iff $\mathcal{X}$ is regular. □

Examples: The known regular generalized polygons are the following. (i) the "usual" (q, q) generalized quadrangle with automorphism group $Sp(4, q)$ (ii) the (q^2, q) generalized quadrangle with automorphism group $O(5, q)$ (iii) the "usual" (q, q) generalized hexagon with automorphism group $G_2(q)$, (iv) the (q^3, q) generalized hexagon with automorphism group ${}^3D_4(q)$ and (v) the (q^2, q) generalized octagon with automorphism group ${}^2F_4(q)$.

It is widely believed that these five series are actually characterized by regularity, and this is actually known in the case of the generalized quadrangles in (i) and (ii), cf. [13]. Thus, Theorem 2.3 is a potential (actual for $n = 4$) coding theoretic characterization of these five series (among generalized polygons of the same parameters). This may be old wine in a new bottle, but it is our paradigm in the search for new coding theoretic characterizations. Notice, however, one peculiarity of generalized polygons revealed here which can not be expected of the 2-designs we study next: Theorem 2.3 is characteristic free; its validity is independent of the choice of the prime p.

3 Dual Codes of 2-Designs

Recall that a $2-(v, k, \lambda)$ design (or 2-design with parameters v, k, λ) is an incidence system with v points such that each block is incident with k points (so k is called the block size) and any two distinct points are together incident with λ common blocks. One classical series of examples is the points versus t-flats incidence system $PG_t(n, \mathbb{F}_q)$ in the n-dimensional projective space $PG(n, \mathbb{F}_q)$ over the finite field $\mathbb{F}_q$ of order q. When $q = p^e$ with p prime (i.e. $p = \text{char}(\mathbb{F}_q)$) the p-ary code (and its dual) of $PG_t(n, \mathbb{F}_q)$ seems to carry a huge amount of information about these designs. In [9], Hamada computed the dimensions of these codes.

Hamada (and independently Sachar) conjectured that $PG_t(n, \mathbb{F}_q)$ maximises the dimension of the dual p-ary code among all 2-designs with the same parameters. This conjecture is still open. However, they also made the stronger conjecture that $PG_t(n, \mathbb{F}_q)$ is actually characterised as the unique maximiser of the number of code words in the dual p-ary code. This stronger conjecture was refuted by Tonchev (cf. [14]) when he observed that there are exactly five distinct quasi-symmetric 2-(31, 7, 3) designs, including $PG_2(4, \mathbb{F}_2)$, and all of them have the same dimension 15 for their dual binary code. However, the strong conjecture may still be valid in some special cases, for instance for the points-versus-hyperplanes designs $PG_{n-1}(n, \mathbb{F}_q)$. In any case, this stronger conjecture is our model in the following discussions.

In [2], we proved a very general (and sometimes tight) lower bound on the minimum weight of the dual p-ary code of arbitrary incidence systems. This is our starting point. To state it, we need:

Definition 3.1. Let $\mathcal{X}_i = (P_i, B_i, I_i)$, $i = 1, 2$, be two incidence systems and let λ be a positive integer. Then the λ-join of $\mathcal{X}_1$ and $\mathcal{X}_2$ is the incidence system $\mathcal{X}$ whose point set is the disjoint union $P_1 \sqcup P_2$, and whose blocks are the blocks of $\mathcal{X}_1$ and $\mathcal{X}_2$ together with λ new blocks, say $(x_1, x_2)_i$, $1 \leq i \leq \lambda$, each incident with x_1 and x_2 and with no other points, for each $(x_1, x_2) \in P_1 \times P_2$.

With this definition, the following is Theorem 2 from [2]:

Theorem 3.2. *(Bagchi and Inamdar): Let n, λ be positive integers and let p be a prime. Let $\mathcal{X}$ be any finite incidence system in which any two distinct points are together incident with at most λ blocks and each point is incident with at least $n + \lambda$ blocks. Then the minimum weight of the dual p-ary code $C_p^{\perp}(\mathcal{X})$ is at least $2(\frac{n}{\lambda} + 1 - \frac{n}{p\lambda})$. Further, a point set S is the support of a word of $C_p^{\perp}(\mathcal{X})$ of this minimum weight if and only if the induced subsystem on S is the λ-join of two $2 - (\frac{n}{\lambda} + 1 - \frac{n}{\lambda p}, p, \lambda)$ designs.*

Notice that if $\mathcal{X}$ is any incidence system, say with point set P, then we may define $r(x)$ to be the number of blocks of $\mathcal{X}$ incident with x and $\lambda(x, y)$ to be the number of blocks of $\mathcal{X}$ incident with both x and y, for $x \neq y$ in P. Then, let us put $\lambda = \max\limits_{x \neq y} \lambda(x, y)$, $r = \min\limits_{x} r(x)$ and $n = r - \lambda$. Then Theorem 2 may be applied to $\mathcal{X}$ with this choice of n and λ, and any prime p. However, if we are looking for instances of equality in this theorem, then the parameters $(\frac{n}{\lambda} + 1 - \frac{n}{\lambda p}, p, \lambda)$ should satisfy the usual divisibility condition for a 2-design, in particular we must choose p to be a prime such that $p \mid n$.

Having said this, we are interested in investigating the cases of equality in Theorem 3.2 for 2-designs $\mathcal{X}$. Recall that if $\mathcal{X}$ is a $2 - (v, k, \lambda)$ design then each point of $\mathcal{X}$ is incident with r blocks, where the *replication number* r of $\mathcal{X}$ is given by $r(k-1) = \lambda(v-1)$. The number $n = r - \lambda$ is called the *order* of the 2-design, and it is known to play a key role in the coding theory of square designs. We introduce:

Definition 3.3. A **subdesign** of a 2-design is a subsystem which is itself a 2-design. If D_1, D_2 are two subdesigns of a 2-design D then we say that D_1 and D_2 are **totally disjoint** if no point of D_1 is incident with any block of D_2 and no point of D_2 is incident with any block of D_1. (In particular, this implies that D_1 and D_2 have disjoint point sets and disjoint block sets. Notice that this definition would be very awkward to formulate if we identified blocks with subsets of the point set!).

In terms of this definition, the specialization of Theorem 3.2 to designs has the following interesting formulation. Unfortunately, the parameter λ of a $2 - (v, k, \lambda)$ design has no standard name. In the following, we use the name *balance* for this parameter.

Theorem 3.4. *Let D be a 2-design of order n and balance λ. Then, for any prime p, the minimum weight of $C_p^{\perp}(D)$ is at least $2(\frac{n}{\lambda} + 1 - \frac{n}{\lambda p})$. Further, a set S of points of D is the support of a word of weight $2(\frac{n}{\lambda} + 1 - \frac{n}{\lambda p})$ in $C_p^{\perp}(D)$ if and only if S is the (disjoint) union of the point sets of two totally disjoint $2 - (\frac{n}{\lambda} + 1 - \frac{n}{p\lambda}, p, \lambda)$ subdesigns of D.*

Proof. The lower bound on minimum weight is from Theorem 3.2. Also, if S is the support of a word of $C_p^{\perp}(D)$ attaining this bound, then by Theorem 3.2, S must be as stated. Conversely, let $S = S_1 \sqcup S_2$ where S_1, S_2 are the point sets of totally disjoint subdesigns D_1, D_2 with the parameters given. Since D_i has the same balance λ as the design D, any block of D incident with two or more points of D_i must be a block of D_i (and hence incident with exactly p points of S_i and no point of S_j, $j \neq i$). Therefore, for any point $x \in S_1$, there are exactly $\lambda(\frac{n}{\lambda} + 1 - \frac{n}{p\lambda}) = n + \lambda - \frac{n}{p}$ blocks incident with x each of which is incident with a unique point in S_2. Since D_i has replication number $\frac{n}{p}$ and D has replication number $n + \lambda$, this shows that the induced subsystem of D on the point-set S is precisely the λ-join of D_1 and D_2. Therefore, by Theorem 3.2, S is the support of a word of $C_p^{\perp}(D)$. □

This has the following interesting consequence.

Corollary 3.5. *With notations as in Theorem 3.4, let $\mathcal{D}$ be the collection of all 2–$(\frac{n}{\lambda} + 1 - \frac{n}{p\lambda}, p, \lambda)$ sub-designs of D. For $D_1, D_2 \in \mathcal{D}$, write $D_1 \sim D_2$ if either $D_1 = D_2$ or D_1 and D_2 are totally disjoint. Then$\sim$is an equivalence relation on $\mathcal{D}$.*

Proof. Only transitively needs proof. So let D_1, D_2, D_3 be distinct members of $\mathcal{D}$ with $D_1 \sim D_2 \sim D_3$. Let P_i be the point-set of D_i, $1 \leq i \leq 3$, and let P be the point-set of D. For $1 \leq i \leq 3$, let $w_i : P \to \mathbb{F}_p$ be the indicator function of P_i. Then by Theorem 3.4, $w_1 - w_2$ and $w_2 - w_3$ are words in $C_p^{\perp}(D)$. Therefore, their sum $w_1 - w_3$ is also in $C_p^{\perp}(D)$. Hence, by Theorem 3.4, $D_1 \sim D_3$. □

Definition 3.6. A **club** in a 2-design D of order n and balance λ is a maximal collection of pairwise totally disjoint $2 - (\frac{n}{\lambda} + 1 - \frac{n}{\lambda p}, p, \lambda)$ subdesigns of D. In other words, it is an equivalence class of the relation $\sim$ in Corollary 3.5. (More precisely, this ought to be called a p-club, but the prime p will be clear from the context.)

Notice that if S is the support of a minimum weight word in a p-ary code C then there are exactly $p - 1$ words of C with support S; they are scalar multiples of each other. This is because if $w_1 \neq w_2$ are any two words with support S, then we can always choose a $\lambda \in \mathbb{F}_p^*$ such that the support of $w_2 - \lambda w_1 \in C$ is properly contained in S and hence $w_2 - \lambda w_1 = 0$. In view of this comment, the following is an obvious consequence of what we have seen so far:

Proposition 3.7. *If the 2-design D has t clubs, say of size $c_1, c_2, \ldots, c_t$, then the number of minimum weight words in $C_p^{\perp}(D)$ is $(p-1)\sum_{i=1}^{t} \binom{c_i}{2}$.*

4 Codes of Finite Projective Spaces

We specialize to the point-line designs $PG_1(n, \mathbb{F}_q)$. Notice that these are Steiner designs, i.e. they have balance $\lambda = 1$. We follow an usual convention and call the blocks of Steiner 2-designs "*lines*". Although we shall soon specialize further to the case $q = p$, we begin with a purely combinatorial characterization of $PG_1(n, \mathbb{F}_q)$, q a power of the prime p. From now on we identify lines of Steiner 2-designs with their shadows.

Definition 4.1. Let D be a Steiner 2-design. A **flat** F in D is a set of points such that F contains the line joining each pair of distinct points in F. It is a proper flat if F is a proper subset of the point set of D. A **hyperplane** in D is a proper flat which meets every line of D.

Lemma 4.2. *The size of any proper flat F in a Steiner 2-design D is at most r, the replication number of D. Equality holds here iff F is a hyperplane of D.*

Proof. Fix a point x outside F. For any point y in F, let xy be the unique line joining x to y. Then $y \mapsto xy$ is a one-one function from F into the set of r lines through x. Hence, $\#(F) \leq r$. Clearly, equality holds in this argument iff F is a hyperplane. □

A 2-design is called trivial if each point is incident with each block; it is non-trivial otherwise. The parameters of any non-trivial 2-design satisfy Fisher's inequality $r \geq k$, equivalently $b \geq v$. Any 2-design satisfying equality here is called a *square* (or symmetric) design. A $2-(v, k, \lambda)$ design D is a square design iff its parameters satisfy $k(k-1) = \lambda(v-1)$. This holds iff the dual incidence system D^* is also a 2-design (necessarily with the same parameters as D) iff any two distinct blocks of D are together incident with exactly λ points. We also recall:

Definition 4.3. A **line** in a $2-(v, k, \lambda)$ design is a set of at least two points which is the intersection of λ distinct blocks. Clearly, any two distinct points are together in a unique line. However, the lines of a 2-design need not form the lines of a Steiner 2-design since in general they may have variable sizes.

Notice that the point-hyperplane design $PG_{n-1}(n, \mathbb{F}_q)$ is self dual; the lines of this square design are precisely the lines of $PG(n, \mathbb{F}_q)$. With this background, we have the following characterization of $PG_{n-1}(n, \mathbb{F}_q)$.

Theorem 4.4. *(Dembowski and Wagner, cf. [11]): Let D be a square 2-design in which every line meets every block. Then either D is a projective plane (i.e. a square Steiner 2-design) or $D = PG_{n-1}(n, \mathbb{F}_q)$ for some $n \geq 3$ and prime power q.*

Now we state and prove a similar characterization of the point-line designs $PG_1(n, \mathbb{F}_q)$. Its proof is crucially dependent on the Dembowski–Wagner theorem.

Theorem 4.5. *Let D be a Steiner 2-design on v points. Then D has at most v hyperplanes. Equality holds here iff either D is a projective plane or $D = PG_1(n, \mathbb{F}_q)$ for some $n \geq 3$ and some prime power q.*

Proof. Let r and k be the replication number and block size of D. Suppose D has v distinct hyperplanes; we must show that it has no more. Let $\mathcal{H}$ be a collection of v hyperplanes of D and let E be the incidence system with the point set of D and with $\mathcal{H}$ as its set of blocks, incidence being set-membership. By Lemma 4.2, E has constant block size r. Also, any $H \in \mathcal{H}$ carries a $2-(r,k,1)$ subdesign D_H of D, with replication number $= \frac{r-1}{k-1}$. If $H' \neq H$ is another member of $\mathcal{H}$, then $H' \cap H$ is a flat of D_H, and hence by Lemma 4.2, $\#(H' \cap H) \leq \frac{r-1}{k-1}$. Thus, each block of E has size r and any two distinct blocks of E have at most $\frac{r-1}{k-1}$ points in common. Therefore, letting e_i denote the number of points which are in exactly i blocks of E, an obvious two-way counting yields:

$$\sum_{i\geq 0} e_i = v,$$

$$\sum_{i\geq 0} i\, e_i = rv,$$

$$\sum_{i\geq 0} i(i-1)e_i \leq v(v-1)\cdot\frac{r-1}{k-1} = r(r-1)v.$$

Therefore, we get:

$$0 \leq \sum_{i\geq 0}(i-r)^2 e_i = \sum_{i\geq 0} i(i-1)e_i - (2r-1)\sum_{i\geq 0} i\, e_i + r^2\sum_{i\geq 0} e_i$$

$$\leq r(r-1)v - (2r-1)rv + r^2 v = 0.$$

Therefore, $\sum(i-r)^2 e_i = 0$. Thus, $e_i = 0$ for $i \neq r$. This means that each point of E is in exactly r blocks of E. Also, the inequalities in the above argument must actually be equalities. Thus, any two distinct blocks of E have exactly $\frac{r-1}{k-1}$ points in common. This means that the dual E^* of E is a $2-(v, r, \frac{r-1}{k-1})$ design. Since this design has $b = v$, it is square. Therefore, E itself is a square design. Now notice that any block of a design is uniquely determined by its remaining blocks (the incidence system consisting of the remaining blocks has two replication numbers $r-1$ and r, and the missing block must consist of the points of replication $r-1$). However, if there is a hyperplane H of D outside $\mathcal{H}$, then replacing any particular block of E by the new block H one obtains another 2-design (by the above argument), contrary to the observation just made.

Thus, if D has at least v hyperplanes then it has exactly v hyperplanes (proving the inequality) and in that case the hyperplanes of D form the blocks of a square design E. From the definition of hyperplane one sees that every line of E contains a (unique) line of D and every line of D meets every block of E. Therefore, every line of E meets every block of E. Therefore, by Theorem 4.4, E is either a projective plane (in which case $D = E$ is a projective plane) or else $E = PG_{n-1}(n, \mathbb{F}_q)$. In the latter case, comparing parameters of D and E, one sees that the lines of D are precisely the lines of $PG_{n-1}(n, \mathbb{F}_q)$, so that $D = PG_1(n, \mathbb{F}_q)$. □

Now we specialize further to the case $q = p$. Let D be a $2-(\frac{p^{n+1}-1}{p-1}, p+1, 1)$ design, p prime. By Definition 3.6, a club in D is a maximal collection of pairwise totally disjoint $2-(p^{n-1},p,1)$ subdesigns of D. Therefore, letting $v=\frac{p^{n+1}-1}{p-1}$. $v_0=p^{n-1}$, we see that any club in D has at most $\lfloor \frac{v}{v_0} \rfloor = p+1$ members. This leads to:

Definition 4.6. A **large club** in a $2-(\frac{p^{n+1}-1}{p-1},p+1,1)$ design is a club with $p+1$ members.

Then we have:

Theorem 4.7. *For a prime p, any $2-(\frac{p^{n+1}-1}{p-1},p+1,1)$ design D has at most as many large clubs as there are lines of D. If $n\geq 3$, equality holds here iff $D=PG_1(n,\mathbb{F}_p)$.*

Proof. Let $D_0,D_1,\ldots,D_p$ be a large club, with corresponding point sets $P_0,P_1,\ldots,P_p$. Let P be the point set of D and put $Q=P\backslash(P_0\sqcup\ldots\sqcup P_p)$. By Theorem 3.4, for $i\neq j$, the subsystem induced by D on $P_i\sqcup P_j$ is the join (i.e. 1-join) of D_i and D_j. It follows that for $0\leq i\leq p$, $H_i=Q\sqcup P_i$ is a hyperplane of D and Q is a flat of size $\frac{p^{n-1}-1}{p-1}$. We call a flat of this size a *coline* of D. Thus, with any large club in D we have associated a coline Q which is contained in $p+1$ hyperplanes.

Now notice that if Q is a coline and $x\notin Q$ a point, then there is at most one hyperplane $H\supseteq Q\cup\{x\}$. H must be the union of the lines joining x to the points of Q. Thus, the hyperplanes through Q are pairwise disjoint outside Q. Hence, any coline is in at most $p+1$ hyperplanes, and equality holds when it is associated with a large club. It follows that any coline is associated with at most one large club. The members D_i of a large club associated with the coline Q must be the subdesigns of D carried by $H_i\backslash Q$, where H_i are the hyperplanes through Q. Thus, the map from large clubs to colines is injective.

Now suppose D has (at least) b large clubs, where b is the number of lines of D. Thus, D has b colines each of which is the intersection of $\binom{p+1}{2}$ pairs of a hyperplanes. Thus, there are (at least) $\binom{p+1}{2}b$ unordered pairs of hyperplanes in D. Letting N denotes the total number of hyperplanes in D, we deduce that $\binom{N}{2}\geq\binom{p+1}{2}b=\binom{v}{2}$, where v is the number of points of D. Therefore, $N\geq v$. Hence, by Theorem 4.2, $N=v$ and (if $n\geq 3$) $D=PG_1(n,\mathbb{F}_p)$. □

Notice that the proof of Theorem 4.7 (applied to $D=PG_1(n,\mathbb{F}_p)$) shows that $C_p^{\perp}(PG_1(n,\mathbb{F}_p))$ has minimum weight $2p^{n-1}$ and its words of minimum weight are precisely the non-zero scalar multiples of the words w_1-w_2 where w_1,w_2 are indicator functions of distinct hyperplanes of $PG(n,\mathbb{F}_p)$. This is a special case of Proposition 2 in [2].

Theorem 4.7 is actually a characterization of $PG_1(n,\mathbb{F}_p)$, $n\geq 3$, by its dual p-ary code, even if this is somewhat obscured by the technical terms. But we would like to prove:

Conjecture 4.8. Let D be a $2-(\frac{p^{n+1}-1}{p-1}, p+1, 1)$ design, $n \geq 3$, p prime. Then $C_p^{\perp}(D)$ has at most $\frac{p}{2}(p^n-1)(p^{n+1}-1)/(p-1)$ words of (minimum) weight $2p^{n-1}$. Equality holds here iff $D = PG_1(n, \mathbb{F}_p)$.

In view of Theorem 4.7, it suffices to show that if D has the right number of totally disjoint pairs of $2-(p^{n-1}, p, 1)$ subdesigns, then there must be the right number of large clubs.

5 Projective Planes of Prime Order

We recall that a projective plane is just a square Steiner 2-design. Thus, a projective plane of order n is nothing but a $2-(n^2+n+1, n+1, 1)$ design. It is a folklore conjecture that, for each prime p, there is a unique projective plane of order p, namely the desarguesian plane $PG_1(2, \mathbb{F}_p)$. Indeed, since the dual p-ary code of any projective plane of order p is easily seen to have dimension $\binom{p+1}{2}$, this conjecture may be seen as a special case of the strong Hamada–Sachar conjecture.

Thus, it is specially interesting to obtain coding theoretic characterizations of $PG_1(2, \mathbb{F}_p)$. Such a characterization will provide a powerful tool to attack the uniqueness conjecture.

Theorem 3.4 from Sect. 3 specializes to the following result, first proved by Inamdar.

Corollary 5.1. *Let π be a projective plane of prime order p. Then the minimum weight of $C_p^{\perp}(\pi)$ is $= 2p$ and, up to multiplication by nonzero scalars, the words of minimum weight in $C_p^{\perp}(\pi)$ are $w_1 - w_2$, where w_1, w_2 are indicator functions of distinct lines.*

Since this holds for all putative projective planes of order p, the minimum weight words of the dual code fail to distinguish them. We suspect that looking at the second minimum weight will suffice. But what is the second minimum weight for the dual p-ary code even for the desarguesian plane $PG_1(2, \mathbb{F}_p)$? Nobody knows! In [6], it was shown that this weight is at least $\frac{5}{2}(p+1)$ for $p \geq 11$. The following result shows that this weight is at most $3p-3$ for every prime $p \geq 5$. Note that when $p = 11$, the upper and lower bounds coincide, so that the second minimum weight is $3p-3$ in this case. Indeed, calculations show that it is $3p-3$ for $5 \leq p \leq 11$.

Theorem 5.2. *For any prime $p \geq 5$, the dual p-ary code of $PG_1(2, \mathbb{F}_p)$ contains at least $\frac{1}{6}\, p^3(p^3-1)(p^2-1)$ words of weight $3p-3$.*

Proof. Take a point x, three distinct lines ℓ_1, ℓ_2, ℓ_3 through x and a line ℓ not passing through x. Put $S = (\ell_1 \cup \ell_2 \cup \ell_3) \backslash (\ell \cup \{x\})$. Thus, S is a set of size $3p-3$. It is enough to construct one word of $C_p^{\perp}$ with support S. To this end, we set up co-ordinates such that ℓ is the line at infinity, x is the origin, and ℓ_1, ℓ_2, ℓ_3 are the lines with equation $Y = 0, X = 0$ and $Y = X$. Notice that, in this co-ordinate system, S consists of $3p - 3$ affine points. Define the word w by setting $w(P) = 0$ for any point P of the projective plane, $P \notin S$, and define it on S by the formula

$$w(\alpha, \beta) = \begin{cases} 1/\alpha & \text{if } \beta = 0, \alpha \neq 0 \\ 1/\beta & \text{if } \alpha = 0, \beta \neq 0 \\ -1/\alpha & \text{if } \alpha = \beta \neq 0. \end{cases}$$

Then it is easy to see that w is a word of $C_p^\perp$, with support S. Therefore, S supports (at least, but one can show exactly) $p - 1$ words of $C_p^\perp$, namely the non-zero scalar multiples of w. Since there are $(p^2 + p + 1) \times p^2 \times \binom{p+1}{3}$ choices for S, we are done. □

Notice that in Theorem 5.2, p could be any prime power. But, when p is not prime, the words obtained are defined only over $\mathbb{F}_p$ and not over its prime subfield. In view of the comments preceding Theorem 5.2, we suspect that when p is prime, the second minimum weight of $C_p^\perp(PG_1(2, \mathbb{F}_p))$ is $3p - 3$ and the words described in that theorem are its only words of weight $3p - 3$. But we go ahead and state a much stronger conjecture (a priori stronger, that is!).

We recall that a *linear space* is an incidence system with exactly one line joining any two distinct points, and each line containing at least two points. We shall say that a linear space is of *order p* if each of its points is in exactly $p + 1$ lines. Clearly, any linear space of order p has at most $p^2 + p + 1$ points, with equality only for projective planes of order p. Thus, there are only finitely many linear spaces of any given order, and the projective planes are the largest among them.

Definition 5.3. An incidence system $\mathcal{X}$ is said to be non-trivial at a prime p if $C_p^\perp(\mathcal{X}) \neq \{0\}$.

Examples of small linear spaces of order p which are non-trivial at the prime p:

(a) Let $\mathcal{X}_p$ be the linear space with two disjoint "long lines" ℓ_1, ℓ_2 of size p each and p^2 short lines of size two each joining a point of ℓ_1 to a point of ℓ_2. The word sending each point of ℓ_1 to +1 and each point of ℓ_2 to $-1$is in $C_p^\perp(\mathcal{X}_p)$. Thus, $\mathcal{X}_p$ is non-trivial at p.

(b) For $p \geq 3$, let $\mathcal{Y}_p$ be the linear space on $3p - 3$ points described as follows. Its point-set is $\mathbb{F}_p^* \times \{1, 2, 3\}$. It has three long lines $\ell_i = \mathbb{F}_p^* \times \{i\}$, $i = 1, 2, 3$, of size $p - 1$ each. It has $(p - 1)(p - 2)$ lines of size 3, namely $\{(\alpha_1, 1), (\alpha_2, 2), (\alpha_3, 3)\}$ where $\alpha_1, \alpha_2, \alpha_3 \in \mathbb{F}_p^*$ with $\alpha_1 + \alpha_2 + \alpha_3 = 0$. It also has $3p - 3$ lines of size 2, namely $\{(\alpha_1, i), (\alpha_2, j)\}$ where $1 \leq i \neq j \leq 3$, $\alpha_1, \alpha_2 \in \mathbb{F}_p^*$ with $\alpha_1 + \alpha_2 = 0$. It is easy to verify that $\mathcal{Y}_p$ is a linear space of order p. The word $w : (\alpha, i) \mapsto \alpha$ is clearly a non-trivial word in $C_p^\perp(\mathcal{Y}_p)$. Thus, $\mathcal{Y}_p$ is non-trivial at p.

Inamdar deduced Corollary 5.1 from the following result by looking at the subsystem of π induced on the support of a minimum weight word in $C_p^{\perp}(\pi)$:

Proposition 5.4. *For primes $p, \mathcal{X}_p$ is the unique smallest linear space of order p which is non-trivial at p. That is, it has the smallest number $(2p)$ of points among all such spaces.*

Proof. This is the special case $n = p, \lambda = 1$ of Theorem 3.2. □

The following conjecture was first formulated by the author in [5]. Here, we have tried to motivate it by putting it in its wider context.

Conjecture 5.5. (The $3p-3$ conjecture): For primes $p \geq 5, \mathcal{Y}_p$ is the unique second smallest linear space of order p which is non-trivial at p.

That is, we conjecture that if Z is a linear space of prime order $p \geq 5$ which is non-trivial at p, then either $Z = \mathcal{X}_p$ or $Z = \mathcal{Y}_p$ or Z has at least $3p - 2$ points.
Clearly, $\mathcal{X}_p$ is obtained as the induced subsystem of $PG_1(2, \mathbb{F}_p)$ on the support of a minimum weight word of its dual code, and $\mathcal{Y}_p$ is obtained as the induced subsystem on one of the words described in Theorem 5.2. This is why they are non-trivial at p. Indeed, in our final result, we shall show that Conjecture 5.5 implies the following coding theoretic characterization of $PG_1(2, \mathbb{F}_p)$ among projective planes of order p:

Conjecture 5.6. Let π be a projective plane of prime order $p \geq 5$. Then the second minimum weight of $C_p^{\perp}(\pi)$ is $\geq 3p - 3$ and there are at most $\frac{1}{6}\, p^3(p^3 - 1)$ $(p^2 - 1)$ words of weight $3p - 3$ in this code. If, further, both π and its dual π^* satisfy equality in the last inequality then $\pi = PG_1(2, \mathbb{F}_p)$.

Theorem 5.7. *Conjecture 5.5 implies Conjecture 5.6.*

Proof. The first statement follows by looking at the induced subsystem of π on the support of words of weight $\leq 3p - 3$. In view of the uniqueness statement in Conjecture 5.5, it also follows that the support S of any word in $C_p^{\perp}(\pi)$ of weight $3p - 3$ is as described in the proof of Theorem 5.2. Hence, the upper bound on the number of weight $3p - 3$ words follows since the bound is just $(p - 1)$ times the number of such sets S in π. If equality holds for π, then the induced subsystem on each such set S is isomorphic to $\mathcal{Y}_p$. This has the following translation. If $S = (\ell_1 \cup \ell_2 \cup \ell_3) \backslash (\ell \cup x)$, then letting σ denote the affine plane obtained from π^* by deleting its "line" x, the Bruck subnet of σ corresponding to the three "points" ℓ_1, ℓ_2, ℓ_3 at infinity is co-ordinatized by the cyclic group of order p. Indeed, any word of $C_p^{\perp}(\pi)$ with support S provides such a co-ordinatization. As this holds for any three concurrent lines ℓ_1, ℓ_2, ℓ_3 of π, and since π^* is also assumed to satisfy the same, standard arguments imply that π must be desarguesian (cf. Proof of Theorem 5.1 in [11]). □

References

1. E.F. Assmus and J.D. Key, *Designs and their codes*, Cambridge Tracts in Math. **103**, Cambridge Univ. Press, Cambridge, 1992
2. B. Bagchi and S.P. Inamdar, *Projective geometric codes*, J. Combin. Theory Series A **99** (2002), 128–142
3. B. Bagchi and N.S.N. Sastry, *Even order inversive planes, generalized quadrangles and codes*, Geom. Dedicata **22** (1987), 137–147
4. B. Bagchi and N.S.N. Sastry, *Codes associated with generalized polygons*, Geom. Dedicata **27** (1988), 1–8
5. B. Bagchi, *A conjecture on linear spaces of prime order, J. Combin. Inform. & System Sciences*, **34** (2009), 23–31
6. V. Fack, S.L. Fancsali, L. Storme, G. Van de Voorde and J. Winne, *Small weight codewords in the codes arising from Desarguesian projective planes*, Designs, Codes and Cryptog. **46** (2008), 25–44
7. W. Feit and G. Higman, *The non-existence of certain generalized polygons*, J. Algebra **1** (1964), 114–131
8. M. Hall, jr., Combinatorial Theory, Second edition, Wiley publishers, New York, 1986
9. N. Hamada, *The rank of the incidence matrix of points and d-flats in finite geometries*, J. Sci. Hiroshima Univ., Ser. A1, **32** (1968), 381–396
10. C.W.H. Lam, L.H. Thiel and S. Swiercz, *The non-existence of finite projective planes of order 10*, Canad. J. Math. **41** (1989), 1117–1123
11. E.S. Lander, Symmetric designs: an algebraic approach, Cambridge Univ. Press, Cambridge, 1983
12. G.E. Moorhouse, *Bruck nets, codes and characters of loops*, Designs, Codes and Cryptog. **1** (1991), 7–29
13. S.E. Payne and J.A. Thas, Finite generalized quadrangles, Pitman advanced publishing program, London, 1984
14. V.D. Tonchev, *Quasi-symmetric 2-(31,7,7) designs and a revision of Hamada's conjecture*, J. Combin. Theory Ser. A, **42** (1986), 104–110

The Geometry of Extremal Elements in a Lie Algebra

Arjeh M. Cohen

Abstract Let L be a simple finite-dimensional Lie algebra over an algebraically closed field of characteristic distinct from 2 and from 3. Then L contains an extremal element, that is, an element x such that $[x,[x,L]]$ is contained in the linear span of x in L. Suppose that L contains no sandwich, that is, no element x such that $[x,[x,L]] = 0$. Then, up to very few exceptions in characteristic 5, the Lie algebra L is generated by extremal elements and we can construct a building of irreducible and spherical type on the set of extremal elements of L. Therefore, by Tits' classification of such buildings, L is determined by a known shadow space of a building. This gives a geometric alternative to the classical classification of finite-dimensional simple Lie algebras over the complex numbers and of classical finite-dimensional simple modular Lie algebras over algebraically closed fields of characteristic ≥ 5. This paper surveys developments pertaining to this kind of approach to classical Lie algebras.

Subject Classifications: AMS classification: 17B05, 17B45, 51B25

Keywords Lie algebra • Lie incidence geometry

1 Extremal Elements in Lie Algebras

Let k be a field and consider a Lie algebra L over k. For each $z \in L$ we denote by ad_z as usual left multiplication by z, so $\mathrm{ad}_z(x) = [z,x]$ for $x \in L$.

If the characteristic of k is distinct from 2, a nonzero element x of L is called *extremal* if $[x,[x,L]]$ is contained in the linear span of x in L. Then, by linearity

A.M. Cohen (✉)
Technische Universiteit Eindhoven, Den Dolech 2, 5612 AZ Eindhoven, The Netherlands
e-mail: amc@win.tue.nl

N.S.N. Sastry (ed.), *Buildings, Finite Geometries and Groups*, Springer Proceedings in Mathematics 10, DOI 10.1007/978-1-4614-0709-6_2,

of $[x,[x,y]]$ in y, an element x of L is extremal if and only if there is a linear functional $g_x : L \to k$ such that, for all $y \in L$,

$$[x,[x,y]] = 2g_x(y)x. \tag{1}$$

Note that $g_x(y) = 0$ if x and y commute.

The two identities of the following lemma were first obtained by Premet through polarization (see [7, 14]).

Lemma 1.1. *Suppose the characteristic of k is not even and let $x \in L$ be extremal. Then, for all $y, z \in L$,*

$$[[x,y],[x,z]] = g_x([y,z])x + g_x(z)[x,y] - g_x(y)[x,z], \tag{2}$$

$$[x,[y,[x,z]]] = g_x([y,z])x - g_x(z)[x,y] - g_x(y)[x,z]. \tag{3}$$

Observe that the number 2 occurring in (1) has disappeared. Accordingly, these identities cannot be derived from the above definition of extremality if the characteristic of k is even. In order to include the proper notion of extremality for even characteristic, we need to take these identities along.

Definition 1.2. A nonzero element x of L is called *extremal* if there is a linear functional $g_x : L \to k$ such that

(i) $[x,[x,y]] = 2g_x(y)x$ for each $y \in L$;
(ii) $[[x,y],[x,z]] = g_x([y,z])x + g_x(z)[x,y] - g_x(y)[x,z]$ for all $y, z \in L$.

The set of extremal elements of L is denoted $\mathrm{E}(L)$, or just E if no confusion is imminent. Likewise, a projective point kx with $x \in L \setminus \{0\}$ is called *extremal* if x is extremal. The set of extremal points is denoted $\mathcal{E}(L)$ or just $\mathcal{E}$.

In view of (i), extremal elements are nilpotent of order 3. If $x \in \mathrm{E}$ is such that Definition 1.2 is satisfied with $g_x = 0$, then we call x a *sandwich*.

By definition, each extremal element satisfies the first Premet identity (2). An extremal element is easily shown to satisfy the second Premet identity (3), and so Lemma 1.1 holds for all characteristics. According to the above discussion, requirement (ii) is not needed if the characteristic of k is not even. In characteristic 2, formulas (i) and (ii) do not uniquely define the linear functional g_x, but the leeway is removed by insisting that $g_x = 0$ whenever x is a sandwich; see [10, Lemma 16] for details. We will adopt this convention. A sandwich $x \in \mathrm{E}$ is nilpotent of order 2 and, due to (3), satisfies $\mathrm{ad}_x \mathrm{ad}_y \mathrm{ad}_x = 0$ for each $y \in L$, which explains to some extent the name.

The fact that extremal elements are special nilpotent elements is exhibited by the direct relation with automorphisms. For $x \in L$ and $t \in k$ define the map $\exp(x,t) : L \to L$ by

$$\exp(x,t)y = y + t[x,y] + t^2 g_x(y)x. \tag{4}$$

Lemma 1.3. *Let $x \in L$ be extremal with functional g_x. Then the map $t \mapsto \exp(x,t)$ is an injective homomorphism of groups $k^+ \to Aut(L)$, where k^+ stands for the additive group of k.*

Here are some elementary facts on extremal elements. Proofs can be found in [10].

Proposition 1.4. *Let $x, y \in \mathrm{E}$ and $z \in L$.*

(i) The subalgebra $\langle \mathrm{E} \rangle$ of L generated by E is linearly spanned by E.
(ii) On $\langle \mathrm{E} \rangle$ there is a unique symmetric bilinear form g with value $g_x(y)$ at (x, y).
(iii) For all $a, b, c \in \langle \mathrm{E} \rangle$, we have $g([a,b],c) = g(a,[b,c])$. In particular, the form g of (ii) is associative.
(iv) Assume that L is generated by E. Then the radical of g, that is, $Rad(g) = \{x \in L \mid g(x, L) = 0\}$, is a nilpotent ideal of L containing all sandwiches of L.

A consequence of the last assertion is that sandwiches in Lie algebras generated by extremal elements disappear in the transition to the quotient by $\mathrm{Rad}(g)$. Moreover, if $x \in \mathrm{E}(L)$ is not a sandwich, then $x + \mathrm{Rad}(g)$ is an extremal element of $L/\mathrm{Rad}(g)$ and is not a sandwich. So, going over from L to $L/\mathrm{Rad}(g)$, we retain all extremal elements that are not a sandwich and remove all sandwiches. For a characterization of simple Lie algebras generated by extremal elements that are not sandwiches, the assertion allows us to assume that they have no sandwiches.

Example 1.5. Let V be a vector space over k. Then each member X of $\mathfrak{gl}(V)$, the Lie algebra of all endomorphisms of V, of rank 1 with $X^2 = 0$ is extremal. The linear 1-space of X determines the incident pair (x, H) of the point that is its image $x = XV$ and the hyperplane $H = \ker X$ that is its kernel.

Suppose that V has finite dimension n. Then the subalgebra $\langle \mathrm{E} \rangle$ coincides with $\mathfrak{sl}(V)$, the matrices of trace 0 in $\mathfrak{gl}(V)$. If the characteristic of k is a prime dividing n, then the identity belongs to $\mathrm{Rad}(g)$. The subalgebra $\mathfrak{n}(V)$ of all upper triangular matrices with respect to a chosen basis of V all of whose main diagonal entries are zero, is easily seen to be generated by $\mathrm{E}(\mathfrak{gl}(V)) \cap \mathfrak{n}(V)$ and to coincide with its own g-radical. If $n = 3$, we refer to this example as the *Heisenberg algebra* and denote it by $\mathfrak{h}$. It is a 3-dimensional Lie algebra with one-dimensional center $Z(\mathfrak{h})$ such that $[\mathfrak{h}, \mathfrak{h}] = Z(\mathfrak{h})$.

Example 1.6. Let k be a field of characteristic 3 and let V be the vector space k^3. We will consider the 7-dimensional Lie algebra $L = \mathfrak{sl}(V)/Z$, where $Z = Z(\mathfrak{sl}(V))$ is the (one-dimensional) center of $\mathfrak{sl}(V)$ spanned by the identity matrix. Besides the images mod Z of the rank 1 matrices X in $\mathfrak{sl}(V)$ with $X^2 = 0$, the images mod Z of the rank 2 matrices Y in $\mathfrak{sl}(V)$ with $Y^2 \neq 0$ and $Y^3 = 0$ also belong to $\mathrm{E}(L)$. The group generated by all $\exp(x,t)$ for $t \in k$ and $x \in \mathrm{E}(L)$ consists of the k-rational points of the split algebraic group of type G_2. The Lie algebra of this algebraic group is 14-dimensional and contains L as an ideal.

Example 1.7. Let V be a vector space over k supplied with a symplectic form f. Let $\mathfrak{sp}(V, f)$ be the Lie algebra of all endomorphisms of V preserving f (so $X \in$

$\mathfrak{sp}(V, f)$ if and only if $X \in \mathfrak{gl}(V)$ and $f(Xa, b) + f(a, Xb) = 0$ for all $a, b \in V$). Then each infinitesimal symplectic transvection, that is, endomorphism of $\mathfrak{sl}(V)$ of the form $x \mapsto f(x, a)a$ for some $a \in V$, is extremal in $\mathfrak{sp}(V, f)$. If $n = 2$ and f is non-degenerate, then $\mathfrak{sp}(V, f) = \mathfrak{sl}(V)$. If $n = 3$ and f has rank 2, then $\mathfrak{sp}(V, f)$ is a semi-direct product of the ideal $\mathrm{Rad}(g)$, which is isomorphic to $\mathfrak{h}$ (of Example 1.5), and $\mathfrak{sl}(k^2)$.

Example 1.8. Let V be a vector space over k supplied with a quadratic form κ. Let $\mathfrak{o}(V, \kappa)$ be the Lie algebra of all endomorphisms of V preserving κ. Denote by f the associated bilinear form: $f(a, b) = \kappa(a + b) - \kappa(a) - \kappa(b)$ for all $a, b \in V$. Then $X \in \mathfrak{o}(V, \kappa)$ if and only if $X \in \mathfrak{gl}(V)$ and $f(Xa, a) = 0$ for all $a \in V$. Each infinitesimal Siegel transvection, that is, endomorphism of $\mathfrak{sl}(V)$ of the form $x \mapsto f(x, a)b + f(x, b)a$ for some $a, b \in V$ with $\kappa(a) = f(a, b) = \kappa(b) = 0$, is extremal in $\mathfrak{o}(V, \kappa)$.

There is a unitary variant of this example, which we denote $\mathfrak{u}(V, \kappa)$, where κ is a unitary form on V.

Example 1.9. Consider an arbitrary k-split connected reductive linear algebraic group G, with reductive rank n and semisimple rank ℓ. Fix a maximal split torus T of G and let Φ be the root system with respect to T, let X be the character group of T, and Y its dual so there is a bilinear pairing $\langle \cdot, \cdot \rangle \to \mathbb{Z}$. Fix a basis $e_1, \ldots, e_n$ of X and a dual basis $f_1, \ldots, f_n$ of Y (so $\langle e_i, f_j \rangle = \delta_{ij}$). There is a coroot system $\Phi^\star$ in Y with a bijective correspondence $\alpha \mapsto \alpha^\star$ between Φ and $\Phi^\star$ such that $\langle \alpha, \alpha^\star \rangle = 2$ for each $\alpha \in \Phi$. Let L be the Lie algebra of G and H the Lie subalgebra of T. Then L has basis elements e_α for $\alpha \in \Phi$ and $h_i \in H$ for $i = 1, \ldots, n$ with structure constants:

$$[h_i, h_j] = 0, \tag{5}$$

$$[e_\alpha, h_i] = \langle \alpha, f_i \rangle \, e_\alpha, \tag{6}$$

$$[e_{-\alpha}, e_\alpha] = \sum_{i=1}^{n} \langle e_i, \alpha^\star \rangle h_i, \tag{7}$$

$$[e_\alpha, e_\beta] = \begin{cases} N_{\alpha\beta}\, e_{\alpha+\beta} & \text{for } \alpha + \beta \in \Phi, \\ 0 & \text{for } \alpha + \beta \notin \Phi, \beta \neq -\alpha, \end{cases} \tag{8}$$

for certain integral constants $N_{\alpha\beta}$ (see [6, 28]). These integers are interpreted as elements of the field k. Such a basis of L is called a *Chevalley basis*. Now fix a long root $\alpha \in \Phi$. Then $e_\alpha \in \mathrm{E}(L)$. For k of characteristic distinct from 2, this can be checked as follows: as α is long, $2\alpha + \beta \in \Phi$ is a root only if $\beta = -\alpha$, so

$$[e_\alpha, [e_\alpha, L]] \subseteq [e_\alpha, [e_\alpha, H]] + \sum_{\beta \in \Phi} [e_\alpha, [e_\alpha, e_\beta]] \subseteq \sum_{\beta \in \Phi : 2\alpha + \beta \in \Phi} k e_{2\alpha+\beta} \subseteq k e_\alpha.$$

Now G acts on L via the adjoint representation and the full G-orbit of ke_α is contained in $\mathcal{E}(L)$. In most cases, this is all of $\mathcal{E}(L)$. This example generalizes Examples 1.5, 1.7, and 1.8. If G is not semisimple, then $\langle E(L)\rangle$ is properly contained in L. If G is semisimple and the characteristic of k is not bad for G, then $\langle E(L)\rangle = L$ (see [13]).

There are five essentially different positions two extremal elements can be in with respect to each other.

Definition 1.10. For $i \in \{-2, -1, 0, 1, 2\}$, define relations E_i on E as follows.

i	$x\mathrm{E}_i y$	Name
-2	$kx = ky$	Identical
-1	$[x, y] = 0, kx + ky \subseteq \mathrm{E} \cup \{0\}, kx \neq ky$	Strongly commuting
0	$[x, y] = 0, kx + ky \not\subseteq \mathrm{E} \cup \{0\}$	Polar
1	$[x, y] \neq 0, g(x, y) = 0$	Special
2	$g(x, y) \neq 0$	Hyperbolic

The name refers to the pair (x, y) in the individual cases. As the relation E_i on E does not depend on the choice of vector in the 1-space spanned by x or y, the names are also valid for the projective pairs (kx, ky) from $\mathcal{E}$. The corresponding projective relations are denoted $\mathcal{E}_i$.

Clearly, the relations E_i are symmetric and partition $\mathrm{E} \times \mathrm{E}$, and the $\mathcal{E}_i$ partition $\mathcal{E} \times \mathcal{E}$.

The following cases of few extremal generators are discussed in [14]; see also [18].

Lemma 1.11. *Let $x, y, z \in \mathrm{E}$.*

(i) The subalgebra of L generated by x and y is at most 3-dimensional. It is commutative if $x\mathrm{E}_i y$ with $i \leq 0$, it is isomorphic to the Heisenberg algebra $\mathfrak{h}$ if $x\mathrm{E}_1 y$, and it is isomorphic to $\mathfrak{sl}(k^2)$ if $x\mathrm{E}_2 y$.

(ii) The subalgebra $\langle x, y, z\rangle$ is at most 8-dimensional. It is nilpotent, isomorphic to (a possibly twisted form of) $\mathfrak{sl}(k^3)$ (like $\mathfrak{u}(\mathbb{R}^3, \kappa)$ where κ is a non-compact unitary form), or an extension of $\mathfrak{sl}(k^2)$ by a nilpotent ideal.

An application in [14] of a beautiful result of Zelmanov and Kostrikin [34] gives that the dimension of a Lie algebra generated by a finite number of extremal elements is finite. Besides, there is always a nilpotent Lie algebra of maximal dimension $d(N)$ for any given finite number N of extremal generators. The above lemma shows that $d(2) = 3$ and $d(3) = 8$ are attained by simple Lie algebras. It is also known that $d(4) = 28$ and $d(5) = 537$ if k has characteristic 0. The former case is not only realized by a nilpotent Lie algebra but also by the 28-dimensional Lie algebra $\mathfrak{o}(k^8, \kappa)$ of Example 1.8, where κ is a quadratic form on k^8 of Witt index 4. The Lie algebras of types E_6, E_7 and E_8 are generated by five extremal elements. This leads to an intriguing question.

Problem 1.12. Is there a 'generic' Lie algebra of maximal possible dimension 537 in the 5 generator case?

The notion of genericity is made precise by Draisma and in't panhuis in [17]. For any given finite graph Γ, they construct an algebraic variety X over k whose k-points parametrize Lie algebras over k generated by a set of extremal elements in bijective correspondence with the vertex set of the graph, with prescribed commutation relations for pairs corresponding to nonedges of Γ. If Γ is a connected, simply laced Dynkin diagram of finite or affine type, then they prove that X is an affine space, and that all points in an open dense subset of X parametrize Lie algebras isomorphic to a single fixed Lie algebra. If Γ is of affine type, then this fixed Lie algebra is the split finite-dimensional simple Lie algebra corresponding to the associated finite-type Dynkin diagram. This gives a new construction of these Lie algebras, in which they come together with interesting degenerations, corresponding to points outside the open dense subset. For graphs Γ on five vertices, the maximal dimensions of Lie algebras realizing Γ have been investigated by Roozemond [26]. In [19], the classical Lie algebras are characterized by series of graphs Γ. The easiest series is the path of length n (Coxeter diagram A_n), which corresponds to $\mathfrak{sp}(k^n, f)$, where f is a symplectic form on k^n of maximal rank. So, if n is even, then f is non-degenerate and the classical Lie algebra is of type $C_{n/2}$ and if n is odd, it is a semi-direct product of $\mathrm{Rad}(g)$ and the classical Lie algebra of type $C_{(n-1)/2}$. If $n = 3$, we recover the semidirect product of $\mathfrak{h}$ and $\mathfrak{sl}(k^2)$ described in Example 1.7.

2 Geometry from Extremal Elements

Throughout this section L will be a Lie algebra over the field k. In [22, 23] Premet proved that extremal elements occur in many simple Lie algebras.

Theorem 2.1. *If k is algebraically closed of characteristic distinct from* 2 *and* 3 *and L is finite-dimensional and simple, then L has an extremal element.*

Using powerful methods, Benkart [1, Theorem 3.2] showed that if a simple Lie algebra over an algebraically closed field of characteristic $p \geq 7$ or $p = 0$ contains a nilpotent element of order at most $p - 1$ and no sandwiches, then it is of classical type. In this section, we survey an attempt to replace this proof by a geometric argument with recourse to Tits' theory of buildings [32]. For the classification of simple modular Lie algebras in characteristic at least 5, the reader is referred to [2, 9, 24, 29].

Our approach will consist of three steps, two of which have been completed. First, in this section, we discuss how one can produce a geometry from L. The geometry will have point set $\mathcal{E}(L)$ and its line set $\mathcal{F}$ will be the projective lines in L fully contained in $\mathcal{E}(L)$. This geometry satisfies properties summarized by the term root filtration space (see Definition 3.1).

Then, in a second step, root filtration spaces are characterized as certain (root) shadow spaces of spherical buildings.

The third and final step will be to show that, given a root filtration space $(\mathcal{E}, \mathcal{F})$ satisfying some mild conditions, then, up to isomorphism, there is at most one simple Lie algebra whose root filtration space is isomorphic to $(\mathcal{E}, \mathcal{F})$. There is an abundance of examples of root filtration spaces that do not arise from Lie algebras.

We now return to the first step. The point set of our geometry will be $\mathcal{E}$. By Theorem 2.1, this set is not empty if k is algebraically closed. This is a good starting point in view of the fact that, at least for k of characteristic distinct from 2, 3, and 5, a self-contained proof by Tange is available in [30]. But we need $\mathcal{E}$ to be more than a singleton.

Example 2.2. Let k be a field of characteristic $p = 5$ and take $W_{1,1}(5)$ to be the vector space over k with basis $z^i\,\partial_z$, for $i = 0, \ldots, 4$. The Lie bracket is defined on two of these elements by

$$[z^i\,\partial_z, z^j\,\partial_z] := (j - i)z^{i+j-1}\partial_z,$$

with the convention that $z^i = 0$ whenever $i \notin \{0, \ldots, 4\}$. The Lie bracket extends bilinearly to a multiplication on $W_{1,1}(5)$ which turns it into a Lie algebra of dimension 5 over k.

Now $x = -z^2\partial_z$ is extremal in $W_{1,1}(5)$. Together with $y = \partial_z$ and $h = 2z\partial_z$ it forms an $\mathfrak{sl}_2$-triple in $W_{1,1}(5)$. But $[y, [y, 2z^4\partial_z]] = x$, so y is not extremal in $W_{1,1}(5)$. In fact, no nilpotent element generating a subalgebra isomorphic to $\mathfrak{sl}(k^2)$ with x is extremal in $W_{1,1}(5)$. So, although the Jacobson-Morozov theorem (see [20, page 98]) holds, it does not provide a second extremal element. It readily follows that $W_{1,1}(5)$ is not generated by $\mathrm{E}(W_{1,1}(5))$.

The failure of Jacobson–Morozov elements to be extremal is exceptional. A relatively succinct proof in [12] shows that L generally has more than one extremal point.

Theorem 2.3. *Let the characteristic of k be distinct from 2 and 3, and let L be simple. Suppose that L contains an extremal element that is not a sandwich. Then either k has characteristic 5 and L is isomorphic to $W_{1,1}(5)$ or L is linearly spanned by extremal elements.*

Note that the field k need not be algebraically closed. If no pair $(x, y) \in \mathrm{E} \times \mathrm{E}$ would lie in E_2, then, in view of Lemma 1.11(i), the Lie algebra L would be nilpotent. So, in simple Lie algebras containing extremal elements, E_2 will be non-empty. In the presence of a hyperbolic pair of extremal elements, the following structural result from [10] is very useful.

Proposition 2.4. *Suppose that $x, y \in E$ satisfy $g_x(y) = 1$ and write $U = \{u \in L \mid g_x(u) = g_y(u) = g_x([y, u]) = 0\}$.*

(i) There is a $\mathbb{Z}$-grading

$$L = L_{-2}(x, y) + L_{-1}(x, y) + L_0(x, y) + L_1(x, y) + L_2(x, y), \qquad (9)$$

with $L_{-2}(x,y) = kx$, $L_2(x,y) = ky$, $L_0(x,y) = N_L(kx) \cap N_L(ky)$, $L_{-1}(x,y) = [x,U]$, *and* $L_1(x,y) = [y,U]$. *Furthermore,* ad_x *induces a linear isomorphism* $L_1(x,y) \to L_{-1}(x,y)$ *with inverse* $-\mathrm{ad}_y$.

(ii) There is a filtration

$$L_{\le -2}(x) \subseteq L_{\le -1}(x) \subseteq L_{\le 0}(x) \subseteq L_{\le 1}(x) \subseteq L_{\le 2}(x) = L, \tag{10}$$

where $L_{\le i}(x) = \sum_{j=-2}^{i} L_j(x,y)$. *Moreover,* $L_{\le 1}(x) = \{z \in L \mid g_x(z) = 0\}$, $L_{\le 0}(x) = N_L(kx)$, *and* $L_{\le -1}(x) = kx + [x, L_{\le 1}(x)]$. *In particular, the subspaces* $L_{\le i}(x)$ *are independent of the choice of* y.

For each $i \in \{-2,-1,0,1,2\}$, the subspace $L_i(x,y)$ is contained in the i-eigenspace of $\mathrm{ad}_{[x,y]}$. If the characteristic is distinct from 2, 3, 5, these eigenspaces coincide with the subspaces of the grading of (i). This illustrates why it is difficult to obtain the result for characteristic 5 and why there is no result for characteristics 2 and 3.

Recall that $\mathcal{F}$ is the set of projective lines of L that lie entirely in $\mathcal{E}$. This implies that two points are collinear if and only if they strongly commute. Being a subspace of the projective space on L, the line space $(\mathcal{E},\mathcal{F})$ is partial linear. It satisfies a host of useful properties. For notions not explained here, see [8] or the beginning of the next section.

Theorem 2.5. *Let* L *be generated by* $\mathrm{E}(L)$ *and suppose* L *has no sandwiches. Consider the space* $(\mathcal{E},\mathcal{F})$ *together with the symmetric relations* $\mathcal{E}_i$ $(i = -2, ..., 2)$ *on* $\mathcal{E}$. *It satisfies the following properties, where we write* $\mathcal{E}_{\le i}$ *for* $\cup_{j \le i}\mathcal{E}_j$.

(A) The relation $\mathcal{E}_{-2}$ *is equality on* $\mathcal{E}$.
(B) The relation $\mathcal{E}_{-1}$ *is collinearity of distinct points of* $\mathcal{E}$.
(C) There is a map $\mathcal{E}_1 \to \mathcal{E}$, *denoted by* $(u,v) \mapsto [u,v]$ *such that, if* $(u,v) \in \mathcal{E}_1$ *and* $x \in \mathcal{E}_i(u) \cap \mathcal{E}_j(v)$, *then* $[u,v] \in \mathcal{E}_{\le i+j}(x)$.
(D) For each $(x,y) \in \mathcal{E}_2$, *we have* $\mathcal{E}_{\le 0}(x) \cap \mathcal{E}_{\le -1}(y) = \emptyset$.
(E) For each $x \in \mathcal{E}$, *the subsets* $\mathcal{E}_{\le -1}(x)$ *and* $\mathcal{E}_{\le 0}(x)$ *are subspaces of* $(\mathcal{E},\mathcal{F})$.
(F) For each $x \in \mathcal{E}$, *the subset* $\mathcal{E}_{\le 1}(x)$ *is a hyperplane of* $(\mathcal{E},\mathcal{F})$.

This is the geometric structure that we will take as a starting point of our geometric characterization attempt for classical Lie algebras.

3 Root Filtration Spaces

We introduce root filtration spaces and derive some of their properties. We begin with some notation for relations on a set $\mathcal{E}$. Let $x \in \mathcal{E}$. For a relation $\mathcal{X}$ on $\mathcal{E}$, we denote by $\mathcal{X}(x)$ the set of all elements $y \in \mathcal{E}$ with $(x,y) \in \mathcal{X}$. If, in addition, $y,z \in \mathcal{E}$ and $Y \subseteq \mathcal{E}$, we write $\mathcal{X}(x,y)$ for $\mathcal{X}(x) \cap \mathcal{X}(y)$, $\mathcal{X}(x,y,z)$ for $\mathcal{X}(x) \cap \mathcal{X}(y) \cap \mathcal{X}(z)$, and $\mathcal{X}(Y)$ for $\bigcap_{y \in Y} \mathcal{X}(y)$, etc.

A *point-line space* (or just *space*) is a pair $(\mathcal{E}, \mathcal{F})$ consisting of a set $\mathcal{E}$ (of points) and a collection $\mathcal{F}$ of subsets of $\mathcal{E}$ of size at least 2 (whose members are called lines). A space is called a *gamma space* if, for each point p and each line l not on p, the set of points on l collinear with p is either empty, a singleton, or all of l. It is called a *partial linear space* if every pair of distinct points is on at most one line, and a *linear space* if every pair of distinct points is on precisely one line. A *subspace* of $(\mathcal{E}, \mathcal{F})$ is a subset of $\mathcal{E}$ containing each line that has at least two points in common with it. The *rank* of a linear space is one less than the length of a maximal chain of proper subspaces; if there is no such finite chain, the rank is said to be ∞. A *singular subspace* of a space is a subspace in which any two points are collinear. The *singular rank* of a partial linear space is the supremum of all ranks of maximal singular subspaces.

We say that a subspace of a point-line space is a *hyperplane* if every line has a non-empty intersection with it. Thus, the whole point set is a hyperplane.

The definitions of *polar space*, *nondegeneracy* of a polar space, and *rank* of a polar space, are as in [8].

Definition 3.1. Let $(\mathcal{E}, \mathcal{F})$ be a partial linear space. For $\{\mathcal{E}_i\}_{-2\leq i\leq 2}$, a quintuple of disjoint symmetric relations partitioning $\mathcal{E}\times\mathcal{E}$, we call $(\mathcal{E}, \mathcal{F})$ a *root filtration space with filtration* $\{\mathcal{E}_i\}_{-2\leq i\leq 2}$ if the properties (A)–(F) of Theorem 2.5 are satisfied.

If, in addition, $(\mathcal{E}, \mathcal{F})$ satisfies the following two conditions, it is called a *non-degenerate* root filtration space.

(G) For each $x \in \mathcal{E}$ the set $\mathcal{E}_2(x)$ is not empty.
(H) The graph $(\mathcal{E}, \mathcal{E}_{-1})$ is connected.

Thus, by definition, Theorem 2.5 can be restated as in the first part of the following result of [10].

Theorem 3.2. *Suppose that L is a Lie algebra over k generated by* E. *Let $\mathcal{F}$ be the set of projective lines all of whose points belong to $\mathcal{E}$. If L does not contain sandwiches then $(\mathcal{E}, \mathcal{F})$ is a root filtration space with filtration $(\mathcal{E}_i)_{-2\leq i\leq 2}$ as defined in Definition 3.1. Furthermore, let $\mathcal{B}_i$ $(i \in I)$ be the connected components of the graph $(\mathcal{E}, \mathcal{E}_2)$. Then each $\mathcal{B}_i$ is either a non-degenerate root filtration space or a root filtration space with an empty set of lines. Furthermore, L is the direct sum of the Lie subalgebras $\langle\mathcal{B}_i\rangle$ generated by $\mathcal{B}_i$ for $i \in I$.*

The second part of the theorem shows that the decomposition of L into direct summands can be recovered from the geometry.

Condition (C) gives a kind of filtration around each extremal point x. According to Lemma 3.3(ii) below, $[u, v]$ is the unique point in $\mathcal{E}_{\leq -1}(u) \cap \mathcal{E}_{\leq -1}(v)$, so the map $[\cdot,\cdot]$ is uniquely determined by the relations $(\mathcal{E}_i)_{-2\leq i\leq 2}$.

Condition (D) is referred to as the *triangle condition on x, y, z* by Timmesfeld [31]. It can be replaced by the statement that, for each $(x, y) \in \mathcal{E}_2$, we have $\mathcal{E}_{\leq i}(x)\cap \mathcal{E}_{\leq j}(y) = \emptyset$ whenever $i + j < 0$.

Condition (E) can be replaced by the statement that $\mathcal{E}_{\leq i}(x)$ is a subspace of $(\mathcal{E}, \mathcal{F})$ for each i (see Lemma 3.3(i) below).

We will use some of the terminology introduced for Lie algebras in Definition 1.10, calling a pair $(x, z) \in \mathcal{E}_i$ *hyperbolic* if $i = 2$, *special* if $i = 1$, *polar* if $i = 0$, *collinear* if $i = -1$ (so collinearity is only used for distinct points). In addition we say (x, z) are *commuting* (notation $[x, z] = 0$) if $i \leq 0$. The following properties of root filtration spaces are derived in [11].

Lemma 3.3. *In a root filtration space $(\mathcal{E}, \mathcal{F})$, the following properties hold.*

(i) For each $i \in \{-2, \ldots, 2\}$ and each $x \in \mathcal{E}$, the subset $\mathcal{E}_{\leq i}(x)$ is a subspace of $(\mathcal{E}, \mathcal{F})$.
(ii) If $(u, v) \in \mathcal{E}_1$, then $[u, v]$ is the unique common neighbor of both u and v in the collinearity graph $(\mathcal{E}, \mathcal{E}_{-1})$ of $(\mathcal{E}, \mathcal{F})$.
(iii) If $(u, v) \in \mathcal{E}_1$, then $\mathcal{E}_0(u) \cap \mathcal{E}_2(v) \subseteq \mathcal{E}_1([u, v])$.
(iv) If $(x, y) \in \mathcal{E}_0$ and $z \in \mathcal{E}_{-1}(y)$, then either $z \in \mathcal{E}_{\leq 0}(x)$, or $z \in \mathcal{E}_1(x)$ and $\mathcal{E}_{-1}(x, y, z) = \{[x, z]\}$.
(v) If (x, q) and (u, z) belong to $\mathcal{E}_1$ whereas $u = [x, q]$ and $q = [u, z]$, then $(x, z) \in \mathcal{E}_2$.
(vi) If P is a pentagon in the collinearity graph $(\mathcal{E}, \mathcal{E}_{-1})$ (that is, the induced subgraph is a pentagon), then each distinct non-collinear pair of points of P is polar.
(vii) If $(u, v) \in \mathcal{E}_1$, then $\mathcal{E}_{-1}(u) \cap \mathcal{E}_0([u, v]) \subseteq \mathcal{E}_1(v)$.
(viii) Let $y \in \mathcal{E}$ and $l \in \mathcal{F}$ be such that $y \in \mathcal{E}_0(l)$. Then $\mathcal{E}_{\leq -1}(y, l)$ is a non-empty singular subspace of $(\mathcal{E}, \mathcal{F})$.

Here are some examples of root filtration spaces.

Example 3.4. Every linear space is a trivial example of a root filtration space with $\mathcal{E}_i = \emptyset$ for $i \geq 0$.

Every space without lines is a trivial example of a root filtration space with $\mathcal{E}_i = \emptyset$ for $-2 < i < 2$ and $\mathcal{E}_2$ the relation of being distinct. Even if we keep $\mathcal{E}_1 = \mathcal{E}_{-1} = \emptyset$ and allow for the relation of being distinct to be partitioned in $\mathcal{E}_0$ and $\mathcal{E}_2$, the result is a root filtration space. For example, if $(\mathcal{E}, \mathcal{L})$ is a polar space, taking $\mathcal{E}_2$ to be the non-collinearity relation, $\mathcal{E}_{-2}$ the identity relation, and $\mathcal{E}_0$ the complement in $\mathcal{E} \times \mathcal{E}$ of $\mathcal{E}_{-2} \cup \mathcal{E}_2$, we obtain a root filtration space $(\mathcal{E}, \emptyset)$ with $\mathcal{E}_{-1} = \mathcal{E}_1 = \emptyset$. Nondegeneracy of this polar space $(\mathcal{E}, \mathcal{L})$ is equivalent to condition (G). This makes clear why some nondegeneracy conditions like (G) and (H) are needed.

Example 3.5. Every generalized hexagon $(\mathcal{E}, \mathcal{F})$ with $\mathcal{E}_i$ for $i = 1, 2$ the set of points at mutual distance $i + 1$, and $[x, y]$ the unique point collinear with both x and y for $(x, y) \in \mathcal{E}_1$, is a non-degenerate root filtration space with $\mathcal{E}_0 = \emptyset$. Conversely, if $(\mathcal{E}, \mathcal{F})$ is a non-degenerate root filtration space with $\mathcal{E}_0 = \emptyset$, then it is a generalized hexagon.

Example 3.6. Let $\mathbb{P}$ and $\mathbb{H}$ be projective spaces in duality: $\mathbb{H}$ is a collection of hyperplanes forming a subspace of the dual of $\mathbb{P}$ annihilating $\mathbb{P}$. The latter means that the intersection of all hyperplanes of $\mathbb{H}$ is empty. If $\mathbb{P}$ has finite rank, this condition forces $\mathbb{H}$ to be the dual of $\mathbb{P}$. Take $\mathcal{E}$ to be the set of incident pairs from $\mathbb{P} \times \mathbb{H}$. The set $\mathcal{F}$ of lines is built up of two kinds: those consisting of all (x, H) with hyperplane $H \in \mathbb{H}$ fixed and x running through the points of a line of $\mathbb{P}$ inside H, and those consisting of all (x, H) with point x fixed and H running through the hyperplanes in $\mathbb{H}$ containing a fixed codimension 2 subspace of $\mathbb{P}$ containing x. So, $((x, H), (y, K)) \in \mathcal{E}_{-1}$ iff $x = y$ or $H = K$ (but not both). Then $(\mathcal{E}, \mathcal{F})$ is a root filtration space with $\mathcal{E}_0, \mathcal{E}_1, \mathcal{E}_2$ defined as follows: $(x, H) \in \mathcal{E}_0((y, K))$ iff $x \in K$ and $y \in H$ and $x \neq y$ and $H \neq K$, $(x, H) \in \mathcal{E}_1((y, K))$ iff $x \in K$ (in which case $[(x, H), (y, K)] = (x, K)$) or $y \in H$ (in which case $[(x, H), (y, K)] = (y, H)$) but not both, and $(x, H) \in \mathcal{E}_2((y, K))$ iff $x \notin K$ and $y \notin H$. We denote this root filtration space by $\mathcal{E}(\mathbb{P}, \mathbb{H})$.

Example 3.7. Let $(\mathcal{P}, \mathcal{E})$ be a non-degenerate polar space. Then the Grassmann space $(\mathcal{E}, \mathcal{F})$ on $(\mathcal{P}, \mathcal{E})$, where $\mathcal{F}$ consists of the pencils of lines on a point in a singular plane, is a root filtration space with $l \in \mathcal{E}_{-1}(m)$ iff l and m span a singular plane, $l \in \mathcal{E}_0(m)$ iff either l and m span a singular subspace not contained in a plane, or l and m intersect but do not span a singular plane, $l \in \mathcal{E}_1(m)$ iff there is a unique line n such that both the span of l and n and the span of n and m are singular planes (in which case $n = [l, m]$, and the span of l and m is not a singular subspace), and $\mathcal{E}_2$ is the complement of $\mathcal{E}_{\leq 1}$ in $\mathcal{E} \times \mathcal{E}$. The same construction for a projective space instead of a polar space leads to a root filtration space with $\mathcal{E}_1 = \mathcal{E}_2 = \emptyset$.

Let $\mathbb{P}$ and $\mathbb{H}$ be as in Example 3.6. Consider the space $(\mathcal{P}, \mathcal{L})$ whose point set $\mathcal{P}$ is the disjoint union of $\mathbb{P}$ and $\mathbb{H}$ and whose line set $\mathcal{L}$ is the union of the line set of $\mathbb{P}$, the line set of $\mathbb{H}$ and the set of all unordered pairs $\{x, H\}$ with $x \in \mathbb{P}$ and $H \in \mathbb{H}$ such that $x \in H$. This is a non-degenerate polar space, called the *dualized projective space* of $\mathbb{P}$ and $\mathbb{H}$. The root filtration space $\mathcal{E}(\mathbb{P}, \mathbb{H})$ defined in Example 3.6 is a subspace of the Grassmann space on the dualized projective space $(\mathcal{P}, \mathcal{L})$.

Example 3.8. Suppose that $(\mathcal{E}^{(1)}, \mathcal{F}^{(1)})$ and $(\mathcal{E}^{(2)}, \mathcal{F}^{(2)})$ are root filtration spaces. Let $\mathcal{E}$ be the disjoint union of $\mathcal{E}^{(1)}$ and $\mathcal{E}^{(2)}$ and let $\mathcal{F}$ be the disjoint union of $\mathcal{F}^{(1)}$ and $\mathcal{F}^{(2)}$. Then $(\mathcal{E}, \mathcal{F})$ is a root filtration space with filtration $\mathcal{E}_0 = \mathcal{E}_0^{(1)} \cup \mathcal{E}_0^{(2)} \cup (\mathcal{E}^{(1)} \times \mathcal{E}^{(2)}) \cup (\mathcal{E}^{(2)} \times \mathcal{E}^{(1)})$ and $\mathcal{E}_i = \mathcal{E}_i^{(1)} \cup \mathcal{E}_i^{(2)}$ for $i \neq 0$.

Under mild conditions, an arbitrary root filtration space can be deconstructed in the vein of Example 3.8.

Lemma 3.9. *If $(\mathcal{E}, \mathcal{F})$ is a root filtration space satisfying (G), then $(\mathcal{E}, \mathcal{F})$ is the disjoint union of connected subspaces $\mathcal{B}_i$ such that $\mathcal{B}_i \times \mathcal{B}_j \subseteq \mathcal{E}_0$ whenever $i \neq j$ unless $\mathcal{B}_i \times \mathcal{B}_j \subseteq \mathcal{E}_2$, in which case $\mathcal{B}_i$ and $\mathcal{B}_j$ are singletons. Moreover, if $x, y \in \mathcal{B}_i$ for some i and $(x, y) \in \mathcal{E}_0$ then $\mathcal{E}_{-1}(x, y) \neq \emptyset$.*

Remark 3.10. The relations $(\mathcal{E}_i)_{-2 \leq i \leq 2}$ and the map $[\cdot, \cdot] : \mathcal{E}_1 \to \mathcal{E}$ of a non-degenerate root filtration space $(\mathcal{E}, \mathcal{F})$ are fully determined by the space $(\mathcal{E}, \mathcal{F})$ itself. For, $\mathcal{E}_{-1}$, $\mathcal{E}_0 \cup \mathcal{E}_1$, and $\mathcal{E}_2$ are the relations of having distance 1, 2, and 3 in the collinearity graph of $(\mathcal{E}, \mathcal{F})$, and, for x, y in $\mathcal{E}$ at mutual distance 2, we have $x \in \mathcal{E}_1(y)$ if and only if x and y have a unique common neighbor (which coincides with $[x, y]$). Therefore, we will often not mention the filtration explicitly when introducing a non-degenerate root filtration space.

Each spherical building whose Coxeter diagram comes from a Dynkin diagram corresponds to a root filtration space. This holds in particular for all thick spherical buildings. To clarify this, we need the following definitions. Recall that a Dynkin diagram is the Coxeter diagram of a Weyl group whose bonds with an even label greater than 2 are directed.

Definition 3.11. Let Y_n be an irreducible Dynkin diagram of rank $n > 1$. We number its nodes with $1, \dots, n$ as in Bourbaki [4]. Denote by X_n the corresponding Coxeter diagram (obtained by removing the arrow of a multiple bond). To avoid confusion between the Dynkin diagram and the Coxeter diagram, we shall write $X_n = (\mathrm{B}|\mathrm{C})_n$ for the Coxeter diagram corresponding to both B_n and C_n.

Let $\widetilde{Y}_n$ be the extended (or affine) Dynkin diagram of Y_n. Its nodes are those of Y_n and an additional node, numbered 0. By J we denote the subset of $\{1, \dots, n\}$ consisting of all nodes of Y_n adjacent to 0. Then $J = \{1, n\}$ if $M = \mathrm{A}_n$, and $J = \{j\}$, where $j = 2$ if $Y_n = \mathrm{B}_n$, $j = 1$ if $Y_n = \mathrm{C}_n$, $j = 2$ if $Y_n = \mathrm{D}_n$ or E_6, $j = 1$ if $Y_n = \mathrm{E}_7$, $j = 8$ if $Y_n = \mathrm{E}_8$, $j = 1$ if $Y_n = \mathrm{F}_4$, $j = 2$ if $Y_n = \mathrm{G}_2$. We shall call J the *root nodes* or, if appropriate, j the *root node* of Y_n.

Let $\mathcal{C}$ be a building of type X_n. Following [33], we view it as a chamber system over $R = \{1, \dots, n\}$. Let J be an arbitrary subset of R. The J-*shadow* of a chamber c of $\mathcal{C}$ is the $(R \setminus J)$-cell containing c. For $j \in J$, we define a j-*line* to be the union of all $(R \setminus J)$-cells containing a chamber from a given j-panel. The pair $(\mathcal{E}, \mathcal{F})$ consisting of the set $\mathcal{E}$ of all J-shadows and the set $\mathcal{F}$ of all j-lines, for $j \in J$, is called the *shadow space* for $\mathcal{C}$ of type $X_{n,J}$. If $J = \{j\}$, we also write $X_{n,j}$ instead of $X_{n,J}$. If J is the set of root nodes of a corresponding Dynkin diagram Y_n, we call $(\mathcal{E}, \mathcal{F})$ the *root shadow space* of $\mathcal{C}$ with respect to Y_n. If X_n has multiple bonds, there are two choices for Y_n, whence two root shadow spaces of $\mathcal{C}$.

Example 3.12. Let $(\mathcal{E}, \mathcal{F})$ be the root shadow space of a spherical building of irreducible Dynkin type Y_n with $n \geq 3$. If $Y_n = \mathrm{C}_n$, then $(\mathcal{E}, \mathcal{F})$ is a non-degenerate polar space. If $Y_n \neq \mathrm{C}_n$, then $(\mathcal{E}, \mathcal{F})$ is a non-degenerate root filtration space.

These results are proved in [10], by means of the following facts, where J is the set of root nodes.

- The various mutual geometric positions two J-cells can be in (in case of a sufficiently transitive automorphism group, such as the examples coming from algebraic groups, this means the orbits of the group on pairs of J-cells) are parameterized by left and right J-reduced elements of the corresponding Coxeter group of type X_n.
- In turn, these reduced elements are parameterized by the inner products between pairs of long roots of the associated root system. The five possible values are the familiar integers $-2, -1, 0, 1, 2$.

Root shadow spaces of buildings of rank at least 3 have been discussed. Those of rank 1 are non-degenerate polar spaces of rank 1 and those of irreducible Dynkin type of rank 2 are either generalized hexagons (cases A_2 and G_2) or generalized quadrangles (cases, B_2 and C_2). Therefore, every root shadow space of a building of irreducible spherical type is a non-degenerate root filtration space or a non-degenerate polar space.

Example 3.5 represents the root shadow space of a building of type G_2, its dual corresponds to the other interpretation of G_2 as a Dynkin diagram. Example 3.6 represents a generalization to arbitrary rank of type A_n, Example 3.7 represents both Dynkin types B_n and D_n, and the polar spaces of Example 3.4 are of Dynkin type C_n.

Timmesfeld's non-degenerate sets of abstract root subgroups are in fact non-degenerate root filtration spaces. To explain this observation, we begin by recalling the notion of abstract root subgroups as appearing in [31, Definition (1.1), Chap. II] (the wording is adjusted to our setting). If A and B are subgroups of a given group G, then $[A, B]$ stands for the subgroup of G generated by all commutators $[a, b] := a^{-1}b^{-1}ab = a^{-1}a^b$ with $a \in A$ and $b \in B$. Similarly for $[a, B]$ and $[A, b]$.

Definition 3.13. Let G be a group. A set $\mathcal{E}$ of abelian non-trivial subgroups of G is called a set of *abstract root subgroups* of G if it satisfies the following two conditions.

(I) $G = \langle \mathcal{E} \rangle$ and $\mathcal{E}^g \subseteq \mathcal{E}$ for each $g \in G$.
(II) For each pair $a, b \in \mathcal{E}$ one of the following cases occurs, where $X = \langle a, b \rangle$:

- (−2) $a = b$, and so $X = a = b$.
- (−1) $[a, b] = 1$, $a \neq b$, and $X \setminus \{1\}$ is partitioned by $c \setminus \{1\}$ for $c \in \mathcal{E}$ with $c \leq X$. Here, we call the *line ab* the set of elements $c \in \mathcal{E}$ with $c \leq ab$. By $\mathcal{F}$ we denote the set of lines.
- (0) $[a, b] = 1$ and $X \setminus \{1\}$ is not partitioned by $c \setminus \{1\}$ for $c \in \mathcal{E}$ with $c \leq X$.
- (1) $[a, b]$ belongs to $\mathcal{E}$ and coincides with $[a_0, b]$ and with $[a, b_0]$ for every nontrivial $a_0 \in a$ and $b_0 \in b$; this subgroup is nontrivial and is contained in $Z(X)$, the center of X.
- (2) For each nontrivial $a_0 \in a$ there exists a nontrivial $b_0 \in b$ such that $a^{b_0} = b^{a_0}$; and similarly with a and b interchanged.

Case (2) above is described as 'X is a rank one group with unipotent subgroups a and b'. Chapter I of [31] is concerned with the structure of such groups. The subgroups a and b as in (II)(2) are X-conjugate and their X-conjugacy class is called a *hyperbolic line*. Typical examples of X are the groups (P)SL$(2, k)$ for a (skewfield) k, in which case the hyperbolic line corresponds to the points of the projective line on which X acts 2-transitively. When viewed as the rational points over a field k of an algebraic group, its Lie algebra corresponds to $\mathfrak{sl}(k^2)$ (up to central isogeny), and a and b to a hyperbolic pair of points therein.

Case (1) is the so-called special case; typical examples are extra-special p-groups of order p^3, often suggestively denoted by p^{1+2}. When viewed as the $(\mathbb{Z}/p\mathbb{Z})$-rational points of the upper triangular group in SL(k^3) for k a field of characteristic p, its Lie algebra is the Heisenberg Lie algebra $\mathfrak{h}$ over k of Example 1.5.

For each $j \in \{-2, -1, 0, 1, 2\}$ we write $\mathcal{E}_j$ to denote the relation on $\mathcal{E}$ expressing that a, b are in case (j). So $(a, b) \in \mathcal{E}_{-2}$ is equivalent to $a = b$, and $a \in \mathcal{E}_{-1}(b)$ means that a and b belong to a line in $\mathcal{F}$. Notice that $\mathcal{E}_{\leq 0}(x)$ is the set of subgroups in $\mathcal{E}$ commuting with x.

The following result shows that the images of the maps in Lemma 1.3 are abstract root subgroups of the algebraic group Aut(L).

Theorem 3.14. *Suppose that L is a Lie algebra over k generated by extremal elements and containing no sandwiches. Then the set of subgroups $\{\exp(x,t) \mid t \in k\}$ of Aut(L) for $x \in E$ is a set of abstract root subgroups of the subgroup of Aut(L) which they generate.*

Example 3.15. Let G be a group. In [31, Definition (1.1)], nondegeneracy of a set $\mathcal{E}$ of abstract root subgroups in G is defined by the condition that $\mathcal{E}_{\leq 0}$, $\mathcal{E}_1$, $\mathcal{E}_2$ be non-empty. Suppose that $\mathcal{E}$ is a non-degenerate set of abstract root subgroups of G such that $(\mathcal{E}, \mathcal{E}_2)$ is connected and $\mathcal{E}_{-1}$ is nonempty. If G has no solvable normal subgroup, then $(\mathcal{E}, \mathcal{F})$, for $\mathcal{F}$ the set of lines of $\mathcal{E}$, is a non-degenerate root filtration space with thick lines. In [31, Sect. II.4], Timmesfeld showed that, when G is generated by a set of abstract root groups, the solvable radical of G is nilpotent of class at most two.

Observe that, for $(a,b) \in \mathcal{E}_1$, we have $[a,b] \in \mathcal{E}$, so we have a map $[\cdot,\cdot] : \mathcal{E}_1 \to \mathcal{E}$ as required in the definition of root filtration space.

For the proof (in [10]) that the set of abstract root subgroups of G gives rise to a root filtration space, some properties from [31] are needed that occur at an early stage.

4 Parapolar Spaces

This section is concerned with a converse of Example 3.12, which states that each non-degenerate root filtration space is the root shadow space of a building of spherical type. The result is proved in [11].

Theorem 4.1. *Let $(\mathcal{E}, \mathcal{F})$ be a non-degenerate root filtration space. If the singular rank of $(\mathcal{E}, \mathcal{F})$ is finite, then $(\mathcal{E}, \mathcal{F})$ is isomorphic to a shadow space of type $\mathrm{A}_{n,\{1,n\}}$ ($n \geq 2$), $(\mathrm{B|C})_{n,2}$ ($n \geq 3$), $\mathrm{D}_{n,2}$ ($n \geq 4$), $\mathrm{E}_{6,2}$, $\mathrm{E}_{7,1}$, $\mathrm{E}_{8,8}$, $\mathrm{F}_{4,1}$, or $\mathrm{G}_{2,2}$.*

The only root shadow spaces missing from this result are polar spaces, that is, the shadow spaces of type $(\mathrm{B|C})_{n,1}$. A characterization of polar spaces in terms of degenerate root filtration spaces is given by Cuypers [15] who extended earlier results of Timmesfeld [31]. See Sect. 5, where it is also described how Cuypers' result can be used in the characterization of classical simple Lie algebras with degenerate root filtration spaces.

We briefly describe the proof of Theorem 4.1. In order to reconstruct the building from a non-degenerate root filtration space, the elements of the building of types other than point and line need to be found. An initial step is to show that the space is a parapolar space. We recall some definitions from [8].

Definition 4.2. A *parapolar space* is a connected partial linear gamma space possessing a collection of geodesically closed subspaces, called *symplecta* (singular: *symplecton*), isomorphic to non-degenerate polar spaces of rank at least 2, with the properties that each line is contained in a symplecton and that each pair of distinct non-collinear points having at least 2 common neighbors is contained in a unique symplecton. (This definition comes from [11] and is a slight improvement of the one in [8], where the existence of symplecta was only required for quadrangles.) If all symplecta are polar spaces of rank k (respectively, of rank at least k) the space is said to have *polar rank* k (respectively, polar rank at least k).

A very useful condition that separates the parapolar spaces of our interest from others is the following condition, which originates from [9].

(K) For each point x and each symplecton S the set of points of S collinear with x is either empty or contains a line.

Note that the set of points inside a symplecton that are collinear with a point outside of it form a singular subspace, so (K) excludes the possibility of a singleton.

Recall the root filtration space $\mathcal{E}(\mathbb{P}, \mathbb{H})$ of Example 3.6. In [11] the following result is established.

Theorem 4.3. *Let* $(\mathcal{E}, \mathcal{F})$ *be a non-degenerate root filtration space.*

(i) If some line in $\mathcal{F}$ *is contained in a unique maximal singular subspace, then one of the following cases prevails.*

  I. The singular rank is 1 *and* $(\mathcal{E}, \mathcal{F})$ *is a root shadow space of type* $\mathrm{G}_{2,1}$.

  II. The singular rank is at least 2*, there is a point that belongs to at least 3 maximal singular subspaces, and* $(\mathcal{E}, \mathcal{F})$ *is a root shadow space of type* $(\mathrm{B}|\mathrm{C})_{3,2}$.

  III. The singular rank is at least 2*, on each point there are precisely 2 maximal singular subspaces, and* $(\mathcal{E}, \mathcal{F})$ *is isomorphic to* $\mathcal{E}(\mathcal{M}, \mathcal{N})$ *for a certain pair* $(\mathcal{M}, \mathcal{N})$ *of projective spaces in duality.*

(ii) If some line in $\mathcal{F}$ *is contained in more than one maximal singular subspace, then* $\mathcal{E}_1 \neq \emptyset$ *and* $(\mathcal{E}, \mathcal{F})$ *is a parapolar space satisfying condition* (K) *whose polar rank is at least* 3.

We briefly discuss the proof. For $(\mathcal{E}, \mathcal{F})$ as in the hypotheses, the set of common neighbors of a polar pair (x, y) is a non-degenerate polar space. In case (ii) this polar space has rank at least 2 and the required properties can be derived by use of Lemma 3.3.

In case (i) it is of rank 1; in other words, the subspace $\mathcal{E}_{-1}(x, y)$ of $\mathcal{E}$ has no lines. It is relatively easy to prove that each line l is in a unique maximal singular subspace $M(l)$. If the singular rank of $(\mathcal{E}, \mathcal{F})$ equals 1, then it can be shown that $\mathrm{E}_2 = \emptyset$, and it follows that $(\mathcal{E}, \mathcal{F})$ is a generalized hexagon, whence case I.

If there is a point in $\mathcal{E}$ that is in at least 3 maximal singular subspaces, then $(\mathcal{E}, \mathcal{F})$ has singular rank 2, and the symplecta of the resulting space can be shown to support the structure of a polar space (in which a line is the collection of symplecta on a line of $(\mathcal{E}, \mathcal{F})$). This leads to case II.

The case remains where each point in $\mathcal{E}$ is in exactly two maximal singular subspaces. Then the collection of maximal singular subspaces of $(\mathcal{E},\mathcal{F})$ can be partitioned into two classes $\mathcal{M}$ and $\mathcal{N}$, such that each of $\mathcal{M}$ and $\mathcal{N}$ carries the structure of a projective space whose lines are the members of the class containing exactly one point of a line in $\mathcal{F}$. Moreover, the space $\mathcal{N}$ can be viewed as a subspace of the dual of $\mathcal{M}$ annihilating $\mathcal{M}$, and, with this identification, $(\mathcal{E},\mathcal{F})$ is isomorphic to $\mathcal{E}(\mathcal{M},\mathcal{N})$; hence, case III.

This ends the discussion of the proof of Theorem 4.3. We proceed with the brief overview of the proof of Theorem 4.1. In case (ii) of Theorem 4.3, the hypotheses of the Kasikova–Shult theorem of [17] hold. Below we state this theorem, which deals with point-line spaces of arbitrary rank. In order to cover the case where the polar rank is big, the following condition is needed.

(L) If all symplecta have rank at least 4, then the singular rank of $(\mathcal{E},\mathcal{F})$ is finite.

The main result of [17, 27] reads as follows.

Theorem 4.4. *Each parapolar space of polar rank at least* 3 *satisfying (K) and (L), and containing a special pair is a shadow space of type* $(\mathrm{B}|\mathrm{C})_{n,2}$ $(n \geq 4)$, $\mathrm{D}_{n,2}$ $(n \geq 4)$, $\mathrm{E}_{6,2}$, $\mathrm{E}_{7,1}$, $\mathrm{E}_{8,8}$, *or* $\mathrm{F}_{4,1}$.

In [27], Shult shows that, if $(\mathcal{E},\mathcal{F})$ is a parapolar space of polar rank at least 3 in which (K) holds, then, for each singular plane π and each line l meeting π at a point x and not lying in a singular subspace containing π, either one or all lines m on x in π have the property that $l \cup m$ belongs to a symplecton. In an earlier version of Theorem 4.4, this property was stated as a condition.

Theorem 4.4 is proved by reconstructing the elements of missing types and invoking [33] to recognize the building. The reconstruction is carried out by deriving conditions needed in earlier theorems characterizing buildings by means of point-line spaces, like [9].

Theorem 4.1 follows from Theorems 4.3 and 4.4. For, the singular rank of $(\mathcal{E},\mathcal{F})$ is finite, so case III of Theorem 4.3 leads to $(\mathcal{E},\mathcal{F})$ being a shadow space of type $\mathrm{A}_{n,\{1,n\}}$.

5 Degenerate Root Filtration Spaces

In this section, we are concerned with root filtration spaces that are polar spaces. In particular, we review how Lie algebras generated by extremal elements in which no strongly commuting pairs occur, have a root filtration space coming from a polar space, as described in Example 3.4.

At the root filtration space level, we have a useful result by Cuypers [15], which generalizes an abstract root group version by Timmesfeld [31]. In order to state it, we need the notion of transversal coclique.

Definition 5.1. A *dual affine plane* is a point-line space obtained from a projective plane by removing a point and all lines containing that point. A *transversal coclique* in a dual affine plane is the set of points forming such a removed line together with the removed point.

Theorem 5.2. *Let* $(\mathcal{E}, \mathcal{H})$ *be a point-line space and let* $\sim$ *denote being collinear and distinct. Write* $\perp$ *for its complement. Suppose* $(\mathcal{E}, \mathcal{H})$ *satisfies the following six conditions.*

(i) $(\mathcal{E}, \mathcal{H})$ *is a connected partial linear space that is not linear.*
(ii) Each line contains at least four points.
(iii) The subspace of $(\mathcal{E}, \mathcal{H})$ *generated by any triple of points* x, y, z *with* $x \sim y \sim z \perp x$ *is a dual affine plane.*
(iv) If a point is not collinear with two points of a transversal coclique in a subspace isomorphic to a dual affine plane, then it is not collinear with any point of that coclique.
(v) If x, y *are points with* $x^{\perp} \subseteq y^{\perp}$, *then* $x = y$.
(vi) $(\mathcal{E}, \perp)$ *is connected.*

Then $(\mathcal{E}, \perp)$ *is the collinearity graph of a non-degenerate polar space.*

The following characterization of the geometry on $\mathcal{E}(L)$ can be found in the PhD thesis [18] of in't panhuis.

Theorem 5.3. *Let* L *be a Lie algebra over a field* k *of size at least* 3 *generated by its extremal elements. Assume the following two conditions hold.*

(i) The sets $\mathcal{E}_{-1}$ *and* $\mathcal{E}_1$ *are empty.*
(ii) The set $\mathcal{E}_0$ *is nonempty and the graph* $(\mathcal{E}, \mathcal{E}_2)$ *is connected.*

Then $(\mathcal{E}, \mathcal{E}_0)$ *is the collinearity graph of a non-degenerate polar space.*

This result suffices for finding the full polar space as lines of a non-degenerate polar space are of the form $\{x, y\}^{\perp\perp}$ for distinct collinear points x and y. The proof of this theorem is based on Cuypers' Theorem 5.2. In the application to L, the members of the line set $\mathcal{H}$ are *hyperbolic lines*, that is, the set of extremal points of a subalgebra of L generated by two extremal elements forming a hyperbolic pair. The relation $\perp$ is just $\mathcal{E}_{-2} \cup \mathcal{E}_0$.

The remaining case, where k has two elements, is harder to characterize in terms of root filtration spaces. It is more easily described when hyperbolic lines are added, as the following result of [18] shows. The theory of abstract root groups was an initiative to view the beautiful results on 3-transpositions by Fischer as a special case of a more general set-up in which the original setting would be the case where a parameterizing field k would have two elements. In this respect, and with the root filtration spaces as geometric generalizations of abstract root groups, it is fitting that Fischer spaces occur as the root filtration spaces for $k = \mathbb{Z}/2\mathbb{Z}$.

Definition 5.4. A *Fischer space* is a partial linear space in which each plane is isomorphic to a dual affine plane of order two or to an affine plane of order three.

Theorem 5.5. *Let L be a Lie algebra over $\mathbb{Z}/2\mathbb{Z}$ generated by extremal elements whose root filtration space satisfies conditions (i) and (ii) of Theorem 5.3. Let $\mathcal{H}$ be the collection of hyperbolic lines of L. Then $(\mathcal{E}, \mathcal{H})$ is a connected Fischer space.*

6 Conclusion

Although there is work in progress, the third step in the characterization of classical Lie algebras by means of root filtration spaces has not yet been finalized and so it is formulated here as a question.

Problem 6.1. Under which conditions (e.g. nondegeneracy) is it true that, for a given root filtration space $(\mathcal{E}, \mathcal{F})$, there is, up to isomorphism, at most one simple Lie algebra whose root filtration space is isomorphic to $(\mathcal{E}, \mathcal{F})$?

This question is more general than the problem posed at the beginning of Sect. 2. If 'classical type' of a simple Lie algebra is interpreted as being the Lie algebra of an algebraic group or a central quotient thereof, Benkart's results are covered by the following proposition.

Proposition 6.2. *Let L be a simple Lie algebra generated by extremal elements over an algebraically closed field k. Then $G = \langle \exp(x,t) \mid x \in \mathrm{E}(L), t \in k \rangle$ is an algebraic subgroup of* $\mathrm{Aut}(L)$ *whose Lie algebra* $\mathrm{Lie}(G)$ *contains an ideal isomorphic to L.*

Proof. As the subgroups U_x for $x \in \mathrm{E}$ are closed connected subgroups of the algebraic group $\mathrm{Aut}(L)$ (cf. Lemma 1.3), the subgroup G of $\mathrm{Aut}(L)$ generated by all of them is also closed in $\mathrm{Aut}(L)$ (see [3, Proposition 2.2]). The Lie algebra $\mathrm{Lie}(G)$ can be viewed as a Lie subalgebra of $\mathrm{Der}(L)$, the subalgebra of all $X \in \mathfrak{gl}(L)$ such that $X[a,b] = [Xa,b] + [a,Xb]$ whenever $a, b \in L$. If $x \in \mathrm{E}$, the derivative of the embedding of U_x in $\mathrm{Aut}(L)$ has image $k\,\mathrm{ad}_x$, so ad_L is a subalgebra of $\mathrm{Lie}(G)$. It is even an ideal of $\mathrm{Lie}(G)$ as it is an ideal of $\mathrm{Der}(L)$, which contains $\mathrm{Lie}(G)$. As L is simple, it is isomorphic to ad_L. We conclude that L is isomorphic to the ideal ad_L of $\mathrm{Lie}(G)$, as required.

The Lie algebras L and $\mathrm{Lie}(G)$ need not coincide as follows from Example 6.3. If k has characteristic 0, all derivations of L are known to be inner (cf. [20, Theorem 6]), and so $\mathrm{Lie}(G)$, being contained in $\mathrm{Der}(\mathrm{ad}_L)$, coincides with ad_L; therefore, L is the Lie algebra of G if k has characteristic 0. This statement is even true if L is not simple provided its center is trivial (and the characteristic is still 0), as it is isomorphic then to the subalgebra ad_L of $\mathfrak{gl}(L)$, which is generated by the algebraic Lie subalgebras $k\,\mathrm{ad}_x$ for $x \in \mathrm{E}(L)$, so [3, Corollary 7.7] applies.

Example 6.3. In Example 1.6, we have seen an exceptional case (with k of characteristic 3) in which the dimension of L is as low as half the dimension of $\mathrm{Lie}(G)$. Take p to be a prime greater than 3 and let k be algebraically closed of characteristic p. Then the analogous construction yields a simple Lie algebra that still deviates from the Lie algebra of the corresponding group although the dimension gap is smaller: put $V = k^p$ and let $L = \mathfrak{sl}(V)/Z$, where $Z = Z(\mathfrak{sl}(V))$ is the (one-dimensional) center of $\mathfrak{sl}(V)$ spanned by the identity matrix. Then L is simple of dimension $p^2 - 2$. So, if H is a connected algebraic subgroup of GL(L) with Lie(H) $= \mathrm{ad}_L$, then it must be simple of dimension $p^2 - 2$. But no simple connected algebraic group has such a dimension (as $p > 4$), so L is not even the Lie algebra of an algebraic group.

In Example 3.4, we have seen that there is little structure on root filtration spaces with $\mathcal{E}_{-1} = \mathcal{E}_1 = \emptyset$, and that polar spaces are among them. There is an abundance of examples of root filtration spaces that do not arise from a Lie algebra. For instance, let k be algebraically closed of characteristic distinct from two and let V be a vector space over k with a non-degenerate quadratic form κ. Define $(\mathcal{E}, \mathcal{F})$ as the polar space of the quadric determined by κ in the projective space on V. If the dimension of V is equal to 3, then the exterior product map leads to a Lie algebra structure on V isomorphic to $\mathfrak{o}(V, \kappa)$. For higher dimensions, no Lie algebra structure (nor any other antisymmetric product) on V exists that is invariant under the orthogonal group on V with respect to κ. But the embedding of $(\mathcal{E}, \mathcal{F})$ in the projective space on V is universal, and so there is no Lie algebra whose root filtration space is isomorphic to $(\mathcal{E}, \mathcal{F})$.

The Lie algebra $\mathfrak{o}(V, \kappa)$ is generated by extremal elements, but, if the dimension is higher than 3, the corresponding root filtration space is not isomorphic to $(\mathcal{E}, \mathcal{F})$. The root filtration space of $\mathfrak{o}(V, \kappa)$ is the Grassmannian of lines on this polar space.

Similarly, not every root filtration space comes from a set of abstract root groups. Another way to understand what is going on, is to consider a thick building of an irreducible spherical type having multiple bonds. Then there are two root shadow spaces of this building, according to two distinct interpretations of the Coxeter diagram as a Dynkin diagram. Except for 'bad characteristics', only one of these choices will lead to a correspondence with long root subgroups, and hence to abstract root subgroups. The number of positions two short root groups can be in is usually larger than five.

In the case where the field has size 2 and the root filtration space of the Lie algebra over $\mathbb{Z}/2\mathbb{Z}$ is degenerate, the hyperbolic lines lead to a Fischer space (see Theorem 5.5). Conversely, there is a universal way to construct a Lie algebra from the geometry such that the points of the geometry map to extremal points of the Lie algebra, which runs as follows. Given a Fischer space $(\mathcal{E}, \mathcal{H})$, let A be the vector space over $\mathbb{Z}/2\mathbb{Z}$ with basis $\mathcal{E}$. Define the multiplication $*$ on A as the one determined by

$$x * y = \begin{cases} x + y + z & \text{if } \{x, y, z\} \in \mathcal{H} \\ 0 & \text{otherwise} \end{cases}.$$

Following [16], we consider the symplectic form f on A satisfying $f(x, y) = 1$ if x and y are distinct and collinear, and $f(x, y) = 0$ otherwise, for $x, y \in \mathcal{E}$. Then

f is associative, in the sense that $f(x, y * z) = f(x * y, z)$ for all $x, y, z \in A$, so $\mathrm{Rad}(f)$ is an ideal of A.

Now the quotient algebra $A/\mathrm{Rad}(f)$ is a Lie algebra if and only if each subset of $\mathcal{E}$ that is an affine plane of $(\mathcal{E}, \mathcal{H})$ belongs to $\mathrm{Rad}(f)$. This Lie algebra $A/\mathrm{Rad}(f)$ may collapse, so that again examples of Fischer spaces without corresponding Lie algebras occur. The Fischer space related to the biggest Fischer group F_{24} is an example where this happens. Also, two nonisomorphic Fischer spaces may lead to isomorphic Lie algebras. In one of these instances, the Lie algebra $A/\mathrm{Rad}(f)$ of the Fischer space related to Fischer's group F_{22} is isomorphic to the Lie algebra over $\mathbb{Z}/2\mathbb{Z}$ of twisted type ${}^2E_6$, which leads to a geometric proof of the existence of an embedding of F_{22} in the twisted Chevalley group ${}^2E_6(2)$. Many more details are to be found in [16, 18].

References

1. G. Benkart, On Inner Ideals and Ad-Nilpotent Elements of Lie Algebras, Trans. Amer. Math. Soc., **232** (1977), 61–81
2. R.E. Block, *On the Mills-Seligman axioms for Lie algebras of classical type*, Trans. Amer. Math. Soc., **121** (1966), 378–392
3. A. Borel, *Linear Algebraic Groups*, 2nd enlarged edition, GTM 126, Springer, 1991
4. N. Bourbaki, *Groupes et Algèbres de Lie, Chapitres 4, 5, et 6*, Hermann, Paris, 1968
5. A.E. Brouwer and A.M. Cohen and A. Neumaier, Distance-regular Graphs, Springer, Berlin 1989
6. R.W. Carter, *Simple groups of Lie type*, Wiley and Sons, New York, 1972
7. V.I. Chernousov, Hasse principle for groups of type $\mathcal{E}_8$, Dokl. Akad. Nauk SSSR, **306**, (1989) 1059–1063
8. A.M. Cohen, Point-line spaces related to buildings, *Handbook of incidence geometry*, pp. 647–737, North-Holland, Amsterdam, 1995
9. A.M. Cohen and B.N. Cooperstein, A characterization of some geometries of exceptional Lie type. Geometriae Dedicata **15** (1983) 73–105
10. A.M. Cohen and G. Ivanyos, Root filtration spaces from Lie algebras and abstract root groups, J. Algebra, **300** (2006), 433–454
11. A.M. Cohen and G. Ivanyos, Root shadow spaces, European J. Combin., **28** (2007), 1419–1441
12. A.M. Cohen and G. Ivanyos and D.A. Roozemond, Simple Lie algebras having extremal elements, Indag. Math. (N.S.), **19** (2008), 177–188
13. A.M. Cohen and D.A. Roozemond, Computing Chevalley bases in small characteristics, J. Algebra, **322** (2009) 703–721
14. A.M. Cohen and A. Steinbach and R. Ushirobira and D.B. Wales, Lie algebras generated by extremal elements, J. Algebra, **236** (2001), 122–154
15. H. Cuypers, The geometry of k-transvection groups, J. Algebra, **300** (2006), 455–471
16. H. Cuypers and M. Horn and J. in't panhuis and S. Shpectorov, Lie algebras and 3-transpositions, arXiv:1104.0536v1[math.GR].
17. J. Draisma and Jos in 't panhuis, Constructing simply laced Lie algebras from extremal elements, Algebra Number Theory, **2** (2008), 551–572
18. J. in't panhuis, *Lie algebras, extremal elements, and geometries*, PhD Thesis, Technische Universiteit Eindhoven, 12 October 2009
19. J. in 't panhuis and E.J. Postma and D.A. Roozemond, Extremal presentations for classical Lie algebras, J. Algebra, **322** (2009), 295–326

20. N. Jacobson, *Lie algebras*, Dover, New York, 1979
21. A. Kasikova and E.E. Shult, Point-line characterizations of Lie geometries, Adv. in Geometry, **2** (2002), 147–188
22. A.A. Premet, Lie algebras without strong degeneration, Mat. Sb. **129** (186), 140–153 (translated in Math. USSR Sbornik, **57** (1987), 151–164)
23. A.A. Premet, Inner ideals in modular Lie algebras, Vestsī Akad. Navuk BSSR Ser. Fīz.-Mat. Navuk, **5** (1986), 11–15
24. A.A. Premet and H. Strade, Simple Lie algebras of small characteristic: I. Sandwich elements, J. Algebra, **189** (1997), 419–480
25. A.A. Premet and H. Strade, Simple Lie algebras of small characteristic: III. Toral Rank 2 Case, J. Algebra, **242** (2001), 236–337
26. D.A. Roozemond, *Lie algebras generated by extremal elements*, Master Thesis, August 2005, Technische Universiteit Eindhoven. http://alexandria.tue.nl/extra2/afstversl/wsk-i/roozemond2005.pdf
27. E.E. Shult, On characterizing the long-root geometries, Adv. Geom., **10** (2010), 353–370
28. T.A. Springer, *Linear algebraic groups, second edition*, Birkhäuser, Basel 1991
29. H. Strade, *Simple Lie algebras over fields of positive characteristic, I. Structure theory*, de Gruyter, Berlin, 2004
30. R. Tange, *Extremal elements in Lie algebras*, Master's thesis, Technische Universiteit Eindhoven, June 2002. http://alexandria.tue.nl/extra2/afstversl/wsk-i/tange2002.pdf
31. F.G. Timmesfeld, *Abstract root subgroups and simple groups of Lie type*, Monographs in Mathematics 95, Birkhäuser Verlag, Basel, 2001
32. J. Tits, *Buildings of spherical type and finite BN-pairs*, Lecture Notes in Math. 386, Springer, Berlin, 1974
33. J. Tits, A local approach to buildings, pp. 317–322 in *The Geometric Vein (Coxeter Festschrift)*, Springer, 1981
34. E.I. Zelmanov and A.I. Kostrikin, A theorem on sandwich algebras (Translated in Proc. Steklov Inst. Math. **1991**, no. 4, 121–126, Galois theory, rings, algebraic groups and their applications (Russian)), Trudy Mat. Inst. Steklov., **183** (1990), 106–111, 225

Properties of a 27-Dimensional Space of Symmetric Bilinear Forms Acted on by E_6

Rod Gow

Abstract We use the 27-dimensional simple exceptional Jordan algebra to construct a 27-dimensional space of symmetric bilinear forms invariant under the action of the simple Lie group E_6. Using the E_6 action on the Jordan algebra, we show that the space contains elements of exactly three non-zero ranks, namely, 10, 18 and 27. Many of the properties of the Jordan algebra are reflected in rank properties of the space of forms. We investigate subspaces containing no elements of rank 27 and also constant rank subspaces. The results vary depending on the nature of the octonion algebra used to define the Jordan algebra.

Subject Classifications: 15A03, 15A33

Keywords Octonions • Jordan algebra • Cubic form • Rank

1 Introduction

The purpose of this paper is to investigate properties of a 27-dimensional space $\mathcal{N}$ of symmetric bilinear forms obtained from the simple exceptional Jordan algebra of dimension 27. This subspace contains only non-zero forms of rank 10, 18 or 27 and it is naturally a module for the group of type E_6, which acts on the Jordan algebra.

Certain properties of $\mathcal{N}$ are independent of the underlying field, but others depend on whether the octonion algebra, on which the Jordan algebra is modelled, is a division algebra or not. A description of much of the information we need about octonion algebras, Jordan algebras and the group E_6 may be found in Chap. 4 of [7].

R. Gow (✉)
School of Mathematical Sciences, University College, Belfield, Dublin 4, Ireland
e-mail: rod.gow@ucd.ie

N.S.N. Sastry (ed.), *Buildings, Finite Geometries and Groups*, Springer Proceedings in Mathematics 10, DOI 10.1007/978-1-4614-0709-6_3,

2 Octonion Algebras

Let K be a field of characteristic different from 2 and let C be an octonion algebra over K. We recall that C is an eight-dimensional non-associative algebra over K with multiplicative identity e. It is equipped with a non-degenerate quadratic form N which permits composition, meaning that

$$N(xy) = N(x)N(y)$$

for all x and y in C. We call N the *norm* on C. It satisfies $N(e) = 1$. We let p denote the associated polarizing symmetric bilinear form.

Let x be an element of C. We define the conjugate $\overline{x}$ of x by

$$\overline{x} = p(x,e)e - x.$$

The conjugation operator is an anti-involution on C satisfying

$$x\overline{x} = \overline{x}x = N(x)e.$$

Provided that $N(x) \neq 0$, x has an inverse x^{-1} given by

$$x^{-1} = \frac{1}{N(x)}\overline{x}.$$

3 The Exceptional 27-Dimensional Jordan Algebra and the Group E_6

We suppose for the rest of this paper that the field K has characteristic different from 2 and 3. As before, let C denote an octonion algebra over K. Let A denote the K-algebra of 3×3 matrices with entries in C which are Hermitian with respect to the conjugation operation in C. We may write the elements of A in the form

$$\begin{pmatrix} \alpha_1 & c & \overline{b} \\ \overline{c} & \alpha_2 & a \\ b & \overline{a} & \alpha_3 \end{pmatrix},$$

where the $\alpha_i \in K$ and $a, b, c \in C$. Note that we have

$$\dim_K A = 27.$$

A is a Jordan algebra, with Jordan product defined by

$$xy = \frac{1}{2}(x \cdot y + y \cdot x)$$

for elements x and y in A. Here, $x \cdot y$ denotes the usual matrix product of x and y.

There is a quadratic form Q defined on A by the formula

$$Q(x) = \frac{1}{2}\,\mathrm{tr}(x^2),$$

where tr denotes the usual trace function defined on 3×3 matrices. Let $\langle\ ,\ \rangle$ denote the symmetric bilinear form on $A \times A$ obtained from polarization of Q. We have

$$\langle x, y\rangle = \mathrm{tr}(xy)$$

for x and y in A.

There is also a cubic form $\det : A \to K$ which can be defined in terms of Q and $\langle\ ,\ \rangle$ by the formula

$$\det(x) = \frac{1}{3}\langle x^2, x\rangle - Q(x)\langle x, e\rangle + \frac{1}{6}\langle x, e\rangle^3.$$

Here, e denotes the identity element of A. If we set

$$\begin{pmatrix} \alpha_1 & c & \overline{b} \\ \overline{c} & \alpha_2 & a \\ b & \overline{a} & \alpha_3 \end{pmatrix},$$

we can also express $\det(x)$ by the formula

$$\det(x) = \alpha_1\alpha_2\alpha_3 - \alpha_1 a\overline{a} - \alpha_2 b\overline{b} - \alpha_3 c\overline{c} + (ab)c + \overline{(ab)c}.$$

This formula shows that if the entries of A lie in a commutative associative subalgebra of C, $\det(x)$ is the usual determinant of a 3×3 matrix. It follows that det maps A onto K.

Each element $x \in A$ satisfies the cubic equation

$$x^3 - \langle x, e\rangle x^2 - \left(Q(x) - \frac{1}{2}\langle x, e\rangle^2\right) x - \det(x)e = 0.$$

This is known as the Cayley–Hamilton equation for A.

Let $\langle\ ,\ ,\ \rangle$ denote the symmetric trilinear form $A \times A \times A \to K$ obtained by polarization of det. Thus

$$\langle x, x, x\rangle = \det(x).$$

This trilinear form can be expressed in terms of the bilinear form $\langle\ ,\ \rangle$, as shown in [5, formula 5.14, p. 121].

Given $x \in A$, we define a symmetric bilinear form $F_x : A \times A \to K$ by

$$F_x(y, z) = \langle x, y, z \rangle$$

for all y and z in A. This form was introduced by Jacobson in [1, p. 364]. As x ranges over A, we obtain a 27-dimensional subspace of symmetric bilinear forms defined on $A \times A$, which we denote by $\mathcal{N}$. In the remaining part of this paper, we shall investigate $\mathcal{N}$.

To this end, we recall that the cross product $x \times y$ of elements x and y is defined by the formula

$$\langle x \times y, z \rangle = 3\langle x, y, z \rangle.$$

The following lemma is clear from this definition.

Lemma 3.1. *The radical of F_x consists of all elements y satisfying $x \times y = 0$.*

Theorem 3.2. *Given $x \in A$, there is a non-zero element $y \in A$ with $x \times y = 0$ if and only if* $\det(x) = 0$.

Proof. Suppose that $\det(x) \neq 0$. Then Lemma 5.9.1 of [5] shows that the only solution y to the equation $x \times y = 0$ is $y = 0$. The converse part of the theorem is a consequence of a general theorem we shall employ later in this section. However, it may be of interest to see a reasonably constructive proof which explains in part some of our later calculations.

Suppose therefore that x is a non-zero element with $\det(x) = 0$. We set $\lambda = \langle x, e \rangle / 2$. Employing Lemma 5.2.1(i) of [5], in conjunction with the Cayley–Hamilton theorem in A, we find that

$$x \times (x^2 - \lambda x) = 0,$$

and thus provided $x^2 - \lambda x \neq 0$, we may take $y = x^2 - \lambda x$.

Suppose now that $x^2 - \lambda x = 0$. If $\lambda = 0$, so that $x^2 = 0$, we obtain $x \times x = 0$, and we may take $y = x$. Finally, suppose that $\lambda \neq 0$ and set

$$w = \frac{1}{\lambda} x.$$

It is clear that $w^2 = w$ and $\langle w, e \rangle = 2$. Furthermore, $x \times y = 0$ if and only if $w \times y = 0$.

We set $F = Ke \oplus Kw$. It is straightforward to see that F is non-degenerate with respect to $\langle\, ,\, \rangle$ and we set $E = F^{\perp}$. Thus $A = E \perp F$. An elementary calculation using Lemma 5.2.1(i) of [5] shows that $w \times F \leq F$, $w \times E \leq E$. Furthermore, $w \times z = 0$ for $z \in F$ implies that $z = 0$. Consequently, any non-zero y with $w \times y = 0$ must be in E.

Suppose now that $y \in E$. Then we find that $w \times y = wy - y$. Thus the y we are seeking are those elements in E with $wy = y$. We set $u = e - w$, so that u is a primitive idempotent in A. Then, working in terms of u, the y we seek are elements in E satisfying $uy = 0$. In the notation of Sect. 3.5.3 of [5], such elements constitute the subspace E_0 of E. We obtain a non-zero solution y provided $E_0 \neq 0$. However, Proposition 5.3.5 of [5] shows that if $E_0 = 0$, then $\dim_K A = 2$, which is absurd. Consequently, E_0 is non-zero and we obtain a non-zero solution for y, as required.

We note that in the final part of the argument above, the radical of S_x is precisely E_0. We will see this fact in evidence in the next lemma.

Next, we consider two idempotents in A whose associated bilinear forms can be investigated in some detail.

Lemma 3.3. *Let w be the element*

$$\begin{pmatrix} 1 & 0 & 0 \\ 0 & 1 & 0 \\ 0 & 0 & 0 \end{pmatrix}$$

and let $u = e - w$. Then the radical of F_w has dimension 9 and the radical of F_u has dimension 17. Thus F_w has rank 18 and F_u has rank 10.

Proof. We note that $w^2 = w$ and $\langle w, e \rangle = 2$. It follows from Theorem 3.2 that the elements y satisfying $w \times y = 0$ are those in the subspace denoted by E_0. Such elements satisfy $\langle y, e \rangle = 0$ and $\langle y, w \rangle = 0$. A straightforward calculation shows that E_0 consists of all elements

$$\begin{pmatrix} \lambda & c & 0 \\ \overline{c} & -\lambda & 0 \\ 0 & 0 & 0 \end{pmatrix},$$

where $\lambda \in K$ and $c \in C$. This subspace is clearly 9-dimensional, as required. (We note that the dimension formulae $\dim E_0 = 9$, $\dim E_1 = 16$ occur on p.174 of [5].)

The element u is also an idempotent and it satisfies $\langle u, e \rangle = 1$. Now $u \times F \leq F$ and $u \times E \leq E$. An easy calculation shows that $u \times y = uy - y/2$ when $y \in E$. Thus, in the notation of Lemma 5.3.1 of [5], the elements $y \in E$ with $u \times y = 0$ are precisely those in the subspace E_1. We also have

$$\dim E_1 = \dim E - \dim E_0 = \dim F^{\perp} - \dim E_0 = \dim A - \dim F - \dim E_0 = 16.$$

Since $u \times u = 0$ and $u \times e \neq 0$, it follows that those z in F with $u \times z = 0$ are the scalar multiples of u. We, thus, see that the subspace of all those y in A with $u \times y = 0$ is 17-dimensional, as required. (Of course, this dimension could also have been obtained by explicit identification of the subspace of those y with $u \times y = 0$.)

The calculation of the rank of F_u is also performed in [1, pp. 367–368].

Let $\mathrm{NP}(A)$ be the subgroup of group of all invertible linear transformations g of A into itself satisfying

$$\det(g(x)) = \det(x)$$

for all $x \in A$. Note that any g in $\mathrm{NP}(A)$ also satisfies

$$\langle g(x), g(y), g(z) \rangle = \langle x, y, z \rangle$$

for all x, y and z in A.

Lemma 3.4. *Let x be an element of A and g an element of G. Then F_x and $F_{g(x)}$ have the same rank.*

Proof. Let $y \in A$ be an element in the radical of F_x. Then we have

$$\langle x \times y, z \rangle = 0$$

for all $z \in A$. Applying g, we obtain

$$\langle g(x) \times g(y), g(z) \rangle = \langle x \times y, z \rangle = 0$$

for all $z \in A$. Since g is invertible, we deduce that $g(y)$ is in the radical of $F_{g(x)}$. Conversely, it is straightforward to prove that if w is in the radical of $F_{g(x)}$, then $g^{-1}(w)$ is in the radical of F_x. This proves what we want.

Suppose now that the octonion algebra C is split. It is known that in this case $\mathrm{NP}(A)$ is either a perfect threefold cover of the simple group $E_6(K)$ of split type over K when K contains an element of order 3, or else is the split group $E_6(K)$ itself. Furthermore, in its action on the one-dimensional subspaces of A, $\mathrm{NP}(A)$ has exactly three orbits, by a theorem of [4]. In fact, we can be more precise, since the orbits of $\mathrm{NP}(A)$ on the non-zero elements of A have representatives

$$\begin{pmatrix} 1 & 0 & 0 \\ 0 & 0 & 0 \\ 0 & 0 & 0 \end{pmatrix}, \quad \begin{pmatrix} 1 & 0 & 0 \\ 0 & 1 & 0 \\ 0 & 0 & 0 \end{pmatrix}, \quad \begin{pmatrix} 1 & 0 & 0 \\ 0 & 1 & 0 \\ 0 & 0 & k \end{pmatrix},$$

where k runs over the non-zero elements of K. An element of the third type above has non-zero determinant k, and thus of course two such elements with different k-values cannot be in the same $\mathrm{NP}(A)$-orbit. For a discussion of this description of the orbits, see for example [3, p. 938]. Theorem 7 of [2] is essentially a proof that elements of the third type are representatives of the $\mathrm{NP}(A)$-orbits on the elements with non-zero determinant. Note that the first two orbit representatives are the elements considered in Lemma 3.3.

There is a concept of *rank* in the theory of exceptional simple Jordan algebras. As we shall use the word rank in its traditional sense in linear algebra, we will refer instead to *Jordan rank*. Elements in the first orbit have Jordan rank 1 and the orbit comprises non-zero elements x satisfying $x \times x = 0$. Elements in the second orbit have Jordan rank 2 and the orbit comprises those elements x that satisfy $\det(x) = 0$, $x \times x \neq 0$. The elements in the third orbit have non-zero determinant and Jordan rank 3. Such elements have a Jordan inverse.

Since we can identify the one-dimensional subspaces of A with the one-dimensional subspaces of the space $\mathcal{N}$ of symmetric bilinear forms, it follows that in its natural action on $\mathcal{N}$, $\mathrm{NP}(A)$ also has three orbits on these subspaces. The orbits can be described in terms of the possible ranks of the elements F_x, as we deduce from Lemma 3.3 and the next simple result.

Corollary 3.5. *Suppose that the octonion algebra C is split. Then the non-zero elements of $\mathcal{N}$ have rank* 10, 18 *or* 27. *The elements of rank* 10 *correspond to elements of Jordan rank* 1, *those of rank* 18 *to elements of Jordan rank* 2, *and those of rank* 27 *to elements of Jordan rank* 3.

Proof. We know from Lemma 3.3 that $\mathcal{N}$ contains elements of rank 10, 18 and 27. Since $\mathrm{NP}(A)$ has exactly three orbits in its action on the one-dimensional subspaces of $\mathcal{N}$, namely the subspaces corresponding to elements of Jordan rank 1, 2 and 3, the result is immediate from Lemma 3.4.

We can extend part of the conclusion of Corollary 3.5 to all octonion algebras by exploiting the algebraic closure of the field K.

Corollary 3.6. *The non-zero elements of $\mathcal{N}$ have rank* 10, 18 *or* 27 *for any octonion algebra C (split or non-split).*

Proof. Let $\overline{K}$ denote an algebraic closure of K and let $\overline{C} = \overline{K} \otimes_K C$. $\overline{C}$ is an octonion algebra over $\overline{K}$, and it is split, since $\overline{K}$ is algebraically closed. Let $\overline{A}$ denote the Jordan algebra over $\overline{K}$ consisting of 3×3 hermitian matrices with entries in $\overline{C}$. A is naturally a K-subspace of $\overline{A}$ and its $\overline{K}$-span is $\overline{A}$. Likewise, let $\overline{\mathcal{N}}$ denote the 27-dimensional subspace of symmetric bilinear forms derived from $\overline{A}$.

$\mathcal{N}$ is a K-subspace of $\overline{\mathcal{N}}$ whose $\overline{K}$-span is $\overline{\mathcal{N}}$. Given a bilinear form F_x, say, in $\mathcal{N}$, we may naturally extend F_x to an element $\overline{F_x}$, say, of $\overline{\mathcal{N}}$ by extension of scalars. The rank of $\overline{F_x}$ is identical with that of F_x. Thus, since $\mathcal{N}$ contains elements of rank 10, 18 and 27, and $\overline{\mathcal{N}}$ contains only elements of these three ranks, the corollary follows.

We have the following result concerning the 27-dimensional fundamental representation of the split group of type E_6.

Theorem 3.7. *Let K be a field of characteristic different from* 2 *or* 3. *Then the fundamental representation of degree* 27 *of the split algebraic group of type E_6 may be realized on a* 27*-dimensional space $\mathcal{N}$ of symmetric bilinear forms whose non-zero elements have rank* 10, 18 *or* 27. *Moreover, in its action on $\mathcal{N}$, the split group of type E_6 acts with exactly three orbits on the one-dimensional subspaces. Representatives for the orbits are the subspaces spanned by any form of rank* 10, *any form of rank* 18 *and any form of rank* 27.

4 Subspaces with Special Rank Properties

Let x be a fixed non-zero element of Jordan rank 1 or 2. and let R be the radical of the symmetric bilinear form F_x. We saw that this consists of all elements y satisfying

$$x \times y = 0.$$

Since $x \times y = y \times x$, it is clear that y is in the radical of F_x if and only if x is in the radical of F_y.

We now set

$$\mathcal{R} = \{\, F_y \in \mathcal{N} : y \in R\,\}.$$

Each non-zero element of $\mathcal{R}$ contains x in its radical and hence has rank 10 or 18. Furthermore

$$\dim \mathcal{R} = 9$$

if x has Jordan rank 2 and

$$\dim \mathcal{R} = 17$$

if x has Jordan rank 1.

Consider the case that x has Jordan rank 1. Up to action by NP(A), we may take x to be

$$\begin{pmatrix} 1 & 0 & 0 \\ 0 & 0 & 0 \\ 0 & 0 & 0 \end{pmatrix}.$$

A straightforward calculation shows that R consists of all elements of the form

$$\begin{pmatrix} \alpha & c & \overline{b} \\ \overline{c} & 0 & 0 \\ b & 0 & 0 \end{pmatrix},$$

where $\alpha \in K, b \in C, c \in C$.

Consider the 16-dimensional subspace S of R consisting of all elements y of the form

$$\begin{pmatrix} 0 & c & \overline{b} \\ \overline{c} & 0 & 0 \\ b & 0 & 0 \end{pmatrix}.$$

Such a non-zero element has Jordan rank 1 or 2. We now check $y \times y$. Since $\langle y, e\rangle = 0$,

$$y \times y = y^2 - \frac{1}{2}\langle y, y\rangle e.$$

We easily calculate that

$$y^2 = \begin{pmatrix} N(c)+N(b) & 0 & 0 \\ 0 & N(c) & \overline{bc} \\ 0 & bc & N(b) \end{pmatrix}.$$

It follows that $\langle y, y \rangle = N(b) + N(c)$ and

$$y \times y = \begin{pmatrix} 0 & 0 & 0 \\ 0 & -N(b) & \overline{bc} \\ 0 & bc & -N(c) \end{pmatrix}.$$

Thus, in order that $y \times y = 0$ (so that y has Jordan rank 1) we require

$$N(b) = N(c) = 0.$$

This can only happen if C is split. Thus, if C is a division algebra, all non-zero elements of S have Jordan rank 2.

Correspondingly, we have the following result about the space $\mathcal{N}$ of symmetric bilinear forms.

Theorem 4.1. *Suppose that C is a division algebra. Let S be the 16-dimensional subspace of C constructed above whose non-zero elements all have Jordan rank 2. Then there is a corresponding 16-dimensional constant rank 18 subspace of $\mathcal{N}$.*

The existence of such a constant rank subspace seem to me to be unusual in the theory of symmetric bilinear forms.

We now want to look at constant rank 10 subspaces of $\mathcal{N}$. In this context, we need to know about when a linear combination of two elements of Jordan rank 1 can also have Jordan rank 1.

Lemma 4.2. *Let x and y be elements of Jordan rank 1 in A and let λ be a non-zero element of K. Then $x + \lambda y$ has Jordan rank 1 if and only if*

$$x \times y = 0.$$

Proof. The element $x + \lambda y$ has Jordan rank 1 if and only if

$$(x + \lambda y) \times (x + \lambda y) = 0.$$

Using the bilinearity of the cross product and the fact that $x \times y = y \times x$, the result follows.

The following is a theorem of Jacobson [2], Proposition 6.

Lemma 4.3. *Suppose that C is a division algebra and let x and y be linearly independent elements of Jordan rank 1 in A. Then $x \times y$ has Jordan rank 1.*

Corollary 4.4. *Suppose that C is a division algebra. Then the maximum dimension of a subspace of A whose non-zero elements all have Jordan rank 1 is 1. Equivalently, the maximum dimension of a constant rank 10 subspace of $\mathcal{N}$ is also 1. Furthermore, if x and y are linearly independent elements of Jordan rank 1, $x + y$ has Jordan rank 2.*

What about subspaces of elements of Jordan rank 1 in the split case? Here, we have information from a paper of Springer and Veldkamp [6].

We suppose that C is split, and let $a \in C$ be a non-zero element with $N(a) = 0$. Then the set of $c \in C$ satisfying

$$ac = 0$$

is a 4-dimensional subspace of C. Consider all elements $z \in A$ with

$$z = \begin{pmatrix} 0 & 0 & 0 \\ 0 & 0 & c \\ 0 & \overline{c} & \mu \end{pmatrix},$$

where μ is an arbitrary element of K. This set is a 5-dimensional subspace in which each non-zero element has Jordan rank 1. Correspondingly, $\mathcal{N}$ contains a 5-dimensional constant rank 10 subspace. Likewise, consider all elements

$$w = \begin{pmatrix} 0 & \lambda a & 0 \\ \lambda \overline{a} & \mu & c \\ 0 & \overline{c} & 0 \end{pmatrix},$$

where λ, μ are arbitrary elements of K and a and c are as before.

This subset is a 6-dimensional subspace of A in which all non-zero elements have Jordan rank 1. Correspondingly, $\mathcal{N}$ contains a 6-dimensional constant rank 10 subspace.

If we understand the Springer–Veldkamp paper correctly, the two constant rank 10 subspaces of $\mathcal{N}$ constructed are maximal with respect to the constant rank property and all other maximal constant rank subspaces are obtained by transitive action of the group $\mathrm{NP}(A)$.

References

1. N. Jacobson, *Structure and representations of Jordan algebras*, American Mathematical Society, Providence, Rhode Island, 1968
2. N. Jacobson, *Some groups of transformations defined by Jordan algebras*. III, J. Reine Angew. Math.**207** (1961), 61–85
3. S. Krutelevich, *Jordan algebras, exceptional groups, and Bhargava compositions*, J. Algebra **314** (2007), 924–977
4. J. G. M. Mars, *Les nombres de Tamagawa de certains groupes excepionnels*, Bull. Soc. Math. France **94** (1966), 97–140

5. T. A. Springer and F. D. Veldkamp, *Octonions, Jordan algebras and exceptional groups*, Springer, Berlin, 2000
6. T. A. Springer and F. D. Veldkamp, *On Hjelmslev-Moufang planes*, Math. Zeit. **107** (1968), 249–263
7. R. A. Wilson, *The finite simple groups*, Springer, London, 2009

On the Geometry of Global Function Fields, the Riemann–Roch Theorem, and Finiteness Properties of S-Arithmetic Groups

Köhl né Gramlich

Abstract In this survey I sketch Behr–Harder reduction theory for reductive groups over global functions fields and briefly describe its applicability in the theory of S-arithmetic groups, notably homological and isoperimetric properties.

Subject Classifications: MSC 2010: Primary 20-02 Secondary 11F75, 14L15, 20G30, 20G35, 20J06, 51E24

Keywords S-arithmetic groups over global function fields • Reduction theory • Bruhat–Tits buildings • Finiteness properties • Isoperimetric properties

1 Introduction

Harder's reduction theory [23, 24] provides filtrations of Euclidean buildings that allow one to deduce cohomological [26] and homological [16, 40] properties of S-arithmetic groups over global function fields. In this survey, I will sketch the main points of Harder's reduction theory, starting from Weil's geometry of numbers and the Riemann–Roch theorem. I will describe a filtration, used for example in [5], that is particularly useful for deriving finiteness properties of S-arithmetic groups. Finally, I will state the recently established rank theorem and some of its earlier partial verifications that do not restrict the cardinality of the underlying field of constants. As a motivation for further research I also state a much more general conjecture on isoperimetric properties of S-arithmetic groups over global fields (number fields or function fields).

K. né Gramlich (✉)
Justus-Liebig-Universität Gießen, Mathematisches Institut,
Arndtstraße 2, 35392 Gießen, Germany
e-mail: ralf.koehl@math.uni-giessen.de; ralf.gramlich@math.uni-giessen.de

N.S.N. Sastry (ed.), *Buildings, Finite Geometries and Groups*, Springer Proceedings in Mathematics 10, DOI 10.1007/978-1-4614-0709-6_4,

2 Projective Varieties

In this section, I give a quick introduction to the concept of projective varieties. For further reading the sources [27, I] and [33, 2] are highly recommended.

Definition 2.1. Let k be a perfect field, let $\overline{k}$ be its algebraic closure, and let $S \subset \overline{k}[X] = \overline{k}[x_0, x_1, ..., x_n]$ be a set of homogeneous polynomials. The set

$$Z(S) = \{P \in \mathbb{P}_n(\overline{k}) \mid f(P) = 0 \text{ for all } f \in S\}$$

is called a **projective algebraic set**. The algebraic set $Z(S)$ is **defined over** k, if S can be chosen to be contained in $k[X]$. The **Zariski topology** on $\mathbb{P}_n(\overline{k})$ is defined by taking the closed sets to be the projective algebraic sets.

A non-empty projective algebraic set in $\mathbb{P}_n(\overline{k})$ is called a **projective variety**, if it is irreducible in the Zariski topology of $\mathbb{P}_n(\overline{k})$, i.e., if it is not equal to the union of two proper closed subsets. It is called a **projective** k**-variety** if it is defined over k. The **dimension** of a non-empty projective variety is defined to be its dimension as a topological space in the induced Zariski topology, i.e., the supremum of all integers n such that there exists a chain $Z_0 \subsetneq Z_1 \subsetneq \cdots \subsetneq Z_n$ of distinct non-empty irreducible closed subsets.

Theorem 2.2 ([33, 2.3.8]). *A projective algebraic set V is a projective variety if and only if the ideal $I(V)$ of $\overline{k}[X]$ generated by the set $\{f \in \overline{k}[X] \mid f$ homogeneous and $f(P) = 0$ for all $P \in V\}$ is a prime ideal of $\overline{k}[X]$.*

Definition 2.3. A non-empty intersection of a projective variety in $\mathbb{P}_n(\overline{k})$ with an open subset of $\mathbb{P}_n(\overline{k})$ is called a **quasi-projective variety**.

Let $V \subseteq \mathbb{P}_n(\overline{k})$ be a quasi-projective variety. For each $P \in V$ there exists a hyperplane $H \subset \mathbb{P}_n(\overline{k})$ with $P \notin H$. Then $P \in V \setminus H = V \cap A_n(\overline{k})$ for $A_n(\overline{k}) := \mathbb{P}_n(\overline{k}) \setminus H$. Let r be a defining relation of the hyperplane H in the variables $x_0, ..., x_n$ (cf. Definition 2.1). A $\overline{k}$-valued function f on V is called *regular at* P, if there exists a neighbourhood N of P in $V \cap A_n(\overline{k})$ such that there exist polynomials $a, b \in \overline{k}[X]/(r)$ with $b(Q) \neq 0$ for all $Q \in N$ and $f_{|N} = \frac{a_{|N}}{b_{|N}}$. The function f is *regular* on a non-empty open subset U of V, if it is regular at every point of U.

Definition 2.4. Let V be a quasi-projective variety and let $P \in V$. Then the **local ring** $\mathcal{O}_P = \mathcal{O}_P(V)$ **at** P is defined as the ring of germs $[f]$ of functions $f : V \to \overline{k}$ which are regular on a neighbourhood of P. In other words, an element of $\mathcal{O}_P$ is an equivalence class of pairs (U, f) where U is an open subset of Y containing P, and f is a regular function on U, and two such pairs (U, f) and (V, g) are equivalent if $f_{|U \cap V} = g_{|U \cap V}$. For a non-empty open subset U of V define $\mathcal{O}_U = \mathcal{O}_U(V) := \bigcap_{P \in U} \mathcal{O}_P(V)$.

The ring $\mathcal{O}_P$ is indeed a local ring in the sense of commutative algebra: its unique maximal ideal $\mathfrak{m}_P$ is the set of germs of regular functions which vanish at P. The ith power $\mathfrak{m}_P^i$ of $\mathfrak{m}_P$ consists of the germs of regular functions whose vanishing order at

P is at least i. Taking these powers $\mathfrak{m}_P^i$ as a neighbourhood basis of $\mathcal{O}_P$, this defines a topology on $\mathcal{O}_P$, the $\mathfrak{m}_P$*-adic topology*. The inverse limit $\hat{\mathcal{O}}_P := \lim_{\leftarrow} \mathcal{O}_P/\mathfrak{m}_P^i$ is the *completion* of $\mathcal{O}_P$; cf. [20, 7.1], [27, p. 33], [50, VIII, Sect. 2]. It is a local ring whose maximal ideal is denoted by $\hat{\mathfrak{m}}_P$.

Example 2.5. The projective line $\mathbb{P}_1(\mathbb{C}) \cong \mathbb{C} \cup \{\infty\}$ is a non-singular projective curve (see Definition 3.1 below). For $P \in \mathbb{C}$ one has

$$\mathcal{O}_P = \left\{\frac{a}{b} \in \mathbb{C}(t) \mid b(P) \neq 0\right\},$$

$$\mathfrak{m}_P = \left\{\frac{a}{b} \in \mathcal{O}_P \mid a(P) = 0\right\},$$

$$\hat{\mathcal{O}}_P = \left\{\frac{a}{b} \in \mathbb{C}((t)) \mid b(P) \neq 0\right\},$$

$$\hat{\mathfrak{m}}_P = \left\{\frac{a}{b} \in \hat{\mathcal{O}}_P \mid a(P) = 0\right\}.$$

Definition 2.6. Let V be a variety defined over k. Then the **field $\overline{k}(V)$ of rational functions of** V consists of the equivalence classes of pairs (U, f) where U is a non-empty open subset of V and f is a regular function on U, and two such pairs (U, f) and (V, g) are equivalent if $f_{|U\cap V} = g_{|U\cap V}$. The **field of k-rational functions of** V/k is

$$K := k(V) := \left\{h \in \overline{k}(V) \mid \sigma(h) = h \text{ for all } \sigma \in \mathrm{Gal}(\overline{k}/k)\right\}.$$

Definition 2.7. For a point $P \in \mathbb{P}_n(\overline{k})$ the set $\{\sigma(P) \mid \sigma \in \mathrm{Gal}(\overline{k}/k)\}$ is called a **k-closed point**. A k-closed point of cardinality 1 is called **k-rational**.

For each $P \in \mathbb{P}_n(\overline{k})$, each homogeneous polynomial $f \in \overline{k}[X]$, and each $\sigma \in \mathrm{Gal}(\overline{k}/k)$ one has $f(P) = 0$ if and only if $\sigma(f(P)) = 0$ if and only if $\sigma(f)(\sigma(P)) = 0$. Hence, given a projective k-variety $V \subseteq \mathbb{P}_n(\overline{k})$, the assertion $P \in V$ is equivalent to the assertion $\{\sigma(P) \mid \sigma \in \mathrm{Gal}(\overline{k}/k)\} \subset V$. It, therefore, makes sense to speak of *k-closed points* of V. The set of k-closed points of V is denoted by $V^\circ = V^\circ/k$. The *degree* $\deg(P)$ of a k-closed point P equals the number of points it contains.

Example 2.8. For an involutive automorphism σ of $\mathbb{C}$, define $\mathbb{K} := \mathrm{Fix}_{\mathbb{C}}(\sigma)$. The set of $\mathbb{K}$-closed points of $\mathbb{P}_1(\mathbb{C})$ equals the set of σ-orbits of $\mathbb{P}_1(\mathbb{C})$, i.e. the σ-fixed/$\mathbb{K}$-rational points of $\mathbb{P}_1(\mathbb{C})$ and the pairs of σ-conjugate points of $\mathbb{P}_1(\mathbb{C})$.

3 Curves Over Finite Fields Considered as Varieties

In this section, I turn to one of the main objects of study of this survey, non-singular projective curves. For further reading, the sources [27, I], [33, 3], and [37, II] or, if one prefers a very algebraic approach, [17, I, II] and [35, 5, 6] are highly recommended.

Definition 3.1. Let k be a perfect field. A projective variety of dimension 1 defined over k is called a **projective curve** over k (cf. Definition 2.1). A projective curve Y is called **non-singular** at $P \in Y$ if $\mathcal{O}_P$ (cf. Definition 2.4) is a discrete valuation ring. It is called **non-singular** if it is non-singular at each of its points.

For Y non-singular at P, the valuation of $\mathcal{O}_P$ is given by

$$\overline{\nu}_P : \mathcal{O}_P \to \mathbb{Z} \cup \{\infty\} : x \mapsto \sup\left\{i \in \mathbb{N} \cup \{0\} \mid x \in \mathfrak{m}_P^i\right\},$$

with the understanding that $\mathfrak{m}_P^0 = \mathcal{O}_P$. This valuation extends to the valuation $\hat{\mathcal{O}}_P \to \mathbb{Z} \cup \{\infty\} : x \mapsto \sup\left\{i \in \mathbb{N} \cup \{0\} \mid x \in \hat{\mathfrak{m}}_P^i\right\}$, as $\hat{\mathfrak{m}}_P \cap \mathcal{O}_P = \mathfrak{m}_P$; cf. [20, 7.1], [27, I.5.4A]. The valuation $\overline{\nu}_P : \mathcal{O}_P \to \mathbb{Z} \cup \{\infty\}$ also extends to a valuation on its field of fractions $\overline{k}(Y)$ via $\overline{\nu}_P(x) := \overline{\nu}_P(a) - \overline{\nu}_P(b)$ for $a, b \in \mathcal{O}_P$ with $\frac{a}{b} = x$.

The geometric intuition is that $\overline{\nu}_P(x)$ indicates whether P is a zero or a pole of x and counts its multiplicity (cf. Example 2.5).

An *algebraic function field* (in one variable over k) is an extension field $\mathbb{F}$ of k that admits an element x that is transcendental over k such that $\mathbb{F}/k(x)$ is a field extension of finite degree.

Theorem 3.2 ([33, 3.2.9]). *There is a one-to-one correspondence between k-isomorphism classes of non-singular projective curves over k and k-isomorphism classes of algebraic function fields of one variable with full constant field k, via the map $Y/k \mapsto k(Y)$ (cf. Definition 2.6).*

Here, the *full constant field* of an algebraic function field K over k is the algebraic closure of k in K. In case the full constant field is finite, K is a *global function field.* For example, $\mathbb{F}_q(t)$ is a global function field with full constant field $\mathbb{F}_q$.

Definition 3.3. Two discrete valuations ν_1 and ν_2 of an algebraic function field K are **equivalent** if there exists a constant $c > 0$ such that $\nu_1(x) = c\nu_2(x)$ for all $0 \neq x \in K$. An equivalence class of discrete valuations is called a **place**. The **degree** of a valuation/place is the degree of the residue class field $\mathcal{O}_P/\mathfrak{m}_P$ over the constant field k. This is always a finite number; cf. [33, 1.5.13].

Theorem 3.4 ([33, 3.1.15]). *Let $Y/\mathbb{F}_q$ be a non-singular projective curve. Then there exists a natural one-to-one correspondence between $\mathbb{F}_q$-closed points of Y and places (cf. Definition 3.3) of the field $K = \mathbb{F}_q(Y)$ of $\mathbb{F}_q$-rational functions (cf. Definition 2.6). Moreover, the degree of an $\mathbb{F}_q$-closed point is equal to the degree of the corresponding place.*

3.5. Given a point P in an $\mathbb{F}_q$-closed point of Y, the valuation $\overline{\nu}_P : \mathcal{O}_P \to \mathbb{Z} \cup \{\infty\}$ (cf. Definition 3.1) restricts to a valuation $\nu_P : \mathcal{O}_{P,K} := \mathcal{O}_P \cap \mathbb{F}_q(Y) \to \mathbb{Z} \cup \{\infty\}$. By [33, 3.1.14] this valuation only depends on the $\mathbb{F}_q$-closed point of Y containing P and not on the particular choice of P inside that $\mathbb{F}_q$-closed point. As before, ν_P extends to the field of $\mathbb{F}_q$-rational functions $K = \mathbb{F}_q(Y)$, the completion $\hat{\mathcal{O}}_{P,K} := \widehat{\mathcal{O}_P \cap K}$, and to the field of fractions K_P of the completion $\hat{\mathcal{O}}_{P,K}$. The field K_P is the **local function field at** P. For example, $\mathbb{F}_q((t))$ is the local function field of the global function field $\mathbb{F}(t)$ at the place corresponding to the irreducible polynomial t.

Definition 3.6. Let $Y/\mathbb{F}_q$ be a non-singular projective curve. The **Weil divisor group** $\mathrm{Div}(Y) = \mathrm{Div}(Y/\mathbb{F}_q)$ is the free abelian group over the set Y° of $\mathbb{F}_q$-closed points of Y (cf. Definition 2.7). An element $D = \sum_{P\in Y^\circ} n_P P \in \mathrm{Div}(Y)$ is called a **Weil divisor** of Y. It is **effective**, in symbols $D \geq 0$, if $n_P \geq 0$ for all $P \in Y^\circ$. For two divisors D_1 and D_2 of Y one writes $D_1 \geq D_2$, if $D_1 - D_2 \geq 0$. The **degree** $\deg(D)$ of a Weil divisor $D = \sum_{P\in Y^\circ} n_P P$ is given by $\deg(D) := \sum_{P\in Y^\circ} n_P \deg(P)$ (cf. Definition 2.7). Also, define $\nu_P(D) := n_P$.

For $0 \neq x \in K = \mathbb{F}_q(Y)$, define the *divisor* $\mathrm{div}(x)$ of x by $\mathrm{div}(x) := \sum_{P\in Y^\circ} \nu_P(x)P$. As, by [33, 3.3.2], [35, 5.1], any $0 \neq x \in K$ admits only finitely many zeros (i.e., points $P \in Y^\circ$ with $\nu_P(x) > 0$) and poles (i.e. points $P \in Y^\circ$ with $\nu_P(x) < 0$), the divisor $\mathrm{div}(x)$ indeed is a Weil divisor. A Weil divisor obtained in this way is *principal*, and has degree 0; cf. [33, 3.4.3], [35, 5.1]. Two Weil divisors that differ by a principal divisor are called *equivalent*.

Definition 3.7. For a divisor D of $Y/\mathbb{F}_q$, the **Riemann–Roch space** $L(D)$ is defined as

$$L(D) := \{0 \neq x \in K = \mathbb{F}_q(Y) \mid \mathrm{div}(x) + D \geq 0\} \cup \{0\}.$$

The Riemann–Roch space $L(D)$ of a divisor D is an $\mathbb{F}_q$-vector space of finite dimension (cf. [33, 3.4.1(iv)]).

Definition 3.8. Define the **ring of repartitions** as

$$\mathbb{A}_K := \left\{ (x_P)_{P\in Y^\circ} \in \prod_{P\in Y^\circ} K \mid x_P \in \mathcal{O}_P \text{ for almost all } P \in Y^\circ \right\}.$$

By [17, p. 25], [33, 3.3.2], the field K embeds diagonally in $\mathbb{A}_K$. For any divisor $D \in \mathrm{Div}(Y)$ define $\mathbb{A}_K(D) := \{x \in \mathbb{A}_K \mid \nu_P(x_P) + \nu_P(D) \geq 0 \text{ for all } P \in Y^\circ\}$. Note that $\mathbb{A}_K(D)$ is an $\mathbb{F}_q$-subvector space of $\mathbb{A}_K$.

Definition 3.9. A **Weil differential** of a non-singular projective curve $Y/\mathbb{F}_q$, resp. of its global function field K, is an $\mathbb{F}_q$-linear map $\omega : \mathbb{A}_K \to \mathbb{F}_q$ such that $\omega_{|\mathbb{A}_K(D)+K} = 0$ for some divisor $D \in \mathrm{Div}(Y)$. Denote by Ω_K the set of all Weil differentials of K and by $\Omega_K(D)$ the set of all Weil differentials of K that vanish on $\mathbb{A}_K(D) + K$.

Note the difference between the ring of repartitions $\mathbb{A}_K$ and the ring of adèles $\hat{\mathbb{A}}_K$ defined in Definition 4.3 below (completions). The concept of a Weil differential can equally well be introduced using the ring of adèles $\hat{\mathbb{A}}_K$, cf. [35].

Observation 3.10. *Let* $D \in \mathrm{Div}(Y)$. *Then* $\Omega_K(D) \cong \mathbb{A}_K/\mathbb{A}_K(D) + K$ *as* $\mathbb{F}_q$-*vector spaces.*

If $0 \neq \omega \in \Omega_K$, then by [33, 3.6.11], [35, 6.8] the set $\{D \in \mathrm{Div}(Y) \mid \omega_{|\mathbb{A}_K(D)+K}=0\}$ has a unique maximal element with respect to $\geq$. This maximal element is called a *canonical divisor* of Y and denoted by (ω). By [33, 3.6.10], [35, 6.10]

$$\dim_K(\Omega_K) = 1 \tag{1}$$

and by [33, 3.6.13], [35, 6.9] the Weil differential $x\omega : \mathbb{A}_K \to \mathbb{F}_q : a \mapsto \omega(xa)$ satisfies

$$(x\omega) = \mathrm{div}(x) + (\omega) \tag{2}$$

for any $0 \neq x \in K$, $0 \neq \omega \in \Omega_K$. Hence, any two canonical divisors of Y are equivalent (cf. Definition 3.6).

Observation 3.11. *Let D be a divisor of $Y/\mathbb{F}_q$ and let $W = (\omega)$ be a canonical divisor of Y. Then*

$$L(W - D) \to \Omega_K(D) : x \mapsto x\omega$$

is an isomorphism of $\mathbb{F}_q$-vector spaces.

Proof. For $0 \neq x \in L(W - D)$ we have[1] $(x\omega) \overset{(2)}{=} \mathrm{div}(x) + (\omega) \geq -(W - D) + W = D$, so that indeed $x\omega \in \Omega_K(D)$. Injectivity and $\mathbb{F}_q$-linearity of the map $x \mapsto x\omega$ are clear. In order to prove surjectivity, let $0 \neq \omega_1 \in \Omega_K(D)$. By (1) there exists $0 \neq x \in K$ such that $\omega_1 = x\omega$. Since we have $\mathrm{div}(x) + W \overset{(2)}{=} (x\omega) = (\omega_1) \geq D$, it follows that $x \in L(W - D)$.

Definition 3.12. Let $Y/\mathbb{F}_q$ be a non-singular projective curve and let W be a canonical divisor of Y. Define the **genus** g of the curve $Y/\mathbb{F}_q$ by

$$g := \dim_{\mathbb{F}_q}(L(W)) \overset{3.11}{=} \dim_{\mathbb{F}_q}(\Omega_K(0)) \overset{3.10}{=} \dim_{\mathbb{F}_q}(\mathbb{A}_K/\mathbb{A}_K(0) + K).$$

Theorem 3.13 (Riemann–Roch Theorem, [17, p. 30], [33, 3.6.14], [35, 5.4]). *Let $Y/\mathbb{F}_q$ be a non-singular projective curve of genus g and let W be a canonical divisor. Then for any divisor $D \in \mathrm{Div}(Y)$ one has*

$$\dim_{\mathbb{F}_q}(L(D)) - \dim_{\mathbb{F}_q}(L(W - D)) = \deg(D) + 1 - g.$$

The Riemann–Roch theorem combined with the study of the zeta function of the global function field $\mathbb{F}_q(Y)$ allows one to establish the following useful result.

Proposition 3.14 ([33, 1.10]). *Let $Y/\mathbb{F}_q$ be a non-singular projective curve. Then there exists a Weil divisor of Y of degree* 1.

Note that this Weil divisor of degree 1 need not be effective. In fact, for $p \geq 5$ prime the non-singular projective curve over $\mathbb{F}_p$ given by $x^{p-1} + y^{p-1} = 3z^{p-1}$ does not

[1] The notation $\overset{(2)}{=}$ means that (2) is a justification for the equality.

admit any $\mathbb{F}_p$-rational point and, thus, there cannot exist an effective Weil divisor of degree 1 of this curve.

Definition 3.15. A **vector bundle** of rank r over a curve Y/k is a variety E/k together with a morphism $\pi : E \to Y$ such that there exists an open covering $\{U_i\}$ of Y and isomorphisms $\phi_i : \pi^{-1}(U_i) \to U_i \times A_r(k)$ (where $A_r(k)$ denotes the affine space over k of dimension r) such that for each pair U_i, U_j the composition $\phi_j \phi_i^{-1}{}_{|U_i \cap U_j}$ equals $(\mathrm{id}, \phi_{i,j})$ for a linear map $\phi_{i,j}$. A **section** of a vector bundle $\pi : E \to Y$ over an open set $U \subset Y$ is a map $s : U \to E$ such that $\pi \circ s = \mathrm{id}$.

4 Geometry of Numbers

In this section, I describe Weil's geometry of numbers. For further reading, the sources [44, 2], [45, VI] are highly recommended.

4.1. Let $Y/\mathbb{F}_q$ be a non-singular projective curve, let P be an $\mathbb{F}_q$-closed point of Y, and let ν_P be the valuation of $K = \mathbb{F}_q(Y)$ discussed in Theorem 3.5. Then

$$|\cdot|_P : K \to \mathbb{R} : x \mapsto (q^{\deg(P)})^{-\nu_P(x)}$$

defines an absolute value on K. The completion of K with respect to this absolute value equals K_P (cf. Theorem 3.5), which is locally compact as the zero neighbourhood $\hat{\mathcal{O}}_{P,K}$ is an inverse limit of finite rings (cf. Definition 2.4), whence compact. On each field K_P define the canonical Haar measure to be the one with respect to which $\hat{\mathcal{O}}_{P,K}$ has volume 1.

Theorem 4.2 ([4]). *Let $Y/\mathbb{F}_q$ be a non-singular projective curve. Then each $0 \neq x \in K$ satisfies $|x|_P = 1$ for almost all $P \in Y^\circ$. Moreover, for each $0 \neq x \in K$,*

$$\prod_{P \in Y^\circ} |x|_P = 1.$$

Definition 4.3. In analogy to Definition 3.8 define the **adèle ring**

$$\hat{\mathbb{A}}_K := \left\{ (x_P)_{P \in Y^\circ} \in \prod_{P \in Y^\circ} K_P \mid x_P \in \hat{\mathcal{O}}_P \text{ for almost all } P \in Y^\circ \right\}.$$

It is a locally compact ring and contains K embedded diagonally as a discrete subring (cf. Theorem 4.2). Define

$$|\cdot| : \hat{\mathbb{A}}_K \to \mathbb{R} : x \mapsto \prod_{P \in Y^\circ} |x|_P.$$

The product measure of the canonical Haar measures on each individual K_P (4.1) yields a Haar measure on the locally compact group $(\hat{\mathbb{A}}_K, +)$. Let $\omega_{\hat{\mathbb{A}}_K}$ denote the corresponding volume form. For any divisor $D \in \mathrm{Div}(Y)$, define $\hat{\mathbb{A}}_K(D) := \{x \in \hat{\mathbb{A}}_K \mid \nu_P(x_P)+\nu_P(D) \geq 0 \text{ for all } P \in Y^\circ\}$. Note that $\hat{\mathcal{O}}_K := \hat{\mathbb{A}}_K(0)$ is the maximal compact subring of $\hat{\mathbb{A}}_K$. It has volume 1 with respect to the chosen Haar measure.

4.4. For a finite subset $S \subset Y^\circ$ define the **ring of S-adèles**

$$\hat{\mathbb{A}}_S := \prod_{P \in S} K_P \times \prod_{P \in Y^\circ \setminus S} \hat{\mathcal{O}}_{P,K}$$

(cf. Theorem 3.5). One has $\hat{\mathbb{A}}_K = \lim_{\rightarrow} \hat{\mathbb{A}}_S$. For K embedded diagonally in $\hat{\mathbb{A}}_K$ as a discrete subring (cf. Definition 4.3) define the **ring of S-integers**

$$\mathcal{O}_S := K \cap \hat{\mathbb{A}}_S = \bigcap_{P \in Y^\circ \setminus S} (K \cap \hat{\mathcal{O}}_{P,K}) \stackrel{3.5}{=} \bigcap_{P \in Y^\circ \setminus S} \mathcal{O}_{P,K} \stackrel{2.4}{=} \mathcal{O}_{Y^\circ \setminus S, K};$$

these are the k-rational functions of Y which are regular outside S. The field K embeds diagonally in $\prod_{P \in S} K_P$ as a dense subring and, hence, also in $\mathbb{A}_S$. Define

$$|\cdot|_S : \prod_{P \in S} K_P \to \mathbb{R} : x \mapsto \prod_{P \in S} |x|_P .$$

4.5. Let G be a unimodular locally compact group, let Γ be a discrete subgroup of G, let H be a compact open subgroup of G, let μ be a Haar measure on G, and assume the double coset space $X \cong H \backslash G / \Gamma$, considered as a system of representatives of the Γ-orbits on $H \backslash G$, is countable. Note that μ is also a Haar measure on H, since H is open, and hence of positive volume. Then, cf. [38, p. 84],

$$\int_{G/\Gamma} d\mu = \sum_{x \in X} \left(\int_{G_x/\Gamma_x} d\mu \right) = \sum_{x \in X} \frac{\int_{G_x} d\mu}{|\Gamma_x|} = \sum_{x \in X} \frac{\int_H d\mu}{|\Gamma_x|} = \int_H d\mu \sum_{x \in X} \frac{1}{|\Gamma_x|}.$$

Here, for choices of $x_0 \in X$ and $(g^x_{x_0})_{x \in X} \in G$ with $g^x_{x_0}(x_0) = x$, the map $\bigsqcup_{x \in X} G_x/\Gamma_x \to G/\Gamma : g\Gamma_x \mapsto g^x_{x_0} g \Gamma$ is an isomorphism of orbit spaces, where G_x and Γ_x denote the stabilizers of x in the respective group.

4.6. One has

$$\begin{aligned} \hat{\mathbb{A}}_K(D) \backslash \hat{\mathbb{A}}_K / K &= \hat{\mathbb{A}}_K / \hat{\mathbb{A}}_K(D) + K \cong \mathbb{A}_K / \mathbb{A}_K(D) + K \\ &\stackrel{3.10}{\cong} \Omega_K(D) \stackrel{3.11}{\cong} L(W - D), \end{aligned} \tag{3}$$

$$K \cap \hat{\mathbb{A}}_K(D) = K \cap \{x \in \hat{\mathbb{A}}_K \mid \nu_P(x) + \nu_P(D) \geq 0 \text{ for all } P \in Y^\circ\} = L(D). \tag{4}$$

The observation that the intersection of the lattice K with the compactum $\hat{\mathbb{A}}_K(D)$ equals $L(D)$ is a classical theme in the geometry of numbers, cf. [3, p. 387].

Proposition 4.7 ([44, 2.1.3], [45, p. 100]). *Using the notation of Definition 4.3, one has*

$$\int_{\hat{\mathbb{A}}_K/K} \omega_{\hat{\mathbb{A}}_K} = q^{g-1}.$$

Proof. For the compact open subgroup $\hat{\mathbb{A}}_K(0)$ of $\hat{\mathbb{A}}_K$ Serre's formula (cf. Theorem 4.5) yields

$$\int_{\hat{\mathbb{A}}_K/K} \omega_{\hat{\mathbb{A}}_K} = \int_{\hat{\mathbb{A}}_K(0)} \omega_{\hat{\mathbb{A}}_K} \sum_{\hat{\mathbb{A}}_K(0)\backslash\hat{\mathbb{A}}_K/K} \frac{1}{|K \cap \hat{\mathbb{A}}_K(0)|} = \frac{|\hat{\mathbb{A}}_K(0)\backslash\hat{\mathbb{A}}_K/K|}{|K \cap \hat{\mathbb{A}}_K(0)|} \int_{\hat{\mathbb{A}}_K(0)} \omega_{\hat{\mathbb{A}}_K}. \tag{5}$$

Therefore, as $\int_{\hat{\mathbb{A}}_K(0)} \omega_{\hat{\mathbb{A}}_K} = 1$ (cf. Definition 4.3), the Riemann–Roch Theorem (cf. Theorem 3.13) or, in fact, its underlying definitions and observations 3.12 and 4.6 imply $\int_{\hat{\mathbb{A}}_K/K} \omega_{\hat{\mathbb{A}}_K} = q^{g-1}$.

Proposition 4.8 ([24, p. 39], [45, p. 98]). *One has*

$$\int_{\hat{\mathbb{A}}_K(D)} \omega_{\hat{\mathbb{A}}_K} = q^{\deg(D)}.$$

Proof. The strategy is to first use Proposition 4.7 and then apply Theorem 4.5 to the compact open subgroup $\hat{\mathbb{A}}_K(D)$ of $\hat{\mathbb{A}}_K$ in analogy to (5) in the proof of Proposition 4.7. This way one computes

$$\begin{aligned} q^{g-1} &\overset{4.7}{=} \int_{\hat{\mathbb{A}}_K/K} \omega_{\hat{\mathbb{A}}_K} \\ &\overset{4.5}{=} \frac{|\hat{\mathbb{A}}_K(D)\backslash\hat{\mathbb{A}}_K/K|}{|K \cap \hat{\mathbb{A}}_K(D)|} \int_{\hat{\mathbb{A}}_K(D)} \omega_{\hat{\mathbb{A}}_K} \\ &\overset{4.6}{=} \frac{|L(W-D)|}{|L(D)|} \int_{\hat{\mathbb{A}}_K(D)} \omega_{\hat{\mathbb{A}}_K}. \end{aligned}$$

Hence, by the Riemann–Roch Theorem (cf. Theorem 3.13), one has $\int_{\hat{\mathbb{A}}_K(D)} \omega_{\hat{\mathbb{A}}_K} = q^{\deg(D)}$.

5 Curves Over Finite Fields Considered as Schemes

In this section, I return to the study of non-singular projective curves, in the more general context of schemes. For further reading the sources [27, II–IV], [32, 2, 6, 7] are highly recommended.

The basic idea of Harder's reduction theory is to apply the geometry of numbers to the groups of $\hat{\mathbb{A}}_K$-rational points of reductive K-isotropic algebraic K-groups, see Theorem 7.2 below. The concept of schemes, which generalizes the concept of varieties, allows one to do so in a very efficient way, as in this language one can consider reductive groups over projective curves, thus making the Riemann–Roch theorem applicable.

Definition 5.1. Let X be a topological space, let $\mathfrak{Top}(X)$ be the category of open subsets of X with the inclusion maps as morphisms, and let $\mathfrak{C}$ be one of the categories $\mathfrak{Ab}$ of abelian groups, $\mathfrak{CRing}$ of commutative rings with 1, or $\mathfrak{Mod}(A)$ of modules over the (commutative) ring A. A contravariant functor $\mathcal{F}$ from $\mathfrak{Top}(X)$ to $\mathfrak{C}$ which satisfies $\mathcal{F}(\emptyset) = 0$ is called a **presheaf** on X with values in $\mathfrak{C}$. For presheaves $\mathcal{F}$, $\mathcal{G}$ on X with values in $\mathfrak{C}$, a natural transformation from $\mathcal{F}$ to $\mathcal{G}$ is called a **morphism** of presheaves.

A presheaf $\mathcal{F}$ is called a **sheaf**, if it satisfies the following conditions: (a) if U is an open subset of X, if $\{V_i\}$ is an open covering of U, and if $s \in \mathcal{F}(U)$ satisfies $s_{|V_i} = 0$ for all i, then $s = 0$; and (b) if U is an open subset of X, if $\{V_i\}$ is an open covering of U, and if the $s_i \in \mathcal{F}(V_i)$ satisfy $s_{i\,|V_i \cap V_j} = s_{j\,|V_i \cap V_j}$ for all i, j, then there exists $s \in \mathcal{F}(U)$ such that $s_{|V_i} = s_i$ for all i.

For a presheaf $\mathcal{F}$ on X and an element $x \in X$, the **stalk** $\mathcal{F}_x$ is defined as the direct limit $\lim_{\rightarrow} \mathcal{F}(U)$ over the open neighbourhoods U of x where $U \le V$ if $U \supseteq V$, i.e., $\mathcal{F}_x$ consists of the germs of elements of $\mathcal{F}(U)$ at x.

Example 5.2. Let V be a quasi-projective variety. The functor $\mathcal{O}_V : U \mapsto \mathcal{O}_U(V)$ (cf. Definition 2.4) is a sheaf on V. The stalk $\mathcal{O}_{V,P}$ at a point P equals the local ring $\mathcal{O}_P$ at that point.

Definition 5.3. Let X be a topological space and let $\mathcal{F}$ be a presheaf on X. The **sheaf associated to the presheaf** $\mathcal{F}$, also called the **sheafification of** $\mathcal{F}$, is a pair $(\mathcal{F}^\dagger, \theta)$ consisting of a sheaf $\mathcal{F}^\dagger$ and a morphism $\theta : \mathcal{F} \to \mathcal{F}^\dagger$ of presheaves that satisfies the following universal property: for every morphism $\alpha : \mathcal{F} \to \mathcal{G}$, where $\mathcal{G}$ is a sheaf, there exists a unique morphism $\tilde{\alpha} : \mathcal{F}^\dagger \to \mathcal{G}$ such that $\alpha = \tilde{\alpha} \circ \theta$.

The sheafification of a presheaf is unique for abstract reasons and exists by [32, 2.2.15]; see also [28, 3.3], [27, II.1.2], [32, Ex. 2.2.3].

Definition 5.4. A subfunctor $\mathcal{F}'$ of a sheaf $\mathcal{F}$ that is itself a sheaf is called a **subsheaf** of $\mathcal{F}$. If $\mathcal{F}$ takes values in $\mathfrak{Ab}$, then the **quotient sheaf** $\mathcal{F}/\mathcal{F}'$ is the sheaf associated to the presheaf $U \mapsto \mathcal{F}(U)/\mathcal{F}'(U)$ (cf. Definition 5.3). If $\mathcal{F}$ is a sheaf on X then, given an open cover $\{U_i\}$ of X, an element of $\mathcal{F}/\mathcal{F}'(X)$ can be described by elements $f_i \in \mathcal{F}(U_i)$ such that $\frac{f_i}{f_j} \in \mathcal{F}'(U_i \cap U_j)$ for all i, j.

Definition 5.5. Let $f : X \to Y$ be a continous map between topological spaces, let $\mathcal{F}$ be a sheaf on X, and let $\mathcal{G}$ be a sheaf on Y. Then $Y \supseteq U \mapsto \mathcal{F}(f^{-1}(U))$ defines a sheaf $f_*\mathcal{F}$ on Y, the **direct image sheaf**. For each $x \in X$ there is a natural map $\epsilon_x : (f_*\mathcal{F})_{f(x)} \to \mathcal{F}_x$. The **inverse image sheaf** $f^{-1}\mathcal{G}$ on X is the sheaf associated to the presheaf $X \supseteq U \mapsto \lim_{V \supseteq f(U)} \mathcal{G}(V)$ where the limit is taken over all open subsets V of Y that contain $f(U)$; cf. Definition 5.3. For each $x \in X$ one has $(f^{-1}\mathcal{G})_x = \mathcal{G}_{f(x)}$ [32, p. 37].

Definition 5.6. A **ringed space** is a pair $(X, \mathcal{O}_X)$ consisting of a topological space X and a sheaf of rings $\mathcal{O}_X$, i.e. a sheaf on X with values in $\mathfrak{CRing}$. It is a **locally ringed space**, if for each point $P \in X$, the stalk $\mathcal{O}_{X,P}$ is a local ring.

Example 5.7. Let A be a commutative ring, define the set Spec(A) to be the set of prime ideals of A, and define a topology on Spec(A) by taking the closed sets to be sets of the type $V(\mathfrak{a}) = \{\mathfrak{p} \in \mathrm{Spec}(A) \mid \mathfrak{a} \subseteq \mathfrak{p}\}$, for arbitrary ideals $\mathfrak{a}$ of A; cf. [27, II.2.1], [32, 2.1.1]. For an open set $U \subseteq \mathrm{Spec}(A)$ define $\mathcal{O}(U)$ to be the set of functions $s : U \to \bigsqcup_{\mathfrak{p}\in U} A_\mathfrak{p}$, where $A_\mathfrak{p}$ denotes the localization of A at $\mathfrak{p}$, such that $s(\mathfrak{p}) \in A_\mathfrak{p}$ and for each $\mathfrak{p} \in U$ there exist a neighbourhood $V \subseteq U$ of $\mathfrak{p}$ and elements $a, f \in A$ with $f \notin \mathfrak{q}$ and $s(\mathfrak{q}) = \frac{a}{f}$ in $A_\mathfrak{q}$ for each $\mathfrak{q} \in V$. The resulting locally ringed space $\mathrm{Spec} A = (\mathrm{Spec}(A), \mathcal{O})$ is called the *spectrum* of A; cf. [27, p. 70], [32, 2.3.2].

Definition 5.8. A **morphism of locally ringed spaces** $(f, f^\sharp) : (X, \mathcal{O}_X) \to (Y, \mathcal{O}_Y)$ consists of a continuous map $f : X \to Y$ and a morphism of sheaves of rings $f^\sharp : \mathcal{O}_Y \to f_*\mathcal{O}_X$ such that for every $x \in X$ the induced map $\mathcal{O}_{Y,f(x)} \stackrel{f^\sharp}{\to} (f_*\mathcal{O}_X)_{f(x)} \stackrel{\epsilon_x}{\to} \mathcal{O}_{X,x}$ is a local homomorphism (cf. Definition 5.5). An **isomorphism** is an invertible morphism. A morphism $(f, f^\sharp) : (Z, \mathcal{O}_Z) \to (X, \mathcal{O}_X)$ of ringed spaces is a **closed immersion** if f is a topological closed immersion and if each $f_x^\sharp$ is surjective.

Definition 5.9. Let $(X, \mathcal{O}_X)$ be a ringed space. An $\mathcal{O}_X$**-module** is a sheaf $\mathcal{F}$ on X with values in $\mathfrak{Ab}$ such that for each open subset $U \subseteq X$ the group $\mathcal{F}(U)$ is an $\mathcal{O}_X(U)$-module and, for each inclusion of open sets $V \subseteq U$, the group homomorphism $\mathcal{F}(U) \to \mathcal{F}(V)$ is compatible with the module structures via the ring homomorphism $\mathcal{O}_X(U) \to \mathcal{O}_X(V)$.

If U is an open subset of X, and if $\mathcal{F}$ is an $\mathcal{O}_X$-module, then $\mathcal{F}_{|U}$ is an $\mathcal{O}_{X|U}$-module. If $\mathcal{F}$ and $\mathcal{G}$ are two $\mathcal{O}_X$-modules, the group of morphisms from $\mathcal{F}$ to $\mathcal{G}$ is denoted by $\mathrm{Hom}_{\mathcal{O}_X}(\mathcal{F}, \mathcal{G})$. The presheaf $U \mapsto \mathrm{Hom}_{\mathcal{O}_{X|U}}(\mathcal{F}_{|U}, \mathcal{G}_{|U})$ is a sheaf, denoted by $\mathcal{H}om_{\mathcal{O}_X}(\mathcal{F}, \mathcal{G})$; it is also an $\mathcal{O}_X$-module.

The **tensor product** $\mathcal{F} \otimes_{\mathcal{O}_X} \mathcal{G}$ of two $\mathcal{O}_X$-modules is defined as the sheaf associated to the presheaf $U \mapsto \mathcal{F}(U) \otimes_{\mathcal{O}_X(U)} \mathcal{G}(U)$ (cf. Definition 5.3).

An $\mathcal{O}_X$-module $\mathcal{F}$ is **free** if it is isomorphic to a direct sum of copies of $\mathcal{O}_X$. It is **locally free** if there exists an open cover $\{U_i\}$ of X for which each $\mathcal{F}_{|U_i}$ is a free $\mathcal{O}_X$-module. In that case the **rank** of $\mathcal{F}$ on such an open set is the number of copies of the structure sheaf $\mathcal{O}_X$ needed. For X irreducible, thus, the notion of rank of a locally free sheaf is well defined. A locally free $\mathcal{O}_X$-module of rank 1 is called an **invertible** $\mathcal{O}_X$**-module**.

Definition 5.10. For any ringed space $(X, \mathcal{O}_X)$, define the **Picard group** Pic(X) of X to be the group of isomorphism classes of invertible $\mathcal{O}_X$-modules under the operation $\otimes_{\mathcal{O}_X}$, cf. [27, II.6.12]. The inverse of an invertible $\mathcal{O}_X$-module $\mathcal{L}$ is $\mathcal{L}^\vee := \mathcal{H}om_{\mathcal{O}_X}(\mathcal{L}, \mathcal{O}_X)$, cf. [27, Ex. II.5.1], [32, Ex. 5.1.12].

Definition 5.11. An **affine scheme** is a locally ringed space $(X, \mathcal{O}_X)$ which is isomorphic to the spectrum of some ring. A locally ringed space $(X, \mathcal{O}_X)$ is a

scheme, if every point of X has an open neighbourhood U such that $(U, \mathcal{O}_{X|U})$ is an affine scheme. Its **dimension** is the dimension of X as a topological space (cf. Definition 2.1) and it is called **irreducible** if X is irreducible.

Definition 5.12. A **morphism** of schemes is a morphism of locally ringed spaces. Likewise, an **isomorphism** of schemes is an isomorphism of locally ringed spaces. A **closed subscheme** $(Z, \mathcal{O}_Z)$ of a scheme $(X, \mathcal{O}_X)$ consists of a closed subset Z of X and a closed immersion $(j, j^\sharp)(Z, \mathcal{O}_Z) \to (X, \mathcal{O}_X)$ where j is the canonical injection. (Cf. Definition 5.8.)

Definition 5.13. Let S be a scheme. An S**-scheme** or a **scheme over** S is a scheme X endowed with a morphism of schemes $\pi : X \to S$. If $S = \operatorname{Spec} A$ for a ring A, then an S-scheme is also called an A**-scheme**.

Example 5.14. Let $(X, \mathcal{O}_X)$ be a scheme, for $x \in X$ let $\mathcal{O}_{X,x}$ be the stalk at x (cf. Definition 5.1) and let $\mathfrak{m}_x$ be its maximal ideal. The *residue field* of x on X is $k(x) := \mathcal{O}_{X,x}/\mathfrak{m}_x$. For any field K, by [27, Ex. II.2.7], there exists a one-to-one correspondence between morphisms from the spectrum of K to $(X, \mathcal{O}_X)$ and pairs of a point $x \in X$ and an inclusion $k(x) \to K$. In other words, any spectrum of a field that contains the residue field of some point $x \in X$ as a subfield can be considered as an X-scheme.

Definition 5.15. Let S be a scheme and let X, Y be S-schemes with respect to the morphisms $\pi_X : X \to S$ and $\pi_Y : Y \to S$. The **fibred product** $X \times_S Y$ of X and Y over S is defined to be an S-scheme together with morphisms $p_X : X \times_S Y \to X$ and $p_Y : X \times_S Y \to Y$ satisfying $\pi_X \circ p_X = \pi_Y \circ p_Y$, such that given any S-scheme Z and morphisms $f : Z \to X$ and $g : Z \to Y$ satisfying $\pi_X \circ f = \pi_Y \circ g$, then there exists a unique morphism $\theta : Z \to X \times_S Y$ satisfying $f = p_X \circ \theta$ and $g = p_Y \circ \theta$.

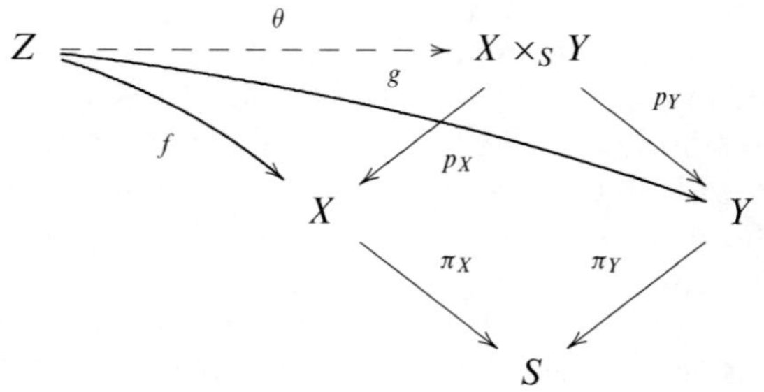

By [27, II.3.3], [32, 3.1.2] the fibred product of two S-schemes exists and is unique up to unique isomorphism.

Example 5.16. Let A be a ring, let $B = \bigoplus_{d \geq 0} B_d$ be a graded A-algebra, and let $B_+ := \bigoplus_{d>0} B_d$. Define the set $\operatorname{Proj}(B)$ to be the set of homogeneous prime ideals of B which do not contain B_+, and define a topology on $\operatorname{Proj}(B)$ by taking the closed sets to be sets of the form $V(\mathfrak{a}) = \{\mathfrak{p} \in \operatorname{Proj}(B) \mid \mathfrak{a} \subseteq \mathfrak{p}\}$, for arbitrary homogeneous ideals $\mathfrak{a}$ of A; cf. [27, II.2.4], [32, p. 50]. For each $\mathfrak{p} \in \mathrm{Proj}(B)$ let $B_{\mathfrak{p}}$ be the ring of elements of degree zero in the localized ring $T^{-1}B$, where T is the multiplicative system of all homogeneous elements of B which are not in $\mathfrak{p}$. For an open set $U \subseteq \mathrm{Proj}(B)$ define $\mathcal{O}(U)$ to be the set of functions $s : U \to \bigsqcup_{\mathfrak{p} \in U} B_{\mathfrak{p}}$ such that $s(\mathfrak{p}) \in B_{\mathfrak{p}}$ and for each $\mathfrak{p} \in U$ there exist a neighbourhood $V \subseteq U$ of $\mathfrak{p}$ and homogeneous elements $a, f \in B$ of the same degree with $f \notin \mathfrak{q}$ for each $\mathfrak{q} \in V$ and $s(\mathfrak{q}) = a/f$ in $B_{\mathfrak{q}}$. The resulting ringed space $\mathrm{Proj}B = (\mathrm{Proj}(B), \mathcal{O})$ is a scheme, in fact an A-scheme, cf. [27, II.2.5], [32, 2.3.38].

Example 5.17. Let A be a ring and let $B = A[x_0, x_1, ..., x_n]$ with grading by degree. Then the scheme $\mathbb{P}_n(A) := \mathrm{Proj}B$ is the *projective n-space over A*.

5.18. The schemes together with their morphisms form a category $\mathfrak{Sch}$. It contains the category $\mathfrak{AffSch}$ of affine schemes as a subcategory. The functor Spec from $\mathfrak{CRing}$ to $\mathfrak{AffSch}$ that sends A to $\mathrm{Spec}A$ and a ring homomorphism $A \to B$ to the corresponding morphism of ringed spaces from the spectrum of B to the spectrum of A (cf. [27, II.2.3]) is a contravariant equivalence (also called a duality) between the category of commutative rings and the category of affine schemes. The global section functor, i.e., the functor from $\mathfrak{AffSch}$ to $\mathfrak{CRing}$ that sends $(\mathrm{Spec}(A), \mathcal{O})$ to $\mathcal{O}(\mathrm{Spec}(A))$, is an inverse functor to Spec; cf. [27, II.2.2, II.2.3].

5.19. The affine scheme $\mathrm{Spec}\mathbb{Z}$ is a final object in both categories of schemes and affine schemes; cf. [27, Ex. II.2.5].

Definition 5.20. Let k be a perfect field. A **projective scheme over** k is a scheme over k that is isomorphic to a closed subscheme of the scheme $\mathbb{P}_n(k)$, for some n (cf. Definition 5.12, Example 5.17). A **projective curve over** k is an irreducible projective scheme of dimension 1. A projective curve Y is called **non-singular** at $P \in Y$, if $\mathcal{O}_P$ (cf. Definition 5.1) is a discrete valuation ring (cf. [32, 4.2.9]). A projective curve is called **non-singular**, if it is non-singular at each of its points.

5.21. Comparing Definition 3.1 with Definition 5.20 it is clear that a non-singular projective curve, defined as a variety, yields a non-singular projective curve if considered as a scheme. In the context of Example 5.2, one can convert locally free $\mathcal{O}_Y$-modules (Definition 5.9) into vector bundles over Y (3.15) and vice versa, cf. [28, p. 61], [27, Ex. II.5.18], [36]: a vector bundle $\pi : E \to Y$ yields a sheaf $\mathcal{L}_E$ by sending an open subset U of Y to the (fibre-wise) $\mathcal{O}_Y(U)$-module of sections $U \to E$ over U.

5.22. Let A be a ring. The quotient of $A[x_0, x_1, ..., x_n]$ by a homogeneous ideal is a **homogeneous** A**-algebra**. By [32, 2.3.41], for every homogeneous A-algebra B, the scheme $\mathrm{Proj}B$ is a projective scheme and, by [32, 5.1.30], conversely every projective scheme over A is isomorphic to $\mathrm{Proj}B$, for some homogeneous A-algebra B.

Definition 5.23. Let X be a scheme. For each open subset U, let $S(U)$ denote the set of elements $s \in \mathcal{O}_X(U)$ such that for each $x \in U$ the germ s_x is not a zero divisor in the stalk $\mathcal{O}_x$. Then the **sheaf** $\mathcal{K}$ **of total quotient rings** of the sheaf $\mathcal{O}_X$ is the sheaf associated to the presheaf $U \mapsto S(U)^{-1}\mathcal{O}_X(U)$ (cf. Definition 5.3). Let $\mathcal{K}^*$ and $\mathcal{O}^*$ denote the sheaves of groups of invertible elements in the sheaves of rings $\mathcal{K}$, resp. $\mathcal{O}_X$. An element of the group $(\mathcal{K}^*/\mathcal{O}^*)(X)$ is called a **Cartier divisor**. A Cartier divisor is **principal** if it is in the image of the natural map $\mathcal{K}^*(X) \to (\mathcal{K}^*/\mathcal{O}^*)(X)$. Two Cartier divisors are **linearly equivalent** if their difference is principal.

Proposition 5.24 ([27, II.6.11], [32, 7.2.16]). *Let Y be a non-singular projective curve over $\mathbb{F}_q$. Then there exists a natural isomorphism between the group* $\mathrm{Div}(Y)$ *of Weil divisors (cf. Definition 3.6) and the group $\mathcal{K}^*/\mathcal{O}^*(Y)$ of Cartier divisors and, furthermore, the principal Weil divisors correspond to the principal Cartier divisors under this isomorphism.*

5.25. In fact, as $\mathcal{O}_Y$ is integral [32, 2.4.17], the sheaf $\mathcal{K}$ is the constant sheaf corresponding to the function field K of Y. As in Definition 5.4, a Cartier divisor is given by a family $\{U_i, f_i\}$ where $\{U_i\}$ is an open cover of Y and $f_i \in \mathcal{K}^*(U_i) = K^*$. For each $P \in Y$ and for all i, j such that $P \in U_i, U_j$ one has $\nu_P(f_i) = \nu_P(f_j)$, as $\frac{f_i}{f_j}$ is invertible on $U_i \cap U_j$. One therefore obtains the well-defined Weil divisor $\sum_{P \in Y^\circ} \nu_P(f_i)P$.

Definition 5.26. Let D be a Cartier divisor on a non-singular projective curve $Y/\mathbb{F}_q$ represented by a family $\{U_i, f_i\}$ where $\{U_i\}$ is an open cover of Y and $f_i \in \mathcal{K}^*(U_i)$. The **sheaf** $\mathcal{L}(D)$ **associated to** D is the subsheaf of $\mathcal{K}$ given by taking $\mathcal{L}(D)$ to be the sub-$\mathcal{O}_Y$-module of $\mathcal{K}$ generated by f_i^{-1} on U_i. Since $\frac{f_i}{f_j}$ is invertible on $U_i \cap U_j$, the elements f_i^{-1} and f_j^{-1} generate the same $\mathcal{O}_Y$-module, and hence $\mathcal{L}(D)$ is well defined.

Proposition 5.27 ([27, II.6.13, II.6.15], [32, 7.1.19]). *Let Y be a non-singular projective curve over $\mathbb{F}_q$. Then for any Cartier divisor D, the sheaf $\mathcal{L}(D)$ is an invertible $\mathcal{O}_Y$-module. Moreover, the map $D \mapsto \mathcal{L}(D)$ induces a surjection from the group of Cartier divisors onto the Picard group* $\mathrm{Pic}(Y)$. *The kernel of this surjection is the group of principal Cartier divisors.*

Definition 5.28. Let $Y/\mathbb{F}_q$ be a non-singular projective curve. Define the **degree** $c(\mathcal{L})$ of an invertible $\mathcal{O}_Y$-module $\mathcal{L}$ to be the degree $\deg(D)$ of the corresponding Weil divisor; cf. Propositions 5.24 and 5.27, Theorem 5.25; [32, 7.3.1, 7.3.2].

Example 5.29. Let $Y = \mathbb{P}_1 = \{(x : y)\}$ and let $U_1 = \{(x : y) \mid y \neq 0\}$ and $U_2 = \{(x : y) \mid x \neq 0\}$ be an open cover of Y with local coordinates $s = \frac{x}{y}$ and $t = \frac{y}{x}$. For $d \in \mathbb{N}$ define the line bundle $L(d) = \{((x : y), (a, b)) \in \mathbb{P}_1 \times A_2 \mid x^d b = y^d a\}$ with canonical projection $\pi : L(d) \to \mathbb{P}_1$. Then

$$U_1 \times A_1 \overset{f_1}{\cong} \pi^{-1}(U_1) = \{((x : y), (a, b)) \in U_1 \times A_2 \mid a = s^d b\}$$
$$(s, b) \mapsto ((s : 1), (s^d b, b))$$

$$U_2 \times A_1 \overset{f_2}{\cong} \pi^{-1}(U_2) = \{((x : y), (a, b)) \in U_1 \times A_2 \mid b = t^d a\}$$
$$(t, a) \mapsto ((1 : t), (a, t^d a)).$$

One has $f_{1|U_1 \cap U_2} = s^d f_{2|U_1 \cap U_2}$, so that the line bundle $L(d)$ corresponds to the Weil divisor $\sum_{P \in Y^\circ} \nu_P(f_i)P$ of degree d (cf. Theorem 5.25).

Theorem 5.30 (Riemann–Roch Theorem [27, IV.1.3], [32, 7.3.26]). *Let $Y/\mathbb{F}_q$ be a non-singular projective curve of genus g. Then for any invertible $\mathcal{O}_Y$-module $\mathcal{L}$*

$$\dim_{\mathbb{F}_q}(H^0(Y,\mathcal{L})) - \dim_{\mathbb{F}_q}(H^1(Y,\mathcal{L})) = c(\mathcal{L}) + 1 - g.$$

Here, $H^i(Y,\mathcal{L})$ denotes sheaf cohomology groups as defined in [27, III], [32, 5.2].

5.31. For a locally free $\mathcal{O}_Y$-module $\mathcal{E}$ of rank n define $c(\mathcal{E}) := \sum_{i=1}^n c(\mathcal{E}_i/\mathcal{E}_{i-1})$ where $0 \subset \mathcal{E}_1 \subset \cdots \subset \mathcal{E}_{n-1} \subset \mathcal{E}_n = \mathcal{E}$ is a filtration such that each $\mathcal{E}_i/\mathcal{E}_{i-1}$ is invertible; cf. [21, 2.1], [23, p. 122]. Then Theorem 5.30 implies $\dim_{\mathbb{F}_q}(H^0(Y,\mathcal{E})) - \dim_{\mathbb{F}_q}(H^1(Y,\mathcal{E})) = c(\mathcal{E}) + (1-g)n$. (Cf. [45, p. 99].)

6 Reduction Theory for Rationally Trivial Group Schemes

In this section, I describe the heart of Harder's reduction theory based on [23]. From this section onwards I will assume the reader has a basic intuition for linear algebraic groups and feels comfortable with some of the standard terminology such as that of Borel subgroups and parabolic subgroups. For introductory reading the sources [11, 29], for further reading the sources [30, 39, 42, 43] are highly recommended. I also recommend [31, Appendix E].

Definition 6.1. An object X of a category $\mathfrak{C}$ is called a **group object**, if for each object Y of $\mathfrak{C}$ the set $\mathrm{Hom}_{\mathfrak{C}}(Y,X)$ of morphisms from Y to X is a group and the correspondence $Y \mapsto \mathrm{Hom}_{\mathfrak{C}}(Y,X)$ is a (contravariant) functor from the category $\mathfrak{C}$ to the category $\mathfrak{Gr}$ of groups; cf. [42, p. 3].

Definition 6.2. A group object in the categories $\mathfrak{Sch}$ of schemes, resp. $\mathfrak{AffSch}$ of affines schemes (cf. Theorem 5.18) is called a **group scheme**, resp. an **affine group scheme**.

6.3. As the categories $\mathfrak{Sch}$ and $\mathfrak{AffSch}$ both have $\mathrm{Spec}\mathbb{Z}$ as a final object (cf. Theorem 5.19) and have finite products, being an (affine) group scheme is an intrinsic property; cf. [42, p. 3]. In particular, any affine group scheme is a group scheme.

6.4. When restricting (and co-restricting accordingly) the functor Spec from $\mathfrak{CRing}$ to $\mathfrak{AffSch}$ (cf. Theorem 5.18) to the subcategory of commutative Hopf algebras, it yields a contravariant equivalence (duality) to the category of affine group schemes; cf. [42, 1.3], [43, p. 9].

Example 6.5. Let $\mathrm{G_a}$ be the covariant functor on the category $\mathfrak{CRing}$ defined by $\mathrm{G_a}(A) = (A,+)$. Since by means of duality (cf. Theorem 6.14)

$$\mathrm{Hom}_{\mathfrak{CRing}}(\mathbb{Z}[T],A) = A \quad \Longleftrightarrow \quad \mathrm{Hom}_{\mathfrak{AffSch}}(\mathrm{Spec}A, \mathrm{Spec}\mathbb{Z}[T]) = A,$$

the functor G_{a}, considered as the contravariant functor on the category $\mathfrak{AffSch}$ that sends $(\mathrm{Spec}(A), \mathcal{O})$ to $(\mathcal{O}(\mathrm{Spec}(A)), +)$, is represented by $\mathrm{Spec}\mathbb{Z}[T]$. Therefore, $\mathrm{Spec}\mathbb{Z}[T]$ is an affine group scheme. By Theorem 6.3, it is also a group scheme, which as a functor sends $(X, \mathcal{O}_X)$ to $(\mathcal{O}_X(X), +)$.

Example 6.6. Let G_{m} be the covariant functor on the category $\mathfrak{CRing}$ defined by $\mathrm{G}_{\mathrm{m}}(A) = (A^*, \cdot)$. Since

$$\mathrm{Hom}_{\mathfrak{CRing}}(\mathbb{Z}[T, T^{-1}], A) = A^* \iff \mathrm{Hom}_{\mathfrak{AffSch}}(\mathrm{Spec}A, \mathrm{Spec}\mathbb{Z}[T, T^{-1}]) = A^*,$$

the functor G_{m}, considered as a contravariant functor on the category $\mathfrak{AffSch}$, is represented by $\mathrm{Spec}\mathbb{Z}[T, T^{-1}]$. Therefore, $\mathrm{Spec}\mathbb{Z}[T, T^{-1}]$ is an affine group scheme and, by Theorem 6.3, also a group scheme.

Example 6.7. Let GL_n be the covariant functor on the category $\mathfrak{CRing}$ defined by

$$A \mapsto \mathrm{GL}_n(A) = \{M \in A^{n \times n} \mid \det(M) \neq 0\}.$$

It is represented by the affine scheme $\mathrm{GL}_n = \mathrm{Spec}\mathbb{Z}[T_{11}, ..., T_{nn}, \det(T_{ij})^{-1}]$ and, hence, is an affine group scheme. By Theorem 6.3, GL_n is also a group scheme, and induces the contravariant functor on $\mathfrak{Sch}$ that sends $(X, \mathcal{O}_X)$ to $\mathrm{GL}_n(\mathcal{O}_X(X))$.

An alternative, basis-free way of defining GL_n is to define a contravariant functor on $\mathfrak{Sch}$ that sends

$$(X, \mathcal{O}_X) \quad \text{to} \quad \mathrm{Aut}_{\mathcal{O}_X(X)}\left(\bigoplus_{i=1}^{n} \mathcal{O}_X(X)\right).$$

Example 6.8. Let SL_n be the covariant functor on the category $\mathfrak{CRing}$ defined by

$$A \mapsto \ker\left(\mathrm{GL}_n(A) \to \mathrm{G}_{\mathrm{m}}(A)\right) = \{M \in A^{n \times n} \mid \det(M) = 1\}.$$

By Yoneda's lemma [43, p. 6] the natural transformation $\mathrm{GL}_n \to \mathrm{G}_{\mathrm{m}}$ can be described by a homomorphism $\mathbb{Z}[T, T^{-1}] \to \mathbb{Z}[T_{11}, ..., T_{nn}, \det(T_{ij})^{-1}]$, the one that sends T to $\det(T_{ij})$. Therefore SL_n is represented by

$$\mathbb{Z}[T_{11}, ..., T_{nn}, \det(T_{ij})^{-1}] \otimes_{\mathbb{Z}[T, T^{-1}]} \mathbb{Z} = \mathbb{Z}[T_{11}, ..., T_{nn}]/(\det(T_{ij}) - 1).$$

In other words, $\mathrm{Spec}\mathbb{Z}[T_{11}, ..., T_{nn}]/(\det(T_{ij}) - 1)$ is an (affine) group scheme.

Definition 6.9. An S-scheme (cf. Definition 5.13) that is an (affine) group scheme is called an **(affine) group S-scheme**. As the categories of (affine) S-schemes have finite products and have the scheme S as a final object, as in Theorem 6.3 being an (affine) group S-scheme is an intrinsic property. In particular, any affine group S-scheme is a group S-scheme.

Example 6.10. Generalizing Example 6.8, let S be a scheme, let G and H be group S-schemes, and let $f : G \to H$ be a homomorphism of group S-schemes, i.e., a morphism of S-schemes from G to H such that for each S-scheme Y the induced map $\mathrm{Hom}_{\mathfrak{C}}(Y, G) \to \mathrm{Hom}_{\mathfrak{C}}(Y, H)$ is a group homomorphism. Define a functor on the category of S-schemes by sending

$$Y \quad \text{to} \quad \ker(\mathrm{Hom}_{\mathfrak{C}}(Y, G) \to \mathrm{Hom}_{\mathfrak{C}}(Y, H)).$$

It is represented by the fibred product S-scheme $G \times_H S$, cf. Definition 5.15.

Example 6.11. Let $(X, \mathcal{O}_X)$ be an (irreducible) scheme and let $\mathcal{E}$ be a locally free $\mathcal{O}_X$-module of finite rank (Definition 5.9). Following [18, I 4.5] define a functor $\mathrm{GL}(\mathcal{E})$ on the category of X-schemes that sends

$$(Y, \mathcal{O}_Y) \quad \text{to} \quad \mathrm{Aut}_{\mathcal{O}_Y(Y)}\left(\left(\mathcal{O}_Y \otimes_{f^{-1}\mathcal{O}_X} f^{-1}\mathcal{E}\right)(Y)\right),$$

where f denotes the morphism of schemes from the X-scheme $(Y, \mathcal{O}_Y)$ to the scheme $(X, \mathcal{O}_X)$ (Definition 5.13) and $f^{-1}\mathcal{O}_X$ and $f^{-1}\mathcal{E}$ denote the respective inverse image sheaves (Definition 5.5). For $\mathcal{E} = \bigoplus_{i=1}^n \mathcal{O}_X$ this is just the affine group X-scheme GL_n/X from Example 6.7. The present more general example is obtained from that example by a twisting process; cf. [23, p. 134], [27, p. 117]. In analogy to Example 6.8 define the (affine) group X-scheme $\mathrm{SL}(\mathcal{E})$ as $\ker(\mathrm{GL}(\mathcal{E}) \to \mathrm{G_m}/X)$ (cf. Example 6.10).

Definition 6.12. Let $Y/\mathbb{F}_q$ be a non-singular projective curve (Theorem 5.21), let $K = \mathbb{F}_q(Y)$ be its field of $\mathbb{F}_q$-rational functions (Definition 2.6) considered as the residue field of Y at its generic point (Example 5.14 and [27, Ex. II.3.6]), let G/Y be an affine group Y-scheme (Definition 6.9), and consider $\mathrm{Spec} K$ as a Y-scheme via the identity map on K (Example 5.14). The scheme G/Y is called a **rationally trivial group** (**Y-**)**scheme** if $G/K := G \times_Y \mathrm{Spec} K$ (Definition 5.15) is a Chevalley scheme [22, I.8.3.1]. It is called **reductive**, if $G/K = G \times_Y \mathrm{Spec} K$ is reductive [19, XIX 1.6].

Proposition 6.13 ([19, XX 1.3, 1.5, 1.16, 1.17; XXII 5.6.5, 5.9.5; XXVI 1.12, 2.1]). *Let $Y/\mathbb{F}_q$ be a non-singular projective curve, let G/Y be a reductive group Y-scheme, and let P/Y be a parabolic subgroup of G/Y (cf. [19, XXVI 1.1]). Then the unipotent radical $R_u(P)$ (cf. [19, XIX 1.2]) admits a filtration $R_u(P) = U_0 \supset U_1 \supset \cdots \supset U_k = \{e\}$ such that each U_i/U_{i+1} is a vector bundle over Y. More precisely, if P_α denotes the vector bundle over Y corresponding to the root space $\mathfrak{g}^\alpha$ (cf. [19, XIX 1.10]), then*

$$U_i = \prod_{\alpha \in \Delta_P^+, l(\alpha) > i} P_\alpha \qquad \text{and} \qquad U_i/U_{i+1} \cong \prod_{\alpha \in \Delta_P^+, l(\alpha) = i+1} P_\alpha.$$

Example 6.14. Let $A_1 \cong \mathcal{O}_Y \cong A_2$, let $\mathcal{E} = A_1 \oplus A_2$ and let $\mathrm{SL}(\mathcal{E}) = \mathrm{SL}_2/Y$ be the group Y-scheme defined in Examples 6.8 and 6.11. Each of $\mathcal{H}om_{\mathcal{O}_Y}(A_1, A_2)$ and $\mathcal{H}om_{\mathcal{O}_Y}(A_2, A_1)$ equals the unipotent radical of a Borel subgroup of $\mathrm{SL}(A_1 \oplus A_2)$ and is an invertible $\mathcal{O}_Y$-module, because $\mathcal{H}om_{\mathcal{O}_Y}(A_i, A_j) \cong A_i^\vee \otimes_{\mathcal{O}_Y} A_j$. By Theorem 5.21, this corresponds to a vector bundle (of dimension 1), confirming Proposition 6.13 for SL_2.

More generally, let $\mathcal{E}$ be a locally free $\mathcal{O}_Y$-module of rank 2 with an invertible $\mathcal{O}_Y$-submodule $\mathcal{L}$ such that $\mathcal{E}/\mathcal{L}$ is also invertible and let $\mathrm{SL}(\mathcal{E})$ be the group Y-scheme defined in Example 6.11. There is a one-to-one correspondence between the Borel subgroups B of $\mathrm{SL}(\mathcal{E})$ and the invertible $\mathcal{O}_Y$-submodules $\mathcal{L}$ such that $\mathcal{E}/\mathcal{L}$ is also invertible, given by $\mathcal{L} \mapsto \mathrm{Stab}_{\mathrm{SL}(\mathcal{E})}(\mathcal{L})$. The unipotent radical $R_u(B)$ is isomorphic to $\mathcal{H}om_{\mathcal{O}_Y}(\mathcal{E}/\mathcal{L}, \mathcal{L})$. For any invertible $\mathcal{O}_Y$-module $\mathcal{H}$, the equality $\mathcal{H}om_{\mathcal{O}_Y}(\mathcal{E}/\mathcal{L} \otimes_{\mathcal{O}_Y} \mathcal{H}, \mathcal{L} \otimes_{\mathcal{O}_Y} \mathcal{H}) = (\mathcal{E}/\mathcal{L} \otimes_{\mathcal{O}_Y} \mathcal{H})^\vee \otimes_{\mathcal{O}_Y} (\mathcal{L} \otimes_{\mathcal{O}_Y} \mathcal{H}) = \mathcal{H}om_{\mathcal{O}_Y}(\mathcal{E}/\mathcal{L}, \mathcal{L})$ implies $\mathrm{SL}(\mathcal{E} \otimes_{\mathcal{O}_Y} \mathcal{H}) \cong \mathrm{SL}(\mathcal{E})$. (Cf. [18, I, 4.5],[19, XX, 5.1].)

Definition 6.15. Let G/Y be a rationally trivial group Y-scheme, let B/Y be a Borel subgroup of G/Y and let $\{\alpha_1, ..., \alpha_r\}$ be the simple roots of B. Using the notation introduced in Theorem 5.21, Definition 5.28 and Proposition 6.13, define

$$n_i(B) := c\left(\mathcal{L}_{B_{\alpha_i}}\right).$$

Note that, since G/Y is rationally trivial, each B_{α_i} is a vector bundle of dimension 1, so that $\mathcal{L}_{B_{\alpha_i}}$ is an invertible $\mathcal{O}_Y$-module.

Theorem 6.16 ([23, 2.2.6]). *Let G/Y be a rationally trivial group Y-scheme of genus g (cf. Definition 3.12). Then there exists a Borel subgroup B/Y such that $n_i(B) \geq -2g$ for all $i \in I = \{1, ..., r\}$.*

Proof. In case G/Y is of type A_1, i.e., if there exists a locally free $\mathcal{O}_Y$-module $\mathcal{E}$ of rank 2 such that $G \cong \mathrm{SL}(\mathcal{E})$, one can proceed as follows. Let $\mathcal{L}$ be an invertible $\mathcal{O}_Y$-submodule such that $\mathcal{E}/\mathcal{L}$ is also invertible and let $B = \mathrm{Stab}_{\mathrm{SL}(\mathcal{E})}(\mathcal{L})$ (cf. Example 6.14).

Then $n_1(B) = c(R_u(B)) = c(\mathcal{H}om_{\mathcal{O}_Y}(\mathcal{E}/\mathcal{L}, \mathcal{L})) = c((\mathcal{E}/\mathcal{L})^\vee \otimes_{\mathcal{O}_Y} \mathcal{L}) = c(\mathcal{L}) - c(\mathcal{E}/\mathcal{L}) = 2c(\mathcal{L}) - c(\mathcal{E})$. By Proposition 3.14 and Definition 5.28 there exists an invertible $\mathcal{O}_Y$-module $\mathcal{H}$ with $c(\mathcal{H}) = 1$. Since $\mathrm{SL}(\mathcal{E} \otimes_{\mathcal{O}_Y} \mathcal{H}^\vee) \cong \mathrm{SL}(\mathcal{E}) \cong \mathrm{SL}(\mathcal{E} \otimes_{\mathcal{O}_Y} \mathcal{H})$ by Example 6.14, the formula $c(\mathcal{E} \otimes_{\mathcal{O}_Y} \mathcal{H}^\vee) = c(\mathcal{E}) - 2c(\mathcal{H}) = c(\mathcal{E}) - 2$ allows us to assume without loss of generality that $2g - 2 < c(\mathcal{E}) \leq 2g$.

By the Riemann–Roch theorem (cf. Theorem 5.31) there exists $0 \neq t \in H^0(Y, \mathcal{E}) \cong \mathcal{E}(Y)$. Since $\mathcal{E}$ is locally free of rank 2 there exists an open cover $\{U_i\}$ of Y such that $\mathcal{E}_{|U_i} \cong \mathcal{O}_{Y|U_i} \oplus \mathcal{O}_{Y|U_i}$ for each U_i. Therefore, t is contained in an invertible $\mathcal{O}_Y$-submodule $\mathcal{L}_t$ such that $\mathcal{E}/\mathcal{L}_t$ is also invertible. As $t \in H^0(Y, \mathcal{L}_t)$, one has $c(\mathcal{L}_t) \geq 0$ by [27, Lemma IV.1.2]. Hence

$$n_1(B_t) = c(\mathcal{H}om_{\mathcal{O}_Y}(\mathcal{E}/\mathcal{L}_t, \mathcal{L}_t)) = 2c(\mathcal{L}_t) - c(\mathcal{E}) \geq -2g,$$

where $B_t = \mathrm{Stab}_{\mathrm{SL}(\mathcal{E})}(\mathcal{L}_t)$ (cf. Example 6.14).

The general case can be reduced to the case A_1 via a local to global argument.

Definition 6.17. Let $c_1 < -2g$. A Borel subgroup B/Y of G/Y is **reduced**, if $n_i(B) \geq c_1$ for all i.

Theorem 6.18 ([23, 2.2.13, 2.2.14]). *Let G/Y be a rationally trivial group Y-scheme. Then there exist constants $c_2 > \gamma > c_1$ such that the following hold: Let B/Y be a reduced Borel subgroup of G/Y and let α_{i_0} be a simple root of B such that $n_{i_0}(B) \geq c_2$. Then each reduced Borel subgroup B' of G/Y satisfies $n_{i_0}(B') \geq \gamma$ and is contained in $P_{i_0}(B)$, where $P_{i_0}(B)$ denotes the maximal parabolic of type α_{i_0} containing B.*

Proposition 6.19 ([23, 2.2.3, 2.2.11]). *To each reduced Borel subgroup B/Y of G/Y, let $I^B := \{i \in I \mid n_i(B) \geq c_2\}$, and let $P^B := \bigcap_{i \in I^B} P_i(B)$. Then*

$$P := \bigcap_{B \subset G/Y \text{ reduced}} P^B$$

either equals G/Y or is a parabolic subgroup of G/Y.

Proof. Let B be a reduced Borel subgroup of G. By Theorem 6.18 each reduced Borel subgroup B' of G is contained in P^B and conversely, again by Theorem 6.18, $B \subset P^{B'}$. Hence, P contains a Borel group and therefore either equals G/Y or is a parabolic subgroup of G/Y.

Theorem 6.18 is the heart of Harder's reduction theory. The only proof of Theorem 6.18 known to me in fact uses Proposition 6.19, as can be guessed from the numbering used in [23]. However, as the length of the proof of Theorem 6.18 by far surpasses anything I can reasonably include in this survey and as I will need to refer to Proposition 6.19 later, I took the liberty of deducing Proposition 6.19 from Theorem 6.18 for the sake of this exposition.

Proposition 6.20 ([23, p. 120], [24, p. 39]). *Let G/Y be a reductive group Y-scheme, let P/Y be a parabolic K-subgroup of G/Y, let $d := \dim(R_u(P))$, and let Δ_P^+ be the set of positive roots of P. Then the character*

$$\chi_P : P \stackrel{\mathrm{Ad}}{\to} \mathrm{GL}\left(\mathrm{Lie}(R_u(P))\right) \stackrel{\det}{\to} \mathrm{GL}\left(\bigwedge^d \mathrm{Lie}(R_u(P))\right) \cong \mathrm{G_m},$$

considered as a character of a maximal split torus contained in P, is given by $\chi_P = \sum_{\alpha \in \Delta_P^+} \dim(P_\alpha)\alpha$.

Definition 6.21. Let B be a minimal parabolic K-subgroup of G/Y, let $\{\alpha_1, ..., \alpha_r\}$ be the simple roots of B, and let $(P_i)_i$ be the maximal parabolic K-subgroups of G of type α_i. Using the notation of Theorems 5.21, 5.31 and Proposition 6.13, define

$$p_i(B) := p(P_i) := \sum_{\alpha \in \Delta_{P_i}^+} c(\mathcal{L}_{P_\alpha}).$$

6.22. Let B be a minimal parabolic K-subgroup of G, let $R(B)$ be the radical of B, let $R_u(B)$ be the unipotent radical of B, let $T = R(B)/R_u(B)$, let $S \subseteq T$ be the maximal K-split subtorus of T, let $\pi = \{\alpha_1, ..., \alpha_r\} \subset X(S)$ be the system of simple roots, and let $X(B) = \mathrm{Hom}_K(B, \mathrm{G_m})$ be the module of K-rational characters of B so that $X(B) \otimes \mathbb{Q} = X(S) \otimes \mathbb{Q}$, let $P_i \supseteq B$ be the maximal K-parabolic of type α_i, let $\chi_{P_i} : P_i \to \mathrm{G_m}$ be the sum of roots of P_i (cf. Proposition 6.20), and let $\chi_i := \chi_{P_i|B}$. The χ_i form a basis of $X(B) \otimes \mathbb{Q}$ and, if $(\cdot, \cdot)$ is a positive definite bilinear form on $X(B) \otimes \mathbb{Q}$ which is invariant under the action of the Weyl group, then

$$(\chi_i, \alpha_j) = 0, \quad \text{if } i \neq j, \text{ and} \tag{6}$$

$$(\chi_i, \alpha_i) > 0 \quad \text{for all } i \in I. \tag{7}$$

If G is rationally trivial, for $m_{i,j} \in \mathbb{Z}$ such that $\chi_i = \sum_{j=1}^r m_{i,j}\alpha_j$, then

$$p_i(B) = \sum_{j=1}^{r} m_{i,j} n_j(B).$$

Furthermore, for $c_{i,j} \in \mathbb{Q}$ such that $\alpha_i = \sum_{j=1}^r c_{i,j}\chi_j$, then

$$n_i(B) = \sum_{j=1}^{r} c_{i,j} p_j(B).$$

7 Reduction Theory for Reductive Groups Over the Adèles

In this section, I describe the interplay between Harder's reduction theory and Weil's geometry of numbers based on [24] in order to arrive at a geometric version of Harder's reduction theory.

7.1. Let G/Y be a reductive group Y-scheme, let P/Y be a parabolic K-subgroup of G/Y, and let ω be a non-trivial volume form on $R_u(P)$ defined over K. This volume form yields a Haar measure of $R_u(P(\hat{\mathbb{A}}_K))$ which is independent of the choice of ω, i.e., for each $0 \neq \lambda \in K$ the volume forms ω and $\lambda\omega$ yield identical Haar measures; cf. [24, p. 37], [44, 2.3.1].

This measure differs from the classical Tamagawa measure by the factor $q^{(1-g)d}$ where $d = \dim(R_u(P))$, cf. [24, p. 38], [41], [42, Sect. 14.4], [44, 2.3], [45, p. 113].

Theorem 7.2 ([24, 1.3.1]). *Let G/Y be a reductive group Y-scheme, let P/Y be a maximal parabolic K-subgroup of G/Y, let ω be a non-trivial volume form of $R_u(P)$ defined over K, and let $\mathfrak{K} := G(\hat{\mathcal{O}}_K)$, cf. Definition 4.3. Then*

$$\int_{R_u(P(\hat{\mathbb{A}}_K))\cap\mathfrak{K}} \omega_{\hat{\mathbb{A}}_K} = q^{p(P)}.$$

Proof. For each $\alpha \in \Delta_P^+$, let $0 \subset P_\alpha^{(1)} \subset \cdots \subset P_\alpha^{(\dim(P_\alpha))} = P_\alpha$ be a filtration such that each $P_\alpha^{(i)}/P_\alpha^{(i-1)}$ is a vector bundle of dimension 1; cf. [21, 2.1], [23, p. 122]. One then computes

$$\begin{aligned}\int_{R_u(P(\hat{\mathbb{A}}_K))\cap\mathfrak{K}} \omega_{\hat{\mathbb{A}}_K} &\overset{6.13}{=} \prod_{\alpha\in\Delta_P^+}\left(\int_{P_\alpha(\hat{\mathbb{A}}_K)\cap\hat{\mathcal{O}}_K^{\dim(P_\alpha)}} \omega_{\hat{\mathbb{A}}_K}\right)\\ &\overset{5.31}{=} \prod_{\alpha\in\Delta_P^+}\prod_{i=1}^{\dim(P_\alpha)}\left(\int_{(P_\alpha^{(i)}/P_\alpha^{(i-1)})(\hat{\mathbb{A}}_K)\cap\hat{\mathcal{O}}_K} \omega_{\hat{\mathbb{A}}_K}\right)\\ &\overset{4.8}{=} \prod_{\alpha\in\Delta_P^+} q^{c(\mathcal{L}_{P_\alpha})}\\ &\overset{6.21}{=} q^{p(P)}.\end{aligned}$$

Theorem 7.3 ([24, 1.3.2]). *Let G/Y be a reductive group Y-scheme, let P/Y be a maximal parabolic K-subgroup, and let $\mathfrak{K} := G(\hat{\mathcal{O}}_K)$. Then, for each $x \in P(\hat{\mathbb{A}}_K)$, one has*

$$\int_{R_u(P(\hat{\mathbb{A}}_K))\cap\mathfrak{K}} \omega_{\hat{\mathbb{A}}_K} = |\chi_P(x)| \int_{R_u(P(\hat{\mathbb{A}}_K))\cap{}^x\mathfrak{K}} \omega_{\hat{\mathbb{A}}_K}.$$

Proof. The absolute value of the determinant of the derivative of conjugation by x

$$\begin{aligned}|\chi_P(\cdot)| : P(\hat{\mathbb{A}}_K) &\overset{\mathrm{Ad}}{\to} \mathrm{GL}\left(\mathrm{Lie}(R_u(P(\hat{\mathbb{A}}_K)))\right) \overset{\det}{\to} \mathrm{GL}\left(\bigwedge^d \mathrm{Lie}(R_u(P(\hat{\mathbb{A}}_K)))\right) \overset{|\cdot|}{\to} \mathbb{R}\\ x &\mapsto |\chi_P(x)|\end{aligned}$$

(cf. Proposition 6.20) measures the ratio of the volumes of $R_u(P(\hat{\mathbb{A}}_K)) \cap \mathfrak{K}$ and of $R_u(P(\hat{\mathbb{A}}_K)) \cap {}^x\mathfrak{K}$.

Definition 7.4. Using the notation of Theorem 6.22, let

$$\pi_i(B, {}^x\mathfrak{K}) := \pi(P_i, {}^x\mathfrak{K}) := \int_{R_u(P_i(\hat{\mathbb{A}}_K))\cap{}^x\mathfrak{K}} \omega_{\hat{\mathbb{A}}_K} \quad \text{and}$$

$$\nu_i(B, {}^x\mathfrak{K}) := \prod_{j=1}^{r} \pi_j(B, {}^x\mathfrak{K})^{c_{i,j}}.$$

Corollary 7.5 ([24, p. 40]). *For each $x \in B(\hat{\mathbb{A}}_K)$, one has $\nu_i(B, \mathfrak{K}) = |\alpha_i(x)|\nu_i(B, {}^x\mathfrak{K})$.*

Observation 7.6. *For each* $x \in G(K)$*, one has* $\pi_i(B,\mathfrak{K}) = \pi_i({}^xB,{}^x\mathfrak{K})$ *and* $\nu_i(B,\mathfrak{K}) = \nu_i({}^xB,{}^x\mathfrak{K})$.

Proof. Conjugation by $x \in G(K)$ maps the K-volume form ω on $R_u(P)$ onto a K-volume form ${}^x\omega$ on $R_u({}^xP)$, whence

$$\pi_i(B,\mathfrak{K}) = \int_{R_u(P_i(\hat{\mathbb{A}}_K))\cap\mathfrak{K}} \omega_{\hat{\mathbb{A}}_K} = \int_{R_u({}^xP_i(\hat{\mathbb{A}}_K))\cap{}^x\mathfrak{K}} {}^x\omega_{\hat{\mathbb{A}}_K} \overset{7.1}{=} \pi_i({}^xB,{}^x\mathfrak{K}).$$

By Definition 7.4 the second identity follows from the first.

Theorem 7.7 ([24, 2.1.1]). *Let* G/Y *be a rationally trivial group* Y*-scheme and let* $\mathfrak{K} := G(\hat{\mathcal{O}}_K)$*. Then there exists a constant* $C_1 > 0$ *such that for each* $x \in G(\hat{\mathbb{A}}_K)$ *there exists a Borel subgroup* B/Y *of* G/Y *such that*

$$\nu_i(B,{}^x\mathfrak{K}) \geq C_1.$$

Proof. By [24, 1.1.2], for each $x \in G(\hat{\mathbb{A}}_K)$ there exists a rationally trivial group scheme $G^{(x)}/Y$ such that ${}^x\mathfrak{K} = G^{(x)}(\hat{\mathcal{O}}_K)$. Let $B^{(x)}/Y$ be a Borel subgroup of $G^{(x)}/Y$. Then, by Theorems 6.22, 7.2, and Definition 7.4,

$$\nu_i(B^{(x)}, G^{(x)}(\hat{\mathcal{O}}_K)) = q^{n_i(B^{(x)})}.$$

Therefore, if $c_1 < -2g$, then by Theorem 6.16 the conclusion of the theorem holds for $C_1 := q^{c_1}$.

Corollary 7.8 ([24, 2.1.2]). *Let* G/Y *be a rationally trivial group* Y*-scheme, let* $\mathfrak{K} := G(\hat{\mathcal{O}}_K)$*, and let* $C_1 = q^{c_1}$ *be a constant for which the conclusion of Theorem 7.7 holds. Then, there exist constants* $C_2 > \Gamma > C_1$ *such that the following hold: let* $x \in G(\hat{\mathbb{A}}_K)$*, let* B/Y *be a Borel subgroup of* G/Y *such that* $\nu_i(B,{}^x\mathfrak{K}) \geq C_1$ *for all* $i \in I$*, and let* α_{i_0} *be a simple root of* B *such that* $\nu_{i_0}(B,{}^x\mathfrak{K}) \geq C_2$*. Then each Borel subgroup* B'/Y *of* G/Y *with* $\nu_i(B',{}^x\mathfrak{K}) \geq C_1$ *for all* $i \in I$ *satisfies* $\nu_{i_0}(B',{}^x\mathfrak{K}) \geq \Gamma$ *and is contained in* $P_{i_0}(B)$.

Proof. By Theorem 6.18 the conclusion holds for $C_2 := q^{c_2}$ and $\Gamma := q^\gamma$.

Corollary 7.9 ([24, 2.3.2]). *Let* G/Y *be a reductive group* Y*-scheme and let* $\mathfrak{K} := G(\hat{\mathcal{O}}_K)$*. Then there exists a constant* $C_1 > 0$ *such that for each* $x \in G(\hat{\mathbb{A}}_K)$ *there exists a minimal* K*-parabolic subgroup* B/Y *of* G/Y *with*

$$\nu_i(B,{}^x\mathfrak{K}) \geq C_1 \qquad \text{for all } i \in I.$$

Definition 7.10. Once and for all fix $C_1 \in (0,1)$ such that the conclusion of Corollary 7.9 holds. A pair consisting of a minimal parabolic subgroup B of G/Y and an element $x \in G(\hat{\mathbb{A}}_K)$ is called **reduced**, if $\nu_i(B,{}^x\mathfrak{K}) \geq C_1$ for all i.

Corollary 7.11 ([24, 2.3.3]). *Let G/Y be a reductive group Y-scheme and let $\mathfrak{K} := G(\hat{\mathcal{O}}_K)$. Then there exist constants $C_2 > \Gamma > C_1$ such that the following hold: let $x \in G(\hat{\mathbb{A}}_K)$, let B/Y be a minimal parabolic K-subgroup of G/Y such that (B, x) is reduced, and let α_{i_0} be a simple root of B such that $\nu_{i_0}(B, {}^x\mathfrak{K}) \geq C_2$. Then each minimal parabolic K-subgroup B'/Y of G/Y with (B', x) reduced satisfies $\nu_{i_0}(B', {}^x\mathfrak{K}) \geq \Gamma$ and $B' \subset P_{i_0}(B)$.*

Note that Corollaries 7.9 and 7.11 follow from Theorem 7.7 and Corollary 7.8 by a standard field extension argument using [24, 2.3.5].

7.12. Let G/Y be a reductive group Y-scheme, let B/Y be a minimal parabolic K-subgroup of G, and let $\mathfrak{K} := G(\hat{\mathcal{O}}_K)$. Since $G(\hat{\mathbb{A}}_K)/B(\hat{\mathbb{A}}_K)$ is compact and $\mathfrak{K}$ is open [24, p. 36], the double coset space $\mathfrak{K}\backslash G(\hat{\mathbb{A}}_K)/B(\hat{\mathbb{A}}_K)$ is finite. As $B(\hat{\mathbb{A}}_K)$ is self-normalizing in $G(\hat{\mathbb{A}}_K)$, one can consider $G(\hat{\mathbb{A}}_K)/B(\hat{\mathbb{A}}_K)$ as the space of conjugates of $B(\hat{\mathbb{A}}_K)$ in $G(\hat{\mathbb{A}}_K)$, and $\mathfrak{K}\backslash G(\hat{\mathbb{A}}_K)/B(\hat{\mathbb{A}}_K)$ as the space of $\mathfrak{K}$-orbits via conjugation on these.

Theorem 7.13 ([24, p. 40]). *Let G/Y be a reductive group Y-scheme, let B/Y be a minimal parabolic K-subgroup of G, let $\mathfrak{K} := G(\hat{\mathcal{O}}_K)$, let $B^{(1)}, \ldots, B^{(t)}$ be a system of representatives of the $\mathfrak{K}$-orbit space $\mathfrak{K}\backslash G(\hat{\mathbb{A}}_K)/B(\hat{\mathbb{A}}_K)$ (cf. Theorem 7.12), let $\xi_s \in G(K)$ with $B^{(s)} = \xi_s^{-1} B^{(1)} \xi_s$, and for $c \in \mathbb{R}$ define $B^{(s)}(c) = \{x \in B^{(s)}(\hat{\mathbb{A}}_K) \mid |\alpha_i(x)| \leq c$ for all $i \in I\}$. Then there exists $r \in \mathbb{R}$ such that*

$$\bigcup_{s=1}^{t} B^{(1)}(r)\xi_s\mathfrak{K}$$

is a fundamental domain for $G(K)\backslash G(\hat{\mathbb{A}}_K)$.

Proof. For $x \in G(\hat{\mathbb{A}}_K)$, by Corollary 7.9, there exists a minimal parabolic K-subgroup B of G such that $\nu_i(B, {}^x\mathfrak{K}) \geq C_1$ for all $i \in I$. Let $B^{(s)}$ be the representative of the $\mathfrak{K}$-orbit of $x^{-1}Bx$ in $G(\hat{\mathbb{A}}_K)/B(\hat{\mathbb{A}}_K)$, let $u \in \mathfrak{K}$ such that $u^{-1}x^{-1}Bxu = B^{(s)}$, let $a \in G(K)$ with $aBa^{-1} = B^{(s)}$, and define $y := axu \in N_{G(\hat{\mathbb{A}}_K)}(B^{(s)}(\mathbb{A}_K)) = B^{(s)}(\mathbb{A}_K)$. Since $a \in G(K)$ one has

$$\nu_i(B, {}^x\mathfrak{K}) \overset{7.6}{=} \nu_i({}^aB, {}^{ax}\mathfrak{K}) = \nu_i(B^{(s)}, {}^y\mathfrak{K}).$$

By Corollary 7.5, for each $i \in I$ one has $|\alpha_i(y)| = \nu_i(B^{(s)}, \mathfrak{K}) \left(\nu_i(B^{(s)}, {}^y\mathfrak{K})\right)^{-1} \leq \nu_i(B^{(s)}, \mathfrak{K})C_1^{-1}$. Hence, for $r := \max\{\nu_i(B^{(s)}, \mathfrak{K})C_1^{-1} \mid 1 \leq s \leq t, i \in I\}$, to each $x \in G(\hat{\mathbb{A}}_K)$ there exists $a \in G(K)$ and $u \in \mathfrak{K}$ such that $ax = yu^{-1} \in B^{(s)}(r)\mathfrak{K} = \xi_s^{-1}B^{(1)}(r)\xi_s\mathfrak{K}$. The claim follows because $a, \xi_s \in G(K)$.

8 Filtrations of Euclidean Buildings

In this section, I translate the geometric version of Harder's reduction theory into the setting of Euclidean buildings based on [26]. From this section onwards the survey is intended for the reader familiar with the concept of Euclidean buildings, as simplicial complexes and as CAT(0) spaces. For both introductory and further reading, the sources [1], [10, 13, 46, 47] are highly recommended.

8.1. Let $Y/\mathbb{F}_q$ be a non-singular projective curve, let $S \subset Y^\circ$ be finite, and let G/Y be a reductive group Y-scheme. Clearly, each of Theorem 7.7, Corollaries 7.8, 7.9, 7.11 holds for $x \in G(\hat{\mathbb{A}}_S) \subset G(\hat{\mathbb{A}}_K)$ (cf. Theorem 4.4). Since

$$G(K \cap \hat{\mathbb{A}}_S)\backslash G(\hat{\mathbb{A}}_S)/G(\hat{\mathcal{O}}_K) \overset{4.4}{\cong} G(\mathcal{O}_S)\Big\backslash G\left(\prod_{P\in S} K_P\right)\Big/ G\left(\prod_{P\in S} \hat{\mathcal{O}}_{P,K}\right)$$
$$= G(\mathcal{O}_S)\Big\backslash \prod_{P\in S} G(K_P)/G(\hat{\mathcal{O}}_{P,K}),$$

the functions $\pi_i(B, {}^x\mathfrak{K})$ and $\nu_i(B, {}^x\mathfrak{K})$ (cf. Definition 7.4) allow one to define $G(\mathcal{O}_S)$-invariant (cf. Observation 7.6) filtrations on the Euclidean building X of $\prod_{P\in S} G(K_P)$. The group $G(\mathcal{O}_S)$ is called an **S-arithmetic group**. The set of special vertices of X is $X_\nu := \prod_{P\in S} G(K_P)/G(\hat{\mathcal{O}}_{P,K})$. The diagonal embedding of K in $\prod_{P\in S} K_P$ (cf. Theorem 4.4) yields a diagonal embedding of the spherical building of $G(K)$ into the spherical building at infinity of X with respect to the complete system of apartments. Note that this embedding is in general not simplicial.

Definition 8.2. Let X be a CAT(0) space, e.g., a Euclidean building, and let $\gamma : [0,\infty) \to X$ be a unit speed geodesic ray, i.e., $d(\gamma(t), \gamma(0)) = t$ for all $t \geq 0$. The function

$$b_\gamma : X \to \mathbb{R} : x \mapsto \lim_{t\to\infty} (t - d(x, \gamma(t)))$$

is the **Busemann function** with respect to γ. Note that $t = d(\gamma(0), \gamma(t)) \leq d(\gamma(0), x) + d(x, \gamma(t))$ and $t - d(x, \gamma(t))$ non-decreasing in t, so that the limit always exists. A linear reparametrization of a Busemann function is called a **generalized Busemann function**. For a (generalized) Busemann function b_γ, a sub-level set of $-b_\gamma$ is a **horoball centred at** $\gamma(\infty)$. The boundary of a horoball is a **horosphere**, centred at $\gamma(\infty)$.

Some sources define the Busemann function with respect to a geodesic ray as $b_\gamma : X \to \mathbb{R} : x \mapsto \lim_{t\to\infty}(d(x, \gamma(t)) - t)$. This does not affect the concept of a horoball, i.e., in that case a horoball is defined as a sub-level set of b_γ.

Theorem 8.3. *Let P be a maximal K-parabolic and let $\mathfrak{K} := G(\mathcal{O}_K)$. Then for each $g \in \prod_{P\in S} P(K_P)$ one has*

$$\log_q(\pi(P, {}^g\mathfrak{K})) = \log_q(\pi(P, \mathfrak{K})) + \sum_{P\in S} \deg(P)\nu_P(\chi_P(g)).$$

In particular, there exists a generalized Busemann function $p(P,\cdot) : X \to \mathbb{R}$ *whose restriction to the set* X_v *of special vertices of* X *equals* $\log_q(\pi(P,\cdot))$.

Proof. One has

$$\begin{aligned}\log_q(\pi(P,{}^g\mathfrak{K})) &\overset{7.3}{=} \log_q(\pi(P,\mathfrak{K})) + \log_q(|\chi_P(g)|^{-1}) \\ &\overset{4.4}{=} \log_q(\pi(P,\mathfrak{K})) + \log_q(|\chi_P(g)|_S^{-1}) \\ &\overset{4.1}{=} \log_q(\pi(P,\mathfrak{K})) + \sum_{P\in S} \deg(P)\nu_P(\chi_P(g)).\end{aligned}$$

The maximal K-parabolic P corresponds to a vertex in the building of $G(K)$ and hence yields an element ξ in the spherical building at infinity of X (cf. Theorem 8.1). As χ_P is a multiple of the fundamental weight corresponding to P, any geodesic ray γ in X with $\gamma(\infty) = \xi$ provides a generalized Busemann function $b_\gamma = p(P,\cdot) : X \to \mathbb{R}$ as claimed.

Definition 8.4. Let B/Y be a minimal parabolic K-subgroup of G/Y, let $\{\alpha_1, ..., \alpha_r\}$ be the simple roots of B, let $(P_i)_i$ be the maximal parabolic K-subgroups of G, of type α_i, and, for each i, let $p(P_i,\cdot) : X \to \mathbb{R}$ be the generalized Busemann function from Theorem 8.3. For $x \in X$ define

$$\begin{aligned}p_i(B,x) &:= p(P_i,x), \\ \overline{p}_i(B,x) &:= c_i\, p_i(B,x), \\ n_i(B,x) &:= \sum_{j=1}^{r} c_{i,j}\, p_j(B,x),\end{aligned}$$

where $(c_i)_{1\le i\le r}$ is a family of positive real numbers such that each $c_i\, p_i(B,\cdot)$ is a Busemann function and $c_{i,j} \in \mathbb{Q}$ are such that $\alpha_i = \sum_{j=1}^{r} c_{i,j}\chi_j$ (cf. Theorem 6.22).

Corollary 8.5. *Let* B/Y *be a minimal parabolic* K*-subgroup of* G/Y*. For each* $1 \le i \le r$ *and each* $g \in \prod_{P\in S} B(K_P)$ *one has*

$$n_i(B,{}^g\mathfrak{K}) = n_i(B,\mathfrak{K}) + \sum_{P\in S} \deg(P)\nu_P(\alpha_i(g)).$$

Proof. Combine Corollary 7.5 with Definition 8.4.

8.6. The chambers of X are compact and pairwise isometric and each contains a special vertex. Hence, there exists $d > 0$ such that for each element $x \in X$ there exists a special vertex $x' \in X$ such that for each minimal parabolic K-subgroup B of G

$$n_i(B,x) \ge c \text{ implies } n_i(B,x') \ge c - d \quad \text{for all } i \in I \text{ and } c \in \mathbb{R} \text{ and,}$$

$$n_i(B,x') \ge c \text{ implies } n_i(B,x) \ge c - d \quad \text{for all } i \in I \text{ and } c \in \mathbb{R}.$$

Theorem 8.7 ([26, 1.4.2]). *There exists $c_1 \in \mathbb{R}$ such that for each element $x \in X$ there exists a minimal parabolic K-subgroup B of G with $n_i(B,x) \geq c_1$ for all $i \in I$.*

Proof. In case one only considers special vertices $x \in X$, such a constant $c_1 := \log_q(C_1)$ exists by Theorem 7.7. If d is a constant for which the conclusion of Theorem 8.6 holds, then replacing c_1 by $c_1 - d$ therefore implies the assertion for arbitrary elements $x \in X$.

Theorem 8.8 ([26, 1.4.4]). *There exist constants $c_2 > \gamma > c_1$ such that for $x \in X$, a minimal K-parabolic B with $n_i(B,x) \geq c_1$ for all $i \in I$, the family $(P_i)_{i \in I}$ of maximal parabolic K-subgroups of G containing B, and $j \in I$ with $n_j(B,x) \geq c_2$, each minimal parabolic K-subgroup B' of G with $n_i(B',x) \geq c_1$ for all $i \in I$ satisfies $n_j(B',x) \geq \gamma$, and is contained in P_j.*

Proof. In case one only considers special vertices $x \in X$, such constants $c_2 := \log_q(C_2)$ and $\gamma := \log_q(\Gamma)$ exist by Corollary 7.11. Choose again a constant d for which the conclusion of Theorem 8.6 holds, define $c_1' := c_1 - d$ and $C_1' := q^{c_1'}$, use this constant C_1' in Definition 7.10 to define reduced pairs, let C_2' and Γ' be constants for which the conclusion of Corollary 7.11 holds for this definition of a reduced pair, and let $c_2' := \log_q(C_2')$. For $c_2 := c_2' + d$ by Theorem 8.6 there exists a special vertex $x' \in X$ such that

$$\begin{aligned} n_i(B,x) \geq c_1 \quad &\text{implies } n_i(B,x') \geq c_1' \text{ for all } i \in I, \\ n_i(B',x) \geq c_1 \quad &\text{implies } n_i(B',x') \geq c_1' \text{ for all } i \in I, \text{ and} \\ n_j(B,x) \geq c_2 \quad &\text{implies } n_j(B,x') \geq c_2'. \end{aligned}$$

We conclude from Corollary 7.11 that B' is contained in P_j and that $n_j(B',x') \geq \log_q(\Gamma')$, whence $n_j(B',x) \geq \log_q(\Gamma') - d =: \gamma$.

Notation 8.9 *Once and for all fix constants $c_1, c_2, \gamma \in \mathbb{R}$ such that c_1 is negative, $c_2 > \gamma > c_1$ and such that the conclusions of Theorems 8.7 and 8.8 hold.*

Definition 8.10. A pair (B,x) consisting of a minimal K-parabolic subgroup B of G and an element $x \in X$ such that $n_i(B,x) \geq c_1$ for all $i \in I$ is called **reduced**. For a minimal parabolic K-subgroup B of G and a maximal K-parabolic $P_j \supseteq B$, following [25, p. 254], an element $x \in X$ is called **close to the boundary of X with respect to P_j**, if (B,x) is a reduced pair and $n_j(B,x) \geq c_2$. An element $x \in X$ is called **close to the boundary of X**, if there exists a maximal K-parabolic P such that x is close to the boundary of X with respect to P. For $x \in X$ close to the boundary of X, define

$$P_x := \bigcap \{P \subset G \mid x \text{ is close to the boundary of } X \text{ with respect to } P\}.$$

By Theorem 8.8 the group P_x is a K-parabolic subgroup of G (cf. Proposition 6.19). Following [23, p. 138], it is called the **isolated parabolic subgroup of G**

corresponding to x. For each K-parabolic $Q \supset P_x$, the element $x \in X$ is called **close to the boundary of** X **with respect to** Q.

Proposition 8.11 ([6, p. 35]). *Let $x \in X$ be close to the boundary of X and let P_x be the corresponding isolated parabolic subgroup of G. Let γ be a geodesic ray in X with $\gamma(0) = x$ and whose end point lies in the simplex of the building at infinity corresponding to P_x. Then each $y \in \gamma([0,\infty))$ is close to the boundary of X. Moreover, one has $P_y = P_x$ and for each minimal K-parabolic B the pair (B, y) is reduced if and only if (B, x) is reduced.*

Proof (Bux, Köhl, Witzel). First, notice that by (6) and (7) for each reduced pair (B, x) and all $i \in I$ one has $n_i(B, y) \geq n_i(B, x)$. In particular, each reduced pair (B, x) gives rise to a reduced pair (B, y) and, moreover, y lies close to the boundary of X with respect to P_x. This implies $P_y \subseteq P_x$.

Conversely, let (B, y) be a reduced pair. Then by Theorem 8.8 one has $B \subseteq P_y \subseteq P_x$. Let $(P_i)_{i \in I}$ be the family of maximal parabolic K-subgroups of G containing B and let $I' \subseteq I$ such that $P_x = \bigcap_{i \in I'} P_i$. If there exists $j \in I$ such that $n_j(B, y) \geq c_2$, but $P_j \not\supseteq P_x$, then $j \in I \backslash I'$. As for each $i \in I \backslash I'$ one has $n_i(B, x) = n_i(B, y)$, in particular $n_j(B, x) = n_j(B, y) \geq c_2$, and in view of Proposition 8.10 the pair (B, x) cannot be reduced. As being a reduced pair is a closed condition (see Definition 8.10), there, therefore, exists a minimal $a \in (0, \infty)$ such that $(B, \gamma(a))$ is reduced. The first paragraph of this proof implies that $P_{\gamma(a)} \subseteq P_x$. Thus, Theorem 8.8 applied to the reduced pair $(B, \gamma(a))$ yields $n_i(B, \gamma(a)) \geq \gamma$, and hence $n_i(B, \gamma(a)) > c_1$ by Notation 8.9, for all $i \in I'$. As the $n_i(B, \gamma(b))$ are continuous in b and as for each $i \in I \backslash I'$ one has $n_i(B, x) = n_i(B, \gamma(b)) = n_i(B, y) \geq c_1$, this contradicts the minimality of a. Therefore, (B, x) has to be reduced. Consequently, each $j \in I$ with the property that $n_j(B, y) \geq c_2$ satisfies $P_j \supseteq P_x$, whence $j \in I'$. Thus we have $P_y = P_x$.

Definition 8.12. For $c \in \mathbb{R}$ define

$$X^n(c) = \{x \in X \mid (B, x) \text{ reduced implies } n_i(B, x) \leq c \text{ for all } i \in I\},$$

$$X^p(c) = \{x \in X \mid (B, x) \text{ reduced implies } p_i(B, x) \leq c \text{ for all } i \in I\}, \quad \text{and}$$

$$X^{\overline{p}}(c) = \{x \in X \mid (B, x) \text{ reduced implies } \overline{p}_i(B, x) \leq c \text{ for all } i \in I\}.$$

These filtrations are $G(\mathcal{O}_S)$-invariant (cf. Observation 7.6) and $G(\mathcal{O}_S)$-cocompact (cf. [24, 2.2.2]). There exists $c_3 \in \mathbb{R}$ such that $X^n(c_2) \subseteq X^{\overline{p}}(c_3)$.

Proposition 8.13 ([8, Sect. 5]). *Let $c \geq c_3$, let $x \in X \backslash X^{\overline{p}}(c)$ and let $\overline{x} \in X^{\overline{p}}(c)$ be an element at which the function $X^{\overline{p}}(c) \to \mathbb{R} : z \mapsto d(x, z)$ assumes a global minimum. Then $P_{\overline{x}} = P_x$. Furthermore, there exists a unique unit speed geodesic ray $\gamma_x^c : [0, \infty) \to X$ with $\gamma_x^c(0) = x$ along which the function $X \backslash X^{\overline{p}}(c) \to \mathbb{R} : x \mapsto d(x, X^{\overline{p}}(c))$ assumes its steepest ascent; its end point lies in the simplex at infinity corresponding to P_x.*

The preceding proposition shows that one can measure the distance from the set $X^{\overline{p}}(c_3)$ in a neat way. It also shows that there is a substantial difference between K-rank 1 and higher K-rank. Indeed, by Proposition 8.13, in the case of K-rank 1 the boundary of $X^{\overline{p}}(c_3)$ consists of hypersurfaces of codimension one which must have pairwise empty intersections, as otherwise one would need non-minimal isolated K-parabolics, which do not exist. The following result makes this heuristic argument more concrete.

Theorem 8.14 ([16, 3.7]). *If* $\mathrm{rk}_K(G) = 1$, *then there exists a collection* $\mathcal{H}$ *of pairwise disjoint horoballs of* X *such that* $X^{\overline{p}}(c_3) = X \backslash \mathcal{H}$ *is* $G(\mathcal{O}_S)$*-invariant and* $G(\mathcal{O}_S)$*-cocompact.*

Proof. Since the functions $\overline{p}$ are Busemann functions (cf. Theorem 8.3, Definition 8.4), there clearly exists a collection $\mathcal{H}$ of horoballs of X such that $X^{\overline{p}}(c_3) = X \backslash \mathcal{H}$. By Definition 8.12 the set $X^{\overline{p}}(c_3)$ is $G(\mathcal{O}_S)$-invariant and $G(\mathcal{O}_S)$-cocompact. It therefore remains to prove that the horoballs in $\mathcal{H}$ can be chosen to be either disjoint or equal. Let $H_1, H_2 \in \mathcal{H}$ have non-trivial intersection and let B_1, B_2 be the minimal K-parabolic subgroups corresponding to their respective centres (cf. Definition 8.2). Then there exists $x \in X$ such that $\overline{p}_\alpha(B_1, x), \overline{p}_\alpha(B_2, x) \geq c_3$, where α denotes the unique simple root. Therefore, by Definition 8.12, $n_\alpha(B_1, x), n_\alpha(B_2, x) \geq c_2$, and by Theorem 8.8 one has $B_1 = B_2$. Hence, $H_1 \subseteq H_2$ or $H_2 \subseteq H_1$. If $H_1 \neq H_2$, one can remove one of the two from $\mathcal{H}$ without changing $X \backslash \mathcal{H}$.

9 Applications and Conjectures

In this section, I sketch the applicability of the geometric version of Harder's reduction theory to the study of finiteness properties of S-arithmetic groups over global function fields and state a very general conjecture on isoperimetric properties of S-arithmetic groups over arbitrary global fields. For reduction theory over number fields, I strongly recommend [34].

9.1 *Finiteness Properties of S-Arithmetic Groups*

Definition 9.1. A group Γ is of **type** F_m, if it admits a free action on a contractible CW complex X with finitely many orbits on the m-skeleton of X.

A group action on a CW complex is called *cellular*, if the action preserves the cell structure, and *rigid*, if each group element that elementwise fixes the skeleton of a cell in fact elementwise fixes the whole cell.

Theorem 9.2 ([12, 1.1]). *Let $m \in \mathbb{N}$, let X be an $(m-1)$-connected CW complex, and let $\Gamma \to \mathrm{Aut}(X)$ act cellularly, rigidly and cocompactly on X such that the stabilizer of each i-cell is of type F_{m-i}. Then Γ is of type F_m.*

Theorem 9.3 ([16, 7.7]). *Let $X = X_1 \times \cdots \times X_t$ be an affine building, decomposed into its irreducible factors, let ∂X be the spherical building at infinity of X, let $\partial^j X$ be the spherical building at infinity of $\prod_{i=1, i \neq j}^{t} X_i$, and let $\xi \in \partial X \setminus \bigcup_{1 \leq j \leq t} \partial^j X$. Then any horosphere centred at ξ (Definition 8.2) is $(\dim(X) - 2)$-connected.*

Theorem 9.4 ([16, 8.1]). *Let K be a global function field, let G be an absolutely almost simple K-group of K-rank 1, let $\emptyset \neq S \subset Y^\circ$ be finite, let X be the Euclidean building of $\prod_{P \in S} G(K_P)$, and let $m = \dim(X) = \sum_{P \in S} \mathrm{rk}_{K_P}(G)$. Then $G(\mathcal{O}_S)$ is of type F_{m-1}, but not F_m.*

Proof. The group $G(\mathcal{O}_S)$ clearly acts rigidly and cellularly on the (contractible) building X. By Theorem 8.14, the group $G(\mathcal{O}_S)$ acts cocompactly on $X^{\overline{P}}(c_3) = X \setminus \mathcal{H}$ and, in view of Theorem 8.1, each of the horoballs $H \in \mathcal{H}$ is centred at some ξ which satisfies the hypothesis of Theorem 9.3. Therefore, by Theorem 9.3, $X^{\overline{P}}(c_3)$ is $(n-2)$-connected. As cell stabilizers in $G(\mathcal{O}_S)$ are finite, whence of type F_∞, the claim follows from Theorem 9.2.

For K-rank greater than 1 this strategy cannot work, as Theorem 8.14 becomes false. It can be adapted, however, which leads to a couple of technical difficulties that have first been overcome in [7, 48] for the S-arithmetic groups $G(\mathbb{F}_q[t])$ and $G(\mathbb{F}_q[t, t^{-1}])$ where G is an absolutely almost simple $\mathbb{F}_q$-group of rank $n \geq 1$. In this situation one can in fact make use of the theory of Euclidean twin buildings, as in [2].

Theorem 9.5 ([7, A], [48, Main Theorem]). *Let G be an absolutely almost simple $\mathbb{F}_q$-group of rank $n \geq 1$. Then $G(\mathbb{F}_q[t])$ is of type F_{n-1}, but not F_n, and $G(\mathbb{F}_q[t, t^{-1}])$ is of type F_{2n-1}, but not F_{2n}.*

Recently, a combination of the ideas developed in [7, 16, 48] with Harder's reduction theory allowed the authors of [8] to prove the following theorem, providing a positive answer to the question asked in [1, 13.20], [5, p. 80], [13, p. 197].

Theorem 9.6 ([8, Rank Theorem]). *Let K be a global function field, let G be an absolutely almost simple K-isotropic K-group, let $\emptyset \neq S \subset Y^\circ$ be finite, let X be the Euclidean building of $\prod_{P \in S} G(K_P)$, and let $m = \dim(X) = \sum_{P \in S} \mathrm{rk}_{K_P}(G)$. Then $G(\mathcal{O}_S)$ is of type F_{m-1}, but not F_m.*

Remark 9.7. Using the notation introduced above, already [15, 1.1] established that $G(\mathcal{O}_S)$ is not of type F_m.

Remark 9.8. S-arithmetic groups over number fields are known to be of type F_∞, cf. [14, Sect. 11].

9.2 Isoperimetric Properties of S-Arithmetic Groups

In the number field case, results similar to Corollaries 7.9 and 7.11 hold, cf. [9], [24, p. 53]. To the best of my knowledge, the question posed by Harder [24, p. 54] whether it is possible to prove the results from [9] using methods similar (or at least closer) to the approach used by Harder is still open. The key, of course, would be to prove Theorem 6.18 for number fields.

As this problem by itself probably will not attract sufficient attention, I will finish this survey by stating a very general conjecture on properties of S-arithmetic groups over arbitrary global fields, i.e., global function fields or number fields, which, if verified, provides another proof of Theorem 9.6.

Definition 9.9. A **coarse n-manifold** Σ in a metric space X is a function from the vertices of a triangulated n-manifold M into X. The homeomorphism type of the manifold M is called the **topological type** of Σ. The **boundary** $\partial\Sigma$ is the restriction to ∂M of the function Σ. The coarse manifold Σ has **scale** $r \in \mathbb{R}_+$, if $d(\Sigma(x), \Sigma(y)) \leq r$ for all adjacent vertices x, y of M. The **volume** $\mathrm{vol}(\Sigma)$ equals the number of vertices in M.

Conjecture 9.10 ([15]). Let K be a global field, i.e., global function field or a number field, let G be an absolutely almost simple K-isotropic K-group, let S be a non-empty finite set of places of K containing all archimedean ones, and let $n < \sum_{P\in S} \mathrm{rk}_{K_P}(G)$. To any $r_1 > 0$ there exists a linear polynomial f and $r_2 > 0$ such that, if $\Sigma \subseteq \prod_{P\in S} G(K_P)$ is a coarse n-manifold of scale r_1 with $\partial\Sigma \subseteq G(\mathcal{O}_S)$, then there exists a coarse n-manifold $\Sigma' \subseteq G(\mathcal{O}_S)$ of scale r_2 and identical topological type as Σ such that $\partial\Sigma' = \partial\Sigma$ and $\mathrm{vol}(\Sigma') \leq f(\mathrm{Vol}(\Sigma))$.

In case $n < |S|$, the existence of a polynomial f of *unspecified degree* and $r_2 > 0$ as in the conjecture have been established by Bestvina, Eskin and Wortman. The precise statement of their result is as follows.

Theorem 9.11 ([49, 2]). *Let K be a global field, i.e., global function field or a number field, let G be an absolutely almost simple K-isotropic K-group, let S be a non-empty finite set of places of K containing all archimedean ones, and let $n < |S|$. To any $r_1 > 0$ there exists a polynomial f and $r_2 > 0$ such that, if $\Sigma \subseteq \prod_{P\in S} G(K_P)$ is a coarse n-manifold of scale r_1 with $\partial\Sigma \subseteq G(\mathcal{O}_S)$, then there exists a coarse n-manifold $\Sigma' \subseteq G(\mathcal{O}_S)$ of scale r_2 and identical topological type as Σ such that $\partial\Sigma' = \partial\Sigma$ and $\mathrm{vol}(\Sigma') \leq f(\mathrm{Vol}(\Sigma))$.*

Note that this result provides an alternative proof that the S-arithmetic group $G(\mathcal{O}_S)$ in Theorem 9.6 is of type $F_{|S|-1}$.

Acknowledgements The author expresses his gratitude to Kai-Uwe Bux, Bernhard Mühlherr and Stefan Witzel for numerous invaluable discussions on the topics of this survey during joint research activities at the Hausdorff Institute of Mathematics at Bonn, the MFO at Oberwolfach, and the University of Bielefeld. The author also thanks Kai-Uwe Bux, Max Horn, Timothée Marquis, Andreas Mars, Susanne Schimpf, Rebecca Waldecker, Markus-Ludwig Wermer, Stefan Witzel and the participants of the reduction theory seminar at Bielefeld during summer 2010, especially Werner Hoffmann and Andrei Rapinchuk, for numerous comments, remarks and suggestions on how to improve the contents and exposition of this survey. Furthermore, the author thanks two anonymous referees for valuable comments and observations.

References

1. Peter Abramenko and Kenneth S. Brown. *Buildings – Theory and Applications*, vol. 248 of *Graduate Texts in Mathematics*. Springer, Berlin. 2008.
2. Peter Abramenko. *Twin buildings and applications to S-arithmetic groups*, vol. 1641 of *Lecture Notes in Mathematics*. Springer, Berlin. 1996.
3. J. Vernon Armitage. Algebraic functions and an analogue of the geometry of numbers: the Riemann–Roch theorem. *Arch. Math.* (1967), pp. 383–393.
4. Emil Artin and George Whaples. Axiomatic characterization of fields by the product formula for valuations. *Bull. Amer. Math. Soc.* (1945), pp. 469–492.
5. Helmut Behr. Arithmetic groups over function fields I. *J. reine angew. Math.* **495** (1998), pp. 79–118.
6. Helmut Behr. Higher finiteness properties of S-arithmetic groups in the function field case I. In *Groups: topological, combinatorial and arithmetic aspects* (edited by Thomas W. Müller), pp. 27–42. Cambridge University Press, Cambridge. 2004.
7. Kai-Uwe Bux and Ralf Gramlich and Stefan Witzel. Finiteness properties of Chevalley groups over a polynomial ring over a finite field. `arXiv:0908.4531`. 2009.
8. Kai-Uwe Bux and Ralf Gramlich and Stefan Witzel. Higher finiteness properties of reductive arithmetic groups in positive characteristic: the rank theorem. `arXiv:1102.0428` 2011.
9. Armand Borel, Harish-Chandra. Arithmetic subgroups of algebraic groups. *Ann. Math.* **75** (1962), pp. 485–535.
10. Martin Bridson, André Haefliger. *Metric spaces of non-positive curvature*. Springer, Berlin. 1999.
11. Armand Borel. *Linear Algebraic Groups*. Springer, Berlin. 1991.
12. Kenneth S. Brown. Finiteness properties of groups. *J. Pure Appl. Algebra* **44** (1987), pp. 45–75.
13. Kenneth S. Brown. *Buildings*. Springer, Berlin. 1989.
14. Armand Borel, Jean-Pierre Serre. Cohomologie d'immeubles et de groupes S-arithmétiques. *Topology* **15** (1976), pp. 211–232.
15. Kai-Uwe Bux, Kevin Wortman. Finiteness properties of arithmetic groups over function fields. *Invent. math.* **167** (2007), pp. 335–378.
16. Kai-Uwe Bux, Kevin Wortman. Connectivity properties of horospheres in Euclidean buildings and applications to finiteness properties of discrete groups. `arXiv:0808.2087v1`. 2008.
17. Claude Chevalley. *Introduction to the theory of algebraic functions of one variable*. American Mathematical Society, New York. 1951.
18. Michel Demazure, Alexandre Grothendieck (editors). *Séminaire de Géométrie Algébrique du Bois Marie – 1962–64 – Schémas en groupes, Tome 1*. Springer, Berlin. 1970.
19. Michel Demazure, Alexandre Grothendieck (editors). *Séminaire de Géométrie Algébrique du Bois Marie – 1962–64 – Schémas en groupes, Tome 3*. Springer, Berlin. 1970.
20. David Eisenbud. *Commutative algebra*. Springer, Berlin. 1995.
21. Alexandre Grothendieck. Sur la classification des fibrés sur la sphère de Riemann. *Amer. J. Math.* **79** (1957), pp. 121–138.
22. Alexandre Grothendieck. Éléments de géométrie algébrique. I. Le langage des schémas. *Inst. Hautes Études Sci. Publ. Math.* **4** (1960), pp. 5–228.
23. Günter Harder. Halbeinfache Gruppenschemata über vollständigen Kurven. *Invent. math.* **6** (1968), pp. 107–149.

24. Günter Harder. Minkowskische Reduktionstheorie über Funktionenkörpern. *Invent. math.* **7** (1969), pp. 33–54.
25. Günter Harder. Chevalley groups over function fields and automorphic forms. *Ann. of Math.* **100** (1974), pp. 249–306.
26. Günter Harder. Die Kohomologie S-arithmetischer Gruppen über Funktionenkörpern. *Invent. math.* **42** (1977), pp. 135–175.
27. Robin Hartshorne. *Algebraic geometry*. Springer, Berlin. 1977.
28. Günter Harder. *Lectures on Algebraic Geometry I*. Vieweg, Wiesbaden. 2008.
29. James E. Humphreys. *Linear Algebraic Groups*. Springer. 1975.
30. Jens Carsten Jantzen. *Representations of algebraic groups*. American Mathematical Society. 2003.
31. Gérard Laumon. *Cohomology of Drinfeld modular varieties. Part II: automorphis forms, trace formulas and Langlands correspondence*. Cambridge University Press, Cambridge. 1997.
32. Qing Liu. *Algebraic Geometry and Arithmetic Curves*. Oxford University Press, Oxford. 2002.
33. Harald Niederreiter and Chaoping Xing. *Algebraic geometry in coding theory and cryptography*. Princeton University Press, Princeton. 2009.
34. Vladimir Platonov and Andrei Rapinchuk. *Algebraic groups and number theory*, vol. 139 of *Pure and Applied Mathematics*. Academic Press Inc., Boston, MA. 1994.
35. Michael Rosen. *Number theory in function fields*. Springer, Berlin. 2002.
36. Jean Pierre Serre. Faisceaux algébriques coherents. *Ann. Math.* **61** (1955), pp. 197–278.
37. Jean Pierre Serre. *Algebraic groups and class fields*. Springer, Berlin. 1988.
38. Jean Pierre Serre. *Trees*. Springer, Berlin. Corrected second printing. 2003.
39. Tonny A. Springer. *Linear Algebraic Groups*. Birkhäuser, Basel. 1998.
40. Ulrich Stuhler. Homological properties of certain arithmetic groups in the function field case. *Invent. Math.* **57** (1980), pp. 263–281.
41. Tsuneo Tamagawa. Adèles. In *Algebraic groups and discontinuous subgroups*, pp. 113–121. American Mathematical Society, Providence. 1966.
42. Valentin Voskresenskiĭ. *Algebraic groups and their birational invariants*, vol. 179 of *Translations of Mathematical Monographs*. American Mathematical Society, Providence, RI. 1998.
43. William C. Waterhouse. *Introduction to affine group schemes*. Springer, Berlin. 1979.
44. André Weil. *Adèles and algebraic groups*. Birkhäuser, Basel. 1982.
45. André Weil. *Basic number theory*. Springer, Berlin. 1995.
46. Richard Weiss. *The structure of spherical buildings*. Princeton University Press, Princeton. 2003.
47. Richard Weiss. *The structure of affine buildings*. Princeton University Press, Princeton. 2009.
48. Stefan Witzel. Finiteness properties of Chevalley groups over a Laurent polynomial ring over a finite field. `arXiv:1007.5216`. 2010.
49. Kevin Wortman. Isoperimetric and finiteness properties of arithmetic groups. *Oberwolfach Reports, Oberwolfach Reports* **7** (2010), pp. 1213–1215.
50. Oscar Zariski and Pierre Samuel. *Commutative algebra, volume 2*. Springer, Berlin. 1975.

Some Remarks on Two-Transitive Permutation Groups as Multiplication Groups of Quasigroups

Gerhard Hiss and Frank Lübeck

Abstract We explore the question of which finite two-transitive permutation groups are multiplication groups of quasigroups.

Subject Classifications: [2010] 20N05, 20B20, 20B40

Keywords Quasigroups • Multiplication groups • Two-transitive groups

1 Introduction

A quasigroup is a finite set Q with a binary operation such that the equation $x \cdot y = z$ in Q has a unique solution, whenever two of x, y, z are given. The quasigroup Q is called a loop, if it has a neutral element, and commutative if $x \cdot y = y \cdot x$ for all $x, y \in Q$. The multiplication table T of $Q = \{q_1, \ldots, q_n\}$ is the $(n \times n)$-matrix whose rows and columns are labelled by $q_1, \ldots, q_n$ (in this order), and the entry in row q_i and column q_j equals the product $q_i \cdot q_j$. The axioms of a quasigroup simply say that T is a Latin square.

Row i of T contains the images of $q_1, \ldots, q_n$ under the left multiplication $L(q_i)$: $q_j^{L(q_i)} = q_i \cdot q_j$. Similarly, column j of T contains the images of $q_1, \ldots, q_n$ under the right multiplication $R(q_j)$: $q_i^{R(q_j)} = q_i \cdot q_j$. The multiplication group $M(Q)$ of Q is, by definition, the permutation group on Q generated by $L(q_i), R(q_i), 1 \leq i \leq n$. It is clearly transitive on Q.

G. Hiss (✉) • F. Lübeck
Lehrstuhl D für Mathematik, RWTH Aachen University, 52056 Aachen, Germany
e-mail: gerhard.hiss@math.rwth-aachen.de; frank.luebeck@math.rwth-aachen.de

N.S.N. Sastry (ed.), *Buildings, Finite Geometries and Groups*, Springer Proceedings in Mathematics 10, DOI 10.1007/978-1-4614-0709-6_5,

It has been shown by Smith (see [16, Theorem 523]) that the multiplication group of a quasigroup is multiplicity free, i.e., the permutation character of $M(Q)$ corresponding to its transitive action on Q is multiplicity free. Equivalently, the centraliser algebra of the permutation representation is commutative.

It is, thus, a natural question to ask which multiplicity free transitive permutation groups are multiplication groups of quasigroups. The diploma thesis of Sebastian Köhler [13] answers this question for all permutation groups up to degree 10. In this survey we restrict this question to the two-transitive permutation groups.

We first collect some basic facts and reductions which are useful to decide if a given permutation group is the multiplication group of a quasigroup. Then we consider the two-transitive permutation groups systematically and explore which of them are multiplication groups of quasigroups. Here, we can settle some previously unknown cases, mainly with computational methods. It is perhaps worth mentioning that we found a commutative loop on 24 points whose multiplication group is the Mathieu group M_{24}. A non-commutative loop with this multiplication group had previously been found by Nagy [15].

Section 2 gives an account of Ihringer's results of [10]. We have reformulated these in terms of permutation groups, since this appears to be more appropriate to our computational approach. Section 3 contains a sufficient condition for a permutation group to be a multiplication group of a quasigroup. Our condition is a consequence of Ihringer's criterion [10, Theorem 1], but easier to check. Section 4 contains a systematic account on the knowledge about two-transitive permutation groups with respect to this question. In the last section, we comment on some of the computational methods we applied.

2 Permutation Groups

Let n be a positive integer and put $Q_n := \{1, \ldots, n\}$. In order to simplify notation, we only consider quasigroups with underlying sets Q_n, usually writing $*$ for the multiplication to avoid confusion with integer multiplication. The multiplication group of a quasigroup $(Q_n, *)$ will thus be a subgroup of S_n, the group of permutations of Q_n.

For $\rho_1, \ldots, \rho_n \in S_n$ consider the matrix

$$T := T_{\rho_1,\ldots,\rho_n} := (i^{\rho_j})_{1 \leq i,j \leq n},$$

i.e., column j of T contains the images of $1, 2, \ldots, n$ under ρ_j.

The following lemma is obvious.

Lemma 2.1. *Let* $\rho_1, \ldots, \rho_n \in S_n$ *and put* $T := T_{\rho_1,\ldots,\rho_n}$.

(a) *The following statements are equivalent.*

  (i) T *is the multiplication table of a quasigroup on* Q_n.
  (ii) *Every row of* T *contains all the numbers* $1, \ldots, n$.
  (iii) *For all* $1 \leq i \neq j \leq n$, *the permutation* $\rho_i \rho_j^{-1}$ *has no fixed points.*

(b) *Suppose* T *satisfies the conditions of* (a). *Let* Q *be the quasigroup on the set* Q_n *with multiplication table* T. *Then*

$$M(Q) = \langle \lambda_1, \ldots, \lambda_n, \rho_1, \ldots, \rho_n \rangle,$$

where for $1 \le i \le n$, λ_i *is left multiplication by* i, *given by*

$$j^{\lambda_i} = i^{\rho_j}, \qquad 1 \le j \le n. \tag{1}$$

An alternative way of phrasing Condition (a)(ii) of Lemma 2.1 is to say that $R = \{\rho_1, \ldots, \rho_n\}$ is a *driving sequence* in the sense of [2]. Similarly, Condition (a)(iii) is the same as saying that R constitutes a *sharply transitive subset* of G, i.e., for every pair (i, j) with $1 \le i, j \le n$, there is a unique $\rho \in R$ with $i^{\rho} = j$.

The direct product $S_n \times S_n \times S_n$ acts on the set of matrices satisfying the conditions of Lemma 2.1(a) as follows. Let $T = T_{\rho_1,\ldots,\rho_n} =: (t_{ij})$ be such a matrix and let $(\pi, \sigma, \tau) \in S_n \times S_n \times S_n$. Then define $T^{\pi,\sigma,\tau} = (t'_{ij})$ by

$$t'_{ij} := t^{\tau}_{i'j'}$$

with $(i', j') := (i^{\pi^{-1}}, j^{\sigma^{-1}})$. Thus, the first two components of (π, σ, τ) act by permuting the rows and columns of T, respectively, and the last component acts by renumbering its entries. We consider the effect of these actions on the multiplication group.

Lemma 2.2. *Let* $\rho_1, \ldots, \rho_n \in S_n$ *such that* $T = T_{\rho_1,\ldots,\rho_n}$ *satisfies the conditions of* Lemma 2.1(a), *and let* $(\pi, \sigma, \tau) \in S_n \times S_n \times S_n$.

Let $\rho'_1, \ldots, \rho'_n, \lambda'_1, \ldots, \lambda'_n \in S_n$ *such that* $T^{\pi,\sigma,\tau} = T_{\rho'_1,\ldots,\rho'_n}$, *and such that* $\lambda'_1, \ldots, \lambda'_n$ *are defined by* (1) *with respect to* $\rho'_1, \ldots, \rho'_n$. *Fix* $1 \le i \le n$.

(a) *If* $\sigma = \tau = 1$, *then* $\rho'_i = \pi^{-1}\rho_i$ *and* $\lambda'_i = \lambda_{i'}$ *with* $i' = i^{\pi^{-1}}$.
(b) *If* $\pi = \tau = 1$, *then* $\rho'_i = \rho_{i'}$ *with* $i' = i^{\sigma^{-1}}$ *and* $\lambda'_i = \sigma^{-1}\lambda_i$.
(c) *If* $\pi = \sigma = 1$, *then* $\lambda'_i = \lambda_i\tau$, *and* $\rho'_i = \rho_i\tau$.

Proof. We only prove (a), since the other parts are proven similarly. Let $T = (t_{ij})$. Putting $i' = i^{\pi^{-1}}$, we have

$$i^{\rho'_j} = t_{i'j} = (i')^{\rho_j} = i^{\pi^{-1}\rho_j}$$

for all $1 \le i, j \le n$. Thus, $\rho'_j = \pi^{-1}\rho_j$ for all $1 \le j \le n$. We also have

$$j^{\lambda'_i} = t_{i'j} = (i')^{\rho_j} = j^{\lambda_{i'}}$$

for all $1 \le i, j \le n$. Thus, $\lambda'_i = \lambda_{i'}$ for all $1 \le i \le n$.

Two quasigroups on Q_n whose multiplication tables are in the same orbit under the action of $S_n \times S_n \times S_n$ are called *isotopic*. This relation can be used to construct inclusions between multiplication groups. The following lemma, which rephrases [10, Proposition 3], provides examples of such inclusions.

Lemma 2.3 ([10, Proposition 3]). *Let $Q = (Q_n, *)$ be a quasigroup. Then there are quasigroups Q' and Q'' on the same underlying set Q_n, isotopic to Q, with $M(Q'') \leq M(Q') \leq M(Q)$ such that Q' has a left unit and Q'' is a loop.*

Proof. For $1 \leq i \leq n$, write ρ_i and λ_i for the right and left multiplication with i, respectively. Let T denote the multiplication table of Q. If $\lambda_1 = 1$, put $Q' = Q$. Otherwise, define Q' by its multiplication table $T^{(1,\lambda_1,1)}$. Then Q' has a left unit, and by Lemma 2.2(b) we have

$$\begin{aligned} M(Q') &= \langle \lambda_1^{-1}\lambda_1, \ldots, \lambda_1^{-1}\lambda_n, \rho_1', \ldots, \rho_n' \rangle \\ &\leq \langle \lambda_1, \ldots, \lambda_n, \rho_1, \ldots, \rho_n \rangle = M(Q), \end{aligned}$$

(where $(\rho_1', \ldots, \rho_n')$ is a permutation of $(\rho_1, \ldots, \rho_n)$).

If $\rho_1' = 1$, put $Q'' = Q'$. Otherwise, define Q'' by its multiplication table $T^{(\rho_1',\lambda_1,1)}$. Then Q'' is a loop and

$$\begin{aligned} M(Q'') &= \langle \lambda_1^{-1}\lambda_1, \ldots, \lambda_1^{-1}\lambda_n, (\rho_1')^{-1}\rho_1', \ldots, (\rho_1')^{-1}\rho_n' \rangle \\ &\leq M(Q') \leq M(Q). \end{aligned}$$

Corollary 2.4 (([10, Theorem 1])). *Let $G \leq S_n$. Then G is the multiplication group of a quasigroup on Q_n, if and only if there is a loop $Q = (Q_n, *)$ and $\rho, \lambda \in S_n$ such that $G = \langle M(Q), \rho, \lambda \rangle$.*

Proof. If G is the multiplication group of a quasigroup Q, let Q'' be as in Lemma 2.3. Then $M(Q) = \langle M(Q''), \rho_1, \lambda_1 \rangle$. Conversely, if $G = \langle M(Q), \rho, \lambda \rangle$, where Q is a loop with multiplication table T, define the quasigroup Q' by its multiplication table $T^{(\rho^{-1},\lambda^{-1},1)}$. By Lemma 2.2, $M(Q') = G$.

Thus, if $G \leq S_n$ does not contain a subgroup which is the multiplication group of a loop on Q_n, then G is not the multiplication group of any quasigroup.

3 A Sufficient Condition

We continue with the notation from Sect. 2. The following corollary contains a sufficient condition for a permutation group to be the multiplication group of a quasigroup. It is a special case of Corollary 2.4, but its purely group theoretical condition is easier to check.

Corollary 3.1. *Let $G \leq S_n$. Suppose that G contains an abelian subgroup H acting regularly on Q_n.*

If $G = \langle H, \rho, \lambda \rangle$ for some $\rho, \lambda \in G$, then G is the multiplication group of a quasigroup on Q_n. If $G = \langle H, \rho \rangle$ for some $\rho \in G$, this quasigroup can be chosen to have a left or right unit.

Proof. Number the elements of H as $\rho_1, \rho_2, \ldots, \rho_n$ such that $1^{\rho_i} = i$ for $1 \leq i \leq n$. Then the multiplication table of H with respect to the ordering $\rho_1, \ldots, \rho_n$ equals $T_{\rho_1,\ldots,\rho_n}$ (identifying an entry i with ρ_i). Since H is abelian, it is equal to the multiplication group $M(Q)$ of the loop on Q_n with multiplication table $T_{\rho_1,\ldots,\rho_n}$. The first statement follows from Corollary 2.4. For the second statement use the construction in the proof of Lemma 2.3.

The condition in Corollary 3.1 is not necessary. For example, the alternating group A_6, acting naturally on six letters, does not have a regular subgroup, but is the multiplication group of a quasigroup (see Sect. 4 below).

In [11, Theorem 1.1], the primitive permutation groups containing an abelian regular subgroup were determined. This gives a wealth of examples for groups satisfying the hypotheses of Corollary 3.1.

If $G \leq S_n$ satisfies the hypothesis of Corollary 3.1, then the Corollary together with the theorem of Smith show that G is a multiplicity free permutation group. This can easily be shown directly.

Lemma 3.2. *Let the assumptions be as in* Corollary 3.1. *Then the permutation character corresponding to the natural action of G on Q_n is multiplicity free.*

Proof. Let V denote the permutation $\mathbb{C}G$-module arising from the embedding $G \to S_n$. We have to show that $\mathrm{End}_{\mathbb{C}G}(V)$ is commutative. Since $\mathrm{End}_{\mathbb{C}G}(V)$ is contained in $\mathrm{End}_{\mathbb{C}H}(V\downarrow_H)$ as a subalgebra, it suffices to show that the latter is commutative. Since H acts regularly on $\{1, \ldots, n\}$, we have $\mathrm{End}_{\mathbb{C}H}(V\downarrow_H) \cong \mathbb{C}H$ as $\mathbb{C}$-algebra, and the result follows.

4 Two-Transitive Groups

Using the classification of the finite simple groups, the two-transitive groups have been enumerated. These groups come in two types: Those with an elementary abelian socle, the affine groups, and those with a nonabelian simple socle, the almost simple groups. The tables in Cameron's book [1, Sects. 7.3, 7.4] contain a complete description of all two-transitive groups. (We thank the referee for pointing out an omission in Cameron's table of the affine groups. A correct list can be found in [12, Appendix 1].)

4.1 The Almost Simple Two-Transitive Groups

We summarize the known results for these groups, enumerating them by their simple socles. We have tried to locate all references to previously known cases. A missing reference for a particular result indicates that we are not aware of a corresponding publication. Most of the cases described below were treated by explicit computations, the strategy of which is explained in Sect. 5 at the end of our paper.

$\boldsymbol{A_n}$ The groups A_n and S_n in their natural permutation representation are multiplication groups of quasigroups (even if they are not two-transitive). This is due to Ihringer [10, Theorem 2]. An n-cycle generates a regular subgroup of S_n. Except in case $n = 2m$, and m odd, A_n has a regular abelian subgroup. In the latter case, Ihringer constructs a commutative loop Q on Q_n with $M(Q) \leq A_n$. Since S_n and A_n are 2-generated, the result follows from Corollary 2.4.

A stronger result is due to Drápal and Kepka. They show in [6, Theorem (4.1)], that with the (true) exception of $n = 4, 5$, there is a loop of order n whose multiplication (left, right ore two-sided) group is isomorphic to A_n.

Our computations suggest that for $n \geq 6$ there is always a commutative loop on Q_n with multiplication group S_n. We are not aware of any general construction or reference supporting this observation.

PSL$(\boldsymbol{d}, \boldsymbol{q})$ If $G = \mathrm{PGL}(d, q)$ acting on the space of lines or on the space of hyperplanes of $\mathbb{F}_q^d$, then G is the multiplication group of a quasigroup. Again, this is due to Ihringer [10, Theorem 2] and follows from the fact that G is 2-generated and a Singer cycle is a regular subgroup.

For $d \geq 3$ and $q^d > 8$ (i.e., $q \neq 2$ if $d = 3$), Nagy has constructed in [15, Theorem 3.1] a loop Q on $\mathbb{P}(\mathbb{F}_q^d)$ such that $\mathrm{PSL}(d, q) \leq M(Q) \leq \mathrm{PGL}(d, q)$. The case $d = 3$ and $q = 2$ is a true exception: There is no quasigroup on 7 points with multiplication group $\mathrm{PSL}(3, 2) = \mathrm{PGL}(3, 2)$. Our computations have shown that there are no quasigroups on the corresponding projective lines with multiplication groups $\mathrm{PSL}(3, 4)$ or $\mathrm{PSL}(3, 7)$, so that in these cases, the multiplication groups of Nagy's loop are $\mathrm{PGL}(3, 4)$ and $\mathrm{PGL}(3, 7)$, respectively.

Drápal [5] has shown that if Q is a loop on the projective line $\mathbb{P}(\mathbb{F}_q)$ with multiplication group contained in $\mathrm{P}\Gamma\mathrm{L}(2, q)$, then $M(Q)$ is isomorphic to a Singer cycle (unless $q = 3$ or 4). In particular, $\mathrm{PGL}(2, q)$ is not the multiplication group of a loop (the cases $q = 2, 3, 4$ do not provide exceptions to this latter statement).

Together with Ihringer's criterion Drápal's result (or rather a weaker form of it) can be used to show that $\mathrm{PSL}(2, q)$ is not the multiplication group of a quasigroup, if q is odd and >3.

Proposition 4.1. *If $q > 3$ is odd,* $\mathrm{PSL}(2, q)$ *is not the multiplication group of a quasigroup.*

Proof. Assume the contrary. By Ihringer's criterion (see Corollary 2.4), there is a loop Q on $\mathbb{P}(\mathbb{F}_q)$ with $M(Q) \leq \mathrm{PSL}(2, q)$. Since $\mathrm{PSL}(2, q)$ is not triply transitive and every element fixes at most 2 points, it follows from [4, Proposition 2.5] that Q is an abelian group. Hence, $Q = M(Q) \leq \mathrm{PSL}(2,q)$ and Q is a regular subgroup. However, $\mathrm{PSL}(2,q)$ does not have a regular subgroup for $q > 3$.

We remark that $\mathrm{PSL}(2,3)$ is an exception to the above proposition. Indeed, $\mathrm{PSL}(2,3)$ is permutation equivalent to A_4 acing on 4 letters. The latter is the multiplication group of a quasigroup by Corollary 3.1.

$\mathrm{Sp}(2d,2)$, $d \geq 2$, $n = 2^{2d-1} \pm 2^{d-1}$ Here we only have information on the smallest cases $d = 2,3$. Suppose first that $d = 2$. We have $\mathrm{Sp}(4,2) \cong S_6$. The action of $\mathrm{Sp}(4,2)$ on $2^{2\cdot 2-1} - 2^{2-1} = 6$ letters is permutation equivalent to the natural action of S_6, and thus $\mathrm{Sp}(4,2)$ is the multiplication group of a quasigroup on 6 points, in fact even of a commutative loop. On the other hand, there is no quasigroup on $2^{2\cdot 2-1} + 2^{2-1} = 10$ points with multiplicatin group $\mathrm{Sp}(4,2)$.

Let us now turn to the case $d = 3$, i.e., to the group $\mathrm{Sp}(6,2)$ acting on 28 or 36 points. Computer calculations have shown that this group is not the multiplication group of any quasigroup on 28 or 36 points.

$\mathrm{PSU}(3,q)$ Acting on $q^3 + 1$ Points The groups $\mathrm{PSU}(3,2)$ and $\mathrm{PGU}(3,2)$ acting on 9 points, have regular abelian normal subgroups. The quotient groups are the quaternion group and $\mathrm{SL}(2,3)$, respectively. Since each of these can be generated by two elements, $\mathrm{PSU}(3,2)$ and $\mathrm{PGU}(3,2)$ are multiplication groups of quasigroups by Corollary 3.1.

The groups $\mathrm{PSU}(3,3)$ and $\mathrm{PSU}(3,4)$ acting on 28 and 65 points, respectively, are not multiplication groups of quasigroups.

$\mathrm{Sz}(q)$, $q = 2^{2m+1}$, $m \geq 0$, Acting on $q^2 + 1$ Points The same argument as for $\mathrm{PSL}(2,q)$, q odd, settles the following case.

Proposition 4.2. *If $q = 2^{2m+1}$, $m \geq 1$, then $\mathrm{Sz}(q)$ is not the multiplication group of a quasigroup.*

Proof. Analogous to the proof of Proposition 4.1.

Note that $\mathrm{Sz}(2)$, being a Frobenius group on 5 points, is an exception to the above proposition.

$\mathrm{Ree}(q)$, $q = 3^{2m+1}$, $m \geq 0$, Acting on $q^3 + 1$ Points We have only investigated the smallest case $q = 3$. Here, $\mathrm{Ree}(3) \cong \mathrm{P}\Gamma\mathrm{L}(2,8)$ acting on 28 points. Neither $\mathrm{P}\Gamma\mathrm{L}(2,8)$ nor its subgroup $\mathrm{PSL}(2,8)$ are multiplication groups of quasigroups in this action.

$\mathrm{PSL}(2,11)$ and M_{11} Acting on 11 Points Each of the groups $\mathrm{PSL}(2,11)$ and M_{11} acting on 11 points contain two conjugacy classes of 11-cycles acting fixed-point freely. In each case, the group is generated by two such 11-cycles. On the other hand, neither $\mathrm{PSL}(2,11)$ nor M_{11} are multiplication groups of loops. The results for M_{11} are due to Ihringer [10, Theorem 2] and Drápal [5, p. 257], respectively.

M_{11} Acting on 12 Points This is not the multiplication group of any quasigroup.

M_{12} Acting on 12 Points This group is the multiplication group of a quasigroup, since it is generated by a regular subgroup and an additional element. It is also the multiplication group of a loop, the arithmetic progression loop constructed by Conway [3, p. 327].

A_7 Acting on 15 Points This is not the multiplication group of any quasigroup.

M_{22} Acting on 22 Points This group is not the multiplication group of any quasigroup (see [5, 14]).

M_{23} Acting on 23 Points This group is the multiplication group of a quasigroup with a left unit, since it is generated by a regular subgroup and an additional element [10]. It is not the multiplication group of a loop [5, 15].

M_{24} Acting on 24 Points This is the multiplication group of a commutative loop. A non-commutative loop with multiplication group M_{24} was found in [15]. We explicitly state one of the commutative loops we have found in Table 1.

PSL(2, 8) Acting on 28 Points This has already been treated in the subsection on the Ree groups.

HS Acting on 176 Points This is not the multiplication group of any quasigroup.

Co_3 Acting on 276 Points This case is still open. By a computer search we could prove, however, that no subgroup of Co_3 is the multiplication group of a commutative loop.

4.2 The Two-Transitive Groups with Elementary Abelian Socle

Let G be a two-transitive group with elementary abelian socle V. Then V acts regularly and G is the semidirect product $G = VH$ with H a point stabilizer. Thus, whenever H can be generated by two elements, G is the multiplication group of a quasigroup by Corollary 3.1.

Hering has classified these two-transitive groups in [8, 9]. The list of occurring point stabilizers H is contained in [8, Sect. 5] and in [12, Appendix 1]. These groups come in three infinite series, and a finite number of exceptional cases.

Using GAP, we have checked that all exceptional point stabilizers occurring in Part IV of Hering's list can be generated by two elements. A point stabilizer H lying in an infinite series has a unique nonabelian composition factor S which is normal in H, and H/S is soluble. It would be a possible but tedious task to check which of the occurring point stabilizers are 2-generated. It is well known that every finite simple groups is 2-generated. Thus, in case $H = S$ is simple, $G = VH$ is indeed the multiplication group of a quasigroup.

Table 1 A commutative loop with multiplication group M_{24}

1	2	3	4	5	6	7	8	9	10	11	12	13	14	15	16	17	18	19	20	21	22	23	24
2	23	16	9	21	7	11	12	8	17	14	19	1	5	6	20	18	4	3	22	13	10	24	15
3	16	24	12	13	17	23	7	19	2	22	20	10	4	21	15	11	5	8	18	14	1	6	9
4	9	12	1	6	10	2	23	3	22	17	15	24	16	14	18	21	19	7	11	5	13	8	20
5	21	13	6	24	1	10	22	16	23	2	18	3	12	8	14	15	20	11	9	4	17	19	7
6	7	17	10	1	24	20	19	23	18	13	11	12	22	3	21	4	9	14	8	15	16	5	2
7	11	23	2	10	20	24	18	17	15	21	3	9	8	1	6	12	16	4	14	22	19	13	5
8	12	7	23	22	19	18	1	10	9	16	14	6	21	17	11	5	24	20	13	2	15	4	3
9	8	19	3	16	23	17	10	1	5	24	2	20	15	18	4	6	13	21	7	12	14	22	11
10	17	2	22	23	18	15	9	5	1	4	13	14	11	7	24	20	12	6	16	19	21	3	8
11	14	22	17	2	13	21	16	24	4	1	7	8	19	20	5	9	10	15	6	23	3	18	12
12	19	20	15	18	11	3	14	2	13	7	23	17	6	4	9	16	21	1	5	8	24	10	22
13	1	10	24	3	12	9	6	20	14	8	17	23	18	11	22	19	15	5	21	7	2	16	4
14	5	4	16	12	22	8	21	15	11	19	6	18	24	13	2	10	3	23	1	20	9	7	17
15	6	21	14	8	3	1	17	18	7	20	4	11	13	23	19	22	2	16	24	9	5	12	10
16	20	15	18	14	21	6	11	4	24	5	9	22	2	19	1	23	8	10	12	3	7	17	13
17	18	11	21	15	4	12	5	6	20	9	16	19	10	22	23	24	7	13	3	1	8	2	14
18	4	5	19	20	9	16	24	13	12	10	21	15	3	2	8	7	23	22	17	6	11	14	1
19	3	8	7	11	14	4	20	21	6	15	1	5	23	16	10	13	22	24	2	17	12	9	18
20	22	18	11	9	8	14	13	7	16	6	5	21	1	24	12	3	17	2	23	10	4	15	19
21	13	14	5	4	15	22	2	12	19	23	8	7	20	9	3	1	6	17	10	24	18	11	16
22	10	1	13	17	16	19	15	14	21	3	24	2	9	5	7	8	11	12	4	18	23	20	6
23	24	6	8	19	5	13	4	22	3	18	10	16	7	12	17	2	14	9	15	11	20	1	21
24	15	9	20	7	2	5	3	11	8	12	22	4	17	10	13	14	1	18	19	16	6	21	23

5 Searching for Quasigroups By Computer

Given a permutation group G on n points we used GAP [7] to search for a quasigroup with G as multiplication group. We either want to find such a quasigroup or prove that such a quasigroup does not exist.

Using Lemma 2.2 there are some easy reductions of the problem: If there is a quasigroup with multiplication group G, then there is a loop with a subgroup of G as multiplication group. And also the converse is true when G is generated by two elements, see Corollary 2.4.

So, we want to search for a quasigroup whose multiplication table is given by row permutations $(\lambda_1, \ldots, \lambda_n)$ which also imply column permutations $(\rho_1, \ldots, \rho_n)$, such that $\lambda_1, \ldots, \lambda_n, \rho_1, \ldots, \rho_n \in G$. We can assume $\lambda_1 = \rho_1 = 1$. Furthermore, using $\pi = \sigma = \tau \in G$ in Lemma 2.2, we need to consider for λ_2 only representatives of conjugacy classes of fixed point free elements of G.

Given partial information $(\lambda_1, \ldots, \lambda_l)$ and $(\rho_1, \ldots, \rho_k)$ of a possible multiplication table we try to extend it as follows: If $l \leq k$ we find all possible λ_{l+1}, otherwise we first change the roles of the λ_i and ρ_i. Such λ_{l+1} must map the points $1, \ldots, k$ to $(l+1)^{\rho_1}, \ldots, (l+1)^{\rho_k}$. GAP can easily find out if such an element exists. If yes, the other such elements are obtained by left multiplication with the elements in the stabilizer of the tuple $(1, \ldots, k)$. For each such candidate for λ_{l+1} we check if $\lambda_{l+1}\lambda_i^{-1}$ is fixed point free for $1 \leq i \leq l$. For each λ_{l+1} fulfilling these conditions we recursively try to extend the table further.

Starting with the various $(\lambda_1 = 1, \lambda_2)$ and $(\rho_1 = 1)$ as described above this backtrack algorithm will either produce multiplication tables of quasigroups we are looking for, or it will show that such quasigroups do not exist.

This algorithm performs quite well in the examples mentioned in the last section. It seems that extending the λ's and ρ's alternately often shows quite early in the recursion that a partial table cannot be extended further.

Compared to the algorithms described in [14] and in [15] our variant needs very little memory and is easy to distribute to several machines. As examples, our program needed less than 4 min to check that the Mathieu group M_{22} on 22 points is not the multiplication group of a quasigroup. The result for HS on 176 points only needed 30 s.

We noticed that many groups with a subgroup which is the multiplication group of a loop also have such a subgroup which is the multiplication group of a commutative loop. By setting $\rho_i = \lambda_i$ for all i in our backtrack search we can restrict the search to commutative loops.

Acknowledgments We thank Barbara Baumeister for bringing reference [11] to our attention. Thanks are also due to Michael Guidici and Cheryl Praeger for discussions on the subject of this paper. Finally, we are indebted to Klaus Lux for suggesting a connection to permutation arrays (which was not pursued further).

References

1. P. J. CAMERON, Permutation groups, London Mathematical Society Student Texts, 45. Cambridge University Press, Cambridge, 1999
2. J. J. CARROLL, G. A. FISHER, A. M. ODLYZKO AND N. J. A. SLOANE, What are the Latin Square Groups? *Amer. Math. Monthly* **80** (1973), 1045–1046
3. J. H. CONWAY, The Golay codes and the Mathieu groups, in: J. H. CONWAY AND N. J. A. SLOANE, Sphere packings, lattices and groups, 3rd ed., Springer, 1999, Chapter 11, pp. 299–330
4. A. DRÁPAL, Multiplication groups of finite loops that fix at most two points, *J. Algebra* **235** (2001), 154–175
5. A. DRÁPAL, Multiplication groups of loops and projective semilinear transformations in dimension two, *J. Algebra* **251** (2002), 256–278
6. A. DRÁPAL AND T. KEPKA, Alternating groups and Latin squares, *European J. Combin.* **10** (1989), 175–180
7. THE GAP GROUP, GAP – Groups, Algorithms, and Programming, Version 4.4.12, 2008 (http://www.gap-system.org)
8. C. HERING, Transitive linear groups and linear groups which contain irreducible subgroups of prime order, *Geom. Dedicata* **2** (1974), 425–460
9. C. HERING, Transitive linear groups and linear groups which contain irreducible subgroups of prime order, II, *J. Algebra* **93** (1985), 151–164
10. T. IHRINGER, On multiplication groups of quasigroups, *European J. Combin.* **5** (1984), 137–141
11. C. H. LI, The finite primitive permutation groups containing an abelian regular subgroup, *Proc. London Math. Soc.* **87** (2003), 725–747
12. M. W. LIEBECK, The affine permutation groups of rank three, *Proc. London Math. Soc.* **54** (1987), 477–516
13. S. KÖHLER, Multiplikationsgruppen von Quasigruppen, Diploma thesis, RWTH Aachen University, 2007
14. P. MÜLLER AND G. P. NAGY, A note on the group of projectivities of finite projective planes, *Innov. Incidence Geom.* **6/7** (2007/08), 291–294
15. G. P. NAGY, On the multiplication groups of semifields, *European J. Combin.* **31** (2010), 18–24
16. J. D. H. SMITH, Representation theory of infinite groups and finite quasigroups, Séminaire de Mathématiques Supérieures, 101. Presses de l'Université de Montréal, Montreal, QC, 1986, 132 pp

Curve Complexes Versus Tits Buildings: Structures and Applications

Lizhen Ji

Abstract Tits buildings $\Delta_{\mathbb{Q}}(\mathbf{G})$ of linear algebraic groups $\mathbf{G}$ defined over the field of rational numbers $\mathbb{Q}$ have played an important role in understanding partial compactifications of symmetric spaces and compactifications of locally symmetric spaces, cohomological properties of arithmetic subgroups and S-arithmetic subgroups of $\mathbf{G}(\mathbb{Q})$. Curve complexes $\mathcal{C}(S_{g,n})$ of surfaces $S_{g,n}$ were introduced to parametrize boundary components of partial compactifications of Teichmüller spaces and were later applied to understand properties of mapping class groups of surfaces and geometry and topology of 3-dimensional manifolds. Tits buildings are spherical buildings, and another important class of buildings consists of Euclidean buildings, for example, the Bruhat–Tits buildings of linear algebraic groups defined over local fields. In this paper, we summarize and compare some properties and applications of buildings and curve complexes. We try to emphasize their similarities but also point out differences. In some sense, curve complexes are combinations of spherical, Euclidean and hyperbolic buildings. We hope that such a comparison might motivate more questions and also suggest methods to solve them at the same time, and furthermore it might introduce buildings to people who study curve complexes and curve complexes to people who study buildings.

2000 Mathematics Subject Classification: 53C35, 30F60, 22E40, 20G15, 57M99.

Keywords Tits building • Curve complex • Spherical building • Euclidean building • Hyperbolic building • Arithmetic groups • Mapping class group • Symmetric space • Teichmüller space • Proper action • Classifying space • Universal space • Locally symmetric space • Mostow strong rigidity • Moduli space • Ending lamination conjecture • Simplicial volume

L. Ji (✉)
Department of Mathematics, University of Michigan, Ann Arbor, MI 48109, USA
e-mail: lji@umich.edu

N.S.N. Sastry (ed.), *Buildings, Finite Geometries and Groups*, Springer Proceedings in Mathematics 10, DOI 10.1007/978-1-4614-0709-6_6,

• Compactification • Boundary • Duality group • Cohomological dimension • Asymptotic cone at infinity • Novikov conjecture • Quasi-isometry • Heegaard splitting

1 Introduction

1.1 Summary

Tits buildings $\Delta_{\mathbb{Q}}(\mathbf{G})$ of linear algebraic groups $\mathbf{G}$ defined over $\mathbb{Q}$ have played an important role in understanding partial compactifications of symmetric spaces and compactifications of locally symmetric spaces, cohomological properties of arithmetic subgroups and S-arithmetic subgroups of $\mathbf{G}(\mathbb{Q})$. On the other hand, curve complexes $\mathcal{C}(S_{g,n})$ of surfaces $S_{g,n}$ were introduced to parametrize boundary components of partial compactifications of Teichmüller spaces and were later applied to understand properties of mapping class groups of surfaces and geometry and topology of 3-dimensional manifolds.

Tits buildings are spherical buildings, and another important class of buildings consists of Euclidean buildings, for example, the Bruhat–Tits buildings of linear algebraic groups defined over local fields. In this paper, we summarize and compare some properties and applications of buildings and curve complexes. We try to emphasize their similarities but also point out differences. In some sense, curve complexes are combinations of spherical, Euclidean and hyperbolic buildings. We hope that such a comparison might motivate more questions and also suggest methods to solve them at the same time, and furthermore it might introduce buildings to people who study curve complexes and curve complexes to people who study buildings.

1.2 The Origin of Tits Buildings

Buildings were originally introduced by Tits [166, 167] in order to realize exceptional Lie groups as the symmetry groups of spaces (or geometry)[1] so that one can construct geometrically analogues of exceptional simple Lie groups over arbitrary fields.[2] See [152] for an overview of motivations and the history of Tits buildings.

[1]Classical simple Lie groups over $\mathbb{C}$ are the symmetry groups of quadratic forms or sesquilinear forms of finite dimensional vector spaces over $\mathbb{C}$. The same construction works for vector spaces over finite fields and produces classical finite groups of Lie type, but this method does not extend to exceptional simple Lie groups.

[2] According to [153, p. 292], "... it is perhaps worth remarking that one of the initial motivations for the theory of buildings, at a time when Chevalley's fundamental "Tohoku paper" had not yet appeared, was the search for a geometric way of obtaining algebraic analogues of the exceptional Lie groups."

Since then, Tits buildings have been applied to many different situations with great success. There are several types of buildings: (1) spherical buildings, for example, Tits buildings, (2) Euclidean buildings, for example, Bruhat–Tits buildings, (3) hyperbolic buildings, (4) $\mathbb{R}$-buildings, (5) topological buildings, (6) twin buildings. We will recall briefly some of these buildings below. See the book [2] for detailed definitions and structures of buildings, and the survey [87] for references on many different applications of buildings.

There are several basic ways in which Tits buildings are used:

1. Buildings describe the large scale geometry or the geometry at infinity of Lie groups, symmetric spaces and locally symmetric spaces.
2. Buildings describe the topology at infinity of partial compactifications of symmetric spaces and of the boundary of compactifications of locally symmetric spaces, and also the topology of the ends of locally symmetric spaces.
3. Buildings provide natural combinatorial and metric spaces on which Lie groups, arithmetic subgroups, p-adic Lie groups, etc., act.
4. Buildings can be used to study cohomological properties of arithmetic groups.

We are mainly interested in infinite buildings in this paper. For finite buildings and their applications in finite groups and their representation theory, see [44].

1.3 The Origin of Curve Complexes

Motivated by the Borel–Serre compactification of locally symmetric spaces [20], Harvey [69, 70, 73] introduced the curve complex $\mathcal{C}(S_g)$ of a compact oriented surface S_g to parametrize the boundary components of partial compactifications of the Teichmüller space $\mathcal{T}_g$, the space of the marked compact Riemann surfaces of genus g which induce compactifications of the moduli space $\mathcal{M}_g$ of Riemann surfaces of genus g. In some sense, it was an exact analogue of the spherical Tits building $\Delta_{\mathbb{Q}}(\mathbf{G})$ of a linear semisimple algebraic group $\mathbf{G}$ defined over $\mathbb{Q}$ which serves as a parameter space for the boundary components of the Borel–Serre partial compactification of the symmetric space $X = G/K$. The same definition works for a more general oriented surface $S_{g,n}$ of genus g with n punctures and gives a curve complex $\mathcal{C}(S_{g,n})$. Motivated by the analogy between arithmetic groups and mapping class groups $\mathrm{Mod}_{g,n} = \mathrm{Diff}^+(S_{g,n})/\mathrm{Diff}^0(S_{g,n})$, the curve complex $\mathcal{C}(S_{g,n})$ was used to study cohomological properties of $\mathrm{Mod}_{g,n}$ [66, 69, 80].

It turns out that the curve complexes $\mathcal{C}(S_{g,n})$ can also be used to study many problems in (lower dimensional) topology and geometry, in particular the ending lamination conjecture of Thurston for 3-dimensional hyperbolic manifolds, quasi-isometric rigidity of mapping class groups $\mathrm{Mod}_{g,n}$ and finiteness of the asymptotic dimension of $\mathrm{Mod}_{g,n}$.

Some applications of curve complexes are motivated by results for Tits buildings, and others are quite different. There are several basic ways in which the curve complexes are used.

1. Curve complexes describe the large scale geometry or the geometry at infinity of Teichmüller spaces and of the moduli spaces of Riemann surfaces.
2. Curve complexes parametrize boundary components of partial compactifications of Teichmüller spaces, they describe the topology at infinity of the partial compactifications of Teichmüller spaces, and they also describe the boundary of compactifications of moduli spaces, and the topology of the ends of moduli spaces.
3. Curve complexes provide natural combinatorial and metric spaces on which mapping class groups act, and structures and representations of the mapping class groups can be understood and constructed from these actions.
4. Curve complexes can be used to study cohomological properties of mapping class groups.
5. Surfaces and simple closed curves occur naturally in the study of 3-dimensional manifolds and hence curve complexes can be used to understand 3-dimensional geometry and topology.

In this paper, we will present some applications of buildings and curve complexes to support these general points. The above two lists suggest many similarities between these two classes of simplicial complexes. On the other hand, there are also dramatic differences between buildings and curve complexes.

Besides the curve complex $\mathcal{C}(S_{g,n})$, there are also several related complexes for surfaces $S_{g,n}$, which are important to understand the geometry and topology of Teichmüller spaces $\mathcal{T}_{g,n}$ and mapping class groups $\mathrm{Mod}_{g,n}$. We will address these issues as well.

This paper can be seen as a sequel of the survey papers [87, 88] in some sense. Since it emphasizes curve complexes and their applications, it can complement the other two papers.

2 Definition of Buildings

In this section, we first introduce the spherical Tits building $\Delta(G)$ of a semisimple Lie group G through a classification of geodesics in the associated symmetric space $X = G/K$ in Sect. 2.1. This justifies the point of view that $\Delta(G)$ describes the geometry at infinity of the symmetric space X. Then in Sect. 2.2, we discuss a more common definition of buildings through a system of apartments. In Sect. 2.3, we introduce Euclidean and hyperbolic buildings and mention other buildings. In Sect. 2.4, we define the rational Tits building $\Delta_{\mathbb{Q}}(\mathbf{G})$ of a linear algebraic group $\mathbf{G}$ defined over $\mathbb{Q}$.

2.1 A Geometric Definition of Tits Buildings via Symmetric Spaces

Let G be a semisimple noncompact Lie group with finitely many connected components, $K \subset G$ be a maximal compact subgroup. Then $X = G/K$ with an invariant metric is a symmetric space of noncompact type, and X is a simply connected nonpositively curved Riemannian manifold. Let $X(\infty)$ be the set of equivalence classes of geodesics in X. Specifically, we assume that all geodesics in X are of unit speed and directed. Two geodesics $\gamma_1(t)$ and $\gamma_2(t)$ in X are defined to be *equivalent* if

$$\limsup_{t \to +\infty} d(\gamma_1(t), \gamma_2(t)) < +\infty,$$

where $d(\cdot,\cdot)$ is the distance function of X. It can be shown that for any point $x_0 \in X$, $X(\infty)$ can be canonically identified with the unit sphere in the tangent space $T_{x_0}X$. It is known that there is a natural topology on $X \cup X(\infty)$ such that it is a compactification of X, called the *geodesic compactification*, or *visual compactification*, and $X(\infty)$ is hence called the *sphere at infinity* (or visual sphere) of X. See [4].

When $X = \mathrm{SL}(2,\mathbb{R})/\mathrm{SO}(2)$ is identified with the upper half place $\mathbf{H}^2$, then $X(\infty) = \mathbb{R} \cup \{i\infty\}$. When $X = \mathrm{SL}(2,\mathbb{R})/\mathrm{SO}(2)$ is identified with the unit disc $D = \{z \in \mathbb{C} \mid |z| < 1\}$, then $X(\infty)$ is equal to the unit circle $S^1 = \partial D$.

Clearly, the isometry action of G on X preserves the equivalence relation between geodesics and hence acts on $X(\infty)$.

A natural question is whether these points in $X(\infty)$, i.e, equivalence classes of geodesics, are the same, i.e., belong to one common G-orbit. If the answer is negative, a natural problem is to parametrize G-orbits in $X(\infty)$.

It turns out that both questions can be answered using parabolic subgroups P of G, i.e., closed subgroups P such that G/P are compact. If $P=G$, it is a parabolic subgroup by definition. Any parabolic subgroup P with $P \neq G$ is called a *proper parabolic subgroup* of G. The set of proper parabolic subgroups of G is a partially ordered set by inclusion, and there are infinitely many maximal and minimal elements in this partially ordered set, which are called *maximal proper parabolic subgroups* and *minimal parabolic subgroups* of G.

Proposition 2.1. *The G-action on X extends to a continuous action on $X \cup X(\infty)$. For any point $z \in X(\infty)$, its stabilizer G_z in G is a proper parabolic subgroup, and every proper parabolic subgroup P of G arises as the stabilizer of a point $z \in X(\infty)$.*

For each proper parabolic subgroup P of G, let σ_P be the set of points in $X(\infty)$ that are fixed by P, equivalently the set of points $z \in X(\infty)$ whose stabilizer G_z in G contains P.

Proposition 2.2. *For every proper parabolic subgroup P, σ_P is a spherical simplex, and its interior σ_P^o, i.e., the open simplex when σ_P is considered as a* *simplex, is equal to the set of points z whose stabilizer is equal to P. The simplex σ_P consists of one point if and only if P is a maximal proper parabolic subgroup. If P is a minimal parabolic subgroup of G, then the dimension of σ_P is equal to $r-1$, where r is the rank of X, i.e., the maximal dimension of totally geodesic flat submanifolds of X. The decomposition of $X(\infty)$ into σ_P,*

$$X(\infty) = \cup_P \sigma_p = \coprod_P \sigma_P^0,$$

gives $X(\infty)$ the structure of an infinite simplicial complex, denoted by $\Delta(X)$.

The simplicial complex $\Delta(X)$ is called the *Tits building* associated with the symmetric space X of noncompact type. If the rank r of X is equal to 1, then $\Delta(X)$ is a 0-simplicial complex, i.e., a disjoint union of points. If $r > 0$, it can be shown that $\Delta(X)$ is connected. This can be seen from the axioms of buildings in the next subsection.

By definition, $\Delta(X)$ classifies geodesics of X into different types.

Proposition 2.3. *Let P_0 be a minimal parabolic subgroup of G. Then every G-orbit in $X(\infty)$ meets σ_{P_0} at exactly one point. G acts on the Tits building $\Delta(X)$ by simplicial maps, and every simplex σ_P is contained in the G-orbit of a face of σ_{P_0}, i.e., $G\backslash\Delta(X)$ can be identified with σ_{P_0}.*

Corollary 2.4. *G acts transitively on $X(\infty)$ if and only if a minimal parabolic subgroup of G is also a maximal parabolic subgroup G, i.e., the rank r of X is equal to 1.*

For more discussion and proofs of the above results, see [60].

2.2 Axioms for Spherical Buildings via Apartments

The Tits building $\Delta(X)$ defined in the previous subsection can be described directly in terms of proper parabolic subgroups of G.

Let $\mathcal{P}$ be the partially ordered set of proper parabolic subgroups of G, where the partial order is given by containment, i.e., the opposite of the inclusion mentioned in the previous subsection. The structure theory of parabolic subgroups of G shows that this poset can be realized by an infinite simplicial complex $\Delta(G)$ satisfying the following conditions:

1. Every parabolic subgroup $P \in \mathcal{P}$ corresponds to a unique simplex σ_P in $\Delta(G)$, and every simplex of $\Delta(G)$ is of this form.
2. For any two parabolic subgroups $P_1, P_2 \in \mathcal{P}$, $P_1 \subset P_2$ if and only if σ_{P_1} contains σ_{P_2} as a face.
3. A simplex σ_P is a point if and only if P is a maximal parabolic subgroup of G. For a non-maximal parabolic subgroup P, the vertices of σ_P correspond to maximal parabolic subgroups that contain P.

This simplicial complex $\Delta(G)$ is called the *Tits building* of G. By Proposition 2.2, $\Delta(G)$ is isomorphic to $\Delta(X)$. These are important examples of spherical Tits buildings.

Definition 2.5. A simplicial complex Δ is called a spherical Tits building if it contains a family of subsets called apartments and satisfies the following conditions:

1. Every apartment is a finite Coxeter complex.
2. Any two simplices are contained in some apartment.
3. Given two apartments Σ and Σ' and simplices $\sigma, \sigma' \in \Sigma \cap \Sigma'$, there exists an isomorphism of Σ onto Σ' which keeps σ, σ' pointwise fixed.

In the above definition, for any finite Coxeter group W, i.e., a finite group generated by reflections with respect to hyperplanes in a fixed vector space, there is a Coxeter complex, which is a finite simplicial complex constructed as follows. Every reflection $\alpha \in W$ fixes a hyperplane $\mathcal{H}_\alpha$. The collection of such hyperplanes $\mathcal{H}_\alpha$ is invariant under W. Connected components of the complement of the union of $\{\mathcal{H}_\alpha\}$ in V are called chambers, which are simplicial cones. The chambers and their faces together give a partition of V into simplicial cones. Let S be the unit sphere in V. Then the intersection of S with these simplicial cones gives a finite simplicial complex, called the *Coxeter complex* of W, whose underlying topological space is S, i.e., the Coxeter complex gives a finite triangulation of the unit sphere.

To see that $\Delta(X)$ is a spherical Tits buildings, we start with the construction of apartments. For any flat totally geodesic submanifold F of X, which is isometric to $\mathbb{R}^r$, its closure in $X \cup X(\infty)$ is equal to $F \cup F(\infty)$, where $F(\infty)$ is homeomorphic to the sphere S^{r-1}. The simplicial complex structure of $X(\infty)$ given by $\Delta(X)$ induces a simplicial complex structure on $F(\infty)$. In fact, $F(\infty)$ is equal to the union of σ_P for some parabolic subgroups P. Denote this finite simplicial complex by Σ_F. Then it is a Coxeter complex associated with the Weyl group of the Lie group G.

Proposition 2.6. *With respect to the collection of finite subcomplexes Σ_F associated with flat subspaces F, the infinite simplicial complex $\Delta(X)$ is a spherical Tits building.*

To understand the spherical building structure of $\Delta(G)$, we need to define apartments in terms of group structure. For any maximal compact subgroup K of G, let $\mathfrak{g} = \mathfrak{k} \oplus \mathfrak{p}$ be the Cartan decomposition. Let $\mathfrak{a} \subset \mathfrak{p}$ be a maximal abelian subalgebra. Let $A = \exp \mathfrak{a}$ be the corresponding Lie subgroup of G. Then there are only finitely many parabolic subgroups P that contain A, and the union of their simplices σ_P gives a triangulation of the unit sphere $\mathfrak{a}(+\infty)$ of $\mathfrak{a}$ and is a Coxeter complex for the Weyl group of G. Denote it by $\Sigma_{\mathfrak{a}}$.

Proposition 2.7. *With respect to the collection of finite subcomplexes $\Sigma_{\mathfrak{a}}$ associated with maximal abelian subspaces, the infinite simplicial complex $\Delta(X)$ is a spherical Tits building.*

Proposition 2.7 is proved using the Bruhat decomposition. To derive Proposition 2.6 from Proposition 2.7, we need to identify their apartments.

The finite simplicial complexes Σ_F and $\Sigma_{\mathfrak{a}}$ can be identified as follows. For any maximal flat totally geodesic submanifold F of X, let $x_0 \in F$ be any point, and $K = G_{x_0}$ be the stabilizer of x_0 in G. Then there exists a maximal abelian subalgebra $\mathfrak{a} \subset \mathfrak{p}$ with respect to the Cartan decomposition of $\mathfrak{g}$ induced by K such that $F = \exp \mathfrak{a}\, x_0$, and Σ_F is identified with $\Sigma_{\mathfrak{a}}$ under the identification between $\Delta(X)$ and $\Delta(G)$. For the proof of the above propositions and more discussion, see the book [60] and references therein.

2.3 *Euclidean and Hyperbolic Buildings*

It will be shown below that the curve complex $\mathcal{C}(S_{g,n})$ shares some properties with spherical buildings, and also with Euclidean buildings and hyperbolic buildings. For convenience, we briefly introduce their definitions here.

First, we define Euclidean reflection groups. Let V be a Euclidean space. An affine reflection group W on V is a group of affine isometries generated by reflections with respect to affine hyperplanes such that the set $\mathcal{H}$ of affine hyperplanes fixed by reflections in W is locally finite. Clearly, a finite reflection group is an affine reflection group. Assume that W is an infinite reflection group and irreducible. Then the hyperplanes in $\mathcal{H}$ divide V into simplices, and W acts simply transitively on the set of simplices. These simplices and their faces form a *Euclidean Coxeter complex.*

Definition 2.8. A simplicial complex Δ is called a Euclidean building if it contains a family of subsets called apartments and satisfies the following conditions:

1. Every apartment is an infinite Euclidean Coxeter complex.
2. Any two simplices are contained in some apartment.
3. Given two apartments Σ and Σ' and simplices $\sigma, \sigma' \in \Sigma \cap \Sigma'$, there exists an isomorphism from Σ onto Σ' which keeps σ, σ' pointwise fixed.

An important source of Euclidean buildings comes from the *Bruhat–Tits building* $\Delta^{\mathrm{BT}}(\mathbf{G})$ associated with a linear semisimple algebraic group $\mathbf{G}$ defined over a non-Archimedean local field k. In this case, the simplices of $\Delta^{\mathrm{BT}}(\mathbf{G})$ are parametrized by parahoric subgroups of $\mathbf{G}(k)$. There is no simple definition or characterization of parahoric subgroups of $\mathbf{G}(k)$ as in the case of parabolic subgroups. When $\mathbf{G}$ is simply connected, any maximal open compact subgroup of $\mathbf{G}(k)$ is a maximal parahoric subgroup, and the converse is also true. See [87, Sect. 3] for more details and references.

Now we introduce hyperbolic buildings. Let $\mathbf{H}^n$ be the real hyperbolic space of dimension n. Let P be a compact convex hyperbolic polyhedron such that the reflections with respect to its codimension 1 faces generate a group W that acts properly on $\mathbf{H}^n$ with P as a fundamental domain. Then the totally geodesic hypersurfaces of $\mathbf{H}^n$ that are fixed by reflections in W are locally finite, and the connected components of the complement of the union of these hyperplanes are hyperbolic polyhedra. These polyhedra and their faces form a polyhedral complex, which is called a *hyperbolic Coxeter complex* associated with P. Hyperbolic Coxeter complexes can only exist for $n \leq 29$.

Definition 2.9. A polyhedral complex Δ is called a hyperbolic building of type P if it contains a family of subcomplexes called apartments and satisfies the following conditions:

1. Every apartment is a hyperbolic Coxeter complex determined by P.
2. Any two polyhedra are contained in some apartment.
3. Given two apartments Σ and Σ' and polyhedra $\sigma, \sigma' \in \Sigma \cap \Sigma'$, there exists an isomorphism of Σ onto Σ' which keeps σ, σ' pointwise fixed.

See [55] for more detail and examples. When $n = 2$, P is given by a compact polygon in the hyperbolic plane $\mathbf{H}^2$ such that the angle at each vertex is equal to π/m for some integer m, and we get a hyperbolic building of Fuchsian type [22].

A natural generalization of spherical buildings is the notion of twin buildings. An important example is a pair of spherical Tits buildings associated to one linear algebraic group, for example, $\mathrm{SL}(n, F_p[t, t^{-1}])$. See [1] for the definition and applications to S-arithmetic subgroups of linear algebraic groups defined over function fields. See [45, 46] for a more geometric group theoretic point of view of buildings and Coxeter complexes.

2.4 Rational Tits Buildings of Linear Algebraic Groups

As explained before, for any semisimple Lie group G, there is a spherical Tits building $\Delta(G)$ that can be related to the geodesic compactification of its symmetric space $X = G/K$. If G is the real locus of a linear algebraic group $\mathbf{G} \subset GL(n, \mathbb{C})$ defined over $\mathbb{Q}$ with positive $\mathbb{Q}$-rank, then there are several other spherical Tits buildings associated with $\mathbf{G}$. They are important for compactifications of locally symmetric spaces and in the study of arithmetic subgroups of $\mathbf{G}(\mathbb{Q})$ and more generally S-arithmetic subgroups of $\mathbf{G}(\mathbb{Q})$.

Let $\mathcal{P}_{\mathbb{Q}}$ be the set of $\mathbb{Q}$-parabolic subgroups of $\mathbf{G}$, i.e., parabolic subgroups defined over $\mathbb{Q}$. Then there is an infinite simplicial complex $\Delta_{\mathbb{Q}}(\mathbf{G})$ whose simplices $\sigma_{\mathbf{P}}$ are parametrized by parabolic subgroups $\mathbf{P}$ in $\mathcal{P}_{\mathbb{Q}}$ and which satisfies the following conditions:

1. When $\mathbf{P}$ is maximal $\mathbb{Q}$-parabolic subgroup, $\sigma_{\mathbf{P}}$ is a point.
2. For two $\mathbb{Q}$-parabolic subgroups $\mathbf{P}_1, \mathbf{P}_2$, the inclusion relation $\mathbf{P}_1 \subset \mathbf{P}_2$ holds if and only if $\sigma_{\mathbf{P}_1}$ contains $\sigma_{\mathbf{P}_2}$ as a face.

The simplicial complex $\Delta_{\mathbb{Q}}(\mathbf{G})$ is called the $\mathbb{Q}$-*Tits building* of $\mathbf{G}$. It cannot be realized as the boundary of a compactification of X as the Tits building $\Delta(G)$, where $G = \mathbf{G}(\mathbb{R})$ is the real locus of $\mathbf{G}$, or as a subset of the boundary of a compactification of X if the $\mathbb{R}$-rank of $\mathbf{G}$ is strictly greater than the $\mathbb{Q}$-rank of $\mathbf{G}$. But $\Delta_{\mathbb{Q}}(\mathbf{G})$ can be realized as the boundary of a partial compactification of X, which gives the Tits compactification of $\Gamma\backslash X$ in [94].

The main application of $\Delta_{\mathbb{Q}}(\mathbf{G})$ is that $\Delta_{\mathbb{Q}}(\mathbf{G})$ naturally parametrizes boundary components of partial compactifications of X whose quotients by arithmetic subgroups $\Gamma \subset \mathbf{G}(\mathbb{Q})$ are compactifications of locally symmetric spaces $\Gamma\backslash X$. See Sect. 6.3.

For every prime p, let $\mathbb{Q}_p$ be the field of p-adic numbers. When $\mathbf{G}$ is considered as a linear algebraic group defined over $\mathbb{Q}_p$, the set of parabolic subgroups of $\mathbf{G}$ defined over $\mathbb{Q}_k$ gives a spherical Tits building $\Delta_{\mathbb{Q}_p}(\mathbf{G})$.

As mentioned before, there is a Euclidean building, the Bruhat–Tits building, associated with the p-adic Lie group $\mathbf{G}(\mathbb{Q}_p)$. Denote this building by $\Delta^{\mathrm{BT}}(\mathbf{G}(\mathbb{Q}_p))$. We will see below that $\Delta^{\mathrm{BT}}(\mathbf{G}(\mathbb{Q}_p))$ can be compactified by adding the spherical building $\Delta_{\mathbb{Q}_p}(\mathbf{G})$ (Proposition 6.25).

3 Definition of Curve Complexes

Let $S_{g,n}$ be an oriented surface of genus g with n punctures (or n boundary components). In the rest of this paper, we assume that the Euler characteristic $\chi(S_{g,n})$ is negative so that $S_{g,n}$ admits complete hyperbolic metrics of finite area.

A simple closed curve c in $S_{g,n}$ is called *essential* if it is not homotopic to a point or a loop around a puncture or a boundary component.

The curve complex $\mathcal{C}(S_{g,n})$ is a simplicial complex such that

1. Its vertices are parametrized by homotopy classes of essential simple closed curves $[c]$ in $S_{g,n}$.
2. Homotopy classes $[c_0], [c_1], \cdots, [c_k]$ form the vertices of a k-simplex if and only if they are pairwise distinct and admit disjoint representatives.

If we put a complete hyperbolic metric of finite area and with geodesic boundary on $S_{g,n}$, then the homotopy class of each essential simple closed curves in $S_{g,n}$ contains a unique simple closed geodesic, and hence the vertices of $\mathcal{C}(S_{g,n})$ corresponds to simple closed geodesics of the hyperbolic metric.

The simplest example is $\mathcal{C}(S_{1,1})$. Let $\Sigma_{1,1}$ be a hyperbolic surface of genus 1 with one puncture. In this case, for every simple closed geodesic of $\Sigma_{1,1}$, there is no other simple closed geodesic that is disjoint from it. Therefore, $\mathcal{C}(S_{1,1})$ is a countable collection of points and can be identified with the rational boundary points $\mathbb{Q}\cup\{\infty\}$ of the upper half-plane $\mathbf{H}^2$. For the surface $S_{1,1}$, a slight modification gives an interesting complex $\mathcal{C}_1(S_{1,1})$. The vertices of $\mathcal{C}_1(S_{1,1})$ still correspond to simple closed geodesics γ in $\Sigma_{1,1}$, and $k+1$ simple closed geodesics $\gamma_1, \cdots, \gamma_{k+1}$ form the vertices of a k-simplex if and only if for every pair of distinct geodesics γ_i and γ_j, the intersection number $\iota(\gamma_i, \gamma_j) = \pm 1$. Then the 1-skeleton of $\mathcal{C}_1(S_{1,1})$ can be identified with the Farey graph, whose vertices are numbers in $\mathbb{Q}\cup\{\infty\}$, the vertex ∞ is joined to every integer n, and two points $\frac{r}{s}$ and $\frac{x}{y}$ in reduced form (i.e., the numerators and denominators do not contain common primes) are connected by an edge if and only if $ry - sx = \pm 1$. The simplicial complex $\mathcal{C}_1(S_{1,1})$ gives an ideal triangulation (or tessellation) of the upper half plane $\mathbf{H}^2$, which is equal to the Farey tessellation, and hence $\mathcal{C}_1(S_{1,1})$ is contractible. (It should be emphasized that a general curve complex $\mathcal{C}_1(S_{g,n})$ is not contractible, and its nontrivial homotopy is used crucially to prove that the mapping class group $\mathrm{Mod}_{g,n}$ is not a virtual Poincaré duality group. See Sect. 7.9 below.)

In the general case, $\mathcal{C}(S_{g,n})$ is fairly complicated and there is no simple geometric model. For example, it is not locally finite in general. It is not locally finite because the complement of a simple closed curve on a surface contains infinitely many homotopy classes of essential simple closed curves except in some special cases (the torus with at most one hole or the sphere with at most four holes). Note that $\mathcal{C}(S_{1,1})$ is a disjoint union of countably many points and hence is locally finite. But the non-locally finiteness phenomenon can be seen in the above example $\mathcal{C}_1(S_{1,1})$ already. One way to see this is to note that $\mathrm{SL}(2,\mathbb{Z})$ acts on the upper half plane, the Farey tessellation and the Farey graph. Each vertex of $\mathcal{C}_1(S_{1,1})$ has an infinite stabilizer in $\mathrm{SL}(2,\mathbb{Z})$, which permutes the 2-simplices that contain the vertex, and hence there are infinitely many edges of $\mathcal{C}_1(S_{1,1})$ out of each vertex, and $\mathcal{C}_1(S_{1,1})$ is not locally finite.

The curve complex is not "homogeneous" in general as in the case of buildings. We will discuss some of its properties later.

Compared with the definition of the spherical Tits building $\Delta(G)$, this suggests that a homotopy class $[c]$ of essential simple closed curves plays a role similar to the one of a maximal parabolic subgroup, and a simplex of $\mathcal{C}(S_{g,n})$ plays the role of a parabolic subgroup. More specifically, we have the following similarities:

1. The boundary components of the Borel–Serre partial compactification of a symmetric space X are parametrized by proper $\mathbb{Q}$-parabolic subgroups of $\mathbf{G}$, and maximal boundary components correspond to maximal $\mathbb{Q}$-parabolic subgroups. On the other hand, the boundary components of an analogous Borel–Serre partial compactification of the Teichmüller space $\mathcal{T}_{g,n}$ or of the completion of $\mathcal{T}_{g,n}$ in the Weil–Petersson metric are parametrized by simplices of $\mathcal{C}(S_{g,n})$, and the maximal boundary components correspond to the vertices. (See Sects. 6 and 7 below for more details).
2. For every $\mathbb{Q}$-parabolic subgroup $\mathbf{P}$ of $\mathbf{G}$, there is an associated Langlands decomposition of X. When $\mathbf{P}$ is a minimal $\mathbb{Q}$-parabolic subgroup, it is reduced to the Iwasawa decomposition of X (or rather of the Lie group G). On the other hand, for every top dimensional simplex of $\mathcal{C}(S_{g,n})$, there is a Fenchel–Nielsen coordinate system (or decomposition) of $\mathcal{T}_{g,n}$.

It is easy to see that the maximal number of homotopy classes of essential simple closed curves with disjoint representatives is equal to $3g-3+n$. Therefore, $\mathcal{C}(S_{g,n})$ has dimension $3g-4+n$. Simplices of $\mathcal{C}(S_{g,n})$ with dimension equal to $3g-4+n$ play a similar role as minimal parabolic subgroups.

There are also several other related complexes.

1. The arc complex for surfaces with at least one puncture or boundary component in [66, 67, 106]. The topology of the arc complex at infinity is closely related to the topology of the curve complex $\mathcal{C}(S_{g,n})$ [67, Theorem 3.4]. The arc complex is used crucially in obtaining a $\mathrm{Mod}_{g,n}$-equivariant cell decomposition and a spine of optimal dimension of $\mathcal{T}_{g,n}$ when $n > 0$ [67, Theorems 1.3 and 2.1].
2. The Torelli curve complex in [51] (see also [101]), its vertices are homotopy classes of separating simple closed curves of S_g and bounding pairs of curves. It plays a similar role for the Torelli group as the curve complex $\mathcal{C}(S_{g,n})$ for the mapping class group $\mathrm{Mod}_{g,n}$, which is equal to $\mathrm{Diff}^+(S_{g,n})/\mathrm{Diff}^0(S_{g,n})$ (see Sect. 7.5 below).
3. The pants complex in [74] whose vertices are pants decompositions of $S_{g,n}$. It was originally used to find generators and relations for the mapping class group $\mathrm{Mod}_{g,n}$. It was also used to describe the coarse geometry of the Weil–Petersson of the Teichmüller space [33]. (Note that the pants complex is a CW-complex but not a simplicial complex.)
4. The train-track complex in [65, 143]. It was used in [65] to prove quasi-isometric rigidity of the mapping class group $\mathrm{Mod}_{g,n}$ (see Proposition 7.15 below).
5. Sub-complexes of $\mathcal{C}(S_{g,n})$ such as the complex of separating curves [101, 115, 145] (see also [72]).
6. The complex of domains and its various subcomplexes [129]. ([129] contains a comprehensive list of complexes associated to surfaces.)
7. A family of complexes related to the curve complex [136].

Remark 3.1. A class of groups closely related to arithmetic groups and mapping class groups consists of outer automorphism groups $\mathrm{Out}(F_n)$ of the free group F_n on n generators. There are several candidates for the analogue of the spherical Tits building $\Delta_{\mathbb{Q}}(\mathbf{G})$ and the curve complex $\mathcal{C}(S_{g,n})$. They are infinite simplicial complexes on which $\mathrm{Out}(F_n)$ acts simplically, and they are also homotopy equivalent to a bouquet of spheres. See [75, 88, 99].

4 Geometric and Topological Properties of Buildings

As mentioned in the introduction, the original motivation of Tits buildings is to give a geometric interpretation of exceptional Lie groups and hence to construct their analogues over finite fields. Their geometric and topological structures have been used for various applications, some of which will be explained in this paper.

First, we define canonical metrics on buildings, their diameters, and curvature properties. Then we state the Solomon–Tits Theorem which determines the topology of buildings. Finally, we explain a relation between Euclidean buildings and spherical buildings.

Recall that a geodesic segment in a metric space (M, d) is an isometric embedding $\gamma : [a, b] \to M$. (M, d) is called *geodesic metric space* if every two points in M are joined by a geodesic segment. M is called *geodesically complete* if every geodesic segment in M, $\gamma : [a,b] \to M$, can be extended infinitely in both directions to a map $\gamma : (-\infty, +\infty)$ such that for any $t \in (-\infty, +\infty)$, when $|t - t'|$ is sufficiently small, $d(\gamma(t'), \gamma(t)) = |t - t'|$, i.e., it is only locally distance minimizing. Note that this is different from the notion of a *complete* geodesic metric space. Clearly, any complete Riemannian manifold is geodesically complete. On the other hand, any Riemannian manifold with nonempty boundary is not geodesically complete, since a geodesic segment perpendicular to the boundary cannot be extended.

Proposition 4.1. *Every spherical Tits building Δ admits a metric such that*

1. *Its restriction to each apartment is isometric to the unit sphere S^{r-1} in $\mathbb{R}^r$, where r is the rank of the building, i.e., the number of vertices of the top dimensional simplices,*
2. *It is a complete geodesic metric space,*
3. *It is geodesically complete.*

This metric is called the Tits metric. If $\Delta = \Delta(G)$, then G acts simplicially and isometrically on $\Delta(G)$.

The idea of the proof is as follows. By definition, each apartment of Δ is a finite Coxeter complex, which is a triangulation of the unit sphere S^{r-1}. The metric on S^{r-1} induces a geodesic metric on each apartment. The axioms for buildings show that the metrics on all the apartments are compatible, for example agree on intersection, and can be patched together to form a metric on Δ. Since all apartments are isometric and can be mapped isometrically from one to another, it can be shown that Δ is a geodesic metric space. Since each apartment is geodesically complete, it can be shown similarly that Δ is also geodesically complete.

Proposition 4.2. *For every spherical Tits building Δ, its diameter with respect to the Tits metric is equal to 2π.*

Proof. It is clear that the diameter of every apartment is equal to 2π. Since every two points of Δ are contained in an apartment, the diameter of Δ is less than or equal to 2π. Since any apartment can be retracted to a fixed apartment, it can be shown that each apartment is a totally geodesic subspace and hence the diameter of Δ is equal to 2π.

Proposition 4.3. *Every Euclidean building Δ admits a metric such that*

1. *Its restriction to each apartment is isometric to the Euclidean space $\mathbb{R}^r$, where r is the rank of the building, i.e., r is equal to the dimension of the top dimensional simplices of Δ, or equivalently, $r + 1$ is the number of vertices of the top dimensional simplices of the Euclidean building Δ,*[3]

[3]It might be worthwhile to emphasize that the rank of a spherical Tits building in Proposition 4.1 is equal to 1 plus the dimension of the top dimensional simplices. This convention is different

2. *It is a complete geodesic metric space,*
3. *It is geodesically complete,*
4. *With respect to this metric, Δ is a CAT(0)-space in the sense that every two distinct points are connected by a unique geodesic, and every triangle in Δ is thinner than the corresponding geodesic in $\mathbb{R}^r$ of the same side lengths.*

This metric is called the Tits metric. If $\Delta = \Delta^{\mathrm{BT}}(\mathbf{G})$, the Bruhat–Tits building of a linear semisimple algebraic group defined over a non-Archimedean local field k, then $G(k)$ acts isometrically on Δ.

See [2] for more details and also [30] for a general discussion on CAT(0)-spaces.

Corollary 4.4. *The diameter of any Euclidean building with respect to the Tits metric is infinite.*

Proof. This follows from the fact that every apartment is a totally geodesic subspace and hence has infinite diameter.

Proposition 4.5. *Every hyperbolic building Δ admits a metric such that*

1. *Its restriction to each apartment is isometric to the hyperbolic space $\mathbf{H}^n$,*
2. *It is a complete geodesic metric space,*
3. *It is geodesically complete,*
4. *With respect to this metric, Δ is a CAT(-1)-space in the sense that every two distinct points are connected by a unique geodesic, and every triangle in Δ is thinner than the corresponding geodesic in $\mathbf{H}^n$ of the same side lengths,*
5. *it has infinite diameter.*

This metric is called the Tits metric on Δ.

Recall that a geodesic metric space is called a *δ-hyperbolic space* if for every triangle in the space, any one side is contained in the δ-neighborhood of the union of the other two sides. It is known that the real hyperbolic space $\mathbf{H}^n$ is δ-hyperbolic. More generally, every simply connected Riemannian manifold with sectional curvature bounded from above is also δ-hyperbolic, and every $CAT(-1)$ geodesic metric space is δ-hyperbolic. Every metric tree is 0-hyperbolic (i.e., δ-hyperbolic, where $\delta = 0$.) Note that every metric space with a finite diameter is automatically hyperbolic for trivial reasons. On the other hand, when $n \geq 2$, $\mathbb{R}^n$ is not δ-hyperbolic.

Corollary 4.6. *Every hyperbolic building is δ-hyperbolic.*

---

from the convention of rank of Euclidean buildings here. Roughly speaking, the reason for these conventions is that the rank of buildings should be equal to the rank of the algebraic groups which define them. For example, a simple algebraic group over $\mathbb{R}$ of rank 1 gives a 0-dimensional Tits building, and a simple algebraic group defined over a p-adic number field of rank 1 gives a 1-dimensional Euclidean building. Both buildings are defined to have rank 1. Another reason for such conventions is that for each Euclidean building of rank r, there is a spherical building of the same rank r which can be added to the infinity of the Euclidean building. See Sect. 6.8 below.

Proposition 4.7. *For a Euclidean building Δ, if its rank is equal to 1, then it is a tree and is a δ-hyperbolic space, otherwise it is not a δ-hyperbolic space.*

Proof. The first statement was mentioned earlier, and the second statement follows from the fact that for any δ-hyperbolic space, any totally geodesic subspace is also δ-hyperbolic. (Recall that a subspace Y of a geodesic space X is called a totally geodesic subspace if for any two points $x, y \in Y$, any geodesic segment in X connecting x and y is contained in Y.)

The topology of spherical buildings is given by the Solomon-Tits Theorem [2].

Proposition 4.8. *Let Δ be a spherical Tits building of rank r. Then Δ is homotopy equivalent to a bouquet of spheres S^{r-1}. If $\Delta = \Delta_{\mathbb{Q}}(\mathbf{G})$, then the bouquet contains infinitely many spheres.*

Basically, fix one simplex σ in Δ, then Δ is the union of apartments containing σ. Each apartment is homotopy equivalent to a sphere, and this union gives the bouquet. When $\Delta = \Delta_{\mathbb{Q}}(\mathbf{G})$, there are infinitely many $\mathbb{Q}$-parabolic subgroups and hence there are infinitely many spheres in the bouquet.

On the other hand, the topology of Euclidean and hyperbolic buildings is trivial, i.e., they are contractible.

Proposition 4.9. *Euclidean buildings and hyperbolic buildings are contractible.*

Proof. By Proposition 4.3, every Euclidean building Δ is a CAT(0)-space. Fix a base point $x_0 \in \Delta$. Then any other point $x \in \Delta$ can be connected to x_0 by a unique geodesic, and deformation retraction along such rays from x_0 shows that Δ is contractible. Since a CAT(-1)-space is also a CAT(0)-space (or by Proposition 4.5), the same proof works for hyperbolic buildings.

5 Geometric and Topological Properties of Curve Complexes

The curve complex $\mathcal{C}(S_{g,n})$ has some properties that are similar to all three types of buildings in the previous subsection.

We can put a metric on $\mathcal{C}(S_{g,n})$ by making each simplex a standard Euclidean one with side length 1.

Proposition 5.1. *$\mathcal{C}(S_{g,n})$ is a complete geodesic space, i.e., it is a geodesic space and is complete as a metric space.*

To prove this proposition, we need mapping class groups and their actions on $\mathcal{C}(S_{g,n})$. Let $\mathrm{Diff}^+(S_{g,n})$ be the group of *orientation preserving* diffeomorphisms of $S_{g,n}$, and $\mathrm{Diff}^0(S_{g,n})$ its identity component, which is a normal subgroup. $\mathrm{Diff}^+(S_{g,n})/\mathrm{Diff}^0(S_{g,n})$ is called the mapping class group of $S_{g,n}$ and denoted by $\mathrm{Mod}_{g,n}$. A closely related group is the *extended mapping class group* $\mathrm{Mod}_{g,n}^{\pm} = \mathrm{Diff}(S_{g,n})/\mathrm{Diff}^0(S_{g,n})$, where $\mathrm{Diff}(S_{g,n})$ is the group of all diffeomorphisms of $S_{g,n}$ including both orientation preserving and orientation reversing diffeomorphisms.

When $3g - 3 + n \geq 2$, for any essential simple closed curve c on $\Sigma_{g,n}$, there are infinitely many essential simple closed curves that are disjoint from c and not homotopy equivalent to c. This implies that the curve complex $\mathcal{C}(S_{g,n})$ is not locally finite. Gluing up the metrics on the simplices certainly gives a metric on the space, but it is not obvious why it is a geodesic metric, i.e., every two points are connected by a geodesic segment. This follows from a theorem on metric simplicial complexes in [29, Theorem 1.1].

Proposition 5.2. *If K is a metric simplicial complex with only finitely many isometry types of simplices, then the metric on K glued up from the metrics of the simplices is a complete geodesic metric space.*

To apply Proposition 5.2 to prove Proposition 5.1, we need to show that $\mathcal{C}(S_{g,n})$ has a large symmetry group.

Proposition 5.3. *The mapping class group* $\mathrm{Mod}_{g,n}$ *and the extended mapping class group* $\mathrm{Mod}_{g,n}^{\pm}$ *act simplicially and isometrically on $\mathcal{C}(S_{g,n})$ with respect to the metric which is glued up from the simplices, and the quotients* $\mathrm{Mod}_{g,n}\backslash\mathcal{C}(S_{g,n})$ *and* $\mathrm{Mod}_{g,n}^{\pm}\backslash\mathcal{C}(S_{g,n})$ *are finite complices.*

Proof. Since diffeomorphisms of $S_{g,n}$ map simple closed curves to simple closed curves, preserve the homotopy classes, and preserve existence of disjoint representatives, it is clear that $\mathrm{Mod}_{g,n}$ acts simplicially on $\mathcal{C}(S_{g,n})$. Since a top dimensional simplex of $\mathcal{C}(S_{g,n})$ corresponds to a pants decomposition of $S_{g,n}$, the statement that $\mathrm{Mod}_{g,n}\backslash\mathcal{C}(S_{g,n})$ is a finite complex is equivalent to the statement that there are only finitely many nonisomorphic pants decompositions of $S_{g,n}$. The latter is reduced to finiteness of homotopy classes of trivalent graphs on the surface. See [40, Sect. 3.5] for detail. The proof for $\mathrm{Mod}_{g,n}^{\pm}$ is the same.

Proof of Proposition 5.1.

By combining Propositions 5.2 and 5.3, we can see that $\mathcal{C}(S_{g,n})$ is a complete geodesic space. (Proposition 5.1 was known and stated in [132, p. 1007].)

Remark 5.4. Since the curve complex $\mathcal{C}(S_{g,n})$ is a thick chamber complex if $g \geq 2$ [69, Proposition on page 266] in the sense that every co-dimension 1 simplex is contained in at least three top dimensional simplices, it might be reasonable to conjecture that $\mathcal{C}(S_{g,n})$ is a geodesically complete metric space, i.e., every geodesic segment in it can be extended infinitely in both directions.

An important metric property of $\mathcal{C}(S_{g,n})$ is the next result [127, Theorem 1.1, Proposition 4.6].

Proposition 5.5. *The curve complex $\mathcal{C}(S_{g,n})$ is a δ-hyperbolic space of infinite diameter.*

For simplified proofs, see [23–25, 63, 64]. As a δ-hyperbolic space, $\mathcal{C}(S_{g,n})$ has a boundary $\partial\mathcal{C}(S_{g,n})$, which is similar to the geodesic boundary $X(\infty)$ of a symmetric space of noncompact type mentioned earlier and consists of equivalence classes of quasi-geodesics. (Note that for a proper δ-hyperbolic space, the boundary can also be defined as the set of equivalence classes of geodesics [98, Proposition 2.10].)

Proposition 5.6. *The boundary $\partial\mathcal{C}(S_{g,n})$ is naturally homeomorphic to the space $\mathcal{EL}(S_{g,n})$ of laminations of $S_{g,n}$ which are filling and minimal (every leaf is dense in the support).*

For every complete hyperbolic metric of finite volume on $S_{g,n}$, a geodesic lamination is a closed subset which is a disjoint union of complete simple geodesics. A *geodesic lamination* is called filling if it intersects every simple closed geodesics. The set of filling geodesic laminations has a natural topology. It is known that different choices of hyperbolic metrics on $S_{g,n}$ give rise to homeomorphic spaces of filling geodesic laminations. Therefore, for the surface $S_{g,n}$ of negative Euler characteristic, there is a well-defined space of filling laminations up to homeomorphism, denoted by $\mathcal{EL}(S_{g,n})$ in the proposition. Proposition 5.6 is due to Klarreich [103]. See also [63] for a proof.

A large scale invariant of a noncompact metric space is the asymptotic dimension introduced by Gromov. Another large scale invariant (or rather property) is Property A, which is a weak amenability-type condition. See [140] for definition.

Proposition 5.7. *The asymptotic dimension of $\mathcal{C}(S_{g,n})$ is finite and has Property A.*

The former was proved in [8] and the latter in [100]. For some results on the quasi-isometry type of $\mathcal{C}(S_{g,n})$, see [148].

The homotopy type of $\mathcal{C}(S_{g,n})$ was determined in [67] (see also [66] and [80]).

Proposition 5.8. *The curve complex $\mathcal{C}(S_{g,n})$ is homotopy equivalent to a bouquet of spheres S^d, where $d = 2g-2 = -\chi(S_g)$ if $n = 0$, $d = -\chi(S_{g,n})-1 = 2g-3+n$ if $g \geq 1$ and $n > 0$, and $d = -\chi(S_{0,n}) - 2 = n - 4$ if $g = 0$.*

The natural question on how many spheres in the bouquet was answered in [86].

Proposition 5.9. *The curve complex $\mathcal{C}(S_{g,n})$ has infinite topology, i.e., the bouquet of spheres in Proposition 5.8 contains infinitely many spheres.*

Propositions 5.8 and 5.9 are an analogue of Solomon-Tits theorem for spherical buildings (Proposition 4.8).

Remark 5.10. Apartments are special and important subcomplexes of buildings and they explain easily the Solomon–Tits theorem. Though the curve complex $\mathcal{C}(S_{g,n})$ share many properties with buildings, one important difference or rather a mystery is that inside $\mathcal{C}(S_{g,n})$, in general, there are no known corresponding distinguished finite subcomplexes whose underlying spaces are spheres or are homotopy equivalent to spheres.

In the special case $g = 2, n = 1$, spheres in $\mathcal{C}(S_{g,n})$ that generate the top dimensional homology group have been constructed in [31, Sect. 4.3]. There are also some candidates of spheres in $\mathcal{C}(S_{g,n})$ when $n = 1$ in [31, Sect. 4.3].

It might be helpful to note that $\mathcal{C}(S_{g,n})$ contains many infinite subcomplexes corresponding to the curve complexes of sub-surfaces of $S_{g,n}$. These subcomplexes have played an important role in understanding the large scale geometry of $\mathcal{C}(S_{g,n})$ and $\mathcal{T}_{g,n}$ in [125, 126].

6 Selected Applications of Buildings

There are many applications of buildings. We select some which are similar to applications of curve complexes discussed below for the purpose of comparison. For more applications of buildings, see [87].

6.1 Automorphism Groups of Buildings

One reason why buildings are useful in geometry and topology is that they are rigid and also admit a large symmetry group.

We state briefly several general results for spherical Tits buildings. For the precise statements, see [166, 5.8, 5.9, 6.3, 6.13, 8.4.5, 9.1, 10.2].

Recall that a building is called *thick* if every simplex of codimension 1 is contained in at least three top dimensional simplices.

Proposition 6.1. *Every thick irreducible spherical Tits building of rank at least 3 is the spherical building of a linear semisimple algebraic group.*

Proposition 6.2. *If $\Delta_k(\mathbf{G})$ and $\Delta_{k'}(\mathbf{G}')$ are irreducible thick spherical Tits buildings of rank at least 2 associated with linear semisimple algebraic groups $\mathbf{G}$ and $\mathbf{G}'$ defined over fields k and k', respectively, then any isomorphism between $\Delta_k(\mathbf{G})$ and $\Delta_{k'}(\mathbf{G}')$ is essentially determined by an isomorphism between the algebraic groups $\mathbf{G}$ and $\mathbf{G}'$ and an isomorphism between k and k'.*

See also [138, Theorem 16.1, Corollary 16.2].

Corollary 6.3. *If $\Delta_k(\mathbf{G})$ is an irreducible thick spherical Tits building of rank at least 2 associated with a semisimple linear algebraic group defined over a field k, then any automorphism of $\Delta_k(\mathbf{G})$ is essentially induced by automorphisms of the group $\mathbf{G}$ and the field k.*

6.2 Mostow Strong Rigidity and Generalizations

One major application of Tits buildings is Mostow strong rigidity (also called Mostow–Prasad strong rigidity) for locally symmetric spaces [138, 144].

Let $X = G/K$ be a symmetric space of noncompact type. Any discrete subgroup $\Gamma \subset G$ acts properly on X and the quotient $\Gamma\backslash X$ is called a locally symmetric space. The invariant metric on X defines an invariant measure on X which induces a measure on $\Gamma\backslash X$. Γ is called a *lattice* if the volume of $\Gamma\backslash X$ with respect to the induced measure is finite. If $G = \mathbf{G}(\mathbb{R})$ is the real locus of a linear semisimple algebraic group and $\Gamma \subset \mathbf{G}(\mathbb{R})$ is an arithmetic subgroup (see the next subsection for a precise definition), then Γ is a lattice in G, by the reduction theory of arithmetic subgroups.

A locally symmetric space $\Gamma\backslash X$ is called *irreducible* if no finite cover splits isometrically as a product.

Theorem 6.4 (Mostow strong rigidity). *Let $\Gamma\backslash X$ and $\Gamma'\backslash X'$ be two irreducible locally symmetric spaces of finite volume. Assume that one of the symmetric spaces X and X' is not isometric to the hyperbolic plane $\mathbf{H}^2$, and Γ and Γ' are isomorphic as abstract groups. Then $\Gamma\backslash X$ and $\Gamma'\backslash X'$ are isometric after suitable scaling of metrics on irreducible factors of X and X'.*

We note that if Γ is torsion-free, then Γ acts on X fixed-point freely, and $\Gamma\backslash X$ is a manifold and its fundamental group $\pi_1(\Gamma\backslash X)$ is equal to Γ. Otherwise, $\Gamma\backslash X$ is an orbifold, and its fundamental group as an orbifold is equal to Γ. Mostow strong rigidity says roughly that if a locally symmetric space of finite volume is not a hyperbolic surface, then its isometry type is determined by its fundamental group.

The proof in [138] depends on the rank of X and works under the assumption that $\Gamma\backslash X$ and $\Gamma'\backslash X'$ are compact. The basic idea is that an isomorphism between Γ and Γ' induces an equivariant quasi-isometry between X and X'. When the rank of X is at least 2, it induces an isomorphism between the Tits buildings $\Delta(X)$ and $\Delta(X')$, which are of rank at least 2, and the rigidity of Tits buildings in Sect. 6.1 implies the desired rigidity. The rank 1 case requires a different proof that involves quasi-conformal maps and ergodic actions of lattices on the sphere $X(\infty)$, which is a real analytic manifold in this case.

The remaining cases where $\Gamma\backslash X$ and $\Gamma'\backslash X'$ are noncompact were proved in [144] when the rank of X is equal to 1, and in [123] when the rank of X is at least 2.

Remark 6.5. Since the original proofs in [138], there have been several different proofs of Mostow strong rigidity. When the locally symmetric spaces are irreducible and the covering symmetric spaces are of rank at least 2, the result also follows from the stronger super-rigidity of Margulis [124]. When $\Gamma\backslash X$ and $\Gamma'\backslash X'$ are compact, there are also proofs of Mostow strong rigidity in some cases by the method of harmonic maps (see [59] for a summary and references).

When the rank of X is equal to 1, in particular when X is the real hyperbolic space, there are at least two completely new proofs:

1. A proof by Gromov using the notion of simplicial volume and the fact that the simplicial volume of a finite volume hyperbolic manifold is proportional to the volume of the hyperbolic metric. See the books [10, 149] for detailed descriptions.

2. A proof in [13] using the notion of barycenter map that characterizes locally symmetric spaces among compact negatively curved Riemannian manifolds in terms of minimal entropy.

There are also proofs in [84,165] of Mostow strong rigidity for hyperbolic spaces using quasi-conformal maps that generalize and simplify the original proof in [138].

Mostow strong rigidity has been generalized in several different ways. One generalization in [4] replaces one locally symmetric space, say $\Gamma'\backslash X'$, by a compact Riemannian manifold M' of nonpositive curvature, and still concludes that if $\Gamma\backslash X$ and M' are homotopy equivalent, $\Gamma\backslash X$ is irreducible and X is of rank at least 2, then M' is also a locally symmetric space isometric to $\Gamma\backslash X$ after a scaling of the metric. A further generalization in [109] replaces the above Riemannian manifold M' by a geodesically complete metric space whose universal covering space is a CAT(0)-space without changing the above conclusion. For this generalization, the following characterization of symmetric spaces of noncompact type and Euclidean buildings was proved in [109, Main Theorem 1.2].

Proposition 6.6. *Let X be a locally compact geodesically complete CAT(0)-space. Let d_{Tits} be the Tits metric on $X(\infty)$. If $(X(\infty), d_{\mathrm{Tits}})$ is isomorphic to a connected thick irreducible spherical building of rank at least 2 with the Tits metric, then X is either a Riemannian symmetric space of noncompact type or a Euclidean building.*

Another type of rigidity problems concerns the characterization of locally symmetric spaces in terms of intrinsic geometric properties. For example, the rank rigidity of nonpositively curved Riemannian manifolds in [3, 38] says that any irreducible nonpositively curved Riemannian manifold M of finite volume with rank at least 2 is a locally symmetric space. The proof of [38] consists of two steps:

1. The construction of a topological Tits building structure on $\widetilde{M}(\infty)$, where $\widetilde{M}$ is the universal covering space of M,
2. The use of rigidity and classification of topological buildings in [39].

All these rigidity results have one thing in common: *the asymptotic geometry of the spaces at infinity is described by Tits buildings, and rigidity of Tits buildings implies the desired rigidity of the spaces.*

6.3 Compactifications of Locally Symmetric Spaces

Let $\mathbf{G} \subset GL(n, \mathbb{C})$ be a linear semisimple algebraic group defined over $\mathbb{Q}$, and $G = \mathbf{G}(\mathbb{R}) = \mathbf{G} \cap GL(n, \mathbb{R})$ the real locus, a real Lie group with finitely many connected components. Let $K \subset G$ be a maximal compact subgroup. Then $X = G/K$ with an invariant metric is a symmetric space of noncompact type. Let $\mathbf{G}(\mathbb{Q}) = \mathbf{G} \cap GL(n, \mathbb{Q})$ be the rational locus of $\mathbf{G}$. A subgroup $\Gamma \subset \mathbf{G}(\mathbb{Q})$ is called an *arithmetic subgroup* if it is commensurable with $\mathbf{G}(\mathbb{Z}) = \mathbf{G}(\mathbb{Q}) \cap GL(n, \mathbb{Z})$, i.e., $\Gamma \cap \mathbf{G}(\mathbb{Z})$ has finite index in both Γ and $\mathbf{G}(\mathbb{Z})$. By the reduction theory of arithmetic subgroups, it is known that $\Gamma\backslash X$ has finite volume and $\Gamma\backslash X$ is compact if and only if the $\mathbb{Q}$-rank $r_{\mathbb{Q}}(\mathbf{G})$ of $\mathbf{G}$ is equal to 0, which is equivalent to the condition that there is no proper $\mathbb{Q}$-parabolic subgroup of $\mathbf{G}$.

For the rest of this subsection, we assume that $\Gamma\backslash X$ is noncompact unless otherwise indicated and let $\mathcal{P}_{\mathbb{Q}}$ be the collection of proper $\mathbb{Q}$-parabolic subgroups of $\mathbf{G}$.

For various applications, we need to compactify $\Gamma\backslash X$. We will discuss one here. See [19] for other motivations and applications.

For any discrete group Γ, there is a *classifying space* $B\Gamma$ which is a CW-complex satisfying the conditions: $\pi_1(B\Gamma) = \Gamma$ and $\pi_i(B\Gamma) = \{1\}$ when $i \geq 2$. Such a space $B\Gamma$ is unique up to homotopy equivalence. Its universal covering space $E\Gamma = \widetilde{B\Gamma}$ is a *universal space for proper and fixed point free actions of* Γ. $E\Gamma$ is a Γ-CW complex characterized by the conditions: (1) Γ acts properly and fixed point freely on $E\Gamma$, and (2) $E\Gamma$ is contractible.

It is known that if Γ contains nontrivial torsion elements, then there does not exist finite dimensional $B\Gamma$ or $E\Gamma$ spaces. In this case, another important space is the *universal space for proper actions of* Γ that is usually denoted by $\underline{E}\Gamma$ and is characterized by the conditions: (1) $\underline{E}\Gamma$ is a Γ-CW complex and Γ acts properly on it, (2) for any finite subgroup $F \subset \Gamma$, the fixed point set $(\underline{E}\Gamma)^F$ is nonempty and contractible. In particular, $\underline{E}\Gamma$ is contractible.

If Γ is torsion-free, then the only finite subgroup is the trivial one, and an $\underline{E}\Gamma$-space is an $E\Gamma$-space.

Definition 6.7. A classifying space $B\Gamma$ is called *finite* if it is a finite CW-complex. Equivalently, $E\Gamma$ is called *cofinite* if the quotient $\Gamma\backslash E\Gamma \cong B\Gamma$ is a finite CW-complex. Similarly, $\underline{E}\Gamma$ is called a *cofinite universal space for proper actions* of Γ if $\Gamma\backslash\underline{E}\Gamma$ is a finite CW-complex.

For topological problems involving Γ, in particular cohomological properties and invariants of Γ, it is important to find explicit and cofinite models of $E\Gamma$ and $\underline{E}\Gamma$ [116].

Proposition 6.8. *Let $\Gamma \subset \mathbf{G}(\mathbb{Q})$ be an arithmetic subgroup as above. Then X is an $\underline{E}\Gamma$-space. It is a cofinite $\underline{E}\Gamma$-space if and only if $\Gamma\backslash X$ is compact.*

Proof. It is known that X is a simply connected and nonpositively curved Riemannian manifold. For any finite subgroup $F \subset \Gamma$, by the Cartan fixed point theorem, the set of fixed points X^F is nonempty. In fact, for any point $x \in X$, the so-called center of gravity of the orbit Fx, which is finite, is fixed by G. Since X^F is a totally geodesic submanifold, it is also contractible.

If X is a cofinite $\underline{E}\Gamma$ space, then $\Gamma\backslash X$ is compact. Conversely, if $\Gamma\backslash X$ is compact, then the existence of an equivariant triangulation implies that X has the structure of a Γ-CW-complex such that $\Gamma\backslash X$ is a finite CW-complex.

When Γ is torsion-free, then $\Gamma\backslash X$ is a $B\Gamma$-space. Assume that $\Gamma\backslash X$ is noncompact. One approach to obtain a compact $B\Gamma$-space is to construct a compactification $\overline{\Gamma\backslash X}$ such that the inclusion $\Gamma\backslash X \to \overline{\Gamma\backslash X}$ is a homotopy equivalence. The Borel–Serre compactification $\overline{\Gamma\backslash X}^{BS}$ in [20] satisfies this property.

For every $\mathbb{Q}$-parabolic subgroup $\mathbf{P} \in \mathcal{P}_{\mathbb{Q}}$, let $P = \mathbf{P}(\mathbb{R})$ be its real locus, $\mathbf{N_P}$ be its unipotent radical, and $N_{\mathbf{P}} = \mathbf{N_P}(\mathbb{R})$ its real locus. Let $\mathbf{A_P}$ be a $\mathbb{Q}$-split component of $\mathbf{P}$ that is stable under the Cartan involution of $\mathbf{G}$ with respect to the maximal compact subgroup K of G, and $A_{\mathbf{P}} = \mathbf{A_P}(\mathbb{R})$ its real locus. Then P admits a *$\mathbb{Q}$-Langlands decomposition*

$$P = N_{\mathbf{P}} A_{\mathbf{P}} M_{\mathbf{P}} \cong N_{\mathbf{P}} \times A_{\mathbf{P}} \times M_{\mathbf{P}}.$$

The subgroup $M_{\mathbf{P}}$ is a reductive subgroup and $M_{\mathbf{P}} \cap K$ is a maximal compact subgroup of $M_{\mathbf{P}}$. The quotient $X_{\mathbf{P}} = M_{\mathbf{P}}/M_{\mathbf{P}} \cap K$ with an invariant metric induced from X is a symmetric space of nonpositive sectional curvature (it might contain a flat factor and may not be of noncompact type), and is called the *boundary symmetric space* of $\mathbf{P}$. The $\mathbb{Q}$-Langlands decomposition of P induces a *horospherical decomposition* of X with respect to the $\mathbb{Q}$-parabolic subgroup $\mathbf{P}$:

$$X \cong N_{\mathbf{P}} \times A_{\mathbf{P}} \times X_{\mathbf{P}}.$$

One example to illustrate this is $\mathbf{G} = \mathrm{SL}(2, \mathbb{C})$, and $\Gamma = \mathrm{SL}(2, \mathbb{Z})$. Then

$$\mathbf{P} = \left\{ \begin{pmatrix} a & b \\ 0 & a^{-1} \end{pmatrix} \mid a \in \mathbb{R}^{\times}, b \in \mathbb{R} \right\}$$

is a $\mathbb{Q}$-parabolic subgroup, and

$$N_{\mathbf{P}} = \left\{ \begin{pmatrix} 1 & b \\ 0 & 1 \end{pmatrix} \mid b \in \mathbb{R} \right\}, \quad A_{\mathbf{P}} = \left\{ \begin{pmatrix} a & 0 \\ 0 & a^{-1} \end{pmatrix} \mid a > 0 \right\},$$

$$M_{\mathbf{P}} = \left\{ \begin{pmatrix} a & 0 \\ 0 & a^{-1} \end{pmatrix} \mid a = \pm 1 \right\}.$$

The boundary symmetric space $X_{\mathbf{P}}$ is one point, and the horospherical decomposition of $X = \mathbf{H}^2 = \mathrm{SL}(2, \mathbb{R})/\mathrm{SO}(2)$ with respect to $\mathbf{P}$ is the (x, y)-horospherical coordinates of the upper half plane $\mathbf{H}^2$. See [19] for more detail.

For every $\mathbb{Q}$-parabolic subgroup $\mathbf{P}$, define

$$e(\mathbf{P}) = N_{\mathbf{P}} \times X_{\mathbf{P}}$$

and call it the *boundary component* of $\mathbf{P}$.

According to the slightly modified procedure in [19], the *Borel–Serre compactification* $\overline{\Gamma\backslash X}^{BS}$ is constructed in the following steps.

1. For every $\mathbf{P} \in \mathcal{P}_{\mathbb{Q}}$, attach the boundary symmetric space $e(\mathbf{P})$ at the infinity of X using the horospherical decomposition of X with respect to $\mathbf{P}$ to obtain a Borel–Serre partial compactification

$$\overline{_{\mathbb{Q}}X}^{\mathrm{BS}} = X \cup \coprod_{\mathbf{P} \in \mathcal{P}_{\mathbb{Q}}} e(\mathbf{P}).$$

2. Show that the Borel–Serre partial compactification $\overline{_{\mathbb{Q}}X}^{\mathrm{BS}}$ is a real analytic manifold with corners.
3. Show that the Γ-action on X extends to a proper real analytic action on $\overline{_{\mathbb{Q}}X}^{\mathrm{BS}}$ with a compact quotient, which is the Borel–Serre compactification of $\Gamma\backslash X$ and denoted by $\overline{\Gamma\backslash X}^{\mathrm{BS}}$.

From the above description, it is clear that the simplices of the Tits building $\Delta_{\mathbb{Q}}(\mathbf{G})$ parametrize the boundary components of the Borel–Serre compactification $\overline{\Gamma\backslash X}^{\mathrm{BS}}$.

Corollary 6.9. *When Γ is torsion-free, $\overline{\Gamma\backslash X}^{\mathrm{BS}}$ is a finite $B\Gamma$-space.*

Proof. When Γ is torsion-free, it acts fixed point freely on $\overline{_{\mathbb{Q}}X}^{\mathrm{BS}}$ and hence $\overline{\Gamma\backslash X}^{\mathrm{BS}}$ is a real analytic compact manifold with corners whose interior is equal to $\Gamma\backslash X$. Then it is clear that the inclusion $\Gamma\backslash X \to \overline{\Gamma\backslash X}^{\mathrm{BS}}$ is a homotopy equivalence by retracting from the boundary faces into the interior. Hence, $\pi_1(\overline{\Gamma\backslash X}^{\mathrm{BS}}) = \Gamma$ and $\pi_i(\overline{\Gamma\backslash X}^{\mathrm{BS}}) = \{1\}$ for $i \geq 2$, and $\overline{\Gamma\backslash X}^{\mathrm{BS}}$ is a $B\Gamma$-space. Since a compact manifold with corners admits a finite triangulation, $\overline{\Gamma\backslash X}^{\mathrm{BS}}$ is a finite $B\Gamma$-space.

When Γ contains torsion elements, the following result was proved in [89].

Proposition 6.10. *The Borel–Serre partial compactification $\overline{_{\mathbb{Q}}X}^{\mathrm{BS}}$ is a cofinite $\underline{E}\Gamma$-space.*

The only non-obvious condition to check is that for any finite subgroup $F \subset \Gamma$, the fixed point set $(\overline{_{\mathbb{Q}}X}^{\mathrm{BS}})^F$ is contractible. The point is that if F fixes a point in the boundary component $N_{\mathbf{P}} \times X_{\mathbf{P}}$, then F is contained in P.

For applications in the next subsection, we note

Proposition 6.11. *The boundary $\partial_{\mathbb{Q}}\overline{X}^{\mathrm{BS}}$ is homotopy equivalent to the spherical Tits building $\Delta_{\mathbb{Q}}(\mathbf{G})$, and hence to a bouquet of infinitely many spheres $S^{r_{\mathbb{Q}}(\mathbf{G})-1}$, where $r_{\mathbb{Q}}(\mathbf{G})$ is the $\mathbb{Q}$-rank of $\mathbf{G}$.*

Proof. Since each boundary space $X_{\mathbf{P}}$ is contractible, and for any two $\mathbb{Q}$-parabolic subgroups $\mathbf{P}_1, \mathbf{P}_2$, the inclusion relation $\mathbf{P}_2 \subset \mathbf{P}_2$ holds if and only if $X_{\mathbf{P}_1}$ is contained in the closure of $X_{\mathbf{P}_2}$, the first statement follows. The second statement follows from Proposition 4.8 and the fact that $\Delta_{\mathbb{Q}}(\mathbf{G})$ is a spherical building of rank $r_{\mathbb{Q}}(\mathbf{G})$.

Remark 6.12. To compare with the two partial compactifications of the Teichmüller space in Sect. 7.5, we mention that in the above construction of the Borel–Serre compactification, if we replace the boundary component $e(\mathbf{P})$ by the boundary symmetric space $X_{\mathbf{P}}$, then we obtain the Reductive Borel–Serre partial compactification

$$\overline{{}_{\mathbb{Q}}X}^{\mathrm{RBS}} = X \cup \coprod_{\mathbf{P}\in\mathcal{P}_{\mathbb{Q}}} X_{\mathbf{P}}.$$

Its boundary decomposes into contractible components $X_{\mathbf{P}}$ and these boundary components are also parametrized by the Tits building $\Delta_{\mathbb{Q}}(\mathbf{G})$ and its boundary $\partial\overline{{}_{\mathbb{Q}}X}^{\mathrm{RBS}}$ is also homotopy equivalent to a bouquet of spheres $S^{r_{\mathbb{Q}}(\mathbf{G})-1}$. The Γ-action on X also extends to a continuous action on $\overline{{}_{\mathbb{Q}}X}^{\mathrm{RBS}}$ with a compact quotient $\Gamma\backslash\overline{{}_{\mathbb{Q}}X}^{\mathrm{RBS}}$. All these results on the boundaries are similar to the Borel–Serre partial compactification $\overline{{}_{\mathbb{Q}}X}^{\mathrm{BS}}$. But the difference is that this extended action on $\overline{{}_{\mathbb{Q}}X}^{\mathrm{RBS}}$ is not proper any more. The inclusion $\Gamma\backslash X \to \Gamma\backslash\overline{{}_{\mathbb{Q}}X}^{\mathrm{RBS}}$ is not a homotopy equivalence. In some sense, $\overline{\Gamma\backslash X}^{\mathrm{BS}}$ is a blow-up of the reductive Borel–Serre compactification $\Gamma\backslash\overline{{}_{\mathbb{Q}}X}^{\mathrm{RBS}}$.

6.4 Cohomological Dimension and Duality Properties of Arithmetic Groups

Let $\Gamma \subset \mathbf{G}(\mathbb{Q})$ be an arithmetic subgroup as in the previous subsection. Once good models of classifying spaces for Γ such as $B\Gamma$ and $\underline{E}\Gamma$ spaces are found, they can be used to study cohomological properties of Γ.

Recall that for any discrete group Γ, the *cohomological dimension* $\mathrm{cd}(\Gamma)$ is defined by

$$\mathrm{cd}(\Gamma) = \sup\{i \mid H^i(\Gamma, M) \neq 0, \text{ for some } \mathbb{Z}\Gamma\text{-module } M\}.$$

It is known that if Γ contains nontrivial torsion elements, then $\mathrm{cd}(\Gamma) = +\infty$.

For any discrete group Γ that is virtually torsion-free, i.e., that contains a finite index torsion-free subgroup Γ', then $\mathrm{cd}(\Gamma')$ is independent of the choice of Γ' and is called the *virtual cohomological dimension* of Γ, denoted by $\mathrm{vcd}(\Gamma)$. It is also known that for any classifying space $B\Gamma'$, $\mathrm{cd}(\Gamma') \leq \dim B\Gamma'$. See [36] for more details.

It is also known that every arithmetic subgroup $\Gamma \subset \mathbf{G}(\mathbb{Q})$ is virtually torsion-free. Since X is an $E\Gamma$-space for a torsion-free arithmetic subgroup, an immediate corollary is the following bound on $\mathrm{vcd}(\Gamma)$:

$$\mathrm{vcd}(\Gamma) \leq \dim X.$$

The precise value of vcd(Γ) was determined in [20] using the Borel–Serre partial compactification $\overline{_{\mathbb{Q}}X}^{BS}$ and the topology of the spherical Tits building $\Delta_{\mathbb{Q}}(\mathbf{G})$.

Proposition 6.13. *The* $\mathrm{vcd}(\Gamma) = \dim X - r_{\mathbb{Q}}(\mathbf{G})$*, where* $r_{\mathbb{Q}}(\mathbf{G})$ *is the* $\mathbb{Q}$*-rank of* $\mathbf{G}$*, i.e., the maximal dimension of* $\mathbb{Q}$*-split tori in* $\mathbf{G}$*.*

This was proved together with a stronger result on duality properties of Γ in [20]. A group Γ is called a *Poincaré duality group of dimension n* if for every $\mathbb{Z}\Gamma$-module M and every i, there is an isomorphism

$$H^i(\Gamma, M) \cong H_{n-i}(\Gamma, M).$$

If Γ admits a $B\Gamma$-space given by a closed orientable manifold of dimension n, then Γ is a Poincaré duality group. But the converse is not true in general. An important question is under what further conditions a Poincaré duality group Γ admits a closed manifold as a $B\Gamma$-space.

More generally, Γ is called a *duality group of dimension n* if there exists a $\mathbb{Z}\Gamma$-module D, called the *dualizing module*, such that for every $\mathbb{Z}\Gamma$-module M and every i, there is an isomorphism

$$H^i(\Gamma, M) \cong H_{n-i}(\Gamma, D \otimes M).$$

In this case, it is known that $\mathrm{cd}(\Gamma) = n$.

Since groups Γ containing nontrivial torsion elements have $\mathrm{cd}(\Gamma) = +\infty$, they cannot be duality groups. On the other hand, if Γ admits a finite-index torsion-free subgroup that is a duality group, it is called a *virtual duality group*. Similarly, the notion of *virtual Poincaré duality group* can be defined.

The stronger result proved in [20] is the following

Proposition 6.14. *Every arithmetic subgroup* $\Gamma \subset \mathbf{G}(\mathbb{Q})$ *as above is a virtual duality group of dimension* $\dim X - r_{\mathbb{Q}}$*, where* $r_{\mathbb{Q}} = r_{\mathbb{Q}}(\mathbf{G})$*. It is a Poincaré duality group if and only if* $r_{\mathbb{Q}} = 0$*, i.e.,* $\Gamma\backslash X$ *is compact. When* $r_{\mathbb{Q}} > 0$*, the dualizing module of* Γ *is equal to the* $\mathbb{Z}\Gamma$*-module* $\overline{H}_{r_{\mathbb{Q}}-1}(\Delta_{\mathbb{Q}}(\mathbf{G}))$*, where* $\overline{H}_*$ *is the reduced homology group.*

The conclusion that Γ is a virtual Poincaré duality group if $\Gamma\backslash X$ is compact is clear since it is a closed orientable manifold if Γ is torsion-free.

Since $\mathbf{G}(\mathbb{Q})$ and hence Γ acts on the set of $\mathbb{Q}$-parabolic subgroups of $\mathbf{G}$ and hence on the spherical Tits building $\Delta_{\mathbb{Q}}(\mathbf{G})$, for every i, $H_i(\Delta_{\mathbb{Q}}(\mathbf{G}))$ is a $\mathbb{Z}\Gamma$-module. By Proposition 4.8, the reduced homology of $\Delta_{\mathbb{Q}}(\mathbf{G})$ is non-zero in only the degree $r_{\mathbb{Q}}(\mathbf{G}) - 1$.

The basic idea of the proof of Proposition 6.14 is as follows. Assume that Γ is torsion-free and $\Gamma\backslash X$ is noncompact. Then $\overline{\Gamma\backslash X}^{BS}$ is a finite $B\Gamma$-space and $\overline{_{\mathbb{Q}}X}^{BS}$ is a cofinite $E\Gamma$-space. By general results for cohomology of groups [36, Propositions 7.5 and 8.2, Theorem 10.1], it suffices to show that $H^i(\Gamma, \mathbb{Z}\Gamma) \cong H^i_c(\overline{_{\mathbb{Q}}X}^{BS}, \mathbb{Z})$ is not zero only when $i = \dim X - r_{\mathbb{Q}}(\mathbf{G})$, where $H^i_c(\overline{_{\mathbb{Q}}X}^{BS}, \mathbb{Z})$

denotes the cohomology group with compact support. By Poincaré-Lefschitz duality,

$$H_c^i(\overline{{}_{\mathbb{Q}}X}^{\mathrm{BS}}, \mathbb{Z}) \cong H_{d-i}(\overline{{}_{\mathbb{Q}}X}^{\mathrm{BS}}, \partial \overline{{}_{\mathbb{Q}}X}^{\mathrm{BS}}),$$

where $d = \dim X$. Since $\overline{{}_{\mathbb{Q}}X}^{\mathrm{BS}}$ is contractible,

$$H_{d-i}(\overline{{}_{\mathbb{Q}}X}^{\mathrm{BS}}, \partial \overline{{}_{\mathbb{Q}}X}^{\mathrm{BS}}) \cong \overline{H}_{d-i-1}(\partial \overline{{}_{\mathbb{Q}}X}^{\mathrm{BS}}),$$

where $\overline{H}_*$ is the reduced homology. By Proposition 6.11, the latter is not zero if and only if $d - i - 1 = r_{\mathbb{Q}}(\mathbf{G}) - 1$, i.e., $i = d - r_{\mathbb{Q}}(\mathbf{G})$. This proves that Γ is a duality group of dimension $d - r_{\mathbb{Q}}(\mathbf{G})$ and the dualizing module is equal to $\overline{H}_{d-r_{\mathbb{Q}}(\mathbf{G})}(\Delta_{\mathbb{Q}}(\mathbf{G}), \mathbb{Z})$.

6.5 *Simplicial Volumes of Locally Symmetric Spaces*

An important homotopy invariant of manifolds is the simplicial volume introduced by Gromov [58].

Suppose that M is a connected oriented compact manifold of dimension n, let $[M]$ be the fundamental class in $H_n(M, \mathbb{R})$, or rather the image of the fundamental class in $H_n(M, \mathbb{Z})$ under the natural map $H_n(M, \mathbb{Z}) \to H_n(M, \mathbb{R})$. For each n-chain $c = \sum_\sigma a_\sigma \sigma$ with $\mathbb{R}$-coefficients, where σ are n-singular simplices, define the simplicial ℓ^1-norm

$$||c||_1 = \sum_\sigma |\sigma|.$$

Then the *simplicial volume* of M, denoted by $||M||$, is defined by

$$||M|| = \inf\{||c||_1 | c \text{ is an } n\text{-chain with } \mathbb{R}\text{-coefficients in the fundamental class } [M]\}.$$

If M is a connected non-orientable manifold, then let $\tilde{M}$ be its double cover and define $||M|| = \frac{1}{2}||\tilde{M}||$. If M is an oriented orbifold, then it has a fundamental class and hence the usual notion of simplicial volume. If M admits a finite smooth cover N, then it also has a *orbifold simplicial volume* $||M||_{\mathrm{orb}} = ||N||/d$, where d is the degree of the covering $N \to M$. It follows from the multiplicative property of simplicial volumes of manifolds that $||M||_{\mathrm{orb}}$ is independent of the choice of a finite smooth cover N. It is known that

$$||M||_{\mathrm{orb}} \geq ||M||,$$

and the strict inequality can occur. See [91] for details.

Assume that M is a connected orientable noncompact manifold of dimension n. Let $H_*^{\mathrm{lf}}(M, \mathbb{R})$ be the locally finite homology group of M. Let $[M]^{\mathrm{lf}}$ be the fundamental class in $H_*^{\mathrm{lf}}(M, \mathbb{R})$. For any locally finite n-chain $c = \sum_\sigma a_\sigma \sigma$, define the simplicial norm $||c||_1 = \sum_\sigma |\sigma|$ as above. Recall that a chain is locally finite means that every compact subset of M meets the images of only finitely many singular simplices σ in the chain. Hence, $||c||_1$ could be equal to infinity.

The simplicial volume $||M||$ of a noncompact manifold M is defined by

$$||M|| = \inf\{||c||_1 \mid c \text{ is a locally finite } n\text{-chain in the fundamental class } [M]^{\mathrm{lf}}\}.$$

One of the motivations of Gromov to introduce the simplicial volume was to give a lower bound on the minimal volume of a manifold M. Consider all complete Riemannian metrics g on M whose sectional curvature $K(g)$ satisfies the bound $|K(g)| \leq 1$ at all points. Let $\mathrm{Vol}(M, g)$ denote the volume of M with respect to the metric g. Then the minimal volume of M is defined by

$$\text{Min-Vol}(M) = \{\inf(\mathrm{Vol}(M, g) \mid \text{all complete metrics } g,\ |K(g)| \leq 1\}.$$

Another major application of simplicial volume is a different proof by Gromov of Mostow strong rigidity for compact hyperbolic spaces of dimension at least 3 as mentioned above. See [10, 149] for detailed discussions.

A basic result in [58, Sect. 0.5] states that there exists a universal constant C_n only depending on the dimension n such that

$$\text{Min-Vol}(M) \geq C_n||M||.$$

Therefore, a natural problem is to understand when the simplicial volume $||M||$ is equal to zero.

It is known that if a compact manifold M admits a self-map of degree greater than or equal to 2, then its simplicial volume $||M|| = 0$. If a noncompact manifold M admits a proper self-map of degree greater than or equal to 2, then $||M|| = 0$ or $||M|| = +\infty$. As a consequence, the simplicial volumes of spheres and tori are equal to zero. It is also known that the simplicial volume of $\mathbb{R}^n$ is equal to 0.

For spaces related to locally symmetric spaces, the following results on simplicial volume are known:

Proposition 6.15. *If M is a complete hyperbolic manifold of finite volume, then $||M|| > 0$. More generally, if M admits a complete metric such that its sectional curvature K is bounded uniformly between two negative constants, then $||M|| > 0$.*

This is a result due to Thurston [58, Sect. 0.3].

Proposition 6.16. *If $M = \Gamma\backslash X$ is a compact locally symmetric space of noncompact type, then $||M|| > 0$.*

This was conjectured by Gromov [58, p. 11] and proved in [37, 107].

Proposition 6.17. *If $M = \Gamma\backslash X$ is an arithmetic locally symmetric space whose $\mathbb{Q}$-rank, denoted by $r_{\mathbb{Q}}(\mathbf{G})$, is greater than or equal to 3, then $||M|| = 0$.*

The vanishing result was proved in [114]. In the proof, Propositions 6.9 and 6.11 are used crucially. Briefly, a vanishing criterion [58, p. 58] was applied. In order to apply this, a suitable covering of $\Gamma\backslash X$ with multiplicity at most $\dim \Gamma\backslash X$ is needed. For this purpose, the conditions that the map $\pi_1^\infty(\Gamma\backslash X) = \pi_1(\overline{\Gamma\backslash X}^{BS}) \to \pi_1(\Gamma\backslash X) = \pi_1(\overline{\Gamma\backslash X}^{BS})$ is injective and that the virtual cohomological dimension of Γ is at most $\dim \Gamma\backslash X - 2$ are needed. Since the virtual cohomological dimension of Γ is equal to $\dim \Gamma\backslash X - r_{\mathbb{Q}}(\mathbf{G})$, the condition that $r_{\mathbb{Q}}(\mathbf{G}) \geq 3$ is more than enough. The assumption that $r_{\mathbb{Q}}(\mathbf{G}) \geq 3$ is needed to show that the boundary $\partial_{\mathbb{Q}} \overline{X}^{BS}$, which is homotopy equivalent to a bouquet of spheres $S^{r_{\mathbb{Q}}(\mathbf{G})-1}$, is simply connected.

If $r_{\mathbb{Q}}(\mathbf{G}) = 1$, this vanishing result does not hold in general. For example, it was proved in [113] that if M is a Hilbert modular variety, then $||M|| > 0$. Note that Hilbert modular varieties are important examples of locally symmetric spaces of $\mathbb{Q}$-rank 1. If the rank of X is equal to 1 and $\Gamma\backslash X$ is noncompact, which implies that $r_{\mathbb{Q}}(\mathbf{G}) = 1$, then the sectional curvature of $\Gamma\backslash X$ is bounded by two negative constants and hence the simplicial volume of $\Gamma\backslash X$ is positive [58, Sect. 0.3].

Remark 6.18. For any topological space M of dimension n that admits a suitable fundamental class in $H_n(M, \mathbb{Z})$ or $H_n^{\mathrm{lf}}(M, \mathbb{Z})$, we can define its simplicial volume. We can show that for any arithmetic locally symmetric space $\Gamma\backslash X$ of $\mathbb{Q}$-rank at least 3, and any irreducible arithmetic locally symmetric space $\Gamma\backslash X$ of $\mathbb{Q}$-rank at least 1 and the rank of X at least 2, the simplicial volumes of the reductive Borel–Serre compactification and the Baily–Borel compactification (if $\Gamma\backslash X$ is Hermitian) also vanish. See [91] for details.

6.6 Asymptotic Cones of Symmetric Spaces and Locally Symmetric Spaces

Buildings also occur naturally in the large scale geometry of symmetric spaces X and locally symmetric spaces $\Gamma\backslash X$.

For any metric space (M, d), choose a basepoint $x_0 \in X$. For any $\varepsilon > 0$, consider the family of pointed metric spaces $(M, \varepsilon d, x_0)$. The limit $\lim_{\varepsilon \to 0}(M, \varepsilon d, x_0)$, if it exists, is called the *asymptotic cone* (or *tangent cone at infinity*) of M and denoted by $\mathrm{Cone}_\infty(M)$. There are always ultra limits, which may not be unique. Any $\mathrm{Cone}_\infty(M)$ is a metric cone and does not depend on the choice of the basepoint x_0.

If (M, d) has finite diameter, then $\mathrm{Cone}_\infty(M)$ consists of one point. If $M = \mathbb{R}^n$ with the standard Euclidean metric, then $\mathrm{Cone}_\infty(\mathbb{R}^n)$ is isomorphic to $\mathbb{R}^n$ due to the scaling invariance of the Euclidean metric. On the other hand, if $M = \mathbf{H}^n$, $n \geq 2$, $\mathrm{Cone}_\infty(M)$ is an $\mathbb{R}$-tree that branches at every point. Recall that a usual simplicial tree branches only at a discrete set of points on every geodesic.

The asymptotic cone of a general symmetric space of noncompact type was determined in [104].

Proposition 6.19. *If X is a symmetric space of noncompact type, then* $\mathrm{Cone}_\infty(X)$ *is an $\mathbb{R}$-Euclidean building.*

One application of this result is the quasi-isometric rigidity of symmetric spaces [48, 104], which is a generalization of Mostow strong rigidity.

Proposition 6.20. *Let X and X' be two symmetric spaces of noncompact type that have no irreducible factors of rank 1. If X and X' are quasi-isometric, then they are isometric up to suitable scaling of the metrics on irreducible factors.*

If $\Gamma \subset \mathbf{G}(\mathbb{Q})$ is an arithmetic subgroup and $\Gamma\backslash X$ is noncompact, then $\mathrm{Cone}_\infty(\Gamma\backslash X)$ was determined in [76, 94, 110].

Proposition 6.21. *If $\Gamma\backslash X$ is a noncompact arithmetic locally symmetric space, then the asymptotic cone at infinity $\mathrm{Cone}_\infty(\Gamma\backslash X)$ is a metric cone over the finite complex $\Gamma\backslash\Delta_{\mathbb{Q}}(\mathbf{G})$.*

Recall that Γ acts on $\Delta_{\mathbb{Q}}(\mathbf{G})$ via the action on the set $\mathcal{P}_{\mathbb{Q}}$ of parabolic subgroups of $\mathbf{G}$. By the reduction theory of arithmetic subgroups, there are only finitely many Γ-conjugacy classes of proper $\mathbb{Q}$-parabolic subgroups and hence $\Gamma\backslash\Delta_{\mathbb{Q}}(\mathbf{G})$ is a finite simplicial complex.

6.7 Applications of Euclidean Buildings

In the previous subsections, we have mainly concentrated on applications of spherical Tits buildings. Now we briefly consider some applications of Euclidean buildings.

A natural generalization of arithmetic subgroups is the class of S-arithmetic subgroups. Let $S = \{p_1, \cdots, p_l, \infty\}$, where $p_1, \cdots, p_l$ are prime distinct numbers, and $\mathbb{Z}[\frac{1}{p_1}, \cdots, \frac{1}{p_l}]$ be the ring of S-integers, denoted by $\mathbb{Z}_S$. For any linear semisimple algebraic group $\mathbf{G} \subset GL(n, \mathbb{C})$ defined over $\mathbb{Q}$, let $\mathbf{G}(\mathbb{Z}_S) = \mathbf{G}(\mathbb{Q}) \cap GL(n, \mathbb{Z}_S)$. Then a subgroup Γ of $\mathbf{G}(\mathbb{Q})$ is called an *S-arithmetic subgroup* if it is commensurable with $\mathbf{G}(\mathbb{Z}_S)$.

It is known that under the diagonal embedding, Γ is a discrete subgroup of $G \times \mathbf{G}(\mathbb{Q}_{p_1}) \times \cdots \times \mathbf{G}(\mathbb{Q}_{p_k})$, where $G = \mathbf{G}(\mathbb{R})$ is the real locus of $\mathbf{G}$.

For each p_i, let Δ_{p_i} be the Bruhat–Tits building of the group $\mathbf{G}(\mathbb{Q}_{p_i})$. Let $X = G/K$ be the symmetric space of noncompact type associated with G as above. Define

$$X_S = X \times \Delta_{p_1} \times \cdots \times \Delta_{p_k}.$$

Then Γ acts properly on X_S, since each $\mathbf{G}(\mathbb{Q}_{p_i})$ acts properly on Δ_{p_i}.

Proposition 6.22. *Any S-arithmetic subgroup Γ acts properly on X_S and X_S is an $\underline{E}\Gamma$-space.*

The reason is that each Δ_{p_i} is a CAT(0)-space and hence the product X_S is also a CAT(0)-space. Since the Cartan fixed point theorem holds for CAT(0)-spaces, for any finite subgroup $F \subset \Gamma$, the fixed point set $(X_S)^F$ is nonempty. It is also a totally geodesic subspace.

If the quotient $\Gamma \backslash X_S$ is compact, then X_S is a cofinite $\underline{E}\Gamma$-space by [93]. Otherwise, we need an analogue of the Borel–Serre partial compactification

$$\overline{_{\mathbb{Q}}X_S}^{\mathrm{BS}} = \overline{_{\mathbb{Q}}X}^{\mathrm{BS}} \times \Delta_{p_1} \times \cdots \times \Delta_{p_k},$$

on which Γ acts properly with a compact quotient. See [89].

Remark 6.23. Bruhat–Tits buildings, or rather the Bruhat–Tits theory, have played an important role in the representation theory of reductive p-adic Lie groups. Due to the lack of knowledge of the author, we only mention one application which is connected with the fact that Bruhat–Tits buildings are CAT(0)-spaces. In [139], it was proved that for any Bruhat–Tits building $\Delta^{\mathrm{BT}}(\mathbf{G}(k))$, and any $g \in \mathrm{Aut}(\Delta^{\mathrm{BT}}(\mathbf{G}(k)))$, the displacement function

$$d_g : \Delta^{\mathrm{BT}}(\mathbf{G}(k)) \to \mathbb{R}_{\geq 0}, x \mapsto \mathrm{dist}(x, gx),$$

is a convex function. This has important applications in representation theory. See [137] for detail.

Remark 6.24. On the set of vertices of a Bruhat–Tits building, there is also a combinatorial distance, which is induced from the distance on the 1-skeleton of the building. This has applications to arithmetic algebraic geometry. In [170], non-Archimedean intersection indices on projective spaces over a non-Archimedean local field of characteristic zero are expressed in terms of the combinatorial distance of the Bruhat–Tits building for $\mathbf{G} = \mathrm{PGL}(n)$.

6.8 Compactifications of Euclidean Buildings

It is clear from the definition that Euclidean buildings are noncompact, since each apartment is a closed noncompact subspace. For many purposes, Euclidean buildings play a similar role for p-adic Lie groups as symmetric spaces of noncompact type for noncompact real Lie groups. There has been a lot of work on compactifications of symmetric spaces motivated by applications in group theory, geometry, topology, and analysis. See [19] for a history, more details and references.

By Proposition 4.3, every Euclidean building Δ is a CAT(0)-space. It is also known that a symmetric space of noncompact type with an invariant Riemannian metric is a complete nonpositively curved simply connected Riemannian manifold and is hence also a CAT(0)-space. This is one important common property shared by both Euclidean buildings and symmetric spaces.

It is known that for any CAT(0)-space Δ, its set of equivalence classes of geodesics $X(\infty)$ forms the geodesic (or visual) boundary of Δ as in the case of a symmetric space in Sect. 2.1. If Δ is a proper metric space, then there is a natural compact topology on $\Delta \cup \Delta(\infty)$ [21,30]. This compactification $X \cup X(\infty)$ is called the geodesic compactification. The following result was also proved in [21].

Proposition 6.25. *(1) If Δ is a Euclidean building, then its geodesic boundary $\Delta(\infty)$ has a natural structure of a spherical Tits building. If the rank of Δ is equal to r, then the rank of $\Delta(\infty)$ is also equal to r. (2) If $\Delta = \Delta^{\mathrm{BT}}(\mathbf{G})$ is the Bruhat–Tits building of an algebraic group $\mathbf{G}$ over a locally compact local field k, then $\Delta = \Delta^{\mathrm{BT}}(\mathbf{G})$ is locally compact, $\Delta(\infty)$ is the spherical Tits building $\Delta_k(\mathbf{G})$, and the geodesic compactification $\Delta^{\mathrm{BT}}(\mathbf{G}) \cup \Delta^{\mathrm{BT}}(\mathbf{G})(\infty)$ is contractible.*

This geodesic compactification explains a close relation between spherical and Euclidean buildings.

Suppose Δ_1, Δ_2 are two locally compact Euclidean buildings. Then there are at least two natural compactifications of the product $\Delta_1 \times \Delta_2$.

The first one is to take the product of the geodesic compactifications $\Delta_1 \cup \Delta_1(\infty)$ and $\Delta_2 \cup \Delta_2(\infty)$,

$$\overline{\Delta_1 \times \Delta_2}^{\mathrm{prod}} = (\Delta_1 \cup \Delta_1(\infty)) \times (\Delta_2 \cup \Delta_2(\infty)).$$

Note that the product $\Delta_1 \times \Delta_2$ is also a CAT(0)-space. Then the geodesic compactiication gives another compactification of $\Delta_1 \times \Delta_2$,

$$\Delta_1 \times \Delta_2 \cup (\Delta_1 \times \Delta_2)(\infty).$$

It is clear that these two compactifications are not isomorphic, and each of them is natural in its own way. It turns out that the second type of compactifications has an application to the Novikov conjecture in geometric topology for certain classes of S-arithmetic subgroups of algebraic groups. Let Γ be an S-arithmetic subgroup of a linear algebraic group as in Sect. 6.7. Assume that Γ acts cocompactly on $X_S = X \times \Delta_{p_1} \times \cdots \Delta_{p_k}$. By Proposition 6.22, X_S is a cocompact universal space for proper actions of Γ, i.e., a cocompact model of $\underline{E}\Gamma$. Since X is also a CAT(0)-space as mentioned above, the product X_S is a CAT(0)-space and hence admits a geodesic compactification. The Γ-action on X_S extends continuously to the compactification. An important point of the geodesic compactification of X_S is that the extended action of Γ on the compactification is small at the boundary (or infinity). Then it follows from a general criterion for validity of Novikov conjectures that the integral Novikov conjecture holds for Γ. See [92,93] for precise statements of the Novikov conjecture and of the criterion and for other details.

On the other hand, if we take the product compactification of X_S induced from compactifications of the factors, i.e., the product of $X \cup X(\infty)$, $\Delta_{p_1} \cup \Delta_{p_1}(\infty), \cdots, \Delta_{p_k} \cup \Delta_{p_k}(\infty)$, the action of Γ might not be small at infinity, and it might not be used for proving the integral Novikov conjecture for Γ.

Motivated by many different compactifications of symmetric spaces of noncompact type (see [19]), there are also corresponding compactifications of

Euclidean buildings. As discussed in Sect. 2.2, one important feature of rich structure of a symmetric space is reflected by the flat subspaces F of maximal dimension in X, which determine the Tits building of X.

There are several ways to understand a compactification $\overline{X}$ of X, and they can be generalized to construct compactifications of Euclidean buildings:

1. Decompose the boundary $\overline{X} - X$ into smaller subspaces which enjoy some natural structures.
2. Determine the closure of maximal flat subspaces F in the compactification $\overline{X}$ and the intersection pattern of these compactificed flat subspaces, and recover the compactification of X from the compactified flats.
3. Interpret X as a moduli space of certain objects and the boundary points $\overline{X} - X$ as degenerate objects.

For the geodesic compactification $X \cup X(\infty)$ in Sect. 2.1, the boundary $X(\infty)$ is naturally decomposed into simplices and can be viewed as the underlying space of the Tits building of X. As pointed out earlier, a geodesic compactification can be constructed for every locally compact Euclidean building.

Another important compactification of X is the maximal Satake compactification $\overline{X}^S_{\max}$, which is the maximal element in the partially ordered finite set of Satake compactifications of X. See [60]. Each maximal flat F of X has a natural decomposition by Weyl chambers and Weyl chamber faces. The closure of F in $\overline{X}^S_{\max}$ is canonically homeomorphic to a polyhedron, its boundary is a finite cell complex dual to the Weyl chamber decomposition, and the compactification $\overline{X}^S_{\max}$ can be constructed by gluing these compactified flats in [60].

It turns out a similar construction works for any Euclidean building Δ by gluing polyhedral compactifications of apartments of Δ. See [108].

To explain Satake compactifications of symmetric spaces, we start with a special Satake compactification of the symmetric space $\mathrm{SL}(n,\mathbb{C})/\mathrm{SU}(n)$. Let $\mathcal{H}_n$ be the real vector space of $(n \times n)$-Hermitian matrices, and $P(\mathcal{H}_n)$ the associated projective space. Then the map $A \mapsto A\overline{A}^t$ defines an embedding of $\mathrm{SL}(n,\mathbb{C})/\mathrm{SU}(n)$ into $P(\mathcal{H}_n)$, and the closure of $\mathrm{SL}(n,\mathbb{C})/\mathrm{SU}(n)$ under this embedding gives a Satake compactification. In this case, every point in $\mathrm{SL}(n,\mathbb{C})/\mathrm{SU}(n)$ corresponds to a positive definite Hermitian matrix of determinant 1 (or equivalently a positive definite Hermitian quadratic form of determinant 1), and the boundary points correspond to projective classes of degenerate positive semi-definite Hermitian matrices.

For a general symmetric space $X = G/K$ of noncompact type, any finite dimensional projectively faithful representation of G gives an isometric embedding of X into $\mathrm{SL}(n,\mathbb{C})/\mathrm{SU}(n)$ for some n, and the closure of X in the above special Satake compactification of $\mathrm{SL}(n,\mathbb{C})/\mathrm{SU}(n)$ gives a Satake compactification of X. Though there are infinitely many representations of G, it turns out that they only give rise to finitely many non-isomorphic Satake compactifications of X.

For the Bruhat–Tits building $\Delta^{\mathrm{BT}}(\mathbf{G})$ of a reductive algebraic group over a local field, the corresponding Satake compactifications have been constructed in [150,

151, 171–173]. The paper [150] constructs the compactification by using Berkovich analytic geometry over complete non-Archimedean fields, and the paper [151] uses irreducible representations of the algebraic group and is more similar to the Satake compactifications of symmetric spaces. The construction in [171] is also similar to the Satake compactifications of symmetric spaces. Compactifications of some special buildings were treated in [172, 173].

The boundary $\partial \overline{X}^S_{\max}$ of the maximal Satake compactification is naturally decomposed into symmetric spaces of noncompact type of smaller dimension, which are naturally parametrized by proper parabolic subgroups of G, where $X = G/K$, or by simplices of the Tits building $\Delta(X)$. Since each boundary symmetric space is a cell, this shows that the boundary $\partial \overline{X}^S_{\max}$ has a natural cell structure which is dual to the Tits building. See [60]. The boundary $\partial \overline{X}^S_{\max}$ can also be constructed in this way by adding these boundary components using the idea of the Borel–Serre compactification of locally symmetric spaces. See [19]. For Bruhat–Tits buildings, a similar construction should also work. For the Bruhat–Tits building of $\mathrm{PGL}(n)$ over a local compact field, such a construction has been carried out in [171].

The maximal Satake compactification $\overline{X}^S_{\max}$ can also be constructed by embedding X into the compact space of closed subgroups of G in [60]. A similar compactification of Bruhat–Tits buildings has been constructed in [61].

An important application of compactifications of symmetric spaces is to harmonic analysis on symmetric spaces, for example, determination of the Poisson boundary. The analogue of the Poisson boundary of certain Bruhat–Tits buildings has been determined in [57].

Remark 6.26. For a hyperbolic building Δ, its geodesic boundary $\Delta(\infty)$ has the structure of a trivial (or rank 1) spherical building structure, i.e., the top dimensional simplices of the building are points. The reason is that the boundary $\mathbf{H}^n(\infty)$ of the real hyperbolic space $\mathbf{H}^n$ has the structure of a rank-1 spherical building.

7 Applications of Curve Complexes

In this section, we discuss some applications of curve complexes briefly. In Sect. 7.1, we discuss the identification of the automorphism group of $\mathcal{C}(S_{g,n})$ with $\mathrm{Mod}_{g,n}$, which is responsible for several rigidity results for the geometry of $\mathcal{T}_{g,n}$ in Sect. 7.2. In Sect. 7.3, we briefly describe the ending lamination conjecture of Thurston on rigidity of 3-dimensional hyperbolic manifolds with a finitely generated fundamental group and its formulation in terms of the boundary $\partial\mathcal{C}(S_{g,n})$ of $\mathcal{C}(S_{g,n})$ as a δ-hyperbolic space. This is probably one of the most striking applications of the curve complex $\mathcal{C}(S_{g,n})$. In Sect. 7.4, we discuss an application to quasi-isometric rigidity of $\mathrm{Mod}_{g,n}$. In Sect. 7.5, we describe several different notions of the Novikov conjectures and an application of the curve complex to finiteness of the asymptotic dimension of $\mathrm{Mod}_{g,n}$. In Sect. 7.6, we mention an application to non-

Gromov-hyperbolicity of the Weil–Petersson metric of $\mathcal{T}_{g,n}$. In Sect. 7.7, we define the Hempel distance of a Heegaard splitting via the image of Heegaard diagrams in $\mathcal{C}(S_g)$. After these applications to 3-dimensional manifolds, we describe the Borel–Serre partial compactification of $\mathcal{T}_{g,n}$ via $\mathcal{C}(S_{g,n})$, the original motivation of introducing $\mathcal{C}(S_{g,n})$. Cohomological properties of $\mathrm{Mod}_{g,n}$ are studied in Sect. 7.9. The asymptotic cone at infinity of $\mathcal{M}_{g,n}$ is described in Sect. 7.10 and the simplicial volume of $\mathcal{M}_{g,n}$ is discussed in Sect. 7.11. In the last Sect. 7.12, we mention applications of $\mathcal{C}(S_{g,n})$ to classification of elements, presentations and unitary representations of $\mathrm{Mod}_{g,n}$.

7.1 Automorphism Groups of Curve Complexes

As mentioned in Sect. 5, the extended mapping class group $\mathrm{Mod}_{g,n}^{\pm}$ acts on $\mathcal{C}(S_{g,n})$ simplicially. This defines a map $\mathrm{Mod}_{g,n}^{\pm} \to \mathrm{Aut}(\mathcal{C}(S_{g,n}))$, where $\mathrm{Aut}(\mathcal{C}(S_{g,n}))$ is the simplicial automorphism group of $\mathcal{C}(S_{g,n})$. It turns out that the image is essentially the whole automorphism group [81, 105], [117, Theorem, p. 7].

Proposition 7.1. *(1) If the dimension of $\mathcal{C}(S_{g,n})$ is at least 1 and $(g,n) \neq (1,2)$, then the map $\mathrm{Mod}_{g,n}^{\pm} \to \mathrm{Aut}(\mathcal{C}(S_{g,n}))$ is surjective. (2) When $(g,n) = (1,2)$, the image of the map $\mathrm{Mod}_{g,n} \to \mathrm{Aut}(\mathcal{C}(S_{g,n}))$ contains all automorphisms of $\mathcal{C}(S_{1,2})$ that preserve vertices represented by separating curves. Furthermore, this map is injective and the image has index 2 in $\mathrm{Aut}(\mathcal{C}(S_{g,n}))$. (3) When $S_{g,n}$ is not a torus with 1 or 2 punctures, or a sphere with at most 4 punctures, or a closed surface of genus 2, then the map $\mathrm{Mod}_{g,n}^{\pm} \to \mathrm{Aut}(\mathcal{C}(S_{g,n}))$ is an isomorphism.*

This is an analogue of Corollary 6.3 for Tits buildings and can be used to prove several rigidity results in the next subsection.

Remark 7.2. There are several related results on rigidity of simplicial maps of $\mathcal{C}(S_{g,n})$ which imply co-Hopfian property of the mapping class group $\mathrm{Mod}_{g,n}$ [9, 79, 162]. Some quasi-isometric embeddings of curve complexes are constructed in [147], which lead to some new quasi-isometric embeddings between mapping class groups.

Remark 7.3. A closely related result on the simplicial automorphism group of an arc and curve complex of a surface is proved in [106]. The automorphism group of a pants complex of a surface is determined in [122].

Remark 7.4. The paper [17] studies the automorphism group $\mathrm{Aut}(\hat{\mathcal{C}}(S_{g,n}))$ of the profinite completion of the complex of curves $\mathcal{C}(S_{g,n})$ and compares it with the Grothendieck–Teichmüller group. It also shows that $\mathrm{Aut}(\hat{\mathcal{C}}(S_{g,n}))$ coincides (apart from two low dimensional exceptions) with the automorphism group of $\hat{\mathrm{Mod}}_{g,n}$, where $\hat{\mathrm{Mod}}_{g,n}$ is the profinite completion of the extended mapping class group $\mathrm{Mod}_{g,n}$.

7.2 Isometry Groups of Teichmüller Spaces

Motivated by the proof of Mostow strong rigidity for irreducible compact locally symmetric spaces of rank at least 2 via rigidity of Tits buildings, Ivanov [82] gave a new proof of the following theorem of Royden [154] by using the rigidity of $\mathcal{C}(S_{g,n})$ in Proposition 7.1.

Proposition 7.5. *The Teichmüller metric d_{Tei} of the Teichmüller space $\mathcal{T}_{g,n}$ is invariant under the natural action of the extended mapping class group $\mathrm{Mod}_{g,n}^{\pm}$. If $g \geq 2$, or $g = 1$ and $n \geq 3$, or $g = 0$ and $n \geq 5$, then $\mathrm{Mod}_{g,n}^{\pm}$ is equal to the full isometry group $\mathrm{Iso}(\mathcal{T}_{g,n}, d_{\mathrm{Tei}})$.*

The Teichmüller metric is historically the first metric defined on $\mathcal{T}_{g,n}$. It is a complete Finsler metric and gives the natural topology on $\mathcal{T}_{g,n}$. The Teichmüller space $\mathcal{T}_{g,n}$ also admits another metric, the Weil–Petersson metric d_{WP}. It is an incomplete Kähler metric. Its isometry group was identified in [128].

Proposition 7.6. *For any Teichmüller space $\mathcal{T}_{g,n}$, the extended mapping class group $\mathrm{Mod}_{g,n}^{\pm}$ is canonically mapped into the isometry group $\mathrm{Iso}(\mathcal{T}_{g,n}, d_{\mathrm{WP}})$. If $g \geq 2$, or $g = 1$ and $n \geq 3$, or $g = 0$ and $n \geq 5$, then $\mathrm{Mod}_{g,n}^{\pm}$ is equal to the full isometry group $\mathrm{Iso}(\mathcal{T}_{g,n}, d_{\mathrm{WP}})$.*

The idea of the proof, simplified in [174], is similar to the previous proposition and can be explained as follows. The completion of $\mathcal{T}_{g,n}$ in the Weil–Petersson metric is the augmented Teichmüller space $\hat{\mathcal{T}}_{g,n}$. The boundary components of $\hat{\mathcal{T}}_{g,n}$ are Teichmüller spaces of stable Riemann surfaces of smaller genus with more punctures (the same Euler characteristic). These boundary components are parametrized by simplices of $\mathcal{C}(S_{g,n})$. Since any isometry of $(\mathcal{T}_{g,n}, d_{\mathrm{WP}})$ extends to its completion, it induces an automorphism of $\mathcal{C}(S_{g,n})$. Proposition 7.1 implies that the isometry is induced by an element of $\mathrm{Mod}_{g,n}$.

Remark 7.7. Teichmüller spaces $\mathcal{T}_{g,n}$ are the counterpart of symmetric spaces in the analogy between Tits buildings and arithmetic groups on the one hand, and curve complexes and mapping class groups on the other hand. The above proposition shows one difference. Any symmetric space is a homogeneous space and has continuous symmetry, but $\mathcal{T}_{g,n}$ has only discrete symmetry, which implies that there is a uniform lower bound on volumes of $\Gamma \backslash \mathcal{T}_{g,n}$. Maybe one result to repair this difference is that for any symmetric space X, there is a uniform lower bound on volumes of all locally symmetric spaces $\Gamma \backslash X$ which are quotients of X.

7.3 Ending Lamination Conjecture of Thurston

Another generalization of Mostow strong rigidity is the ending lamination conjecture of Thurston.

A special case of Mostow strong rigidity concerns 3-dimensional hyperbolic manifolds of finite volume. Let $\Gamma_1\backslash\mathbf{H}^3$ and $\Gamma_2\backslash\mathbf{H}^3$ be two hyperbolic manifolds of finite volume. If $\Gamma_1 \cong \Gamma_2$, then $\Gamma_1\backslash\mathbf{H}^3$ and $\Gamma_2\backslash\mathbf{H}^3$ are isometric.

On the other hand, the conclusion does not hold if the volumes of $\Gamma_1\backslash\mathbf{H}^3$ and $\Gamma_2\backslash\mathbf{H}^3$ are not finite. For example, let $\Gamma \subset \mathrm{SL}(2,\mathbb{R})$ be a torsion-free cocompact Fuchsian group, i.e., a discrete subgroup such that $\Gamma\backslash\mathbf{H}^2$ is a compact hyperbolic surface Σ. Consider Γ as a discrete subgroup of $\mathrm{SL}(2,\mathbb{C})$. Then $\Gamma\backslash\mathbf{H}^3$ is a 3-dimensional hyperbolic manifold of infinite volume and is diffeomorphic to $\Sigma \times \mathbb{R}$. There are injective morphisms $\rho : \Gamma \to \mathrm{SL}(2,\mathbb{C})$ such that $\rho(\Gamma)\backslash\mathbf{H}^3$ is not isometric to $\Gamma\backslash\mathbf{H}^3$, but is still diffeomorphic to $\Sigma\times\mathbb{R}$. A deformation $\rho(\Gamma)$ of Γ is called a *quasi-Fuchsian deformation* if $\rho(\Gamma)$ is a quasi-Fuchsian group in the sense that the limit set of $\rho(\Gamma)$ in $\mathbf{H}^3(\infty)$ is a Jordan curve and the complement of the limit set consists of two simply connected domains Ω_1 and Ω_2. Many quasi-Fuchsian deformations of Γ have the property that $\rho(\Gamma)\backslash\mathbf{H}^3$ is not isometric to $\Gamma\backslash\mathbf{H}^3$.

In this case, clearly Γ and $\rho(\Gamma)$ are isomorphic. Therefore, a direct generalization of Mostow strong rigidity does not hold for infinite volume hyperbolic manifolds of dimension 3, i.e., if $\Gamma_1\backslash\mathbf{H}^3$ and $\Gamma_2\backslash\mathbf{H}^3$ are both of infinite volume, then $\Gamma_1 \cong \Gamma_2$ does not imply that $\Gamma_1\backslash\mathbf{H}^3$ and $\Gamma_2\backslash\mathbf{H}^3$ are isometric. (Note that by [119, Theorem 8.3], if one of the manifolds $\Gamma_1\backslash\mathbf{H}^3$ and $\Gamma_2\backslash\mathbf{H}^3$ has finite volume and $\Gamma_1 \cong \Gamma_2$, then the other manifold also has finite volume and hence $\Gamma_1\backslash\mathbf{H}^3$ and $\Gamma_2\backslash\mathbf{H}^3$ are isometric.)

For each quasi-Fuchsian deformation $\rho(\Gamma)$, the noncompact infinite volume 3-dimensional hyperbolic manifold $\rho(\Gamma)\backslash\mathbf{H}^3$ has two ends[4] and admits a natural compactification $\overline{\rho(\Gamma)\backslash\mathbf{H}^3}$ by adding the two Riemann surfaces $\rho(\Gamma)\backslash\Omega_1$ and $\rho(\Gamma)\backslash\Omega_2$ as the boundary. Therefore, the Riemann surfaces $\rho(\Gamma)\backslash\Omega_1$ and $\rho(\Gamma)\backslash\Omega_2$ are called the *Riemann surfaces at infinity* of the ends of $\rho(\Gamma)\backslash\mathbf{H}^3$.

The following fact establishes a connection between the Riemann surfaces at infinity and the geometry of the interior.

Proposition 7.8. *Two quasi-Fuchsian deformations $\rho_1(\Gamma)$ and $\rho_2(\Gamma)$ of Γ give rise to isometric 3-dimensional hyperbolic manifolds $\rho_1(\Gamma)\backslash\mathbf{H}^3$ and $\rho_2(\Gamma)\backslash\mathbf{H}^3$ if and only if they have the same Riemann surfaces at infinity of the ends. Furthermore, all possible conformal structures on the Riemann surfaces at infinity can arise.*

This is a proper generalization of Mostow strong rigidity for the class of quasi-Fuchsian hyperbolic manifolds, and it gives a complete classification of quasi-Fuchsian hyperbolic manifolds of dimension 3. See [12] [121, Chap. 5] for more details.

Quasi-Fuchsian deformations are important examples of geometrically finite hyperbolic manifolds. Recall that a three-dimensional hyperbolic manifold $\Gamma\backslash\mathbf{H}^3$

---

[4]By an end of a 3-dimensonal manifold, we mean a connected unbounded component of the complement of a sufficiently large compact subset. Strictly speaking, an end is a limit of such unbounded connected components over an exhausting family of compact subspaces. For a precise definition, see [159].

is called *geometrically finite* if Γ admits a finite-sided convex fundamental domain (see [26] for discussion of several equivalent definitions), which is also equivalent to the fact that $\Gamma\backslash\mathbf{H}^3 \cup \Omega(\Gamma)$ is homeomorphic to $M - P$ where M is a compact 3-manifold with boundary and P is a finite collection of disjoint annuli and tori in ∂M, and $\Omega(\Gamma)$ is the largest open subset of $\mathbf{H}^3(\infty)$ where Γ acts properly [43, p. 139].

A similar rigidity result and classification of hyperbolic metrics holds for geometrically finite hyperbolic manifolds M of dimension 3. Roughly speaking, for each end of M, there is a Riemann surface at infinity. Such a collection of Riemann surfaces at infinity together with the topology of the manifold uniquely determines the hyperbolic manifold [12, 120] [43, Sect. 7.3] [119, Theorem 8.1] [121, Chap. 5] [132, p. 188]. We note that the topology of a quasi-Fuchsian hyperbolic manifold is determined by any of the two Riemann surfaces at infinity, and hence the condition on the topology of the manifold is contained in the Riemann surfaces at infinity.

For a finitely generated torsion-free subgroup $\Gamma \subset \mathrm{SL}(2,\mathbb{C})$, the hyperbolic manifold $\Gamma\backslash\mathbf{H}^3$ has only finitely many ends. This case is more complicated than the geometrically finite case. For simplicity, in the following discussion, we assume that Γ *does not contain any nontrivial parabolic element* (see [121, Sect. 5.5] for the complication caused by the presence of parabolic elements.)

A three-dimensional hyperbolic manifold $M = \Gamma\backslash\mathbf{H}^3$ is called *geometrically tame* if each of its ends is either geometrically finite or simply degenerate (defined below). It is called *topologically tame* if M is homeomorphic to the interior of a compact manifold with boundary. It is known that if Γ is finitely generated, then $\Gamma\backslash\mathbf{H}^3$ is topologically tame if and only if it is geometrically tame [18, 41]. Therefore, the solution of Marden's tameness conjecture implies that $\Gamma\backslash\mathbf{H}^3$ is geometrically tame. See [42] for a precise description of these two notions of tameness and a history of Marden's tameness conjecture.

Thurston associated an invariant to an incompressible end of $\Gamma\backslash\mathbf{H}^3$, and an end invariant for a general end was defined in [41] (see also [18]).

If an end is geometrically finite and of infinite volume, then there is a Riemann surface at infinity which gives the end invariant. For a geometrically finite end of finite volume, there is no invariant.

For a simply degenerate end, there is a filling lamination on an associated surface, which is an invariant of the end. For simplicity, we assume that $\Gamma\backslash\mathbf{H}^3$ has no cuspidal ends. Then for each end, there is a surface $S_{g,n}$ such that the end is homeomorphic to $S_{g,n} \times (0, +\infty)$. For every simple closed curve α in $S_{g,n}$, there is a well-defined homotopy class of simple closed curves in the end of $\Gamma\backslash\mathbf{H}^3$ and hence in $\Gamma\backslash\mathbf{H}^3$. Let α^* be the unique geodesic representative in $\Gamma\backslash\mathbf{H}^3$. Then the end is called *simply degenerate* if there is a sequence of simple closed curves α_i on $S_{g,n}$ such that their geodesic representatives α_i^* exit the end, i.e., they are eventually contained in any small neighborhood of the end.

The family of curves α_i in $S_{g,n}$ converges to a lamination, called the *ending lamination* of the end. It can also be interpreted through the curve complex of a surface in the end. The sequence of simple closed curves α_i gives a sequence of vertices in the curve complex $\mathcal{C}(S_{g,n})$. It is known that they converge to a boundary point $\partial\mathcal{C}(S_{g,n})$, which is a filling lamination on $S_{g,n}$ by Proposition 5.6.

Thurston made the following *Ending lamination conjecture.*

Conjecture 7.9. A three dimensional hyperbolic manifold $\Gamma\backslash\mathbf{H}^3$ with finitely generated fundamental group Γ is determined by its topological type and its end invariants.

If $\Gamma\backslash\mathbf{H}^3$ has finite volume, then each end has finite volume and is a cusp, the end invariant is trivial, and the space $\Gamma\backslash\mathbf{H}^3$ is determined by its topology, i.e., Γ as the fundamental group, up to isometry by Mostow strong rigidity.

This ending lamination conjecture has been proved by Brock–Canary–Minsky (see [34, 130, 131]). Slightly different proofs have also be given. See [27, 28]. A crucial step in the proof is to build a bi-Lipschitz model of each of the ends of $\Gamma\backslash\mathbf{H}^3$. (In [27, 28], a weaker Lipschitz model of each end is needed and constructed.) The large scale geometry of the curve complex $\mathcal{C}(S_{g,n})$ and its connection with the Teichmüller metric of $\mathcal{T}_{g,n}$, in particular results in [126, 127], were used crucially for this purpose. (It seems that the papers [126, 127] were motivated by the ending lamination conjecture.) One result in [126] (see [64, Theorem 4.1]) says that when $\mathcal{T}_{g,n}$ is given the Teichmüller metric d_{Tei}, there is a $\mathrm{Mod}_{g,n}$-coarsely-equivariant map $\psi : (\mathcal{T}_{g,n}, d_{\mathrm{Tei}}) \to \mathcal{C}(S_{g,n})$ which is quasi-Lipschitz in the sense:

1. There exists a constant $c > 1$ such that for every point $x \in \mathcal{T}_{g,n}$ and $\gamma \in \mathrm{Mod}_{g,n}$,

$$d(\psi(\gamma x), \gamma\psi(x)) \leq c,$$

2. For any two points $x, x' \in \mathcal{T}_{g,n}$,

$$d(\psi(x), \psi(x')) \leq c\, d_{\mathrm{Tei}}(x, x') + c.$$

This is an important instance of the philosophy that $\mathcal{C}(S_{g,n})$ describes the large scale geometry of $\mathcal{T}_{g,n}$. For detailed descriptions of the ending lamination conjecture and methods to prove it, see [27, 132–134].

Remark 7.10. Besides giving a complete classification of 3-dimensional hyperbolic manifolds, the ending lamination conjecture has applications to the local connectivity of limit sets of Kleinian groups and the Cannon–Thurston map. See [135].

Remark 7.11. See [146] for related results on a model for the Teichmüller metric in terms of combinatorial information on short curves on the hyperbolic surfaces that correspond to points on a Teichmüller geodesic. The paper [102] compares the Thurston boundary of $\mathcal{T}_{g,n}$ and the boundary $\partial\mathcal{C}(S_{g,n})$.

7.4 Quasi-Isometric Rigidity of Mapping Class Groups

In this subsection, we discuss another important application of curve complexes to quasi-isometric rigidity of $\mathrm{Mod}_{g,n}$. As mentioned before, the notion of quasi-isometry was used in the proof of Mostow strong rigidity.

Recall that two metric spaces (M_1, d_1) and (M_2, d_2) are called *quasi-isometric* if there are positive constants C, D and a map $f : M_1 \to M_2$ such that for every two points $p, q \in M_1$,

$$C^{-1}d_2(f(x_1), f(x_2)) - D \le d_1(x_1, x_2) \le Cd_2(f(x_1), f(x_2)) + D,$$

and every point of M_2 lies in the D-neighborhood of $f(M_1)$.

It is clear that if a finitely generated group Γ acts properly and isometrically on a metric space X with a compact quotient, then Γ with any word metric is quasi-isometric to X. The quasi-isometry type of the group Γ does not depend on the choice of a word metric.

A finitely generated group Γ is called *quasi-isometrically rigid* if the following statement holds: for any finitely generated group Γ', if Γ' is quasi-isometric to Γ, then there exists a finite-index subgroup $\Gamma'' \subset \Gamma'$ and a homomorphism $\Gamma'' \to \Gamma$ with finite kernel and finite cokernel, i.e., Γ'' and Γ are virtually isomorphic.

The three propositions below are combinations of results in several papers including [47–49, 54, 104, 160, 161]. See [50] for the history, references and more detailed statements of these propositions. There are also other rigidity results on mapping class groups. See [80].

Proposition 7.12. *If $\Gamma \subset G$ is an irreducible* **non-uniform** *lattice of a semisimple Lie group G, then Γ is quasi-isometrically rigid.*

The assumption that Γ is not uniform is necessary. All uniform lattices of one semisimple Lie group G are quasi-isometric, but they are not necessarily commensurable up to conjugation by elements of G, and hence they are not necessarily isomorphic up to finite index subgroups and quotients. On the other hand, this class of uniform lattices is quasi-isometrically rigid.

Proposition 7.13. *If a finitely generated group Γ' is quasi-isometric to a uniform lattice in a semisimple Lie group G, then there exists a finite index subgroup Γ'' and a homomorphism $\rho : \Gamma'' \to G$ with a finite kernel such that its image is a uniform lattice in G.*

Proposition 7.14. *Non-uniform lattices of* $\mathrm{SL}(2, \mathbb{R})$ *form one quasi-isometry class of groups.*

Quasi-isometric rigidity and quasi-isometric classification of lattices of semisimple Lie groups were obtained by joint efforts of many people. See [50] for the history and more detailed statements of the above three propositions.

In the analogy between the mapping class groups $\mathrm{Mod}_{g,n}$ and lattices of semisimple Lie groups, $\mathrm{Mod}_{g,n}$ correspond to non-uniform lattices. Therefore, the next proposition on the quasi-isometrical rigidity of the mapping class group $\mathrm{Mod}_{g,n}$, proved in [6, 65], is expected.

Proposition 7.15. *Assume that* $3g-3+n \geq 2$. *Then* $\mathrm{Mod}_{g,n}$ *is quasi-isometrically rigid.*

One ingredient used in the proof in [6] is hyperbolicity of the curve complex $\mathcal{C}(S_{g,n})$ in the sense of Gromov, and one ingredient used in the proof of [65] is the train track complex. Results on the asymptotic cones at infinity $\mathrm{Cone}_\infty(\mathrm{Mod}_{g,n})$ in [7] (see Proposition 7.42 below) were used crucially.

The papers [6, 65] also prove the following result on quasi-isometries of $\mathrm{Mod}_{g,n}$.

Proposition 7.16. *Assume that* $3g - 3 + n \geq 2$ *and* $(g, n) \neq (1, 2)$, *then quasi-isometries of* $\mathrm{Mod}_{g,n}$ *are uniformly close to isometries induced by left-multiplication of elements of* $\mathrm{Mod}_{g,n}$.

7.5 Finite Asymptotic Dimension of Mapping Class Groups and Novikov Conjectures

In this subsection, we briefly describe another important application of the hyperbolicity of the curve complex $\mathcal{C}(S_{g,n})$ to prove that the asymptotic dimension of the mapping class group $\mathrm{Mod}_{g,n}$ is finite. We also explain its applications to several different versions of the Novikov conjecture.

For any noncompact metric space (M, d), an important large scale geometry invariant is the asymptotic dimension of M, denoted by $\mathrm{asd}(M)$. It is defined to be the smallest integer n, which could be ∞, such that for every $r > 0$, there exists a cover $\mathcal{C} = \{U_i\}, i \in I$, of M by uniformly bounded sets U_i with r-multiplicity less than or equal to $n + 1$, i.e., every ball in M of radius r intersects at most $n + 1$ sets in $\mathcal{C}$.

For any finitely generated group Γ, its asymptotic dimension $\mathrm{asd}(\Gamma)$ is defined to the asymptotic dimension of Γ endowed with any word metric.

Finiteness of $\mathrm{asd}(\Gamma)$ has applications to the Novikov conjectures for Γ, including various versions of the integral Novikov conjectures. For the convenience of the reader, we briefly recall several versions of the Novikov conjectures, both the integral and rational versions. See [92] for more discussions, details and proper references for the conjectures and results stated below.

To motivate the Novikov conjectures, we first recall the Hirzebruch index theorem. Let M^{4k} be a compact oriented manifold (without boundary) of dimension $4k$. The cup product defines a non-degenerate quadratic form on the middle dimension cohomology group:

$$Q : H^{2k}(M, \mathbb{Q}) \times H^{2k}(M, \mathbb{Q}) \to H^{4k}(M, \mathbb{Q}) = \mathbb{Q}. \tag{1}$$

This quadratic form can be diagonalized over $\mathbb{R}$ to the form Diag$(1, \cdots, 1; -1, \cdots, -1)$, and the number of $+1$'s minus the number of -1's is called the signature of M and denoted by Sgn(M). Since the identification $H^{4k}(M, \mathbb{Q}) = \mathbb{Q}$ depends on the orientation of M, the signature Sgn(M) depends on the orientation and is an oriented homotopy invariant of M.

The Hirzebruch class $\mathcal{L}(M)$ is a power series in Pontrjagin classes $P_1, P_2, \cdots$, with rational coefficients,

$$\mathcal{L}(M) = 1 + L_1 + L_2 + \cdots,$$

where $L_1 = \frac{1}{3}P_1, \quad L_2 = \frac{1}{45}(7P_2 - P_1^2), \cdots$.

Then the *Hirzebruch index theorem* is the following equality:

$$\mathrm{Sgn}(M) = \langle \mathcal{L}(M), [M] \rangle, \tag{2}$$

where the right hand side is the evaluation of $\mathcal{L}(M)$ on the fundamental class $[M]$.

The Hirzebruch class $\mathcal{L}(M)$ depends on the characteristic classes of the tangent bundle of M and a priori depends on the differential structure of M. (In fact, these rational classes in $H^*(M, \mathbb{Q})$ are homeomorphism invariants of M). As pointed out earlier, the left-hand side is an oriented homotopy invariant, and hence the above equality shows that $\langle \mathcal{L}(M), [M] \rangle$ only depends on the oriented homotopy type of M.

To get more homotopy invariants, Novikov introduced the higher signatures. Let $\Gamma = \pi_1(M)$. Let $B\Gamma$ be a classifying space of the discrete group Γ, i.e., a $K(\Gamma, 1)$-space,

$$\pi_1(B\Gamma) = \Gamma, \quad \pi_i(B\Gamma) = \{1\}, \quad i \geq 2.$$

The universal covering space $E\Gamma$ of $B\Gamma$ is contractible and admits a free Γ-action. Equivalently, we can reverse this process and define first $E\Gamma$ as a contractible space with a free Γ-action, and then define $B\Gamma$ as the quotient $\Gamma \backslash E\Gamma$. For example, when $\Gamma = \mathbb{Z}$, it acts freely by translation on $\mathbb{R}$ and hence $E\Gamma = \mathbb{R}$ and $B\Gamma = \mathbb{R}/\mathbb{Z} = S^1$.

For each group Γ, the spaces $E\Gamma$ and $B\Gamma$ are unique up to homotopy. The universal covering map $\tilde{M} \to M$ determines a classifying map $f : M \to B\Gamma$, which is unique up to homotopy.

For any $\alpha \in H^*(B\Gamma, \mathbb{Q})$, $f^*\alpha \in H^*(M, \mathbb{Q})$, and define a *higher signature*

$$Sgn_\alpha(M) = \langle f^*\alpha \cup \mathcal{L}(M), [M] \rangle. \tag{3}$$

The original Novikov conjecture is stated as follows:

Conjecture 7.17 (Novikov conjecture). For any $\alpha \in H^*(B\Gamma, \mathbb{Q})$, the higher signature $Sgn_\alpha(M)$ is an oriented homotopy invariant of M, i.e., if N is another oriented manifold and $g: N \to M$ is an orientation preserving homotopy equivalence, then

$$\langle (g \circ f)^*\alpha \cup \mathcal{L}(N), [N] \rangle = \langle f^*\alpha \cup \mathcal{L}(M), [M] \rangle.$$

The Novikov conjecture can be reformulated in terms of rational injectivity of the assembly map in surgery theory, or L-theory.

The surgery obstruction groups $L_*(\mathbb{Z}[\Gamma])$, or L-groups of $\mathbb{Z}[\Gamma]$, are briefly defined as follows. For $m = 2k$, $L_m(\mathbb{Z}[\Gamma])$ is the Witt group of stable isomorphism classes of $(-1)^k$-quadratic forms on finitely generated free modules over the group ring $\mathbb{Z}[\Gamma]$, and $L_{2k+1}(\mathbb{Z}[\Gamma])$ is a stable automorphism group of hyperbolic $(-1)^k$-quadratic forms on finitely generated free modules over $\mathbb{Z}[\Gamma]$. Since $(-1)^k$ is 4-periodic in m, the groups $L_m(\mathbb{Z}[\Gamma])$ are 4-periodic in m.

Let $\mathbb{L}(\mathbb{Z})$ be the surgery spectrum:

$$\pi_m(\mathbb{L}(\mathbb{Z})) = L_m(\mathbb{Z}), \quad m \in \mathbb{Z}.$$

The spectrum $\mathbb{L}(\mathbb{Z})$ defines a general homology theory with coefficient in $\mathbb{L}(\mathbb{Z})$. For any topological space X, there are general homology groups $H_*(X;\mathbb{L}(\mathbb{Z})) = \pi_*(X_+ \wedge \mathbb{L}(\mathbb{Z}))$, where X_+ is the disjoint union of X and a point.

There is an important notion of assembly map:

$$A : H_*(X;\mathbb{L}(\mathbb{Z})) \to L_*(\mathbb{Z}[\pi_1(X)]). \tag{4}$$

Proposition 7.18. *The Novikov conjecture, i.e., the oriented homotopy invariance of the higher signatures in Conjecture 7.17, is equivalent to the rational injectivity of the assembly map in (4), i.e., the following rational assembly map is injective:*

$$A \otimes \mathbb{Q} : H_*(B\Gamma;\mathbb{L}(\mathbb{Z})) \otimes \mathbb{Q} \to L_*(\mathbb{Z}[\Gamma]) \otimes \mathbb{Q}.$$

The injectivity of the map $A \otimes \mathbb{Q}$ is called *the rational Novikov conjecture.*

Conjecture 7.19 (Integral Novikov conjecture). If Γ is torsion free, then the assembly map $A : H_*(B\Gamma;\mathbb{L}(\mathbb{Z})) \to L_*(\mathbb{Z}[\Gamma])$ is injective.

This conjecture is also called *the L-theory (or surgery theory) integral Novikov conjecture.*

In the integral Novikov conjecture, the torsion free assumption on Γ is important. In fact, it is known that the conjecture is often false for finite groups.

Clearly, the integral Novikov conjecture implies the rational Novikov conjecture and gives an integral version of homotopy invariance of higher signature. There are also several other reasons to consider the integral, rather than the original (rational) Novikov conjecture:

1. Relation to the rigidity of manifolds, in particular, the Borel conjecture for rigidity of aspherical manifolds, which says that two closed aspherical manifolds with the same fundamental group are homeomorphic.
2. Computation of the L-groups $L_*(\mathbb{Z}[\Gamma])$ in terms of a generalized homology theory, i.e., the injectivity of the assembly map in (4) shows that the left-hand side is a summand of the groups $L_*(\mathbb{Z}[\Gamma])$.

Once formulated in terms of the assembly map, there are also other versions of the Novikov conjecture. For any associative ring with unit R, there is a family of algebraic K-groups $K_i(R)$, $i \in \mathbb{Z}$. For example, $K_0(R)$ is defined to be stable equivalence classes of finitely generated projective modules, and $K_1(R) = GL(R)/[GL(R), GL(R)]$. The higher K-groups $K_i(R)$, $i \geq 2$, are defined to be the homotopy groups of the space $BGL(R)^+$, where $BGL(R)$ is the classifying space of $GL(R)$ considered as a discrete group, and $BGL(R)^+$ is the space obtained by applying the Quillen +-construction to the perfect subgroup $E(R) = [GL(R), GL(R)]$, in particular, the homology groups of $BGL(R)^+$ and $BGL(R)$ are equal to each other under inclusion. The K-theory spectrum $\mathbb{K}(R)$ with $\pi_i(\mathbb{K}(R)) = K_i(R), i \in \mathbb{Z}$, is given by the delooping of the infinite loop space $BGL(R)^+ \times K_0(R)$.

Let Γ be a group as above, and $H_*(B\Gamma; \mathbb{K}(\mathbb{Z}))$ the generalized homology of $B\Gamma$ with coefficients in $\mathbb{K}(R)$. There is also an assembly map

$$A : H_*(B\Gamma; \mathbb{K}(\mathbb{Z})) \to K_*(\mathbb{Z}[\Gamma]). \tag{5}$$

Conjecture 7.20 (Integral Novikov conjecture in algebraic K-theory). Assume that Γ is torsion free. Then the assembly map

$$A : H_*(B\Gamma; \mathbb{K}(\mathbb{Z})) \to K_*(\mathbb{Z}[\Gamma]) \tag{6}$$

is injective.

There is also a rational version of the Novikov conjecture in algebraic K-theory, i.e., the induced map $A \otimes \mathbb{Q} : H_*(B\Gamma; \mathbb{K}(\mathbb{Z})) \otimes \mathbb{Q} \to K_*(\mathbb{Z}[\Gamma]) \otimes \mathbb{Q}$ is injective.

One fruitful approach to prove the Novikov conjecture for a group Γ is to show that the asymptotic dimension of Γ is finite.

Theorem 7.21. *If a finitely generated group Γ has finite asymptotic dimension,* asdim $\Gamma < \infty$, *and has finite $B\Gamma$, i.e., its classifying space $B\Gamma$ can be realized as a finite CW-complex, then the integral Novikov conjectures in K-theory and L-theory hold for Γ.*

We note that the existence of a finite model of $B\Gamma$ implies that Γ is torsion-free. For groups containing torsion elements, there is also a modified version of the integral Novikov conjecture. See [92].

In [14], the following result was proved.

Proposition 7.22. *The asymptotic dimension of* $\mathrm{Mod}_{g,n}$ *is finite.*

For any finite index torsion-free subgroup Γ of $\mathrm{Mod}_{g,n}$, the quotient by Γ of the thick part of the Teichmüller space $\mathcal{T}_{g,n}(\varepsilon)$ gives a cofinite model of $B\Gamma$-spaces by [95]. One corollary of the above proposition is the following conclusion.

Corollary 7.23. *For any finite index subgroup Γ of* $\mathrm{Mod}_{g,n}$, *the rational Novikov conjecture in L-theory holds for Γ.*

This result was proved earlier in [62, 100]. Even though it was not stated in [14], the above discussion shows that the following stronger corollary also holds.

Proposition 7.24. *For any finite index torsion-free subgroup Γ of* $\mathrm{Mod}_{g,n}$, *the integral Novikov conjecture in L- and K-theories holds for Γ.*

This shows that the results of [14] imply a stronger version of the Novikov conjecture than what was known previously.

7.6 Non-hyperbolicity of Weil–Petersson Metric of Teichmüller Space

As mentioned before, the Teichmüller space $\mathcal{T}_{g,n}$ admits a $\mathrm{Mod}_{g,n}$-equivariant Kähler metric, the Weil–Petersson metric. It is an incomplete metric with strictly negative sectional curvature. It is known that if any simply connected Riemannian manifold with strictly negative sectional curvature admits a compact quotient, then it is a hyperbolic space in the sense of Gromov. The quotient $\mathrm{Mod}_{g,n}\backslash\mathcal{T}_{g,n}$ is non-compact, and one question raised by Bowditch was whether the Weil–Petersson metric is hyperbolic in the sense of Gromov.

This problem was solved in [32], and the hyperbolicity of the curve complex $\mathcal{C}(S_{g,n})$ was used in the proof.

Proposition 7.25. *The Weil–Petersson metric on $\mathcal{T}_{g,n}$ is Gromov-hyperbolic if and only if $3g-3+n \leq 2$.*

Besides the curve complex $\mathcal{C}(S_{g,n})$, the pants complex and its quasi-isometry with $\mathcal{T}_{g,n}$ endowed with the Weil–Peterson metric was also used crucially in the proof. See [32] for details. For another application of the curve complex to Weil-Petersson geodesics, see [35].

7.7 Heegaard Splittings and Hempel Distance of 3-Manifolds

Another important application of the curve complex $\mathcal{C}(S_g)$ in 3-dimensional topology is the Hempel distance for Heegaard splittings of 3-dimensional manifolds.

A handlebody is a submanifold with boundary of $\mathbb{R}^3$ with only one connected boundary component, which is a compact orientable surface S_g.

Any closed oriented three dimensional manifold M^3 can be written as the union of two handlebodies V_1, V_2 of the same genus glued along their boundary,

$$M = V_1 \cup_{S_g} V_2, \quad \partial V_1 \cong S_g, \quad \partial V_2 \cong S_g.$$

Such a decomposition of M^3 is called a *Heegaard splitting*, and S_g is called the *splitting surface*. It is not unique and there are infinitely many different isotropy classes of Heegaard splittings of M.

A simple closed curve in S_g is called *essential* if it does not bound a disc. The identification of S_g with the boundary of a handlebody V_1 is determined by a maximal set Δ_1 of disjoint non-homotopic essential simple closed curves in S_g that bound essential discs (or meridians) in the handlebody V_1. Let Δ_2 be a corresponding set for the identification of S_g as the boundary of V_2. The pair (Δ_1, Δ_2) is called a *Heegaard diagram* and corresponds to a pair of simplices in $\mathcal{C}(S_g)$.

For each Heegaard splitting $V_1 \cup_{S_g} V_2$ with S_g as the splitting surface, the set of all possible simplices Δ_1 and their faces form a subcomplex K_1 of $\mathcal{C}(S_g)$ (in other words, the subcomplex spanned by the vertices corresponding to all essential simple closed curves of S_g that bound some meridians of the handlebody V_1). Similarly, there is a subcomplex K_2 of $\mathcal{C}(S_g)$ for the other handlebody V_2.

The Hempel distance for the Heegaard splitting $V_1 \cup_{S_g} V_2$ of M^3 is the distance $d(K_1, K_2)$ between K_1 and K_2 with respect to the geodesic length function of $\mathcal{C}(S_g)$.

This defines a numerical invariant of the Heegaard splitting of M and is closely related to topological properties of the manifold M and of the Heegaard splitting. For detail, see [77]. The paper [137] discusses behavior of the Hempel distance under stabilization of splitting.

Remark 7.26. In [96], the curve complex $\mathcal{C}(S_{g,n})$ is replaced by the pants complex of $S_{g,n}$ and a similar distance function is introduced. [97] contains some results relating the bridge number of hyperbolic knots and an invariant defined in terms of the distance function on $\mathcal{C}(S_{g,n})$. The paper [155] applies the idea of Hempel distance to knots and defines a distance for certain knots in lens spaces.

Remark 7.27. The paper [68] shows that for a closed orientable 3-manifold containing an incompressible surface of genus g, any Heegaard splitting has Hempel distance at most $2g$. See also [112, 157, 158, 168] for related results on the Hempel distance.

Remark 7.28. Based on the Hempel distance, the paper [156] defines a translation distance of open book decompositions of three dimensional manifolds, and relates it to the topological properties of the manifolds.

7.8 *Partial Compactifications of Teichmüller Spaces and Their Boundaries*

Recall that $\mathcal{T}_{g,n}$ is the Teichmüller space of marked complex structures on $S_{g,n}$, where a marking is a choice of a set of generators of $S_{g,n}$. Then the mapping class group $\mathrm{Mod}_{g,n}$ acts on $\mathcal{T}_{g,n}$ by changing the markings, and the quotient $\mathrm{Mod}_{g,n}\backslash\mathcal{T}_{g,n}$ is the moduli space $\mathcal{M}_{g,n}$ of Riemann surfaces of genus g with n punctures.

It is known that $\mathcal{T}_{g,n}$ is a complex manifold of complex dimension $3g - 3 + n$ and diffeomorphic to $\mathbb{R}^{6g-6+2n}$, and $\mathrm{Mod}_{g,n}$ acts holomorphically and properly on $\mathcal{T}_{g,n}$. In particular, $\mathcal{M}_{g,n}$ has a natural structure of a complex orbifold.

Proposition 7.29. *For any torsion-free subgroup $\Gamma \subset \mathrm{Mod}_{g,n}$, $\Gamma\backslash\mathcal{T}_{g,n}$ is a $B\Gamma$-space.*

Proof. Since Γ acts properly on $\mathcal{T}_{g,n}$ and is torsion-free, Γ acts fixed-point freely on $\mathcal{T}_{g,n}$. Since $\mathcal{T}_{g,n}$ is contractible, it is an $E\Gamma$-space, and $\Gamma\backslash\mathcal{T}_{g,n}$ is a $B\Gamma$-space.

A more general result is true for subgroups of $\mathrm{Mod}_{g,n}$ containing torsion elements.

Proposition 7.30. *For any subgroup $\Gamma \subset \mathrm{Mod}_{g,n}$, $\mathcal{T}_{g,n}$ is an $\underline{E}\Gamma$-space.*

We only need to check that for any finite subgroup $F \subset \Gamma$, the set of fixed points $\mathcal{T}_{g,n}^F$ is nonempty and contractible. The former follows from the solution of the Nielsen realization problem, and the latter from either the existence of unique earthquake between any two points or the convexity of the Weil–Petersson metric. See [95] for more details.

It is known that $\mathcal{M}_{g,n} = \mathrm{Mod}_{g,n}\backslash\mathcal{T}_{g,n}$ is noncompact. Therefore, $\mathcal{T}_{g,n}$ is not a cofinite $\underline{E}\mathrm{Mod}_{g,n}$-space. To see this, we assume for simplicity that $2g - 2 + n > 0$. Then every Riemann surface in $\mathcal{M}_{g,n}$ admits a unique hyperbolic metric of finite area which is conformal to the complex structure. By pinching a simple closed geodesic on a hyperbolic surface $\Sigma_{g,n}$, we obtain a family of Riemann surfaces in $\mathcal{M}_{g,n}$ that has no limit.

For a non-uniform torsion-free arithmetic subgroup of a linear semisimple algebraic group $\mathbf{G}$, $\Gamma\backslash X$ is a non-cofinite $B\Gamma$-space. To obtain a cofinite $B\Gamma$-space, this problem was solved by the Borel–Serre compactification of $\Gamma\backslash X$, which is a quotient of the Borel–Serre partial compactification of the associated symmetric space $X = G/K$.

In order to construct an analogue of the Borel–Serre partial compactification of $\mathcal{T}_{g,n}$, Harvey [69, 70, 73] introduced the curve complex $\mathcal{C}(S_{g,n})$.

To explain the relation between partial compactifications of $\mathcal{T}_{g,n}$ and simplices of $\mathcal{C}(S_{g,n})$, we note the following compactness criterion of Mumford.

Proposition 7.31. *A sequence of Riemann surfaces Σ_j in $\mathcal{M}_{g,n}$ has no accumulation point in $\mathcal{M}_{g,n}$ if and only if there is a disjoint collection of simple closed geodesics $\gamma_{j,1}, \cdots, \gamma_{j,r_j}$ with respect to the hyperbolic metric of Σ_j whose lengths go to 0 as $j \to +\infty$, i.e., that can be simultaneously pinched.*

Note that in the above proposition, the geodesics must be disjoint since the collar theorem for hyperbolic surfaces implies that two sufficiently short geodesics are disjoint and hence simultaneously pinched geodesics must be disjoint.

Conversely, a collection of disjoint simple closed geodesics in a hyperbolic surface $\Sigma_{g,n}$ can be pinched simultaneously if and only if they are disjoint. This can be seen from pants decompositions of hyperbolic surfaces.

Given the above discussion, a natural compactification of $\mathcal{M}_{g,n}$ is obtained by adding degenerate Riemann surfaces. This is the Deligne–Mumford compactification $\overline{\mathcal{M}_{g,n}}^{\mathrm{DM}}$ of $\mathcal{M}_{g,n}$.

As a topological space, $\overline{\mathcal{M}_{g,n}}^{\mathrm{DM}}$ can be constructed as a quotient of the augmented Teichmüller space $\hat{\mathcal{T}}_{g,n}$. For every simplex σ of $\mathcal{C}(S_{g,n})$, there is a boundary Teichmüller space $\mathcal{T}_{g,n;\sigma}$, which is the Teichmüller space of the surface obtained from $S_{g,n}$ by cutting along the curves in σ. In terms of the hyperbolic metric, every homotopy class of curves in σ determines a unique simple closed geodesic in every marked hyperbolic surface in $\mathcal{T}_{g,n}$, and degenerate hyperbolic surfaces in the boundary $\mathcal{T}_{g,n;\sigma}$ are obtained by pinching *exactly* the geodesics in σ. Then the augmented Teichmüller space $\hat{\mathcal{T}}_{g,n}$ is

$$\hat{\mathcal{T}}_{g,n} = \mathcal{T}_{g,n} \cup \coprod_{\sigma \in \mathcal{C}(S_{g,n})} \mathcal{T}_{g,n;\sigma}.$$

The action of $\mathrm{Mod}_{g,n}$ on $\mathcal{T}_{g,n}$ extends to a continuous action on $\hat{\mathcal{T}}_{g,n}$ and the quotient $\mathrm{Mod}_{g,n}\backslash\hat{\mathcal{T}}_{g,n}$ is homeomorphic to $\overline{\mathcal{M}_{g,n}}^{\mathrm{DM}}$.

An important point here is that the curve complex $\mathcal{C}(S_{g,n})$ is a parameter space for the boundary components of the augmented Teichmüller space $\hat{\mathcal{T}}_{g,n}$ [69, 174]. Since each boundary Teichmüller space $\mathcal{T}_{g,n;\sigma}$ is contractible, the boundary $\partial\hat{\mathcal{T}}_{g,n}$ is homotopy equivalent to $\mathcal{C}(S_{g,n})$ and hence is homotopy equivalent to a bouquet of spheres.

Proposition 7.32. *The action of* $\mathrm{Mod}_{g,n}$ *on* $\hat{\mathcal{T}}_{g,n}$ *is not proper. For every point in a boundary Teichmüller space* $\mathcal{T}_{g,n;\sigma}$*, its stabilizer in* $\mathrm{Mod}_{g,n}$ *contains an infinite subgroup, which is generated by the Dehn twists along the curves in* σ *and is isomorphic to* $\mathbb{Z}^{\sigma}$*, as a subgroup of finite index.*

This implies that for any finite index subgroup $\Gamma \subset \mathrm{Mod}_{g,n}$, the inclusion $\Gamma\backslash\mathcal{T}_{g,n} \to \Gamma\backslash\hat{\mathcal{T}}_{g,n}$ is not a homotopy equivalence. This is similar to the situation with the reductive Borel–Serre compactification of $\Gamma\backslash X$ in Sect. 6.3.

To overcome the problem of infinite stabilizers $\mathbb{Z}^{\sigma}$ for points in $\mathcal{T}_{g,n;\sigma}$, we need to enlarge the boundary component to $\mathbb{R}^{\sigma}\times\mathcal{T}_{g,n;\sigma}$. The resulting space is a Borel–Serre partial compactification $\overline{\mathcal{T}_{g,n}}^{\mathrm{BS}}$ of $\mathcal{T}_{g,n}$ constructed in [70, 73, 83]:

$$\overline{\mathcal{T}_{g,n}}^{\mathrm{BS}} = \mathcal{T}_{g,n} \cup \coprod_{\sigma \in \mathcal{C}(S_{g,n})} \mathbb{R}^{\sigma} \times \mathcal{T}_{g,n;\sigma}.$$

Proposition 7.33. *The action of* $\mathrm{Mod}_{g,n}$ *on* $\mathcal{T}_{g,n}$ *extends to a continuous and proper action on* $\overline{\mathcal{T}_{g,n}}^{\mathrm{BS}}$ *with a compact quotient* $\mathrm{Mod}_{g,n}\backslash\overline{\mathcal{T}_{g,n}}^{\mathrm{BS}}$*. The boundary of the Borel–Serre partial compactification* $\overline{\mathcal{T}_{g,n}}^{\mathrm{BS}}$ *is homotopy equivariant to* $\mathcal{C}(S_{g,n})$ *and hence to a bouquet of spheres* S^d*, where* $d = 2g-2 = -\chi(S_g)$ *if* $n = 0$*,* $d = -\chi(S_{g,n})-1 = 2g-3+n$ *if* $g \geq 1$ *and* $n > 0$*, and* $d = -\chi(S_{0,n})-2 = n-4$ *if* $g = 0$*.*

Corollary 7.34. *For any finite index torsion-free subgroup $\Gamma \subset \mathrm{Mod}_{g,n}$, $\overline{\mathcal{T}_{g,n}}^{BS}$ is a cofinite $\underline{E}\Gamma$-space and the quotient $\Gamma\backslash\overline{\mathcal{T}_{g,n}}^{BS}$ is a finite $B\Gamma$-space.*

Proof. Since $\overline{\mathcal{T}_{g,n}}^{BS}$ is a manifold with corners whose interior is equal to $\mathcal{T}_{g,n}$, it is contractible. Since Γ is torsion-free, it acts fixed point freely on $\overline{\mathcal{T}_{g,n}}^{BS}$, and the quotient $\Gamma \subset \mathrm{Mod}_{g,n}$ is a compact manifold with corners. Therefore, the quotient $\Gamma\backslash\overline{\mathcal{T}_{g,n}}^{BS}$ is a finite $B\Gamma$-space.

On the other hand, for $\mathrm{Mod}_{g,n}$ or its finite index subgroups Γ which contain torsion elements, it is not clear whether $\overline{\mathcal{T}_{g,n}}^{BS}$ is an $\underline{E}\Gamma$-space.

To solve this problem, we can take suitable subspaces of $\mathcal{T}_{g,n}$ instead of a compactification. For any sufficiently small $\varepsilon > 0$, define the ε-thick part

$$\mathcal{T}_{g,n}(\varepsilon) = \{\Sigma \in \mathcal{T}_{g,n} \mid \text{for any simple closed geodesic } \gamma \subset \Sigma, \ell(\gamma) \geq \varepsilon\},$$

where $\ell(\gamma)$ is the length of the geodesic γ.

Proposition 7.35. *The thick part $\mathcal{T}_{g,n}(\varepsilon)$ is a submanifold with corners and invariant under the action of $\mathrm{Mod}_{g,n}$ with a compact quotient. The boundary faces of $\mathcal{T}_{g,n}(\varepsilon)$ are also parametrized by the simplices of $\mathcal{C}(S_{g,n})$ and contractible, and the boundary $\partial\mathcal{T}_{g,n}(\varepsilon)$ is homotopy equivalent to a wedge of spheres S^d, where $d = 2g-2 = -\chi(S_g)$ if $n = 0$, $d = -\chi(S_{g,n}) - 1 = 2g-3+n$ if $g \geq 1$ and $n > 0$, and $d = -\chi(S_{0,n}) - 2 = n-4$ if $g = 0$.*

In [95], we proved

Proposition 7.36. *There is a $\mathrm{Mod}_{g,n}$-equivariant deformation retraction of $\mathcal{T}_{g,n}$ to $\mathcal{T}_{g,n}(\varepsilon)$, and hence $\mathcal{T}_{g,n}(\varepsilon)$ is a cofinite $\underline{E}\mathrm{Mod}_{g,n}$-space.*

If $n > 0$, there is another equivariant deformation of $\mathcal{T}_{g,n}$ which is of the smallest possible dimension, i.e., the virtual cohomological dimension of $\mathrm{Mod}_{g,n}$ in [66].

Problem 7.37. When $n = 0$, construct a $\mathrm{Mod}_{g,0}$-deformation retract of $\mathcal{T}_{g,0}$ of dimension equal to the virtual cohomological dimension of Mod_g, which is equal to $4g-5$.

This problem seems to be completely open. It is not clear what subspaces of $\mathcal{T}_{g,n}$ are possible candidates for such a retract. It is not abvious if such a retract exists.

7.9 Cohomological Dimension and Duality Properties of Mapping Class Groups

As discussed earlier, one important application of the Borel–Serre partial compactification of a symmetric space concerns the virtual cohomological dimension and duality properties of arithmetic subgroups. The analogous partial compactifications of $\mathcal{T}_{g,n}$ in the previous subsection have similar applications.

Proposition 7.38. *When the Euler characteristic $\chi(S_{g,n})$ is positive,* $\mathrm{Mod}_{g,n}$ *is a virtual duality group of dimension d, where $d = 4g - 5 = \dim \mathcal{T}_g + \chi(S_g) - 1$ if $n = 0$, and $d = 4g - 4 + n = \dim \mathcal{T}_{g,n} + \chi(S_{g,n})$ if $g \geq 1$ and $n > 0$, and $d = n - 3 = \dim \mathcal{T}_{0,n} + \chi(S_{0,n}) + 1$. The dualizing module is equal to the only nonzero reduced homology group of $\mathcal{C}(S_{g,n})$, called the Steinberg module of* $\mathrm{Mod}_{g,n}$.

The proof is similar to the proof for arithmetic groups using results on the Borel–Serre partial compactification of $\mathcal{T}_{g,n}$ in the previous subsection. See [66, 67, 69, 80] for details.

In [86], using Proposition 5.9 on the topology of $\mathcal{C}(S_{g,n})$, we proved

Proposition 7.39. $\mathrm{Mod}_{g,n}$ *is not a virtual Poincaré duality group.*

This proposition is equivalent to the fact that when $\dim \mathcal{T}_{g,n} > 0$, the Steinberg module of $\mathrm{Mod}_{g,n}$ is of infinite rank. See [31] for more information on this module.

Since $\mathrm{Mod}_{g,n} \backslash \mathcal{T}_{g,n}$ is noncompact, Proposition 7.39 is consistent with the result that an arithmetic subgroup Γ is not a virtual Poincaré duality group if and only if $\Gamma \backslash X$ is noncompact (Proposition 6.14).

The Euler characteristic of $\mathrm{Mod}_{g,n}$ is given by special values of the Riemann zeta function, and the homology groups of $\mathrm{Mod}_{g,n}$ in low degrees are also known. See [66] and references there.

Problem 7.40. Compute $H_i(\mathrm{Mod}_{g,n}, \mathbb{Z})$ in all degrees. Maybe compute them for special values of g and n, and also for special values of i first.

7.10 Tangent Cones at Infinity of Teichmüller Spaces, Moduli Spaces, and Mapping Class Groups

As discussed earlier, the asymptotic cones at infinity of symmetric spaces and locally symmetric spaces are given by buildings. A natural problem is to consider the asymptotic cones of the Teichmüller space $\mathcal{T}_{g,n}$ and its quotient $\mathcal{M}_{g,n} = \mathrm{Mod}_{g,n} \backslash \mathcal{T}_{g,n}$.

Proposition 7.41. *With respect to the Teichmüller metric, for any finite index subgroup $\Gamma \subset \mathrm{Mod}_{g,n}$, the asymptotic cone $\mathrm{Cone}_\infty(\Gamma \backslash \mathcal{T}_{g,n})$ is equal to the metric cone over $\Gamma \backslash \mathcal{C}(S_{g,n})$.*

This result was proved in [53, 111]. There are some results on $\mathrm{Cone}_\infty(\mathrm{Mod}_{g,n})$ and $\mathrm{Cone}_\infty(\mathcal{T}_{g,n}, d_{\mathrm{WP}})$, where d_{WP} is the Weil–Petersson metric.

By [5], for any asymptotic cone $\mathrm{Cone}_\infty(\mathrm{Mod}_{g,n})$, every point is a global cut-point, i.e., $\mathrm{Cone}_\infty(\mathrm{Mod}_{g,n})$ is tree-graded.

Since the quotient $\mathrm{Mod}_{g,n} \backslash \mathcal{T}_{g,n}$ is noncompact, it is reasonable to expect that $\mathrm{Mod}_{g,n}$ is not quasi-isometric to $\mathcal{T}_{g,n}$, and that $\mathrm{Cone}_\infty(\mathcal{T}_{g,n})$ and $\mathrm{Cone}_\infty(\mathrm{Mod}_{g,n})$ will be different. The following bounds on the dimension of $\mathrm{Cone}_\infty(\mathrm{Mod}_{g,n})$ and $\mathrm{Cone}_\infty(\mathcal{T}_{g,n}, d_{\mathrm{WP}})$ were proved in [7].

Proposition 7.42. *The maximal dimension of locally compact subsets of the asymptotic cone* $\mathrm{Cone}_\infty(\mathrm{Mod}_{g,n})$ *is equal to the maximal rank of free abelian subgroups of* $\mathrm{Mod}_{g,n}$, *which is equal to* $3g-3+n$. *Consequently, the geometric rank of* $\mathrm{Mod}_{g,n}$ *is equal to the maximal rank of free abelian subgroups of* $\mathrm{Mod}_{g,n}$.

The geometric rank of a group Γ is defined as the largest integer n for which there exists a quasi-isometric embedding $\mathbb{Z}^n \to \Gamma$. This result was used to prove the quasi-isometric rigidity of $\mathrm{Mod}_{g,n}$ in [6, 65]. See Sect. 7.4.

Proposition 7.43. *The maximal dimension of locally compact subsets of the asymptotic cone* $\mathrm{Cone}_\infty(\mathcal{T}_{g,n}, d_{\mathrm{WP}})$ *is equal to* $[\frac{3g+p-2}{2}]$, *that is, the integral part of* $\frac{3g+p-2}{2}$, *and hence the geometric rank of* $(\mathcal{T}_{g,n}, d_{\mathrm{WP}})$ *is equal to* $[\frac{3g+p-2}{2}]$.

The geometric rank of a metric space is the maximal dimension of quasi-flats in the metric space. The geometric ranks of some complexes related to the curve complex were determined in [136].

Problem 7.44. Understand global structures of the asymptotic cones at infinity, $\mathrm{Cone}_\infty(\mathrm{Mod}_{g,n})$ and $\mathrm{Cone}_\infty(\mathcal{T}_{g,n}, d_{\mathrm{WP}})$, and determine the maximal dimension of compact subsets of the asymptotic cone at infinity $\mathrm{Cone}_\infty(\mathcal{T}_{g,n}, d_{\mathrm{Tei}})$, where d_{Tei} is the Teichmüller metric.

There are some other results that connect the geometry and topology of $\mathcal{M}_{g,n}$ at infinity with the curve complex $\mathcal{C}(S_{g,n})$.

Proposition 7.45. *Assume that* $\dim \mathcal{T}_{g,n} \geq 2$. *Then for any finite index subgroup* $\Gamma \subset \mathrm{Mod}_{g,n}$, *the quotient* $\Gamma\backslash\mathcal{T}_{g,n}$ *has only one end.*

Briefly, the reason is that under the assumption that $\dim \mathcal{T}_{g,n} \geq 2$, $S_{g,n}$ contains at least two simple closed essential curves, and $\mathcal{C}(S_{g,n})$ is connected. Now a suitable neighborhood of the infinity of $\Gamma\backslash\mathcal{T}_{g,n}$ is homotopy equivalent to $\Gamma\backslash\mathcal{C}(S_{g,n})$ and is hence connected.

Proposition 7.46. *For any finite index subgroup* $\Gamma \subset \mathrm{Mod}_{g,n}$, *the quotient* $\Gamma\backslash\mathcal{T}_{g,n}$ *has a compactification* $\Gamma\backslash\mathcal{T}_{g,n} \cup \Gamma\backslash\mathcal{C}(S_{g,n})$ *whose boundary is equal to the finite complex* $\Gamma\backslash\mathcal{C}(S_{g,n})$.

This is similar to the Tits compactification of locally symmetric spaces in [94]. See [90, Proposition 12.6].

7.11 Simplicial Volumes of Moduli Spaces

As recalled in Sect. 6.5, for an arithmetic locally symmetric space $\Gamma\backslash X$, whether its simplicial volume vanishes or not depends on the size of its $\mathbb{Q}$-rank, which is related to the dimension of the spheres that determines the homotopy type of the Tits building $\Delta_{\mathbb{Q}}(\mathbf{G})$. It turns out that similar results hold for $\mathcal{M}_{g,n}$ [91].

Proposition 7.47. *The orbifold simplicial volume of $\mathcal{M}_g$ is equal to zero if and only if $g \geq 2$. The simplicial volume of $\mathcal{M}_g$ vanishes for all $g \geq 1$.*

It is known that $\mathrm{Mod}_{g,n}$ admits finite index torsion-free subgroups and hence $\mathcal{M}_{g,n}$ admits finite smooth covers. Since the orbifold simplicial volume is greater than or equal to the simplicial volume [91], it suffices to show that the orbifold simplicial volume of $\mathcal{M}_g$ is equal to zero if and only if $g \geq 2$, and the simplicial volume of $\mathcal{M}_1$ is equal to 0.

For the first statement, the proof is similar to the proof of Proposition 6.17 by replacing the Borel–Serre partial compactification of the symmetric space by the thick part $\mathcal{T}_g(\varepsilon)$ of the Teichmüller space $\mathcal{T}_g$, and the rational Tits building $\Delta_{\mathbb{Q}}(\mathbf{G})$ by the curve complex $\mathcal{C}(S_g)$. Since $\mathcal{C}(S_g)$ is homotopy equivalent to a bouquet of spheres of dimension $2g-2$ for $g \geq 2$, it is simply connected. This is a crucial ingredient of the proof of the above proposition. The nonvanishing of the orbifold simplicial volume of $\mathcal{M}_1 \cong \mathrm{SL}(2,\mathbb{Z})\backslash\mathbf{H}^2$ follows from the result of Thurston for finite volume hyperbolic manifolds (Proposition 6.16).

To prove that the simplicial volume of $\mathrm{SL}(2,\mathbb{Z})\backslash\mathbf{H}^2$ vanishes, we note that $\mathrm{SL}(2,\mathbb{Z})\backslash\mathbf{H}^2$ is homeomorphic to $\mathbb{R}^2$. Since the orbifold simplicial volume and hence the simplicial volume of $\mathrm{SL}(2,\mathbb{Z})\backslash\mathbf{H}^2$ is finite, this implies that the simplicial volume of $\mathbb{R}^2$ is finite. Since $\mathbb{R}^2$ admits proper self-maps of degree greater than 2, its simplicial volume is 0. Hence, the simplicial volume of $\mathrm{SL}(2,\mathbb{Z})\backslash\mathbf{H}^2$ is 0. This also gives a new proof of the fact that the simplicial volume of $\mathbb{R}^{2n}$ vanishes. See [91] for details.

More generally, for the general moduli space $\mathcal{M}_{g,n}$, we have

Proposition 7.48. *When $g \geq 1$ and $n > 0$, both the simplicial volume and orbifold simplicial volume of $\mathcal{M}_{g,n}$ are equal to zero if $-\chi(S_{g,n}) - 1 \geq 2$, i.e., $2g+n \geq 5$; and when $g = 0$ and $n \geq 4$, the simplicial volume of $\mathcal{M}_{0,n}$ if $-\chi(S_{0,n}) - 2 \geq 2$, i.e., $n \geq 6$.*

It is known that $\mathcal{M}_{1,1} \cong \mathcal{M}_{1,0} = \mathrm{SL}(2,\mathbb{Z})\backslash\mathbf{H}^2$, which is homeomorphic to $\mathbb{R}^2$. Since the simplicial volume of $\mathbb{R}^2$ is equal to 0, the simplicial volume of $\mathcal{M}_{1,1}$ and $\mathcal{M}_{1,0}$ is also equal to 0. On the other hand, Proposition 6.16, for any finite smooth cover of $\mathcal{M}_{1,1}$ and $\mathcal{M}_{1,0}$ have positive simplicial volume. Therefore, the orbifold simplicial volume of $\mathcal{M}_{1,1}$ and $\mathcal{M}_{1,0}$ is positive.

Since $\mathcal{T}_{0,4} \cong \mathcal{T}_{1,1} \cong \mathbf{H}^2$ [142] and $\mathrm{Mod}_{0,4}$ is commensurable with $\mathrm{SL}(2,\mathbb{Z})$, it follows from that for $(g,n) = (1,0), (1,1), (0,4)$, the orbifold simplicial volume $\mathcal{M}_{g,n}$ is positive.

Problem 7.49. Decide whether the simplicial volume and orbifold simplicial volume of $\mathcal{M}_{g,n}$ vanish or not for $(g,n) = (1,2), (0,5)$.

7.12 *Action of* $\mathbf{Mod}_{g,n}$ *on* $\mathcal{C}(S_{g,n})$ *and Applications*

As mentioned before, $\mathrm{Mod}_{g,n}$ acts on $\mathcal{C}(S_{g,n})$, and this action can be used to understand structures and representations of $\mathrm{Mod}_{g,n}$.

One of the early applications of the curve complex by Harvey [71] concerns a simplified proof of the classification of elements of $\mathrm{Mod}_{g,n}$ by Thurston [164].

Recall that the group $\mathrm{SL}(2,\mathbb{R})$ acts on $\mathbf{H}^2$ by Mobius transformations and any element $\gamma \in \mathrm{SL}(2,\mathbb{R})$ belongs to one of the following types:

1. γ is elliptic and is characterized by the property that it fixes a point in $\mathbf{H}^2$,
2. γ is parabolic and is characterized by the property that it fixes exactly one point in the boundary $\mathbf{H}^2(\infty) \cong S^1$,
3. γ is hyperbolic and is characterized by the property that it fixes exactly two points in the boundary $\mathbf{H}^2(\infty) \cong S^1$.

Since $\mathbf{H}^2 \cup \mathbf{H}^2(\infty)$ is a closed ball and γ acts on it continuously, it must fix some point in $\mathbf{H}^2 \cup \mathbf{H}^2(\infty)$, and the three possibilities above cover all cases.

For any discrete subgroup $\Gamma \subset \mathrm{SL}(2,\mathbb{R})$, any element $\gamma \in \Gamma$ belongs to one of the following types:

1. γ is periodic, i.e., there exists some integer n such that $\gamma^n = Id$,
2. γ is parabolic and fixes a boundary point in $\mathbf{H}^2(\infty)$,
3. γ is hyperbolic and fixes exactly two points in the boundary $\mathbf{H}^2(\infty) \cong S^1$.

Thurston [164] defined a compactification $\overline{\mathcal{T}_{g,n}}^{\mathrm{Th}}$ of $\mathcal{T}_{g,n}$ through projective measured laminations such that

1. $\overline{\mathcal{T}_{g,n}}^{\mathrm{Th}}$ is homeomorphic to a closed ball of dimension $6g-6+n$,
2. the action of $\mathrm{Mod}_{g,n}$ extends to a continuous action on $\overline{\mathcal{T}_{g,n}}^{\mathrm{Th}}$.

Using this action, he classified elements $\gamma \in \mathrm{Mod}_{g,n}$ into corresponding three types:

1. γ is periodic, i.e., there exists some integer n such that $\gamma^n = Id$, and is characterized by the condition that γ fixes a point in $\mathcal{T}_{g,n}$,
2. γ is reducible, i.e., γ leaves invariant a collection of disjoint simple closed essential curves of $S_{g,n}$,
3. γ is pseudo-Anosov and is characterized by the condition that γ fixes exactly two points in the boundary $\partial\overline{\mathcal{T}_{g,n}}^{\mathrm{Th}}$.

Remark 7.50. The action of $\mathrm{Mod}_{g,n}$ on $\mathcal{C}(S_{g,n})$ can be used to define a translation length of an element γ of $\mathrm{Mod}_{g,n}$: for any simple closed curve α in $S_{g,n}$, the translation length of γ is defined by

$$\liminf_{j\to+\infty} \frac{d_{\mathcal{C}}(\alpha, f^j(\alpha))}{j},$$

where $d_{\mathcal{C}}(\alpha, f^j(\alpha))$ is the distance between the two vertices α and $f^j(\alpha)$ of $\mathcal{C}(S_{g,n})$. It was proved in [125] that an element γ is pseudo-Anosov if and only if its translation length is positive. See [52,56] for some estimates on the translation length.

Remark 7.51. An important application of the action of $\mathrm{Mod}_{g,n}$ on the curve complex $\mathcal{C}(S_{g,n})$ is to give an explicit presentation of $\mathrm{Mod}_{g,n}$ in [78, 169] (see also [72, 163]). Another presentation was constructed via the action of $\mathrm{Mod}_{g,n}$ on a variant of the curve complex in [11].

Remark 7.52. The action of $\mathrm{Mod}_{g,n}$ on the curve complex $\mathcal{C}(S_{g,n})$ is also used essentially in the proof of the Tits alternative for $\mathrm{Mod}_{g,n}$ [16, 85].

Remark 7.53. The action of $\mathrm{Mod}_{g,n}$ on the pants complex [74] was also used to construct a presentation of $\mathrm{Mod}_{g,n}$.

Remark 7.54. Another application of the action of $\mathrm{Mod}_{g,n}$ on $\mathcal{C}(S_{g,n})$ is to construct irreducible unitary representations of $\mathrm{Mod}_{g,n}$. It was shown in [141] that the action of $\mathrm{Mod}_{g,n}$ on $\mathcal{C}(S_{g,n})$ has non-commensurable stabilizers and a general method was used to produce irreducible unitary representations.

Remark 7.55. By [125], there is a relative metric on $\mathrm{Mod}_{g,n}$, which is quasi-isometric to $\mathcal{C}(S_{g,n})$. (A relative metric on a group is a word metric on the group with respect to an infinite generating set, consisting of a finite generating set and a finite collection of subgroups.) The paper [118] shows that a random walk on $\mathrm{Mod}_{g,n}$ makes linear progress in this relative metric. For other applications of $\mathrm{Mod}_{g,n}$ on curve complexes, see [15].

Acknowledgements This paper is an expanded version of the lecture notes of an invited talk at *International Conference on Buildings, Finite geometries and Groups* in Bangalore, India, August 29–31, 2010. I would like to thank the organizers for their invitation.

I would also like to thank Dick Canary for many helpful suggestions and references, Bill Harvey for sending valuable reprints, Feng Luo, Juan Souto and Ralf Spatzier for helpful correspondences, Misha Kapovich and B. Sury for helpful comments, and Athanase Papadopoulos for reading several versions of the paper very carefully and for many helpful comments. I would also like to thank an anonymous referee for helpful suggestions and comments. A part of the work in this paper was done during a visit to MSC, Tshinghua University, Beijing, in 2010 and I would like to thank them for providing a stimulating environment. This work was partially Supported by NSF grant DMS 0905283.

References

1. P. Abramenko, Twin buildings and applications to S-arithmetic groups, Lecture Notes in Mathematics, 1641. Springer, Berlin, 1996. x+123 pp
2. P. Abramenko, K. Brown, *Buildings, Theory and applications*, Graduate Texts in Mathematics, 248. Springer, New York, 2008. xxii+747 pp
3. W. Ballmann, Nonpositively curved manifolds of higher rank, *Ann. of Math.* (2) 122 (1985), no. 3, 597–609
4. W. Ballmann, M. Gromov, V. Schroeder, *Manifolds of nonpositive curvature*, Progress in Mathematics, 61. Birkhäuser Boston, Inc., Boston, MA, 1985. vi+263 pp
5. J. Behrstock, Asymptotic geometry of the mapping class group and Teichmüller space, *Geom. Topol.* 10 (2006), 1523–1578

6. J. Behrstock, B. Kleiner, Y. Minsky, L. Mosher, Geometry and rigidity of mapping class groups, arXiv:0801.2006
7. J. Behrstock, Y. Minsky, Dimension and rank for mapping class groups, *Ann. of Math.* (2) 167 (2008), no. 3, 1055–1077
8. G. Bell, K. Fujiwara, The asymptotic dimension of a curve graph is finite, *J. Lond. Math. Soc.* (2) 77 (2008), no. 1, 33–50
9. R. Bell, D. Margalit, Injections of Artin groups, *Comment. Math. Helv.* 82 (2007), no. 4, 725–751
10. R. Benedetti, C. Petronio, *Lectures on hyperbolic geometry*, Universitext. Springer, Berlin, 1992. xiv+330 pp
11. S. Benvenuti, Finite presentations for the mapping class group via the ordered complex of curves, *Adv. Geom.* 1 (2001), no. 3, 291–321
12. L. Bers, The moduli of Kleinian groups. *Russian Math. Surveys* 29 (1974), no. 2, 88–102
13. G. Besson, G. Courtois, S. Gallot, Entropies et rigidités des espaces localement symétriques de courbure strictement négative, *Geom. Funct. Anal.* 5 (1995), no. 5, 731–799
14. M. Bestvina, K. Bromberg, K. Fujiwara, The asymptotic dimension of mapping class groups is finite, arXiv:1006.1939
15. M. Bestvina, K. Fujiwara, Bounded cohomology of subgroups of mapping class groups, *Geom. Topol.* 6 (2002), 69–89
16. J. Birman, A. Lubotzky, J. McCarthy, *Abelian and solvable subgroups of the mapping class groups. Duke Math. J.* 50 (1983), no. 4, 1107–1120
17. M. Boggi, P. Lochak Profinite complexes of curves, their automorphisms and anabelian properties of moduli stacks of curves, arXiv:0706.0859
18. F. Bonahon, Bouts des variétés hyperboliques de dimension 3. *Ann. of Math.* (2) 124 (1986), no. 1, 71–158
19. A. Borel, L. Ji, *Compactifications of symmetric and locally symmetric spaces*, Mathematics: Theory & Applications. Birkhäuser Boston, Inc., Boston, MA, 2006. xvi+479 pp
20. A. Borel, J. P. Serre, Corners and arithmetic groups, *Comment. Math. Helv.* 48 (1973), 436–491
21. A. Borel, J. P. Serre, Cohomologie d'immeubles et de groupes S-arithmétiques, *Topology* 15 (1976), no. 3, 211–232
22. M. Bourdon, Sur les immeubles fuchsiens et leur type de quasi-isométrie, *Ergodic Theory Dynam. Systems* 20 (2000), no. 2, 343–364
23. B. Bowditch, Intersection numbers and the hyperbolicity of the curve complex, *J. Reine Angew. Math.* 598 (2006), 105–129
24. B. Bowditch, Tight geodesics in the curve complex, *Invent. Math.* 171 (2008), no. 2, 281–300
25. B. Bowditch, Length bounds on curves arising from tight geodesics, *Geom. Funct. Anal.* 17 (2007), no. 4, 1001–1042
26. B. Bowditch, Geometrical finiteness for hyperbolic groups, *J. Funct. Anal.* 113 (1993), no. 2, 245–317
27. B. Bowditch, Geometric models for hyperbolic 3-manifolds, preprint
28. B. Bowditch, End invariants of hyperbolic 3-manifolds, preprint
29. M. Bridson, Geodesics and curvature in metric simplicial complexes, in *Group theory from a geometrical viewpoint*, 373–463, World Sci. Publ., River Edge, NJ, 1991
30. M. Bridson, A. Haefliger, *Metric spaces of non-positive curvature*, Grundlehren der Mathematischen Wissenschaften, 319. Springer, Berlin, 1999. xxii+643 pp
31. N. Broaddus, Homology of the curve complex and the Steinberg module of the mapping class group, arXiv:0711.0011
32. J. Brock, B. Farb, Curvature and rank of Teichmüller space. *Amer. J. Math.* 128 (2006), no. 1, 1–22
33. J. Brock, The Weil-Petersson metric and volumes of 3-dimensional hyperbolic convex cores, *J. Amer. Math. Soc.* 16 (2003), no. 3, 495–535
34. J. Brock, R. Canary, Y. Minsky, The classification of Kleinian surface groups, II: The Ending Lamination Conjecture, arXiv:math/0412006v1

35. J. Brock, H. Masur, Y. Minsky, Asymptotics of Weil-Petersson geodesic. I. Ending laminations, recurrence, and flows. *Geom. Funct. Anal.* 19 (2010), no. 5, 1229–1257
36. K. Brown, *Cohomology of groups*, Corrected reprint of the 1982 original. Graduate Texts in Mathematics, 87. Springer, New York, 1994. x+306 pp
37. M. Bucher-Karlsson, Simplicial volume of locally symmetric spaces covered by $SL_3\mathbb{R}/SO(3)$. *Geom. Dedicata* 125 (2007), 203–224
38. K. Burns, R. Spatzier, Manifolds of nonpositive curvature and their buildings, *Inst. Hautes Études Sci. Publ. Math.* No. 65 (1987), 35–59
39. K. Burns, R. Spatzier, On topological Tits buildings and their classification, *Inst. Hautes Études Sci. Publ. Math.* No. 65 (1987), 5–34
40. P. Buser, *Geometry and spectra of compact Riemann surfaces*, Progress in Mathematics, 106. Birkhäuser Boston, Inc., Boston, MA, 1992. xiv+454 pp
41. D. Canary, Ends of hyperbolic 3-manifolds, *J. Amer. Math. Soc.* 6 (1993), no. 1, 1–35
42. D. Canary, Marden's tameness conjecture: history and applications, in *Geometry, analysis and topology of discrete groups*, 137–162, Adv. Lect. Math. (ALM), 6, Int. Press, Somerville, MA, 2008
43. R. Canary, D. McCullough, *Homotopy equivalences of 3-manifolds and deformation theory of Kleinian groups*, Mem. Amer. Math. Soc. 172 (2004), no. 812, xii+218 pp
44. R. Carter, *Finite groups of Lie type. Conjugacy classes and complex characters*, Pure and Applied Mathematics (New York). A Wiley-Interscience Publication. John Wiley & Sons, Inc., New York, 1985. xii+544 pp
45. M. Davis, *The geometry and topology of Coxeter groups*, London Mathematical Society Monographs Series, 32. Princeton University Press, Princeton, NJ, 2008. xvi+584 pp
46. M. Davis, Examples of buildings constructed via covering spaces, *Groups Geom. Dyn.* 3 (2009), no. 2, 279–298
47. A. Eskin, Quasi-isometric rigidity of nonuniform lattices in higher rank symmetric spaces. *J. Amer. Math. Soc.* 11 (1998), no. 2, 321–361
48. A. Eskin, B. Farb, Quasi-flats and rigidity in higher rank symmetric spaces, *J. Amer. Math. Soc.* 10 (1997), no. 3, 653–692
49. J. Essert, Buildings, group homology and lattices, arXiv:1008.4908
50. B. Farb, The quasi-isometry classification of lattices in semisimple Lie groups, *Math. Res. Lett.* 4 (1997), no. 5, 705–717
51. B. Farb, N. Ivanov, The Torelli geometry and its applications: research announcement, *Math. Res. Lett.* 12 (2005), no. 2-3, 293–301
52. B. Farb, C. Leininger, D. Margalit, The lower central series and pseudo-Anosov dilatations. *Amer. J. Math.* 130 (2008), no. 3, 799–827
53. B. Farb, H. Masur, Teichmüller geometry of moduli space. II. $\mathcal{M}(S)$ seen from far away, *In the tradition of Ahlfors-Bers*, V, 71–79, Contemp. Math., 510, Amer. Math. Soc., Providence, RI, 2010
54. B. Farb, R. Schwartz, The large-scale geometry of Hilbert modular groups. *J. Differential Geom.* 44 (1996), no. 3, 435–478
55. D. Gaboriau, F. Paulin, Sur les immeubles hyperboliques, *Geom. Dedicata* 88 (2001), no. 1-3, 153–197
56. V. Gadre, C. Tsai, Minimal pseudo-Anosov translation lengths on the complex of curves, arXiv:1101.2692
57. P. Gérardin, Harmonic functions on buildings of reductive split groups. in *Operator algebras and group representations*, Vol. I (Neptun, 1980), 208–221, Monogr. Stud. Math., 17, Pitman, Boston, MA, 1984
58. M. Gromov, Volume and bounded cohomology, *Inst. Hautes Études Sci. Publ. Math.* No. 56 (1982), 5–99 (1983)
59. M. Gromov, P. Pansu, Rigidity of lattices: an introduction, in *Geometric topology: recent developments* (Montecatini Terme, 1990), 39–137, Lecture Notes in Math., 1504, Springer, Berlin, 1991

60. Y. Guivarc'h, L. Ji, J. C. Taylor, *Compactifications of symmetric spaces*, Progress in Mathematics, 156. Birkhäuser Boston, Inc., Boston, MA, 1998. xiv+284 pp
61. Y. Guivarc'h, B. Rémy, Group-theoretic compactification of Bruhat-Tits buildings, *Ann. Sci. École Norm. Sup.* 39 (2006), 871–920
62. U. Hamenstädt, Geometry of the mapping class groups. I. Boundary amenability, *Invent. Math.* 175 (2009), no. 3, 545–609
63. U. Hamenstädt, Geometry of the complex of curves and of Teichmüller space, in *Handbook of Teichmüller theory*, Vol. I, 447–467, IRMA Lect. Math. Theor. Phys., 11, Eur. Math. Soc., Zürich, 2007
64. U. Hamenstädt, Train tracks and the Gromov boundary of the complex of curves, in *Spaces of Kleinian groups*, 187–207, London Math. Soc. Lecture Note Ser., 329, Cambridge Univ. Press, Cambridge, 2006
65. U. Hamenstädt, Geometry of the mapping class groups III: Quasi-isometric rigidity, arXiv:math/0512429v2
66. J. Harer, The cohomology of the moduli space of curves, in *Theory of moduli (Montecatini Terme, 1985)*, pp. 138–221, Lecture Notes in Math., 1337, Springer, Berlin, 1988
67. J. Harer, The virtual cohomological dimension of the mapping class group of an orientable surface, *Invent. Math.* 84 (1986), no. 1, 157–176
68. K. Hartshorn, Heegaard splittings of Haken manifolds have bounded distance, *Pacific J. Math.* 204 (2002), no. 1, 61–75
69. W. Harvey, Geometric structure of surface mapping class groups, in *Homological group theory (Proc. Sympos., Durham, 1977)*, pp. 255–269, London Math. Soc. Lecture Note Ser., 36, Cambridge Univ. Press, Cambridge-New York, 1979
70. W. Harvey, Boundary structure of the modular group, in *Riemann surfaces and related topics: Proceedings of the 1978 Stony Brook Conference (State Univ. New York, Stony Brook, N.Y., 1978)*, pp. 245–251, Ann. of Math. Stud., 97, Princeton Univ. Press, Princeton, N.J., 1981
71. W. Harvey, Remarks on the curve complex: classification of surface homeomorphisms, in *Kleinian groups and hyperbolic 3-manifolds (Warwick, 2001)*, 165–179, London Math. Soc. Lecture Note Ser., 299, Cambridge Univ. Press, Cambridge, 2003
72. W. Harvey, Modular groups–geometry and physics. in *Discrete groups and geometry* (Birmingham, 1991), 94–103, London Math. Soc. Lecture Note Ser., 173, Cambridge Univ. Press, Cambridge, 1992
73. W. Harvey, Modular groups and representation spaces. in *Geometry of group representations* (Boulder, CO, 1987), 205–214, Contemp. Math., 74, Amer. Math. Soc., Providence, RI, 1988
74. A. Hatcher, W. Thurston, A presentation for the mapping class group of a closed orientable surface, *Topology* 19 (1980), no. 3, 221–237
75. A. Hatcher, K. Vogtmann, The complex of free factors of a free group. *Quart. J. Math. Oxford Ser.* (2) 49 (1998), no. 196, 459–468
76. T. Hattori, Asymptotic geometry of arithmetic quotients of symmetric spaces, *Math. Z.* 222 (1996), 247–277
77. J. Hempel, 3-manifolds as viewed from the curve complex, *Topology* 40 (2001), no. 3, 631–657
78. S. Hirose, A complex of curves and a presentation for the mapping class group of a surface, *Osaka J. Math.* 39 (2002), no. 4, 795–820
79. E. Irmak, Superinjective simplicial maps of complexes of curves and injective homomorphisms of subgroups of mapping class groups, *Topology* 43 (2004), no. 3, 513–541
80. N. Ivanov, Mapping class groups, in *Handbook of geometric topology*, pp. 523–633, North-Holland, Amsterdam, 2002
81. N. Ivanov, Automorphism of complexes of curves and of Teichmüller spaces, *Internat. Math. Res. Notices* 1997, no. 14, 651–666
82. N. Ivanov, Isometries of Teichmüller spaces from the point of view of Mostow rigidity, in *Topology, ergodic theory, real algebraic geometry*, 131–149, Amer. Math. Soc. Transl. Ser. 2, 202, Amer. Math. Soc., Providence, RI, 2001

83. N. Ivanov, Attaching corners to Teichmüller space, *Leningrad Math. J.* 1 (1990), no. 5, 1177–1205
84. N. Ivanov, Action of Möbius transformations on homeomorphisms: stability and rigidity, *Geom. Funct. Anal.* 6 (1996), no. 1, 79–119
85. N. Ivanov, *Subgroups of Teichmüller modular groups*. Translations of Mathematical Monographs, 115. American Mathematical Society, Providence, RI, 1992. xii+127 pp
86. N. Ivanov, L. Ji, Infinite topology of curve complexes and non-Poincaré duality of Teichmüller modular groups, *Enseign. Math.* (2) 54 (2008), no. 3-4, 381–395
87. L. Ji, Buildings and their applications in geometry and topology, *Asian J. Math.* 10 (2006), no. 1, 11–80
88. L. Ji, From symmetric spaces to buildings, curve complexes and outer spaces, *Innov. Incidence Geom.* 10 (2009), 33–80
89. L. Ji, Integral Novikov conjectures and arithmetic groups containing torsion elements, *Comm. Anal. Geom.* 15 (2007), no. 3, 509–533
90. L. Ji, Arithmetic groups, mapping class groups, related groups, and their associated spaces, in *Proceedings of the Fourth International Congress of Mathematicians*, AMS/IP Stud. Adv. Math., vol. 48, pp. 127-175, Amer. Math. Soc., Providence, RI, 2010
91. L. Ji, Simplicial volume of moduli spaces of Riemann surfaces, preprint 2011
92. L. Ji, Large scale geometry, compactifications and the integral Novikov conjectures for arithmetic groups. in *Third International Congress of Chinese Mathematicians*. Part 1, 2, 317–344, AMS/IP Stud. Adv. Math., 42, pt. 1, 2, Amer. Math. Soc., Providence, RI, 2008
93. L. Ji, The integral Novikov conjectures for S-arithmetic groups. I. *K-Theory* 38 (2007), no. 1, 35–47
94. L. Ji, R. MacPherson, Geometry of compactifications of locally symmetric spaces, *Ann. Inst. Fourier (Grenoble)* 52 (2002), no. 2, 457–559
95. L. Ji, S. Wolpert, A cofinite universal space for proper actions for mapping class groups, *In the tradition of Ahlfors-Bers. V*, 151–163, Contemp. Math., 510, Amer. Math. Soc., Providence, RI, 2010
96. J. Johnson, Heegaard splittings and the pants complex, *Algebr. Geom. Topol.* 6 (2006), 853–874
97. J. Johnson, Bridge Number and the Curve Complex, arXiv:math/0603102
98. I. Kapovich, N. Benakli, Boundaries of hyperbolic groups. in *Combinatorial and geometric group theory*, 39–93, Contemp. Math., 296, Amer. Math. Soc., Providence, RI
99. I. Kapovich, M. Lustig, Geometric intersection number and analogues of the curve complex for free groups. *Geom. Topol.* 13 (2009), no. 3, 1805–1833
100. Y. Kida, *The mapping class group from the viewpoint of measure equivalence theory*, Mem. Amer. Math. Soc. 196 (2008), no. 916, viii+190 pp
101. Y. Kida, Automorphisms of the Torelli complex and the complex of separating curves, arXiv:0909.4718
102. Y. Kim, The Thurston boundary of Teichmüller space and complex of curves, *Topology Appl.* 154 (2007), no. 3, 675–682
103. E. Klarreich, The boundary at infinity of the curve complex and the relative Teichmüller space, preprint, http://www.msri.org/people/members/klarreic/curvecomplex.ps
104. B. Kleiner, B. Leeb, Rigidity of quasi-isometries for symmetric spaces and Euclidean buildings, *Inst. Hautes Études Sci. Publ. Math.* No. 86 (1997), 115–197
105. M. Korkmaz, Automorphisms of complexes of curves on punctured spheres and on punctured tori, *Topology Appl.* 95 (1999), no. 2, 85–111
106. M. Korkmaz, A. Papadopoulos, On the arc and curve complex of a surface, *Math. Proc. Cambridge Philos. Soc.* 148 (2010), no. 3, 473–483
107. J. Lafont, B. Schmidt, Simplicial volume of closed locally symmetric spaces of non-compact type, *Acta Math.* 197 (2006), no. 1, 129–143
108. E. Landvogt, A compactification of the Bruhat-Tits building, Lecture Notes in Math. 1619, Springer, 1996, viii+152 pp

109. B. Leeb, A characterization of irreducible symmetric spaces and Euclidean buildings of higher rank by their asymptotic geometry, Bonner Mathematische Schriften, 326. Universität Bonn, Mathematisches Institut, Bonn, 2000. ii+42 pp
110. E. Leuzinger, Tits geometry, arithmetic groups, and the proof of a conjecture of Siegel, *J. Lie Theory* 14 (2004), no. 2, 317–338
111. E. Leuzinger, Reduction theory for mapping class groups and applications to moduli spaces, arXiv:0801.1589
112. T. Li, Saddle tangencies and the distance of Heegaard splittings, *Algebr. Geom. Topol.* 7 (2007), 1119–1134
113. C. Löh, R. Sauer, Simplicial volume of Hilbert modular varieties, *Comment. Math. Helv.* 84 (2009), no. 3, 457–470
114. C. Löh, R. Sauer, Degree theorems and Lipschitz simplicial volume for nonpositively curved manifolds of finite volume, *J. Topol.* 2 (2009), no. 1, 193–225
115. E. Looijenga, Connectivity of complexes of separating curves, arXiv:1001.0823
116. W. Lück, Survey on classifying spaces for families of subgroups, in *Infinite groups: geometric, combinatorial and dynamical aspects*, 269–322, Progr. Math., 248, Birkhäuser, Basel, 2005
117. F. Luo, Automorphisms of the complex of curves, *Topology* 39 (2000), no. 2, 283–298
118. J. Maher, Linear progress in the complex of curves, *Trans. Amer. Math. Soc.* 362 (2010), no. 6, 2963–2991
119. A. Marden, The geometry of finitely generated kleinian groups, *Ann. of Math.* (2) 99 (1974), 383–462
120. A. Marden, Deformations of Kleinian groups, in *Handbook of Teichmüller theory*. Vol. I, 411–446, IRMA Lect. Math. Theor. Phys., 11, Eur. Math. Soc., Zürich, 2007
121. A. Marden, *Outer circles. An introduction to hyperbolic 3-manifolds*, Cambridge University Press, Cambridge, 2007. xviii+427 pp
122. D. Margalit, Automorphisms of the pants complex, *Duke Math. J.* 121 (2004), no. 3, 457–479
123. G. Margulis, Non-uniform lattices in semisimple algebraic groups, in *Lie groups and their representations* (Proc. Summer School on Group Representations of the Bolyai János Math. Soc., Budapest, 1971), pp. 371–553. Halsted, New York, 1975
124. G. Margulis, *Discrete subgroups of semisimple Lie groups*, Ergebnisse der Mathematik und ihrer Grenzgebiete (3), 17. Springer, Berlin, 1991. x+388 pp
125. H. Masur, Y. Minsky, Quasiconvexity in the curve complex, *In the tradition of Ahlfors and Bers, III*, pp. 309–320, Contemp. Math., 355, Amer. Math. Soc., Providence, RI, 2004
126. H. Masur, Y. Minsky, Geometry of the complex of curves. II. Hierarchical structure, *Geom. Funct. Anal.* 10 (2000), no. 4, 902–974
127. H. Masur, Y. Minsky, Geometry of the complex of curves. I. Hyperbolicity, *Invent. Math.* 138 (1999), no. 1, 103–149
128. H. Masur, M. Wolf, The Weil-Petersson isometry group, *Geom. Dedicata* 93 (2002), 177–190
129. J. D. McCarthy, A. Papadopoulos, Simplicial actions of mapping class groups, *Handbook of Teichmüller theory*, vol III, IRMA Lect. Math. Theor. Phys., Eur. Math. Soc., Zürich, 2011
130. Y. Minsky, The classification of Kleinian surface groups. I. Models and bounds, *Ann. of Math.* (2) 171 (2010), no. 1, 1–107
131. Y. Minsky, Curve complexes, surfaces and 3-manifolds, in *International Congress of Mathematicians. Vol. II*, pp. 1001–1033, Eur. Math. Soc., Zürich, 2006
132. Y. Minsky, End invariants and the classification of hyperbolic 3-manifolds, on *Current developments in mathematics*, 2002, 181–217, Int. Press, Somerville, MA, 2003
133. Y. Minsky, Combinatorial and geometrical aspects of hyperbolic 3-manifolds, in *Kleinian groups and hyperbolic 3-manifolds*, 3–40, London Math. Soc. Lecture Note Ser., 299, Cambridge Univ. Press, Cambridge, 2003
134. Y. Minsky, Curve complexes, surfaces and 3-manifolds, in *International Congress of Mathematicians*, Vol. II, 1001–1033, Eur. Math. Soc., Zürich, 2006
135. M. Mj, Cannon-Thurston maps for pared manifolds of bounded geometry. *Geom. Topol.* 13 (2009), no. 1, 189–245

136. M. Mj, Mapping class groups and interpolating complexes: rank. *J. Ramanujan Math. Soc.* 24 (2009), no. 4, 341–357
137. M. Moore, M. Rathbun, High distance knots in closed 3-manifolds, arXiv:0911.3675
138. G. Mostow, *Strong rigidity of locally symmetric spaces*, Annals of Mathematics Studies, No. 78. Princeton University Press, Princeton, N.J.; University of Tokyo Press, Tokyo, 1973. v+195 pp
139. A. Moy, Displacement functions on the Bruhat-Tits building, In *The mathematical legacy of Harish-Chandra*, 483–499, Proc. Sympos. Pure Math., 68, Amer. Math. Soc., Providence, RI, 2000
140. P. Nowak, G. Yu, What is ... property A?, *Notices Amer. Math. Soc.* 55 (2008), no. 4, 474–475
141. L. Paris, Actions and irreducible representations of the mapping class group, *Math. Ann.* 322 (2002), no. 2, 301–315
142. D. Patterson, The TeichmÏler spaces are distinct, *Proc. Amer. Math. Soc.* 35 (1972), 179–182
143. R. Penner, J. Harer, *Combinatorics of train tracks.* Annals of Mathematics Studies, 125. Princeton University Press, Princeton, NJ, 1992. xii+216 pp
144. G. Prasad, Strong rigidity of $\mathbb{Q}$-rank 1 lattices, *Invent. Math.* 21 (1973), 255–286
145. A. Putman, A note on the connectivity of certain complexes associated to surfaces, *Enseign. Math.* (2) 54 (2008), no. 3-4, 287–301
146. K. Rafi, A combinatorial model for the Teichmüller metric, *Geom. Funct. Anal.* 17 (2007), no. 3, 936–959
147. K. Rafi, S. Schleimer, Covers and the curve complex, *Geom. Topol.* 13 (2009), no. 4, 2141–2162
148. K. Rafi, S. Schleimer, Curve complexes are rigid, arXiv:0710.3794
149. J. Ratcliffe, *Foundations of hyperbolic manifolds*, Second edition. Graduate Texts in Mathematics, 149. Springer, New York, 2006. xii+779 pp
150. B. Rémy, A. Thuillier, A. Werner, Bruhat-Tits theory from Berkovich's point of view. I. Realizations and compactifications of buildings. *Ann. Sci. Éc. Norm. Supér.* (4) 43 (2010), no. 3, 461–554
151. B. Rémy, A. Thuillier, A. Werner, Bruhat-Tits theory from Berkovich's point of view. II. Satake compactifications, arXiv:0907.3264
152. M. Ronan, From Galois and Lie to Tits buildings,, in *The Coxeter legacy*, 45–62, Amer. Math. Soc., Providence, RI, 2006
153. M. Ronan, J. Tits, Building buildings, *Math. Ann.* 278 (1987), no. 1-4, 291–306
154. H.L. Royden, Automorphisms and isometries of Teichmüller space, in *Advances in the Theory of Riemann Surfaces* (Proc. Conf., Stony Brook, N.Y., 1969), pp. 369–383, Ann. of Math. Studies, No. 66., Princeton Univ. Press
155. T. Saito, Genus one 1-bridge knots as viewed from the curve complex, *Osaka J. Math.* 41 (2004), no. 2, 427–454
156. T. Saito, R. Yamamoto, Complexity of open book decompositions via arc complex, *J. Knot Theory Ramifications* 19 (2010), no. 1, 55–69
157. M. Scharlemann, Proximity in the curve complex: boundary reduction and bicompressible surfaces, *Pacific J. Math.* 228 (2006), no. 2, 325–348
158. M. Scharlemann, M. Tomova, Alternate Heegaard genus bounds distance, *Geom. Topol.* 10 (2006), 593–617
159. P. Scott, C. T. Wall, Topological methods in group theory. in *Homological group theory* (Proc. Sympos., Durham, 1977), pp. 137–203, London Math. Soc. Lecture Note Ser., 36, Cambridge Univ. Press, Cambridge-New York, 1979
160. R. Schwartz, The quasi-isometry classification of rank one lattices, *Inst. Hautes Études Sci. Publ. Math.*, 82 (1995), 133–168
161. R. Schwartz, Quasi-isometric rigidity and Diophantine approximation. *Acta Math.* 177 (1996), no. 1, 75–112
162. K. Shackleton, Combinatorial rigidity in curve complexes and mapping class groups, *Pacific J. Math.* 230 (2007), no. 1, 217–232

163. B. Szepietowski, A presentation for the mapping class group of a non-orientable surface from the action on the complex of curves, *Osaka J. Math.* 45 (2008), no. 2, 283–326
164. W. Thurston, On the geometry and dynamics of diffeomorphisms of surfaces, *Bull. Amer. Math. Soc. (N.S.)* 19 (1988), no. 2, 417–431
165. P. Tukia, A rigidity theorem for Möbius groups, *Invent. Math.* 97 (1989), no. 2, 405–431
166. J. Tits, Buildings of spherical type and finite BN-pairs, Lecture Notes in Mathematics, Vol. 386. Springer, Berlin-New York, 1974. x+299 pp
167. J. Tits, On buildings and their applications, *Proceedings of the International Congress of Mathematicians (Vancouver, B. C., 1974)*, Vol. 1, pp. 209–220. Canad. Math. Congress, Montreal, Que., 1975
168. M. Tomova, Distance of Heegaard splittings of knot complements, *Pacific J. Math.* 236 (2008), no. 1, 119–138
169. B. Wajnryb, A simple presentation for the mapping class group of an orientable surface. *Israel J. Math.* 45 (1983), no. 2-3, 157–174
170. A. Werner, Non-Archimedean intersection indices on projective spaces and the Bruhat-Tits building for PGL, *Ann. Inst. Fourier (Grenoble)* 51 (2001), no. 6, 1483–1505
171. A. Werner, Compactifications of Bruhat-Tits buildings associated to linear representations. *Proc. Lond. Math. Soc.* (3) 95 (2007), no. 2, 497–518
172. A. Werner, Compactification of the Bruhat–Tits building of PGL by lattices of smaller rank, *Doc. Math.* 6 (2001) 315–341
173. A. Werner, Compactification of the Bruhat–Tits building of PGL by seminorms, *Math. Z.* 248 (2004) 511–526
174. S. Wolpert, Geometry of the Weil-Petersson completion of Teichmüller space, in *Surveys in differential geometry*, Vol. VIII (Boston, MA, 2002), 357–393, Surv. Differ. Geom., VIII, Int. Press, Somerville, MA, 2003

On Isotypies Between Galois Conjugate Blocks

Radha Kessar

Abstract We show that between any pair of Galois conjugate blocks of a finite group, there is an isotypy with all signs positive.

Subject Classifications: 20C05, 20C20

Keywords Galois conjugate blocks • Centers • Isotypy

1 Introduction

Let p be a prime number, let k be an algebraic closure of the field of p elements and let G be a finite group. Let

$$\sigma : k \to k, \ (\lambda \to \lambda^p, \ \lambda \in k)$$

be the Frobenius homomorphism of k and we write also $\sigma : kG \to kG$ for the ring automorphism induced by σ. This is defined by

$$\sigma\left(\sum_{g \in G} \alpha_g g\right) = \sum_{g \in G} \alpha_g^p g.$$

The map σ is a ring automorphism but not a k-algebra automorphism. Since σ is a ring automorphism, σ induces a permutation of the blocks of kG. Here, by a block of a ring A, we mean a primitive idempotent of the center of A. Blocks b and c of kG are said to be *Galois conjugate* if $c = \sigma^n(b)$ for some natural number n.

R. Kessar (✉)
Institute of Mathematics, Fraser Noble Building, King's College,
University of Aberdeen, Aberdeen AB24 3UE, UK
e-mail: r.kessar@abdn.ac.uk

N.S.N. Sastry (ed.), *Buildings, Finite Geometries and Groups*, Springer Proceedings in Mathematics 10, DOI 10.1007/978-1-4614-0709-6_7,

Let $(K, \mathcal{O}, k)$ be a p-modular system and let $^{-} : \mathcal{O} \to k$ be the canonical quotient mapping. Let $^{-} : \mathcal{O}G \to kG$ be the induced $\mathcal{O}$-algebra homomorphism of the group algebras. In particular, $b \to \bar{b}$ induces a bijection between the set of blocks of $\mathcal{O}G$ and the set of blocks of kG. Blocks b and c of $\mathcal{O}G$ are said to be *Galois conjugate* if $\bar{b}$ and $\bar{c}$ are Galois conjugate blocks of kG.

Our main result is the following.

Theorem 1.1. *With the notation above, suppose that K contains a primitive $|G|$-th root of unity. Then any two Galois conjugate blocks of $\mathcal{O}G$ are isotypic.*

As a corollary, we obtain a new proof of the following result of Cliff et al. [7].

Corollary 1.2. *Let G be a finite group and b be a block of kG. Then the center $Z(kGb)$ of kGb has an $\mathbb{F}_p$-form.*

We recall that a finite dimensional k-algebra A is said to have an $\mathbb{F}_p$-form if there exists an $\mathbb{F}_p$-algebra A_0 such that $A \cong k \otimes_{\mathbb{F}_p} A_0$ as k-algebra.

The notion of isotypic blocks (see Definition 2.2) is due to Michel Broué [6, Definition 4.3]. Isotypies between blocks are interesting as they are often the character theoretic shadow of a derived or Morita equivalence between the k-linear module categories of the corresponding block algebras. In [2], it was shown that there exist pairs of Galois conjugate blocks which are not derived equivalent as k-algebras. Thus, the blocks studied in [2] provide examples of pairs of blocks which are isotypic and isomorphic as rings but not derived equivalent as k-algebras. However, it is conjectured that the number of Morita equivalence classes of algebras in any set of Galois conjugate blocks is bounded by a number which depends only on the defect of the blocks and which is independent of G. This conjecture is related to the Donovan finiteness conjectures in block theory [8].

For a non-negative integer d and a finite dimensional commutative k-algebra A, we will say that A occurs as the center of a d-block if there exists a finite group H and a block c of kH with defect d such that $A \cong Z(kHc)$ as k-algebras. Combining Corollary 1.2 with a theorem of Brauer and Feit [4] gives a proof of the following finiteness result. The result itself is known to experts, but as far as I am aware has not appeared before in the literature.

Corollary 1.3. *Let d be a non-negative integer and set $m := \frac{1}{4}p^{2d} + 1$. Up to isomorphism there are at most p^{m^3} k-algebras that occur as centers of d-blocks.*

The paper is divided into three sections. In Sect. 2, we set up notation and recall the definitions of perfect isometries and isotypies. Section 3 contains the proofs of Theorem 1.1, Corollaries 1.2 and 1.3.

2 Notation and Definitions

Throughout, G is a finite group, k is algebraically closed of characteristic p and $(K, \mathcal{O}, k)$ is a p-modular system such that K contains a primitive $|G|$th root of unity. For an element g of G, we will denote by g_p the p-part of g and by $g_{p'}$-the p'-part of G. Denote by G_p the set of elements of G of p-power order and by $G_{p'}$ the set of elements of G of p'-order. For a natural number n, let n_p denote the p-part n and $n_{p'}$ denote the p'-part of n.

2.1 Generalized Decomposition Maps

Let b be a central idempotent of $\mathcal{O}G$. Denote by $\mathrm{CF}(G, K)$ the K-space of K-valued class functions on G. Identifying K-valued class functions on G with their canonical K-linear extensions to K-valued functions on KG, denote by $\mathrm{CF}(G, b, K)$ the K-subspace of class functions ϕ in $\mathrm{CF}(G, K)$ such that $\phi(gb) = \phi(g)$ for all $g \in G$. Denote by $\mathrm{CF}_{p'}(G, K)$ (respectively, $\mathrm{CF}_{p'}(G, b, K)$) the K-subspaces of $\mathrm{CF}(G, K)$ (respectively, $\mathrm{CF}(G, b, K)$) of class functions which vanish on the p-singular classes of G; $\mathrm{Irr}(G, K)$ (respectively, $\mathrm{Irr}(G, b, K)$) the subset of $\mathrm{CF}(G, K)$ (respectively, $\mathrm{CF}(G, b, K)$) of irreducible K-valued characters of G; by $\mathrm{IBr}(G, K)$ (respectively, $\mathrm{IBr}(G, b, K)$) the set of irreducible Brauer characters of G (respectively, irreducible Brauer characters of G in b). For $\phi \in \mathrm{IBr}(G, K)$, let $\phi_0 \in \mathrm{CF}_{p'}(G, K)$ be defined by $\phi_0(y) = \phi(y)$ if $y \in G_{p'}$. Denote by $\mathrm{IBr}(G, K)_0$ the set of class functions ϕ_0, for $\phi \in \mathrm{IBr}(G, K)$ and by $\mathrm{IBr}(G, b, K)_0$ the subset of $\mathrm{IBr}(G, K)_0$ consisting of those ϕ_0 for which $\phi \in \mathrm{IBr}(G, b, K)$.

Let x be a p-element of G. The generalized decomposition map

$$d_G^x : \mathrm{CF}(G, K) \to \mathrm{CF}_{p'}(C_G(x), K)$$

is defined by $d_G^x(\alpha)(y) = \alpha(xy)$ for $\alpha \in \mathrm{CF}(G, K)$, $y \in C_G(x)_{p'}$. If $\chi \in \mathrm{Irr}(G, K)$, $x \in G_p$, then

$$d_G^x(\chi) = \sum_{\phi \in \mathrm{IBr}(C_G(x), K)} \delta_{(G,\chi)}^{(x,\phi)} \phi_0$$

for uniquely determined elements $\delta_{(G,\chi)}^{(x,\phi)}$ of K; $\delta_{(G,\chi)}^{(x,\phi)}$ is called the generalized decomposition number associated to the triple (χ, ϕ, x).

Let e be a central idempotent of $\mathcal{O}C_G(x)$. The map

$$d_G^{(x,e)} : \mathrm{CF}(G, K) \to \mathrm{CF}_{p'}(C_G(x), e, K)$$

is defined by $d_G^{(x,e)}(\alpha)(y) = \alpha(xey)$ for $\alpha \in \mathrm{CF}(G, K)$, $y \in C_G(x)_{p'}$.

2.2 Local Structure of Blocks

Let R denote one of the rings $\mathcal{O}$ or k. For Q a p-subgroup of G, let $(RG)^Q$ denote the R-subalgebra ofRG consisting of the elements of RG which are fixed by the conjugation action of Q. The Brauer homomorphism $\mathrm{Br}_Q : (RG)^Q \to kC_G(Q)$ is the map defined by

$$\mathrm{Br}_Q\left(\sum_{g\in g}\alpha_g g\right) = \sum_{g\in C_G(Q)} \bar{\alpha}_g g.$$

Here, if $R = k$, then $\bar{\alpha}_g$ is to be interpreted as α_g. If b is a block of $\mathcal{O}G$, then a $\bar{b}$-Brauer pair is a pair $(Q, \bar{e})$, where Q is a p-subgroup of G and e is a block of $\mathcal{O}C_G(Q)$ such that $\mathrm{Br}_Q(b)\bar{e} = \bar{e}$. Let $(P, \bar{e}_P)$ be a maximal b-Brauer pair under the Alperin–Broué inclusion of Brauer pairs [1, Definition 1.1], and for each subgroup Q of P, let e_Q be the unique block of $\mathcal{O}C_G(Q)$ such that $(Q, \bar{e}_Q) \le (P, \bar{e}_P)$.

Let $\mathcal{F}_{(P,\bar{e}_P)}(G, \bar{b})$ denote the category whose objects are the subgroups of P and whose morphisms are defined as follows: For Q, R subgroups of P, the set of $\mathcal{F}_{(P,\bar{e}_P)}(G, \bar{b})$ morphisms from Q to R is the set of those group homomorphisms $\varphi : Q \to R$ for which there exists an element g of G such that $\varphi(x) = gxg^{-1}$ for $x \in Q$ and such that $({}^gP, {}^g\bar{e}_Q) \le (R, \bar{e}_R)$; composition of morphisms is the usual composition of functions.

2.3 *Perfect Isometries and Isotypies*

Let H be a finite group such that K contains a primitive $|H|$th root of unity. Let b be a central idempotent of $\mathcal{O}G$ and c a central idempotent of $\mathcal{O}H$.

Definition 2.1. A perfect isometry between b and c is a K-linear map

$$I : \mathrm{CF}(G, b, K) \to \mathrm{CF}(H, c, K)$$

such that the following holds. For each $\chi \in \mathrm{Irr}(G, b, K)$, there exists an $\epsilon_\chi \in \{\pm 1\}$ such that the map $\chi \to \epsilon_\chi I(\chi)$ is a bijection between $\mathrm{Irr}(G, b, K)$ and $\mathrm{Irr}(H, c, K)$ and such that setting

$$\mu := \sum_{\chi\in\mathrm{Irr}(G,b,K)} \chi \times I(\chi),$$

the class function on $G \times H$ which sends an element (x, y) of $G \times H$ to the element $\sum_{\chi\in\mathrm{Irr}(G,b,K)} \chi(x)I(\chi)(y)$ of $\mathcal{O}'$ the following holds:

(a) For each $x \in G$, $y \in H$, $\frac{\mu(x,y)}{|C_G(x)|} \in \mathcal{O}$.
(b) If $x \in G$, $y \in H$ are such that exactly one of x and y is p-singular, then $\mu(x, y) = 0$.

If I as in the above definition is a perfect isometry between b and c, then I induces by restriction a map

$$I_{p'} : \mathrm{CF}_{p'}(G, b, K) \to \mathrm{CF}_{p'}(H, c, K).$$

Definition 2.2. Let b be a block of $\mathcal{O}G$ and c a block of $\mathcal{O}H$. Then b and c are isotypic if the following conditions hold:

(a) There exists a p-group P and inclusions $P \hookrightarrow G$, $P \hookrightarrow H$ such that identifying P with its image in G and in H, there exists a block e_P of $\mathcal{O}C_G(P)$ such that $(P, \bar{e}_P)$ is a maximal $\bar{b}$-Brauer pair and a block f_P of $\mathcal{O}C_H(P)$ such that $(P, \bar{f}_P)$ is a maximal $\bar{c}$-Brauer pair and such that

$$\mathcal{F}_{(P,\bar{e}_P)}(G, \bar{b}) = \mathcal{F}_{(P,\bar{f}_P)}(H, \bar{c}).$$

(b) For each cyclic subgroup Q of P, there exists a perfect isometry

$$I^Q : \mathrm{CF}(C_G(Q), e_Q, K) \to \mathrm{CF}(C_H(Q), f_Q, K)$$

where e_Q (respectively, f_Q) is the unique block of $\mathcal{O}C_G(Q)$ (respectively, $\mathcal{O}C_H(Q)$) with $(Q, \bar{e}_Q) \leq (P, \bar{e}_P)$ (respectively, $(Q, \bar{f}_Q) \leq (P, \bar{f}_P)$) such that for every generator x of Q, we have

$$I^Q_{p'} \circ d_G^{(x,e_Q)} = d_H^{(x,f_Q)} \circ I^{\{1\}}.$$

3 Proofs

Keep the notation and hypothesis of Theorem 1.1. In addition, let $W(k)$ be the unique absolutely non-ramified complete discrete valuation ring having k as residue field, and identify $W(k)$ with its image under the canonical injective homomorphism $W(k) \to \mathcal{O}$ (see [11, Chap. 2, Sect. 5, Theorems 3 and 4]). There is a unique ring automorphism $\sigma_{W(k)} : W(k) \to W(k)$ such that $\overline{\sigma_{W(k)}(\eta)} = \sigma(\bar{\eta})$ for all $\eta \in W(k)$. Further, note that for any p'-root of unity η in K, $\eta \in W(k)$ and $\sigma_{W(k)}(\eta) = \eta^p$.

Let K_0 be the algebraic closure of $\mathbb{Q}$ in K. Choose a field automorphism $\sigma_{K_0} : K_0 \to K_0$ such that if $\eta \in K$ is any $|G|$-th root of unity, then $\sigma_{K_0}(\eta) = \eta^p$ if the order of η is relatively prime to p and $\sigma_{K_0}(\eta) = \eta$ if η is a power of p. Then, $K_0 \cap W(k)$ contains a primitive $|G|_{p'}$-root of unity and σ_{K_0} and $\sigma_{W(k)}$ coincide on $\mathbb{Q}[\eta] \cap W(k)$ for any $|G|_{p'}$-root of unity η. Note that we are not claiming that σ_0 extends to an automorphism of $\mathcal{O}$ which agrees with σ_{K_0} on restriction to $\mathbb{Q}[\eta] \cap \mathcal{O}$ for any $|G|$th root of unity η.

Denote by $\sigma_{W(k)}$ (respectively σ_{K_0}) the natural extension of $\sigma_{W(k)}$ (respectively σ_{K_0}) to $W(k)G$ (respectively K_0G).

Recall from [10, Chap. 3, Theorem 6.22 (ii)] that any block of $\mathcal{O}H$, for H a finite group is an $\mathcal{O}$-linear combination of p'-elements of H.

Lemma 3.1. *Let η be a primitive $|G|_{p'}$ root of unity in K and let $H \leq G$. Let*

$$c = \sum_{g \in H_{p'}} \alpha_g g, \alpha_g \in \mathcal{O}$$

be a block of $\mathcal{O}H$. *Then,*

(i) $\alpha_g \in \mathbb{Q}[\eta] \cap W(k)$ *for all* $g \in H_{p'}$.
(ii) $\sigma_{W(k)}(c)$ *is a block of* $\mathcal{O}H$, $\sigma_{W(k)}(c) = \sigma_{K_0}(c)$ *and*

$$\overline{\sigma_{W(k)}(c)} = \sigma(\bar{c}).$$

Proof. (i) By idempotent lifting, the canonical quotient map $^- : \mathcal{O} \to k$ induces a bijection between the set of central idempotents of $\mathcal{O}H$ and kH. Similarly, the restriction of $^-$ to $W(k)$ induces a bijection between the set of central idempotents of $W(k)H$ and kH. Since a central idempotent of $W(k)H$ is a central idempotent of $\mathcal{O}H$, it follows that $\mathcal{O}H$ and $W(k)H$ have the same central idempotents. In particular, $c \in W(k)H$. So, $\alpha_g \in W(k)$ for all $g \in H_{p'}$. On the other hand, by [10, Chap. 3, Theorem 6.22], for $g \in H_{p'}$, α_g is a $\mathbb{Q}$-linear combination of $|g|$-th roots of unity whence $\alpha_g \in \mathbb{Q}[\eta]$. This proves (i).

(ii) As shown above, the set of blocks of $\mathcal{O}H$ is the same as the set of blocks of $W(k)H$ and $\sigma_{W(k)}$ is an automorphism of $W(k)H$. This proves the first assertion. The others are immediate from (i).

Definition 3.2. For H a subgroup G, let

$$I^H : \mathrm{CF}(H, K) \to \mathrm{CF}(H, K)$$

denote the K-linear map defined by

$$I^H(\phi)(x) = \phi(x_p x_{p'}^p), \quad \text{for } \phi \in \mathrm{CF}(H, K),\ x \in G.$$

If an element $\phi \in \mathrm{CF}(H, K)$ takes values in K_0, denote by $\sigma(\phi)$ the element of $\mathrm{CF}(H, K)$ which sends $g \in H$ to $\sigma_{K_0}(\phi(g))$. Similarly, if ϕ takes values in $W(k)$, then $\sigma_{W(k)}(\phi)$ will denote the element of $\mathrm{CF}(H, K)$ which sends $g \in H$ to $\sigma_{W(k)}(\phi(g))$. We use the same conventions on K-valued class functions defined on $H_{p'}$.

Lemma 3.3. *Let* H *be a subgroup of* G *and let* c *be a block of* H.

(i) For any $\chi \in \mathrm{Irr}(H, K)$,

$$I^H(\chi) = \sigma_{K_0}(\chi) \in \mathrm{Irr}(H, K).$$

(ii) The map

$$\chi \to \sigma_{K_0}(\chi)$$

is a bijection from $\mathrm{Irr}(H, c, K)$ *to* $\mathrm{Irr}(H, \sigma_{K_0}(c), K)$.
(iii) The restriction of I^H *to* $\mathrm{CF}(H, c, K)$ *induces a perfect isometry between* c *and* $\sigma_{K_0}(c)$.
(iv) For any $\phi \in \mathrm{IBr}(H, K)$ *and any* $\chi \in \mathrm{Irr}(G, K)$,

$$I^H(\phi_0) = \sigma_{K_0}(\phi)_0 \in \mathrm{IBr}(H, K)_0$$

and

$$\delta^{(\{1\},\sigma_{K_0}(\phi))}_{(H,\sigma_{K_0}(\chi))} = \delta^{(\{1\},\phi)}_{(H,\chi)}.$$

Proof. (i) Let $\chi \in \mathrm{Irr}(H, K)$, and let $\rho : H \to GL_n(K_0)$ be a representation affording χ. Such a ρ exists by Brauer's splitting field theorem (see for example [10, Chap. 3, Theorem 4.11]). Denoting also by σ_{K_0} the automorphism (as abstract group) of $GL_n(K_0)$ induced by σ_{K_0}, it follows that $\sigma_{K_0} \circ \rho$ is an irreducible representation of H with character $\sigma_{K_0}(\chi)$. We show that $I^H(\chi) = \sigma_{K_0}(\chi)$. Let $h = h_p h_{p'} \in H$. Since K_0 contains all the eigen values of $\rho(h)$, by replacing ρ with an equivalent representation if necessary, we may assume that $\rho(h)$ is a diagonal matrix. Since h_p and $h_{p'}$ are powers of h, it follows that $\rho(h_p)$ is a diagonal matrix $\mathrm{diag}(\zeta_1, \cdots, \zeta_n)$, $\rho(h_{p'})$ is a diagonal matrix $\mathrm{diag}(\eta_1, \cdots, \eta_n)$, where each ζ_i is a $|H|_p$ root of unity and each η_j is an $|H|_{p'}$ root of unity, $\rho(h)$is the diagonal matrix $\mathrm{diag}(\zeta_1\eta_1, \cdots, \zeta_n\eta_n)$ and $\rho(h_p h_{p'}^p)$ is the diagonal matrix $\mathrm{diag}(\zeta_1\eta_1^p, \cdots, \zeta_n\eta_n^p)$. Thus,

$$I^H(\chi)(h) = \sum_{1\le i\le n} \zeta_i \eta_i^p = \sigma_{K_0}\left(\sum_{1\le i\le n} \zeta_i \eta_i\right) = \sigma_{K_0}(\chi)(h).$$

(ii) Let $\chi \in \mathrm{Irr}(H, K)$ and let

$$e_\chi = \frac{\chi(1)}{|H|} \sum_{h\in H} \chi(h)h^{-1}$$

be the primitive central idempotent of KH corresponding to χ. Then,

$$e_{\sigma_{K_0}(\chi)} = \frac{\chi(1)}{|H|} \sum_{h\in H} \sigma_{K_0}(\chi(h))h^{-1} = \sigma_{K_0}(e_\chi).$$

From this it is immediate that if $\chi \in \mathrm{Irr}(H, c, K)$ then $\sigma_{K_0}(\chi) \in \mathrm{Irr}(H, \sigma_{K_0}(c), K)$.

(iii) By (i) and (ii) it suffices to prove that the function

$$\mu := \sum_{\chi\in\mathrm{Irr}(H,c,K)} \chi \times I^H(\chi),$$

satisfies conditions (a) and (b) of Definition 1.1.

Set

$$\iota := \sum_{\chi\in\mathrm{Irr}(H,K,c)} \chi \times \chi$$

and let $x, y \in H$.

Then

$$\mu(x, y) = \iota(x, y_p y_{p'}^p).$$

Since the identity map on $\mathrm{CF}(H, c, K)$ is a perfect isometry with ι the corresponding (virtual) character of $H \times H$, we see that $\mu(x, y)$ is divisible in $\mathcal{O}$ by both $|C_G(x)|$ and by $|C_G(y_p y_{p'}^p)|$ and that $\mu(x, y)$ is 0 if exactly one of x and $y_p y_{p'}^p$ is p-singular. But, clearly $C_G(y_p y_{p'}^p) = C_G(y)$ and $y_p y_{p'}^p$ is p-singular if and only if y is p-singular. This proves (iii).

(iv) Let $\phi \in \mathrm{IBr}(H, K)$ and let $\tau : H \to GL_n(k)$ be an irreducible representation of H affording the Brauer character ϕ. Then $\sigma \circ \tau$ is an irreducible representation of H, where again we denote by σ the induced automorphism of $GL_n(k)$. Let ϕ' be the Brauer character associated to $\sigma \circ \tau$ and let $h \in H_{p'}$. Let Λ be the multiset of eigen values of $\tau(h)$. The multiset of eigen values of $\sigma \circ \tau(h)$ is $\{\lambda^p : \lambda \in \Lambda\}$, hence $\phi'(h) = \sigma_{W(k)}(h)$. Further, since $\{\lambda^p : \lambda \in \Lambda\}$ is also the multiset of eigen values of $\tau(h^p)$, $I^H(\phi_0)(h) = \phi(h^p) = \sigma_{W(k)}(h)$ for all $h \in H_{p'}$. Thus,

$$I^H(\phi_0) = \sigma_{W(k)}(\phi)_0 = \phi'_0 \in \mathrm{IBr}(H, K)_0.$$

Since ϕ takes values in $\mathbb{Z}[\eta]$, with η a primitive $|H|_{p'}$-root of unity, $\sigma_{W(k)}(\phi)_0 = \sigma_{K_0}(\phi)_0$ whence $I^H(\phi_0) = \sigma_{K_0}(\phi)_0 \in \mathrm{IBr}(H, K)_0$.

The compatibility of decomposition numbers is clear from the fact that for any $\chi \in \mathrm{Irr}(H, K)$,

$$\mathrm{Res}|_{H_{p'}} \chi = \sum_{\phi \in \mathrm{IBr}(H,K)} \delta_{H,\chi}^{1,\phi} \phi$$

and that $\delta_{H,\chi}^{1,\phi} \in \mathbb{Z}$ for all $\phi \in \mathrm{IBr}(H, K)$.

Proof of Theorem 1.1. Let b be a block of $\mathcal{O}G$. By Lemma 3.1 it suffices to show that b and $\sigma_{K_0}(b)$ are isotypic. Let P a p-subgroup of G and e_P be a block of $\mathcal{O}C_G(P)$ such that $(P, \bar{e}_P)$ is a maximal $\bar{b}$-Brauer pair. For each $Q \le P$, let e_Q be the unique block of $\mathcal{O}C_G(Q)$ such that $(Q, \bar{e}_Q) \le (P, \bar{e}_P)$. For any p-subgroup Q of G and any $a \in (kG)^Q$, $\sigma(\mathrm{Br}_P(a)) = \mathrm{Br}_P(\sigma(a))$. So, the map $(Q, f) \to (Q, \sigma(f))$ is an isomorphism from the G-poset of $\bar{b}$-Brauer pairs to the G-poset of $\sigma(\bar{b})$-Brauer pairs. By Lemma 3.1, $\sigma(\bar{b}) = \overline{\sigma_{K_0}(b)}$ and $\sigma(\bar{f}) = \overline{\sigma_{K_0}(f)}$ for any block f of $kC_G(Q)$, Q a p-subgroup of G. Thus, $(P, \overline{\sigma_{K_0}(e_P)})$ is a maximal $\overline{\sigma_{K_0}(b)}$-Brauer pair; for every subgroup Q of P, $\sigma_{K_0}(e_Q)$ is the unique block of $\mathcal{O}C_G(Q)$ such that $(Q, \overline{\sigma_{K_0}(e_Q)}) \le (P, \overline{\sigma_{K_0}(e_P)})$, and $\mathcal{F}_{(P, \overline{\sigma_{K_0}(e_P)})}(G, \overline{\sigma_{K_0}(b)}) = \mathcal{F}_{(P, \bar{e}_P)}(G, \bar{b})$.

We will use the maps $I^{C_G(Q)}$ of Definition 3.2 to produce an isotypy between b and $\sigma_{K_0}(b)$. For $Q \le P$, let I^Q be the restriction of $I^{C_G(Q)}$ to $\mathrm{CF}(C_G(Q), e_Q, K)$. We will show that the maps I^Q, as Q runs over the cyclic subgroups of P defines an isotypy between b and c. So, let $Q \le P$ be a cyclic group. By Lemma 3.3,

$$I^Q : \mathrm{CF}(C_G(Q), e_Q, K) \to \mathrm{CF}(C_G(Q), \sigma_{K_0}(e_Q), K)$$

is a perfect isometry. It remains only to check the compatibility condition (b) of Definition 2.2.

Set $H = C_G(Q)$ and let $Q = \langle x \rangle$. We claim that

$$\delta_{(G,\chi)}^{(x,\phi)} = \delta_{(G,\sigma_{K_0}(\chi))}^{(x,\sigma_{K_0}(\phi))}$$

for all $\chi \in \mathrm{Irr}(G, K)$ and all $\phi \in \mathrm{IBr}(H, K)$.

Indeed, for $\tau \in \mathrm{Irr}(H, K)$, let $\zeta_{(\tau,x)} \in K$ be such that $\rho(x)$ is the scalar matrix $(\zeta_{(\tau,x)}, \cdots, \zeta_{(\tau,x)})$ in any representation ρ of H affording τ. Since the order of $\zeta_{(\tau,x)}$ is a divisor of $|G|_p$,

$$\zeta_{(\sigma_{K_0}(\tau),x)} = \sigma_{K_0}(\zeta_{(\tau,x)}) = \zeta_{(\tau,x)}.$$

Also, for $\chi \in \mathrm{Irr}(G, K)$, $\phi \in \mathrm{IBr}(H, K)$,

$$\delta_{(G,\chi)}^{(x,\phi)} = \sum_{\tau \in \mathrm{Irr}(H,K)} \langle \mathrm{Res}|_H \chi, \tau \rangle \zeta_{(\tau,x)} \delta_{(H,\tau)}^{(\{1\},\phi)}.$$

The claim follows since for all $\tau \in \mathrm{Irr}(H, K)$ and all $\phi \in \mathrm{IBr}(H, K)$, $\zeta_{(\sigma_{K_0}(\tau),x)} = \zeta_{(\tau,x)}$, $\delta_{(H,\sigma_{K_0}\tau))}^{(\{1\},\sigma_{K_0}(\phi))} = \delta_{(H,\tau)}^{(\{1\},\phi)}$ by Lemma 3.3 (iv) and $\langle \mathrm{Res}|_H \chi, \tau \rangle \in \mathbb{Z}$.

For $\chi \in \mathrm{Irr}(G, K)$, $\phi \in \mathrm{IBr}(H, K)$,

$$d_G^{(x,e_Q)}(\chi) = \sum_{\phi \in \mathrm{IBr}(H,e_Q,K)} \delta_{(G,\chi)}^{(x,\phi)} \phi_0.$$

The compatibility condition (b) of Definition 2.2 is easily seen to follow from the linearity of the maps I^Q, the claim and Lemma 3.3:

$$\begin{aligned} I_{p'}^Q \circ d_G^{(x,e_Q)}(\chi) &= \sum_{\phi \in \mathrm{IBr}(H,e_Q,K)} \delta_{(G,\chi)}^{(x,\phi)} I^Q(\phi_0) \\ &= \sum_{\phi \in \mathrm{IBr}(H,e_Q,K)} \delta_{(G,\sigma_{K_0}(\chi))}^{(x,\sigma_{K_0}(\phi))} I^Q(\phi_0) \\ &= \sum_{\phi \in \mathrm{IBr}(H,e_Q,K)} \delta_{(G,\sigma_{K_0}(\chi))}^{(x,\sigma_{K_0}(\phi))} \sigma_{K_0}(\phi)_0 \\ &= \sum_{\phi' \in \mathrm{IBr}(H,\sigma_{K_0}(e_Q),K)} \delta_{(G,\sigma_{K_0}(\chi))}^{(x,\phi')} \phi'_0 \\ &= d_G^{(x,\sigma_{K_0}(e_Q))} I^{\{1\}}(\chi). \end{aligned}$$

□

Proof of Corollary 1.2. Let $(K, \mathcal{O}, k)$ be a p-modular system such that k is an algebraic closure of $\mathbb{F}_p$ and K contains a primitive $|G|$-th root of unity. By the theorem, there is a perfect isometry between $\mathcal{O}Gb$ and $\mathcal{O}G\sigma_{K_0}(b)$. Hence, by [5, Théorème 1.4], there is an $\mathcal{O}$-algebra isomorphism $f : Z(\mathcal{O}G\sigma_{K_0}(b)) \to Z(\mathcal{O}Gb)$. This induces a k-algebra isomorphism $\bar{f} : Z(kG\sigma(\bar{b})) \to Z(kG\bar{b})$. On the other hand, σ induces a ring isomorphism $\sigma : Z(kG\bar{b}) \to Z(kG\sigma(\bar{b}))$ such that for all $a \in Z(kG\bar{b})$ and all $\lambda \in k$, $\sigma(\lambda a) = \lambda^p \sigma(a)$. Thus, $\sigma \circ f : Z(kG\sigma(b)) \to Z(kG\sigma(b))$ is a ring automorphism which satisfies $\sigma \circ f(\lambda a) = \lambda^p \sigma \circ f(a)$ for all $a \in Z(kG\sigma(\bar{b}))$ and all $\lambda \in k$. By [8, Lemma 2.1], the fixed points $(Z(kG\sigma(\bar{b})))^{\sigma \circ f}$ of $Z(kG\sigma(\bar{b}))$ under $\sigma \circ f$ are an $\mathbb{F}_p$-subspace of $Z(kG\sigma(\bar{b}))$ such that $Z(kG\sigma(\bar{b})) \cong k \otimes_{\mathbb{F}_p} (Z(kG\sigma(\bar{b})))^{\sigma \circ f}$ as k-vector spaces. Since $\sigma \circ f$ is a homomorphism of rings, $(Z(kG\sigma(\bar{b})))^{\sigma \circ f}$ is an $\mathbb{F}_p$-algebra. Thus, $Z(kG\sigma(\bar{b}))$ and hence $Z(kGb)$ has an $\mathbb{F}_p$-form. □

Proof of Corollary 1.3. Let b be a block of kG with defect d. Since $\dim_k(Z(kGb)) = \dim_K(Z(KGb)) = |\mathrm{Irr}(G, b, K)|$, by [4, Theorem 1] the k-dimension of $Z(kGb)$ is bounded by m. By Corollary 1.2, $Z(kGb)$ has a k-basis such that the multiplicative constants of $Z(kGb)$ with respect to this basis are all in $\mathbb{F}_p$. Thus there are at most p^{m^3} possibilities for the isomorphism type of $Z(kGb)$. □

Remark 3.4. (i) The proof of Theorem 1.1 can be readily adapted to prove that between any pair of Galois conjugate blocks there is a global isotypy in the sense of [3, 1.9]. It is not known whether there is a p-permutation equivalence (cf. [3, Definition 1.3], [9, Definition 1.3]) between any pair of Galois conjugate blocks.

(ii) By Lemma 3.3, the isometries between various blocks in Theorem 1.1 all appear without signs, which seems to render even more surprising the fact that Galois conjugate blocks need not be Morita equivalent [2].

References

1. J. Alperin and M. Broué, Local methods in block theory, *Ann. of Math* **110** (1979), 143–157
2. D. J. Benson and R. Kessar, Blocks inequivalent to their Frobenius twists, *J. Algebra* **315** (2007), no. 2, 588–599
3. R. Boltje, B. Xu, On p-permutation equivalences: between Rickard equivalences and isotypies. *Trans. Amer. Math. Soc.* **360** (2008), no. 10, 5067–5087
4. R. Brauer, W. Feit, On the number of irreducible characters of finite groups in a given block *Proc. Nat. Acad. Sci. U.S.A.* **45** (1959), 361–365
5. M. Broué, Blocs, Isométries parfaites, catégories dériveés *C.R. Acd. Sci. Paris* **307**, Serie I (1988), 13–18
6. M. Broué, Isométries parfaites, types de blocs, catégories dériveés, *Astérisque* **181–182** (1990), 61–91
7. G. Cliff, W. Plesken. A. Weiss, Order-theoretic properties of the center of a block. The Arcata Conference on Representations of Finite Groups (Arcata, Calif., 1986), 413–420, Proc. Sympos. Pure Math., 47, Part 1, Amer. Math. Soc., Providence, RI, 1987
8. R. Kessar, A remark on Donovan's conjecture, *Arch. Math (Basel)* **82** (2004), 391–394
9. M. Linckelmann, Trivial Source bimodule rings for blocks and p-permutation equivalences *Trans. Amer. Math. Soc.* **361** (2009), 3, 1279–1316
10. H. Nagao, Y. Tsushima, Representations of finite groups, Academic Press, 1988
11. J. P. Serre, Corps locaux, Hermann, 1968

Representations of Unitriangular Groups

Tung Le and Kay Magaard

Abstract We survey recent results concerning the representation theory of unitriangular groups.

Subject Classifications: MSC 2000: 20C99

Keywords Irreducible characters • Groups of Lie type • Unipotent groups

1 Introduction

Let p be a prime, $q = p^a$ and $X_r(q)$ be a finite quasisimple group of untwisted rank r defined over the field $\mathbb{F}_q$. By $UX_r(q)$ we denote a Sylow p-subgroup of $X_r(q)$. For example, if $X_r(q) = \mathrm{SL}_{r+1}(q)$, then with respect to a suitable basis of the natural module of $\mathrm{SL}_{r+1}(q)$ the group $USL_{r+1}(q)$ is the set of all upper triangular $(r+1) \times (r+1)$ matrices over the finite field $\mathbb{F}_q$ with diagonal entries all equal to 1. Now $UX_r(q)$ acts faithfully on the Lie algebra of $X_r(q)$ of dimension d. So $UX_r(q)$ may be obtained as an intersection of a $\mathrm{GL}_d(q)$-conjugate of $U\mathrm{SL}_d(q)$ with the image of $X_r(q)$ in its adjoint action.

We explore the question of what is known about $\mathrm{Irr}(UX_r(q))$, and more generally the character table of $UX_r(q)$? Specifically, we will explore the following questions.

Question 1: Is it true that $\chi(1) = q^a$ for $a \in \mathbb{Z}^+$ for all $\chi \in \mathrm{Irr}(UX_r(q))$?
Question 2: Is it true that $|\mathrm{Irr}(UX_r(q))| \in \mathbb{Z}[q]$?

T. Le (✉)
Mathematical Sciences, King's College, University of Aberdeen, Aberdeen AB24 3UE, UK
e-mail: t.le@abdn.ac.uk

K. Magaard
School of Mathematics, University of Birmingham, Edgbaston, Birmingham B15 2TT, UK
e-mail: k.magaard@bham.ac.uk

N.S.N. Sastry (ed.), *Buildings, Finite Geometries and Groups*, Springer Proceedings in Mathematics 10, DOI 10.1007/978-1-4614-0709-6_8,

Question 1, for the case $X_r(q) = \mathrm{SL}_{r+1}(q)$, is an old question going back to the 1960s, whereas Higman [15] conjectured in 1960 that if $X_r(q) = \mathrm{SL}_{r+1}(q)$, then the answer to Question 2 is yes. Subsequently, Higman's conjecture was strengthened by Lehrer [20] who conjectured that the numbers $N_{n,e}(q)$ of irreducible characters of $UA_{n-1}(q)$ of degree q^e are given by a polynomial in q with integer coefficients. Recently, I.M. Isaacs suggested a strengthened form of Lehrer's conjecture stating that $N_{n,e}(q)$ is given by a polynomial in $q-1$ with nonnegative integer coefficients. So, Isaacs' conjecture implies Higman's and Lehrer's conjectures.

2 Character Degrees of $UX_r(q)$

In this section, we consider Question 1. We consider the case $X = SL_n(q)$ first.

Define $\mu_{\mathrm{SL}}(n) := m^2$ if $n = 2m+1$ and $\mu_{\mathrm{SL}}(n) := m(m-1)$ if $n = 2m$. The following two theorems settle the question of character degrees completely for the case $X = SL_n(q)$. In 1973, Gutkin [13] claimed to have proved the result below, however his "proof" turned out to be incorrect. In 1995, Isaacs proved:

Theorem 2.1 (Isaacs). *If $\chi \in \mathrm{Irr}(U\mathrm{SL}_n(q))$, then $\chi(1) = q^a$ where $0 \le a \le \mu_{\mathrm{SL}}(n)$.*

Theorem 2.2 (Huppert). *Moreover, for every $0 \le a \le \mu_{\mathrm{SL}}(n)$ there exists $\Psi \in \mathrm{Irr}(USL_n(q))$ with $\Psi(1) = q^a$.*

For a general group of Lie type $X_r(q)$ define $\mu_X(r)$ to be the maximum over $\chi \in \mathrm{Irr}(UX_r(q))$ of $\log_q(\chi(1))$. For classical groups we have.

Theorem 2.3 (Sangroniz, Szegedy). *If $X_r(q)$ is a classical group of Lie rank r defined over $\mathbb{F}_q$ with q odd and $\chi \in \mathrm{Irr}(UX_r(q))$, then $\chi(1) = q^a$ where $0 \le a \le \mu_X(r)$.*

Independently, Sangroniz, André and Neto determined the exact value of $\mu_X(q)$ whenever X is classical using the Kirillov orbit method. The precise statement is as follows:

Theorem 2.4 (Sangroniz, André and Neto). *If q is odd and $X_r(q)$ is a classical group but not of type A then the values of $\mu_X(r)$ are in $\mathbb{Z}$ and their values are independent of q. In particular, we have*

1. $\mu_{^2A}(m) = \mu_A(m)$.
2. $\mu_B(m) = m(m-1)/2$.
3. $\mu_D(m) = m(m-2)/2$ *if m is even and* $(m-1)^2/2$ *if m is odd.*
4. $\mu_{^2D}(m) = \mu_D(m)$.
5. $\mu_C(m) = m(m-1)/2$.

We conjecture that

Conjecture 2.5 (Le, Magaard). The following are true:

1. If q is odd, then $\mu_F(4) = 10$.
2. $\mu_{^2E}(6) = 16$.
3. $\mu_E(6) = 16$
4. $\mu_E(7) = 28$
5. $\mu_E(8) = 56$

Le's construction, in Sect. 4 below, gives characters of claimed degree. What remains to be shown is that there are no characters of larger degree.

For X classical and q even the situation with regards to character degrees is quite different.

Theorem 2.6 (Gow, Marjoram, Previtali). *If q is even, then for all $0 \leq i \leq [m/2]$ the exists $\chi_i \in \mathrm{Irr}(USp_{2m}(q))$ such that $\chi_i(1) = 2^{-i}q^{m(m-1)/2}$.*

These authors, as well as Sangroniz, show that similar statements are true for orthogonal groups as well.

For low rank classical groups complete descriptions of the irreducible characters have been given. Two recent examples are $U\mathrm{SL}_{13}(q)$ which is due to Evseev [10] and $UD_4(q) = UO_8^+(q)$ which is due to Himstedt et al. [16]. The results concerning $UD_4(q) = UO_8^+(q)$ are tabulated in Table 1. To explain the table we need some background and notation.

Let Φ be a root system of type D_4 in some Euclidean space, with basis $\Delta = \{\alpha_1, \alpha_2, \alpha_3, \alpha_4\}$ of simple roots such that $\alpha_1, \alpha_2, \alpha_4$ are orthogonal to each other. The Dynkin diagram of Φ is

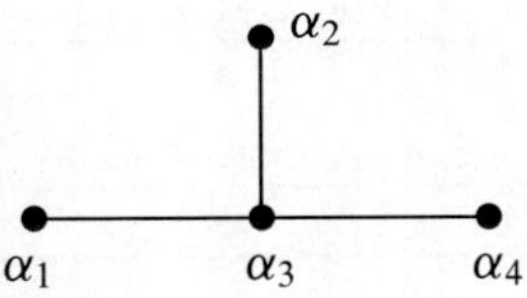

The positive roots are those roots which can be written as linear combinations of the simple roots $\alpha_1, \alpha_2, \alpha_3, \alpha_4$ with nonnegative coefficients and we write Φ_+ for the set of positive roots. We use the notation $\begin{smallmatrix} & 1 & \\ 1 & 2 & 1 \end{smallmatrix}$ for the root $\alpha_1 + \alpha_2 + 2\alpha_3 + \alpha_4$ and we use a similar notation for the remaining positive roots. The *height* of a root is the sum of the coefficients with respect to the basis consisting of the simple roots. The 12 positive roots of Φ are given in Table 2.

The numbering $\alpha_1, \alpha_2, \ldots, \alpha_{12}$ of these roots is in accordance with the output produced by the CHEVIE [11] command

```
CoxeterGroup("D", 4);
```

The labeling of the families $\mathcal{F}_{i,j,k}$ is designed so as to identify the kernel and and the centrally-acting root groups of $UD_4(q)$ of each member of the family.

Table 1 The irreducible characters of $UD_4(q)$

Family	Notation	Parameter set	Number	Degree
$\mathcal{F}_{12}$	χ_{12,a,b_1,b_2,b_3}	$\mathbb{F}_q^\times \times \mathbb{F}_q \times \mathbb{F}_q \times \mathbb{F}_q$	$q^3(q-1)$	q^4
$\mathcal{F}_{11}$	$\chi_{11,a,b_1,b_2,b_3,b_4}$	$\mathbb{F}_q^\times \times \mathbb{F}_q \times \mathbb{F}_q \times \mathbb{F}_q \times \mathbb{F}_q$	$q^4(q-1)$	q^3
$\mathcal{F}_{8,9,10}^{\text{odd}}$	$\chi_{8,9,10,a_1,a_2,a_3,b}$	$\mathbb{F}_q^\times \times \mathbb{F}_q^\times \times \mathbb{F}_q^\times \times \mathbb{F}_q$	$q(q-1)^3$	q^3
$\mathcal{F}_{8,9,10}^{\text{even}}$	$\chi_{8,9,10,q^3,a_1,a_2,a_3}$	$\mathbb{F}_q^\times \times \mathbb{F}_q^\times \times \mathbb{F}_q^\times$	$(q-1)^3$	q^3
	$\chi_{8,9,10,\frac{q^3}{2},x,a_1,a_2,a_3,a_4}$	$\mathbb{Z}_4 \times \mathbb{F}_q^\times \times \mathbb{F}_q^\times \times \mathbb{F}_q^\times \times \mathbb{F}_q^\times$	$4(q-1)^4$	$\frac{q^3}{2}$
$\mathcal{F}_{8,9}$	$\chi_{8,9,q^3,a_1,a_2,a_3}$	$\mathbb{F}_q^\times \times \mathbb{F}_q^\times \times \mathbb{F}_q^\times$	$(q-1)^3$	q^3
	$\chi_{8,9,q^2,a_1,a_2,b_1,b_2}$	$\mathbb{F}_q^\times \times \mathbb{F}_q^\times \times \mathbb{F}_q \times \mathbb{F}_q$	$q^2(q-1)^2$	q^2
$\mathcal{F}_{8,10}$	$\chi_{8,10,q^3,a_1,a_2,a_3}$	$\mathbb{F}_q^\times \times \mathbb{F}_q^\times \times \mathbb{F}_q^\times$	$(q-1)^3$	q^3
	$\chi_{8,10,q^2,a_1,a_2,b_1,b_2}$	$\mathbb{F}_q^\times \times \mathbb{F}_q^\times \times \mathbb{F}_q \times \mathbb{F}_q$	$q^2(q-1)^2$	q^2
$\mathcal{F}_{9,10}$	$\chi_{9,10,q^3,a_1,a_2,a_3}$	$\mathbb{F}_q^\times \times \mathbb{F}_q^\times \times \mathbb{F}_q^\times$	$(q-1)^3$	q^3
	$\chi_{9,10,q^2,a_1,a_2,b_1,b_2}$	$\mathbb{F}_q^\times \times \mathbb{F}_q^\times \times \mathbb{F}_q \times \mathbb{F}_q$	$q^2(q-1)^2$	q^2
$\mathcal{F}_8$	χ_{8,q^3,a_1,a_2}	$\mathbb{F}_q^\times \times \mathbb{F}_q^\times$	$(q-1)^2$	q^3
	χ_{8,q^2,a_1,b_1,b_2}	$\mathbb{F}_q^\times \times \mathbb{F}_q \times \mathbb{F}_q$	$q^2(q-1)$	q^2
$\mathcal{F}_9$	χ_{9,q^3,a_1,a_2}	$\mathbb{F}_q^\times \times \mathbb{F}_q^\times$	$(q-1)^2$	q^3
	χ_{9,q^2,a_1,b_1,b_2}	$\mathbb{F}_q^\times \times \mathbb{F}_q \times \mathbb{F}_q$	$q^2(q-1)$	q^2
$\mathcal{F}_{10}$	χ_{10,q^3,a_1,a_2}	$\mathbb{F}_q^\times \times \mathbb{F}_q^\times$	$(q-1)^2$	q^3
	$\chi_{10,q^2,a_1,b_1,b_2}$	$\mathbb{F}_q^\times \times \mathbb{F}_q \times \mathbb{F}_q$	$q^2(q-1)$	q^2
$\mathcal{F}_{5,6,7}$	$\chi_{5,6,7,a_1,a_2,a_3,b_1,b_2}$	$\mathbb{F}_q^\times \times \mathbb{F}_q^\times \times \mathbb{F}_q^\times \times \mathbb{F}_q \times \mathbb{F}_q$	$q^2(q-1)^3$	q
$\mathcal{F}_{5,6}$	$\chi_{5,6,a_1,a_2,b_1,b_2}$	$\mathbb{F}_q^\times \times \mathbb{F}_q^\times \times \mathbb{F}_q \times \mathbb{F}_q$	$q^2(q-1)^2$	q
$\mathcal{F}_{5,7}$	$\chi_{5,7,a_1,a_2,b_1,b_2}$	$\mathbb{F}_q^\times \times \mathbb{F}_q^\times \times \mathbb{F}_q \times \mathbb{F}_q$	$q^2(q-1)^2$	q
$\mathcal{F}_{6,7}$	$\chi_{6,7,a_1,a_2,b_1,b_2}$	$\mathbb{F}_q^\times \times \mathbb{F}_q^\times \times \mathbb{F}_q \times \mathbb{F}_q$	$q^2(q-1)^2$	q
$\mathcal{F}_5$	χ_{5,a,b_1,b_2}	$\mathbb{F}_q^\times \times \mathbb{F}_q \times \mathbb{F}_q$	$q^2(q-1)$	q
$\mathcal{F}_6$	χ_{6,a,b_1,b_2}	$\mathbb{F}_q^\times \times \mathbb{F}_q \times \mathbb{F}_q$	$q^2(q-1)$	q
$\mathcal{F}_7$	χ_{7,a,b_1,b_2}	$\mathbb{F}_q^\times \times \mathbb{F}_q \times \mathbb{F}_q$	$q^2(q-1)$	q
$\mathcal{F}_{\text{lin}}$	$\chi_{\text{lin},b_1,b_2,b_3,b_4}$	$\mathbb{F}_q \times \mathbb{F}_q \times \mathbb{F}_q \times \mathbb{F}_q$	q^4	1

Table 2 Positive roots of the root system Φ of type D_4

Height	Roots			
5	$\alpha_{12} := \begin{smallmatrix} & 1 & \\ 1 & 2 & 1 \end{smallmatrix}$			
4	$\alpha_{11} := \begin{smallmatrix} & 1 & \\ 1 & 1 & 1 \end{smallmatrix}$			
3	$\alpha_8 := \begin{smallmatrix} & 1 & \\ 1 & 1 & 0 \end{smallmatrix}$	$\alpha_9 := \begin{smallmatrix} & 0 & \\ 1 & 1 & 1 \end{smallmatrix}$	$\alpha_{10} := \begin{smallmatrix} & 1 & \\ 0 & 1 & 1 \end{smallmatrix}$	
2	$\alpha_5 := \begin{smallmatrix} & 0 & \\ 1 & 1 & 0 \end{smallmatrix}$	$\alpha_6 := \begin{smallmatrix} & 1 & \\ 0 & 1 & 0 \end{smallmatrix}$	$\alpha_7 := \begin{smallmatrix} & 0 & \\ 0 & 1 & 1 \end{smallmatrix}$	
1	α_1	α_2	α_3	α_4

Specifically, if $\ell > \max(i,j,k)$, then $X_\ell \subset \ker(\chi)$ for all $\chi \in \mathcal{F}_{i,j,k}$ and if $\ell \in \{i,j,k\}$, then for all $\chi \in \mathcal{F}_{i,j,k}$ we have $X_\ell \subset Z(\chi)$ but $X_\ell \not\subset \ker(\chi)$.

Remark. We construct all characters explicitly by inducing linear characters of certain subgroups.

The prime 2 is bad for classical groups which are not of type A. Recall that p *is a bad prime for* $X_r(q)$ means that when expressing the highest root $\alpha \in \Phi^+$ as a linear combination of simple roots, then at least one of the coefficients exceeds $p-1$. It is worth noting that in $UD_4(q)$ only the character degrees of the family $\mathcal{F}_{8,9,10}$ witness the difference between q even and q odd.

In a related project, Goodwin and Röhrle [12] enumerate all conjugacy classes of $X_r(q)$ where $r \leq 6$ and q is good for $X_r(q)$.

In an ongoing project Goodwin, Himstedt, Le, and Magaard are constructing the generic character table for $UD_4(q)$ and put it into CHEVIE. One intended application of this project is to help find the decomposition numbers for groups of type $D_4(q)$ in characteristics dividing $q+1$.

It is also interesting to note that while the character degrees for the family $\mathcal{F}_{11}$ do not depend on q, the formulae for the character values detects the parity of q. So the exceptional behavior of the prime 2 for UD_4 is witnessed by the character degrees of the family $\mathcal{F}_{8,9,10}$ and by certain character values of members of the family $\mathcal{F}_{11}$.

Finally, we remark that $UD_4^k(q)$ is a homomorphic image of $UD_{4k}(q)$. So for q even, using the family $\mathcal{F}_{8,9,10}$ we can construct characters of degree $2^{-i}q^{3k}$ for all $0 \leq i \leq k$, and thus we see a proliferation of evidence for the exceptional behavior of the prime 2 in terms of the behavior of the character degrees for the groups $UD_n(q)$ as n increases. Similarly we see, using the family $\mathcal{F}_{11}$, that the exceptional behavior for the prime 2 will increasingly be evident in the formulae for the character values of $UD_n(q)$ as well. Of course using the results of Sangroniz and of Anrdé and Neto we can witness similar behavior in the character degrees of the other classical groups.

3 The Number of Characters of a Given Degree

We now turn to Question 2. As we mentioned above in the 1960s, Higman conjectured that $|\mathrm{Irr}(USL_n(q))| \in \mathbb{Z}[q]$.

Definition 3.1. $N_{X,r,a}(q)$ is the number of irreducible characters of $UX_r(q)$ of degree q^a.

In the 1970s, Lehrer conjectured that $N_{A,n-1,a}(q) \in \mathbb{Z}[q]$. Recently, Isaacs conjectured that $N_{A,n-1,a}(q) \in \mathbb{N}[(q-1)]$. As

$$|\mathrm{Irr}(USL_n(q))| = \sum_{a=1}^{\mu(n)} N_{A,n-1,a}(q)$$

both conjectures imply Higman's conjecture. What evidence is there for the conjectures, and do they generalize to other groups of Lie type?

In the 1970s, Lehrer showed that

$$N_{A,2m-1,\mu_A(2m-1)}(q) \geq q(q-1)^{m-1}$$

and

$$N_{A,2m,\mu_A(2m)}(q) \geq (q-1)^m.$$

Later, Isaacs [18] showed that the inequalities are in fact equalities. More recently similar results have been obtained for classical groups which we summarize in the following theorem.

Theorem 3.2 (Sangroniz, André and Neto). *If q is odd and $X_r(q)$ is a classical group but not of type A then the number of characters of $UX_r(q)$ of maximal degree is in $\mathbb{N}[(q-1)]$. In particular, we have*

1. $N_{^2A,m,\mu_{^2A}(m)}(q) = N_{A,m,\mu_A(m)}(q)$.
2. $N_{B,m,\mu_B(m)}(q) = (q-1)^{m/2}q^{(m-2)/2}(q+1)$ *if m is even and* $(q-1)^{(m-1)/2}q^{(m+1)/2}$ *if m is odd.*
3. $N_{D,m,\mu_D(m)}(q) = (q-1)^{(m-2)/2}q^{(m+2)/2}$ *if m is even and* $(q-1)^{(m-1)/2}q^{(m+1)/2}$ *if m is odd.*
4. $N_{^2D,m,\mu_{^2D}(m)}(q) = N_{D,m,\mu_D(m)}(q)$.
5. $N_{C,m,\mu_C(m)}(q) = q(q-1)^{m-1}$.

We also conjecture the following.

Conjecture 3.3 (Le, Magaard). The following are true:

1. If q is odd, then $N_{F,4,10}(q) = (q-1)^3q$.
2. $N_{^2E,6,16}(q) = (q-1)^3q$.
3. $N_{E,6,16}(q) = (q-1)^3q$
4. $N_{E,7,28}(q) = (q-1)^3q^4$
5. $N_{E,8,56}(q) = (q-1)^4q^4$

Le's construction, in Sect. 4 below, gives characters of claimed degree and multiplicity. What remains to be shown is that there are no characters of larger degree.

In his thesis, Le proves recursions for the number of charaters of $UA_{n-1}(q)$ of second and third highest degree.

Theorem 3.4. *The following are true*

$$N_{A,n-1,\mu_A(n)-1}(q) = (q-1)N_{A,n-3,\mu_A(n-3)-1}(q) + q(q-1)^2N_{A,n-5,\mu_A(n-5)}(q)$$

if $n \geq 5$, and

$$\begin{aligned} N_{A,n-1,\mu_A(n)-2}(q) &= (q-1)N_{A,n-3,\mu_A(n-3)-2}(q) + q(q-1)N_{A,n-5,\mu_A(n-5)-1}(q) \\ &\quad + (2q^2(q-1)^3 + (q-1)^4)N_{A,n-7,\mu_A(n-7)}(q) \end{aligned}$$

if $n \geq 7$.

Using these recursions one can easily deduce the following result of Isaacs [17, 18].

Table 3 Numbers and degrees of the irreducible characters of $UD_4(q)$. The numbers of the irreducible characters are given as polynomials in $v = q - 1$

Degree	Number of irreducible characters	Comments
q^4	$v^4 + 3v^3 + 3v^2 + v$	
q^3	$v^5 + 5v^4 + 10v^3 + 7v^2 + v$	If q is odd
	$v^5 + 4v^4 + 10v^3 + 7v^2 + v$	If q is even
$\frac{q^3}{2}$	$4v^4$	Only if q is even
q^2	$3v^4 + 9v^3 + 9v^2 + 3v$	
q	$v^5 + 5v^4 + 10v^3 + 9v^2 + 3v$	
1	$v^4 + 4v^3 + 6v^2 + 4v + 1$	

Corollary 3.5. *The following is true:*

$$N_{A,n-1,\mu_A(n)-1}(q) = q(q-1)^{m-1}(m(q-1)+1)$$

if $n = 2m + 1$ *and*

$$N_{A,n-1,\mu_A(n)-1}(q) = q(q-1)^{m-1}((m-1)q+1)$$

if $n = 2m$.

For low rank groups of type A of rank up to 12 the conjectures of Higman, Lehrer and Isaacs have been confirmed by direct computation. For the ranks 9–12 results see Evseev [10]. For ranks ≤ 8 see Isaacs [18, 22, 23]. The character degrees of groups of type $UC_2(q)$ were computed by Lusztig [21]. The character degrees of groups of type $U^2G_2(q)$ were computed by Eaton [9]. The results of [16] from the previous section are summarized in the following table. The results show that Higman's conjecture does not generalize to situations where the prime is bad for $X_r(q)$ (Table 3).

4 Construction of Characters

Starting with André's PhD thesis [2] systematic constructions of the characters of the unitriangular groups have been undertaken. André's construction is based on studying orbits of the group on the dual of the adjoint representation of the associated Lie algebra, or co-adjoint orbits for short. This approach is also known as the Kirillov orbit method. This approach required a restriction on the characteristic of the field $\mathbb{F}_q$. However, for the case $UA_n(q)$ André in [3] was able to modify his construction so as to remove the restriction on the characteristic.

While the computation of the character table of $UA_r(q)$ has not been achieved to to date, and may never be as this is a provably wild problem, several important milestones have been achieved. It was shown by Diaconis and Isaacs [7] that the characters, which André [3] calls basic, and certain unions of conjugacy classes, now called superclasses, form a partition of the $\mathrm{Irr}(UA_r(q))$, respectively, the conjugacy classes of $UA_r(q)$ in such a way that the basic characters are constant on superclasses. Thus, the basic characters together with the superclasses are an example of a supercharacter table in the sense of Diaconis and Isaacs [7]. Most recently, André and Neto [4, 5] computed a supercharacter table for symplectic and orthogonal groups defined over fields of odd characteristic.

The entries in the supercharacter table of $UA_r(q)$ can be interpreted in a combinatorial way. This has lead to connections to other areas of mathematics, for example Hopf algebras. For development of the combinatorial aspects see Diaconis and Thiem [8] and the references contained therein. For the most recent developments concerning the connection to Hopf algebras see [1].

Still another approach to the characters of unitriangular groups, suggested by Drinfeld, which is based on a geometric interpretation of the Kirillov orbit method has been explored by Boyarchenko [6].

In his thesis, Le took a more group/Lie -theoretic approach which we now describe.

Given a root system Φ, all of whose roots have the same length, for a group of Lie type X and a root $\alpha \in \Phi^+$ define the *hook*

$$h(\alpha) := \{\beta \in \Phi^+ \ : \ \alpha - \beta \in \Phi^+ \cup \{0\}\}.$$

Correspondingly the subgroup

$$H_\alpha := \prod_{\gamma \in h(\alpha)} X_\gamma \subseteq U$$

is called the *hook subgroup* of UX at α.

Remark. If X is quasisimple, α is the highest root of Φ^+ and UX is the unipotent radical of the standard Borel subgroup, then H_α is the unipotent radical of the parabolic subgroup $N_X(X_\alpha)$.

Remark. If α is not the highest root of Φ^+, then H_α is a subgroup of the intersection of the unipotent radical of $C_X(X_\alpha)$ with UX.

For all $\alpha \in \Phi^+$ and all $\lambda \in Hom(\mathbb{F}_q, \mathbb{C})^*$, we can define a unique $\chi_{\alpha,\lambda} \in \mathrm{Irr}(UX)$ of degree $q^{(|h_\alpha|-1)/2}$.

For $X = SL_n(q)$ André calls these elementary characters. For general X, Le calls them *midafi* for "minimal degree almost faithful irreducible". For types D and B some of André and Neto's elementary characters are not irreducible whereas midafi characters always are.

Definition 4.1. Let $X = \mathrm{SL}_n(q)$ and Φ^+ be the set of positive roots. We call a subset $\mathcal{S} \subset \Phi^+$ admissible if for all $\alpha, \beta \in \mathcal{S}$ we have $\langle \alpha, \beta \rangle \ \leq \ 0$; i.e. if no two positions ever lie on the same row or column.

For an admissible set $\mathcal{S} = \{\alpha_1, \ldots, \alpha_k\}$ and $\Lambda \in (Hom(\mathbb{F}_q, \mathbb{C})^*)^k$ we define

$$T(\mathcal{S}, \Lambda) := \otimes_{i=1}^k \chi_{\alpha_i, \lambda_i}.$$

Example: If α_h is the highest root of Φ^+, $n = 2m + 1$, and $\mathcal{S}$ is

$$\{\alpha_h,\ \alpha_h - \alpha_1 - \alpha_{n-1},\ \alpha_h - \alpha_1 - \alpha_2 - \alpha_{n-2} - \alpha_{n-2}, \ldots, \alpha_m + \alpha_{m+1}\}$$

then $T(\mathcal{S}, \Lambda)$ is irreducible of maximal degree, namely $q^{\mu(n)}$ for all Λ. Moreover the the group of diagonal matrices T of $\mathrm{GL}_n(q)$ normalizing $USL_n(q)$ acts transitively on the set of characters of maximal degree, showing that $N_{A,2m,\mu_a(2m+1)}(q) \geq (q-1)^m$

Remark 4.2. More generally, for fixed $\mathcal{S}$ the action of T on the set of $T(\mathcal{S}, \Lambda)$ is transitive.

Remark 4.3. Every linear character is equal to some $T(\mathcal{S}, \Lambda)$ where $\mathcal{S} \subset \{\alpha_1, \alpha_2, \ldots, \alpha_{n-1}\}$.

Theorem 4.4. *For every $\chi \in \mathrm{Irr}(USL_n(q))$ there exists a unique admissible set $\mathcal{S}$ and a unique Λ such that $\langle \chi, T(\mathcal{S}, \Lambda) \rangle \neq 0$.*

There are several proofs of this result in the literature; see for example André [3], Diaconis and Isaacs [7], and Le's thesis [19].

The characters $T(\mathcal{S}, \Lambda)$ are the basic characters defined by André and are examples of supercharacters.

We define

$$i(\mathcal{S}) := \sum_{\alpha \neq \beta \in \mathcal{S}} |h(\alpha) \cap h(\beta)|.$$

Theorem 4.5. $\langle T(\mathcal{S}, \Lambda), T(\mathcal{S}, \Lambda) \rangle = q^{i(\mathcal{S})}$ *and the degree of every constituent is divisible by $T(\mathcal{S}, \Lambda)(1)/q^{i(\mathcal{S})}$*

The problem that remains is how do the characters $T(\mathcal{S}, \Lambda)$ decompose? To this end we make the following

Definition 4.6. An additively closed subset of Φ^+ is called a *pattern*. Subgroups generated by the root subgroups indexed by a pattern are called *pattern subgroups*.

Starting from $\mathcal{S}$ Le defines subgroups $D \cong USL_{|\mathcal{S}|}(q)$, and R and C, where R and C are pattern subgroups of D, and a subgroup K of $USL_n(q)$.

Theorem 4.7 (Le). *The constituents of $\Lambda|_{Z(R/K)} \uparrow^{R/K}$ are in one to one correspondence with the constituents of $T(\mathcal{S}, \Lambda)$.*

For example, if $\Lambda|_{Z(R/K)} = 1$, then $\Lambda|_{Z(R/K)} \uparrow^{R/K}$ is the regular character.

Using this Le proves the formulae for $N_{n,\mu(n)-1}(q)$, and $N_{n,\mu(n)-2}(q)$ by showing that characters of large degree must be consituents of $T(\mathcal{S}, \Lambda)$ where $i(\mathcal{S}) \leq 3$.

We now give the definitions for the relevant pattern subgroups defined by an admissible set $\mathcal{S}$. Suppose that $\mathcal{S} = \{\alpha_{i,j}\}$, where $\alpha_{i,j} := \epsilon_i - \epsilon_j$. The set

$$\mathcal{T}_{\alpha_{i,j}} := \{\alpha_{i,s} \ : \ s < i\}$$

and

$$\mathcal{T} := \bigcup_{\alpha \in \mathcal{S}} \mathcal{T}_\alpha.$$

Now define

$$K := \langle X_\alpha \ : \ \alpha \in \Phi^+ \setminus \mathcal{T} \rangle.$$

Set

$$r(\mathcal{S}) := \{i \ : \ \alpha_{i,j} \in \mathcal{S}\},$$

and define

$$D := \langle X_{\alpha_{g,h-1}} \ : \ g, h \in r(\mathcal{S}) \rangle.$$

Note that if $X_{\alpha_{g,h}} \subset D$, then $\alpha_{g,k}, \alpha_{h,\ell} \in \mathcal{S}$. So define

$$R := \langle X_{\alpha_{g,h-1}} \ : \ k > \ell \rangle$$

and

$$C := \langle X_{\alpha_{g,h-1}} \ : \ k < \ell \rangle.$$

We illustrate the theorem with an example: If $\mathcal{S} := \{\alpha_{1,5}, \alpha_{2,7}, \alpha_{3,6}, \alpha_{4,8}\}$, then $T(\mathcal{S}, \Lambda) = q^{12}$ and $i(\mathcal{S}) = 5$.

Here, $D := \langle X_{\alpha_{i,j}} \ : \ 1 \leq i < j \leq 4 \rangle$, $C = X_{\alpha_{2,3}}$ and R is a special group of order q^{1+4}. Thus, the regular character of R is a sum of $(q-1)$ characters of degree q^2 each occurring with multiplicity q^2 and q^4 characters of degree 1 each occurring with multiplicity 1. Thus, $T(\mathcal{S}, \Lambda)$ decomposes into $(q-1)$ characters of degree q^9 each with multiplicity q^2 and q^4 characters of degree q^7 each with multiplicity 1.

5 Outlook

While Le's thesis sheds light on the decomposition of $T(\mathcal{S}, \Lambda)$ a proof of Higman's conjecture remains elusive. Why? What is needed is an understanding of $\mathrm{Irr}(R/K)$, where R is a pattern group. However, Halasi [14] showed that Higman's conjecture does not hold for all pattern groups.

However, the pattern subgroups that need to be considered for the purpose of decomposing $T(\mathcal{S}, \Lambda)$ enjoy the additional property that they are *complemented*, that is that both R and C are pattern subgroups such that $R \cap C = 1$ and $D = \langle R, C \rangle$. It remains an open question how this extra property can be used.

In closing we mention some open questions and ongoing projects:

- Following Tits' philosophy Himstedt, Le, and Magaard are constructing all characters of $UX_r(q)$ where $X_r(q)$ is exceptional.
- Explain why the primes 3, respectively, 5 are bad for E_6, respectively, E_8 in terms of the representation theory of UE_6 and UE_8.
- The supercharacter theory developed for the classical groups by André is obtained by restricting supercharacters of USL_n down to UX where X is classical. Is it possible to define a supercharacter theory that is intrinsic to UX, where X is classical?
- Apart from the supercharacter theory that Diaconis and Isaacs [7] define for algebra groups, do there exist other "natural" supercharacter theories for pattern subgroups, or for complemented pattern subgroups?
- How can the fact that the pattern subgroups that govern the tensor product decomposition of $T(S, \Lambda)$ are complemented be used?
- How much does the representation theory of UX determine the cross characteristic representations of X?

References

1. Aguiar, Marcelo; André, Carlos; Benedetti, Carolina; Bergeron, Nantel; Chen, Zhi; Diaconis, Persi; Hendrickson, Anders; Hsiao, Samuel; Isaacs, I. Martin; Jedwab, Andrea; Johnson, Kenneth; Karaali, Gizem; Lauve, Aaron; Le, Tung; Lewis, Stephen; Li, Huilan; Magaard, Kay; Marberg, Eric; Novelli, Jean-Christophe; Pang, Amy; Saliola, Franco; Tevlin, Lenny; Thibon, Jean-Yves; Thiem, Nathaniel; Venkateswaran, Vidya; Vinroot, C. Ryan; Yan, Ning; Zabrocki, Mike, Supercharacters, symmetric functions in noncommuting variables, and related Hopf algebras, http://adsabs.harvard.edu/abs/2010arXiv1009.4134A
2. C. André, Irreducible characters of the unitriangular group and coadjoint orbits, PhD thesis, University of Warwick, 1992
3. C. André, The basic character table of the unitriangular group, *J. Algebra* **241** (1) (2001), 437–471
4. C. André, A. Neto, Supercharacters of the Sylow p-subgroups of the finite symplectic and orthogonal groups, *Pacific J. Math.* **239** (2)(2009), 201–230
5. C. André, A. Neto, A supercharacter theory for the Sylow p-subgroups of the finite symplectic and orthogonal groups, *J. Algebra* **322** (2009), 1273–1294
6. M. Boyarchenko, Characters of Unipotent Groups over Finite Fields, **arXiv:0712.2614v1 [math.RT] 17 Dec 2007**
7. P. Diaconis, I. M. Isaacs, Supercharacters and superclasses for algebra groups, *Trans. Amer. Math. Soc.*, **360** (2008), 2359–2392
8. P. Diaconis, N. Thiem, Supercharacter formulas for pattern groups. *Trans. Amer. Math. Soc.* **361** (2009), no. 7, 3501–3533
9. C. Eaton, Dade's Inductive Conjecture for the Ree Groups of Type G_2 in Defining Characteristic, *J. Algebra* **226** (2000), 614–620
10. A. Evseev, Reduction for characters of finite algebra groups, *J. Algebra* **325** (2011), 321–351
11. M. Geck, G. Hiß, F. Lübeck, G. Malle, G. Pfeiffer, CHEVIE – A system for computing and processing generic character tables for finite groups of Lie type, Weyl groups and Hecke algebras, *Appl. Algebra Engrg. Comm. Comput.*, **7** (1996), 175–210

12. S. Goodwin, G. Röhrle, Calculating conjugacy classes in sylow p-subgroups of finite chevalley groups, *J. Algebra*, **321** (2009), 3321–3334
13. E. Gutkin, Representations of algebraic unipotent groups over a self-dual field, Funkts. Analiz i Fgo Prilozheniya **7** (1973), 80
14. Zoltán Halasi, On representations of solvable linear groups, PhD Thesis, Central European University, Budapest, Hungary (2009)
15. G. Higman, Enumerating p-groups. I. Inequalities, Proc. London Math. Soc. (3) 10 (1960) 24–30
16. F. Himstedt, T. Le, K. Magaard, Characters of the Sylow p-subgroups of the Chevalley groups $D_4(p^n)$., Journal of Algebra **332** (2011), 414–427
17. I. M. Isaacs, Characters of groups associated with finite algebras, *J. Algebra*, **177** (1995), 708–730
18. I. M. Isaacs, Counting characters of upper triangular groups, *J. Algebra*, **315** (2007), 698–719
19. T. Le, Irreducible Characters of the Unitriangular Groups, Thesis, Wayne State University, Detroit, MI 2008
20. G. Lehrer, Discrete series and the unipotent subgroup, *Compos.Math.*, **28** (1974), 9–19
21. G. Lusztig, Character sheaves and generalizations, in "The unity of mathematics" (edited by P. Etinghof, V. Retakh, I. M. Singer), 443–455, Progress in Math. **244**, Birkhäuser Boston, MA 2006
22. J. Sangroniz, Character degrees of the Sylow p-subgroups of classical groups. Groups St. Andrews 2001 in Oxford. Vol. II, 487–493, London Math. Soc. Lecture Note Ser., 305, Cambridge Univ. Press, Cambridge, 2003
23. B. Szegedy, Characters of the Borel and Sylow subgroups of classical groups, *J. Algebra*, **267** (2003), 130–136

Hermitian Veronesean Caps

J. Schillewaert and H. Van Maldeghem

Abstract In [J. Schillewaert and H. Van Maldeghem, Quadric Veronesean caps, *Discrete Mathematics*], a characterization theorem for Veronesean varieties in $\mathsf{PG}(N, \mathbb{K})$, with $\mathbb{K}$ a skewfield, is proved. This result extends the theorem for the finite case proved in [J. A. Thas and H. Van Maldeghem, Quart. J. Math. 55 (2004), 99–113]. In this paper, we prove analogous results for Hermitian varieties, extending the results obtained in the finite case in [B. Cooperstein, J. A. Thas and H. Van Maldeghem, Forum Math. 16 (2004), 365–381] in a non-trivial way.

Subject Classifications: 51A45

Keywords Veronesean variety • Hermitian variety • Veronesean embedding

1 Introduction

In [4], we showed that a Veronesean cap endowed with its ovals is a projective space (using Veblen's axiom) and as a corollary we obtained a characterization for quadric Veronesean varieties as a representation of a projective space in another projective space where lines of the former are ovals in the latter. In this paper, we consider Hermitian Veronesean varieties. Here, a characterization as a union of ovoids with

J. Schillewaert (✉)
Department of Mathematics and Statistics, University of Canterbury, New Zealand

Département de Mathématique, Université Libre de Bruxelles, U.L.B., CP 216, Bd du Triomphe, 1050 Bruxelles, Belgique
e-mail: jschille@ulb.ac.be

H. Van Maldeghem
Department of Mathematics, Ghent University, Krijgslaan 281-S22, 9000 Ghent, Belgium
e-mail: hvm@cage.ugent.be

N.S.N. Sastry (ed.), *Buildings, Finite Geometries and Groups*, Springer Proceedings in Mathematics 10, DOI 10.1007/978-1-4614-0709-6_9,

an additional assumption on the tangent planes can be proved in much the same spirit as for quadric Veroneseans, except that, after using Veblen's axiom, one needs an argument to show that all ovoids are isomorphic elliptic quadrics (in the finite case there is only one isomorphism class of elliptic quadrics in projective 3-space over a fixed finite field, since Galois extensions of degree 2 are unique for finite fields). Then we obtain a characterization of Hermitian Veroneseans as a representation of a projective space in another one where the lines of the former are ovoids of the latter. In the finite case, the key lemma is Lemma 3.2 of [7], which is proved with a typical finiteness argument, using counting and dividing. Moreover, later in the proof, in order to arrive at the André representation of an affine plane, one heavily uses the fact that the number of points is already correct. These two facts seemed, up to now, too heavy obstacles for the infinite case. However, in the present paper we use entirely different ideas to generalize Lemma 3.2 of [7]. The proof we provide is more geometric and also largely holds in the finite case, and it provides enough insight in the matter to analyze the smallest case, which is an exception to the theorem, see Remark 4.2. We will describe this case in detail.

We end the paper with an easy application, providing a simple alternative geometric definition of the Hermitian Veronesean. This application generalizes a result of Lunardon [3].

In Sect. 2, we introduce the necessary notions: we review the Veblen–Young theorem [8], which is crucial in our arguments, define Hermitian Veroneseans, and state our main results.

2 Notation and Main Results

2.1 Axiomatization of Projective Spaces

A good exposition on the foundations of projective and polar spaces can be found on Peter Cameron's website, and the paragraph below is based on these lecture notes. At the end of the nineteenth century a lot of work was done on the axiomatization of projective spaces, starting with Pasch. This work culminated in 1910 when Veblen and Young provided a beautiful characterization of projective spaces [8] based on the following axiom.

Veblen's Axiom

If a line intersects two sides of a triangle but does not contain their intersection then it also intersects the third side.

Theorem 2.1 (Veblen–Young theorem). *Let $(X, \mathcal{L})$ be a thick linear space satisfying Veblen's axiom. Then one of the following holds:*

(1) $X = \mathcal{L} = \emptyset$.
(2) $|X| = 1$, $\mathcal{L} = \emptyset$.

(3) $\mathcal{L} = \{X\}, |X| \geq 3.$
(4) $(X, \mathcal{L})$ *is a projective plane.*
(5) $(X, \mathcal{L})$ *is a projective space over a skew field, not necessarily of finite dimension.*

2.2 Quadric Veronesean Caps

Since we will need our results of [4], we briefly recall them here.

Let X be a spanning point set of $\mathrm{PG}(N, \mathbb{K})$, with $\mathbb{K}$ any skew field, and let Π be a collection of planes of $\mathrm{PG}(N, \mathbb{K})$ such that, for any $\pi \in \Pi$, the intersection $\pi \cap X$ is an oval $X(\pi)$ in π (and then, for $x \in X(\pi)$, we denote the tangent line at x to $X(\pi)$ by $T_x(X(\pi))$ or sometimes simply by $T_x(\pi)$). We call X a *Veronesean cap* if the following conditions hold:

(V1) Any two points x and y of X lie in a unique element of Π, denoted by $[x, y]$.
(V2) If $\pi_1, \pi_2 \in \Pi$, with $\pi_1 \neq \pi_2$, then $\pi_1 \cap \pi_2 \subset X$.
(V3) If $x \in X$ and $\pi \in \Pi$, with $x \notin \pi$, then each of the lines $T_x([x, y])$, $y \in \pi \cap X$, is contained in a fixed plane of $\mathrm{PG}(N, \mathbb{K})$, denoted by $T(x, \pi)$.

Then we have the following result.

Theorem 2.2 (Schillewaert and Van Maldeghem [4]). *Let X be a Veronesean cap in $\Pi := \mathrm{PG}(N, \mathbb{K})$. Then $\mathbb{K}$ is a field and there exists a natural number $n \geq 2$ (called the index of X), a projective space $\Pi' := \mathsf{PG}(n(n+3)/2, \mathbb{K})$ containing Π, a subspace R of Π' skew to Π, and a quadric Veronesean $\mathcal{V}_n$ of index n in Π', with $R \cap \mathcal{V}_n = \emptyset$, such that X is the (bijective) projection of $\mathcal{V}_n$ from R onto Π. The subspace R can be empty, in which case X is projectively equivalent to $\mathcal{V}_n$.*

As an application, we showed the following characterization, which basically replaces Condition (V2) with a dimension restriction, and (V3) with the condition that the geometry of points and ovals is a projective space.

Theorem 2.3 (Schillewaert and Van Maldeghem [4]). *Let X be a spanning set of points in the projective space $\mathsf{PG}(d, \mathbb{K})$, with $\mathbb{K}$ any skew field of order at least 3. Suppose that*

(V1*) *For any pair of points $x, y \in X$, there is a unique plane denoted $[x, y]$ such that $[x, y] \cap X$ is an oval, denoted $X([x, y])$;*
(V2*) *The set X endowed with all subsets $X([x, y])$, has the structure of the point-line geometry of a projective space $\mathsf{PG}(n, \mathbb{F})$, for some skew field $\mathbb{F}$, $n \geq 3$, or of any projective plane Π (and we put $n = 2$ in this case);*
(V3*) $d \geq \frac{1}{2}n(n+3)$.

Then $d = \frac{1}{2}n(n+3)$ and X is the point set of a quadric Veronesean of index n. In particular, $\mathbb{F} \cong \mathbb{K}$ if $n \geq 3$, and Π is isomorphic to $\mathsf{PG}(2, \mathbb{K})$ if $n = 2$.

2.3 Hermitian Veronesean caps

An *ovoid* O in a 3-dimensional projective space Σ is a set of points of Σ such that no line of Σ intersects O in at least 3 points, and for every point $x \in O$, there is a unique plane π through x intersecting O in only x and containing all lines through x that meet O in only x. The plane π is called the *tangent plane* at x to O and denoted $T_x(O)$.

Let X be a spanning point set of $\mathrm{PG}(N,\mathbb{K})$, with $\mathbb{K}$ any skew field, and let Ξ be a collection of 3-dimensional projective subspaces of $\mathrm{PG}(N,\mathbb{K})$, called the elliptic spaces of X, such that, for any $\xi \in \Xi$, the intersection $\xi \cap X$ is an ovoid $X(\xi)$ in ξ (and then, for $x \in X(\xi)$, we sometimes denote $T_x(X(\xi))$ simply by $T_x(\xi)$). We call X a *Hermitian Veronesean cap* if the following conditions hold:

(H1) Any two points x and y of X lie in a unique element of Ξ, denoted by $[x,y]$.
(H2) If $\xi_1,\xi_2 \in \Xi$, with $\xi_1 \neq \xi_2$, then $\xi_1 \cap \xi_2 \subset X$.
(H3) If $x \in X$ and $\xi \in \Xi$, with $x \notin \xi$, then each of the planes $T_x([x,y])$, $y \in \xi \cap X$, is contained in a fixed 4-dimensional subspace of $\mathrm{PG}(N,\mathbb{K})$, denoted by $T(x,\xi)$.

In [1], it was shown that the following are examples of Hermitian Veronesean caps.

Hermitian Veronesean Varieties

Let n be a positive integer, let $\mathbb{L}$ be a quadratic extension of a field $\mathbb{K}$, and consider the projective spaces $\mathsf{PG}(n,\mathbb{L})$ and $\mathsf{PG}(N,\mathbb{K})$ with $N = n(n+2)$. Let $r \in \mathbb{L}\setminus\mathbb{K}$ be arbitrary. Then the *Hermitian Veronesean variety of index n* is the image of the map $\pi : \mathsf{PG}(n,\mathbb{L}) \to \mathsf{PG}(N,\mathbb{K})$

$$\pi(\langle(x_0,x_1,\ldots,x_n)\rangle) = \langle(y_{i,j})_{0\le i,j\le n}\rangle,$$

with $y_{i,i} = x_i\bar{x}_i$, $y_{i,j} = x_i\bar{x}_j + \bar{x}_i x_j$ (for $i < j$), and $y_{i,j} = r x_i\bar{x}_j + \bar{r}\bar{x}_i x_j$ (for $i > j$), where $\bar{x}$ denotes the conjugate of x.

Moreover, in the finite case, it is proved in [1] that every Hermitian Veronesean cap is a suitable projection of some Hermitian Veronesean variety. Below we generalize this to the case of finite-dimensional projective spaces over infinite fields.

The following is our main result.

Theorem 2.4. *Let X be a Hermitian cap in $\Pi = \mathsf{PG}(N,\mathbb{K})$, $N > 3$, with corresponding set Ξ of elliptic spaces. Then $\mathbb{K}$ is commutative. Also, X endowed with all $X(\xi)$, for $\xi \in \Xi$, is the point-line structure of a projective space $\mathsf{PG}(n,\mathbb{L})$, with $\mathbb{L}$ a quadratic extension of $\mathbb{K}$, and X is projectively equivalent to a quotient of a Hermitian Veronesean variety of index n. If $n = 2,3$, then $N = n(n+2)$ and X is projectively equivalent to a Hermitian Veronesean variety of index n.*

As an application, we also show the following elegant characterization, which basically replaces Condition (H2) with a dimension restriction, and (H3) with the condition that the geometry of points and ovoids is a projective space.

Theorem 2.5. *Let X be a spanning set of points in the projective space $\mathsf{PG}(d,\mathbb{K})$, with $\mathbb{K}$ any skew field of order at least 3. Suppose that*

(H1*) *For any pair of points $x, y \in X$, there is a unique 3-dimensional subspace denoted $[x, y]$ such that $[x, y] \cap X$ is an ovoid, denoted $X([x, y])$;*
(H2*) *The set X endowed with all subsets $X([x, y])$, has the structure of the point-line geometry of a projective space $\mathsf{PG}(n,\mathbb{L})$, for some skew field $\mathbb{L}$, $n \geq 3$, or of any projective plane Π (and we put $n = 2$ in this case);*
(H3*) $d \geq n(n+2)$.

Then $d = n(n+2)$ and X is the point set of a Hermitian Veronesean variety of index n. In particular, $\mathbb{L}$ is a quadratic extension of $\mathbb{K}$ if $n \geq 3$, and Π is isomorphic to $\mathsf{PG}(2,\mathbb{F})$ if $n = 2$, where $\mathbb{F}$ is a quadratic extension of $\mathbb{K}$.

3 Hermitian Veronesean Caps

3.1 The Projective Space Associated with the Cap

Let $\mathcal{H} = (X, \Theta)$ be a Hermitian Veronesean cap, where X is a set of points in $\mathsf{PG}(N,\mathbb{K})$, for some skew field $\mathbb{K}$, and Θ is a set of elliptic spaces satisfying (H1), (H2) and (H3) introduced before.

Associated with $\mathcal{H}$ we can consider the geometry $\mathcal{P}$ having point set X and line set the set Ξ, endowed with the natural incidence.

Lemma 3.1. *$\mathcal{P}$ is a projective space.*

Proof. First of all, $\mathcal{P}$ is a linear space by (H1).

Let x_{12}, x_{23} and x_{13} be three points of X and denote $O_1 = X([x_{12}, x_{13}])$, $O_2 = X([x_{12}, x_{23}])$ and $O_3 = X([x_{13}, x_{23}])$. Let O_4 be an oval intersecting O_1 in a point x_{14} and O_2 in a point x_{24}, both different from x_{12}. Our purpose is to show that Veblen's axiom holds, which means that we have to show that O_4 intersects O_3. Of course, we may assume that $O_3 \neq O_4$ and that O_4 does not contain x_{13} nor x_{23}. Clearly, $6 \leq \dim V \leq 8$, with $V := \langle O_1, O_2, O_3 \rangle$, and we claim that V contains O_4.

Indeed, let us first show that V contains O_4. Since both $T_{x_{13}}(O_3)$ and $T_{x_{13}}(O_1)$ belong to $\langle O_1, O_3 \rangle \subseteq V$, also $T_{x_{13}}([x_{13}, x_{24}])$ does by applying (H3) with as point x_{13} and as ovoid O_2, and hence $[x_{13}, x_{24}] = \langle T_{x_{13}}([x_{13}, x_{24}]), x_{24} \rangle$ is contained in V. Likewise, applying (H3) to x_{24} and O_1 and reasoning as above yields that O_4 is contained in V.

If V were 6-dimensional, then O_4 and O_3 would meet, and Veblen's axiom would follow automatically.

Next, suppose that $\dim V = 8$. Now we project $V \setminus \langle O_2 \rangle$ from $\langle O_2 \rangle$ onto a four-dimensional space Π of V disjoint from $\langle O_2 \rangle$. The ovoids O_3 and O_4 together with their tangent planes at their intersection point with O_2 are mapped onto two full planes of Π, say Π_3 and Π_4, respectively. Note that Π_3 and Π_4 generate Π, as one has similarly as above that O_1 is contained in $\langle O_2, O_3, O_4 \rangle$, and so the latter is 8-dimensional. Let x be the unique intersection point of Π_3 and Π_4. There are basically four different possibilities.

(1) *There is a point x_i of $O_i \setminus O_2$ projected onto x from $\langle O_2 \rangle$, for $i = 3, 4$, and $x_3 \neq x_4$.*
In this case, since the space $\langle x_3, x_4, O_2 \rangle = \langle x, O_2 \rangle$ is 4-dimensional, the line $\langle x_3, x_4 \rangle$ meets the elliptic space $\langle O_2 \rangle$ in a point y. This implies that the elliptic space $[x_3, x_4]$ intersects $\langle O_2 \rangle$ in y, implying $y \in X$, contradicting $X([x_3, x_4])$ being an ovoid.

(2) *There is a point x_3 of $O_3 \setminus O_2$ projected onto x from $\langle O_2 \rangle$, and the tangent plane $T_{x_{24}}(O_4)$ to O_4 at x_{24} projects from $\langle O_2 \rangle$ onto a line L_4 through x.*
In this case, clearly $\langle T_{x_{24}}(O_4) \rangle$ is contained in $\langle O_2, L_4 \rangle$, which also contains $T_{x_{24}}(O_2)$. Hence, by our axioms, the 5-space $\langle O_2, L_4 \rangle$ also contains $T_{x_{24}}([x_{13}, x_{24}])$ (since the ovoids O_2, O_4 and $X([x_{13}, x_{24}])$ all intersect O_1). Similarly, since the ovoids $X([x_{13}, x_{24}]), O_2$ and $X([x_3, x_{24}])$ all meet the ovoid O_3, the plane $T_{x_{24}}([x_3, x_{24}])$ belongs to $\langle O_2, L_4 \rangle$, which implies that $[x_3, x_{24}]$ belongs to the 5-space $\langle O_2, L_4 \rangle$ and so $[x_3, x_{24}]$ meets $\langle O_2 \rangle$ in a line, contradicting our axioms.

(3) *The tangent plane $T_{x_{2i}}(O_i)$ to O_i at x_{2i}, $i = 3, 4$, projects from $\langle O_2 \rangle$ onto a line L_i through x.*
In this case, as above, the 5-space $\langle O_2, L_4 \rangle$ contains $T_{x_{24}}([x_{13}, x_{24}])$. It follows that the 7-space $U := \langle O_2, L_3, L_4, x_{13} \rangle$ contains $X([x_{13}, x_{24}]), O_2$ and O_3. But, as above, using (H3) with x_{13} and O_2 one easily deduces that U also contains O_1, and so U coincides with V, a contradiction.

(4) The only remaining possibility is that *there is a point z of $(O_3 \cap O_4) \setminus O_2$ projected onto x from $\langle O_2 \rangle$*. But then $O_3 \cap O_4$ is nonempty, and that is exactly what we had to prove.

Finally, suppose that $\dim V = 7$. Now we project $V \setminus \langle O_2 \rangle$ from O_2 onto a three-dimensional space Σ of V disjoint from $\langle O_2 \rangle$. The ovoids O_3 and O_4 together with their tangent planes at their intersection point with O_2 are mapped onto two full planes of Σ, say Π_3 and Π_4, respectively. Let L be the intersection line of Π_3 and Π_4. We distinguish three cases.

(1) *Both tangent planes $T_{x_{23}}(O_3)$ and $T_{x_{24}}(O_4)$ are projected from O_2 onto lines L_3 and L_4 which are distinct from L.* Then we can pick a point y on L not contained in L_3 nor L_4 and continue as in (1) of the case $\dim V = 8$.

(2) *The tangent plane $T_{x_{24}}(O_4)$ is projected onto L from O_2 and $T_{x_{23}}(O_3)$ is projected onto a line $L_3 \neq L$.* Then choose a point x_3 of O_3 that is projected onto a point of L and reason as in (2) of the case $\dim V = 8$ to conclude that $X([x_3, x_{24}])$ belongs to the 5-space $\langle O_2, L \rangle$ and so $[x_3, x_{24}]$ meets $\langle O_2 \rangle$ in a line, contradicting our axioms.

(3) *Both tangent planes* $T_{x_{23}}(O_3)$ *and* $T_{x_{24}}(O_4)$ *are projected onto* L *from* O_2. Then arguing as in (3) of the case $\dim V = 8$ yields that $V = \langle O_2, L, x_{13}\rangle$, a contradiction since $\langle O_2, L, x_{13}\rangle$ is 6-dimensional.

Hence, we have shown that Veblen's axiom holds.

Note that $\mathcal{P}$ is not necessarily finite-dimensional at this stage. If $\mathcal{P}$ has finite dimension n, then we say that the Hermitian Veronesean cap has *index* n.

3.2 The Basic Step

Assume that our Hermitian Veronesean cap has index 2. Then we have the following result.

Theorem 3.2. *If* $\mathcal{H} = (X, \Theta)$ *is a Hermitian cap of index* 2 *in* $\mathsf{PG}(N, \mathbb{K})$*, then* $N = 8$ *and* X *is projectively equivalent to a Hermitian Veronesean variety of index* 2.

Proof. We choose an arbitrary elliptic space ξ and corresponding ovoid $O := X(\xi)$. Assume, by way of contradiction, that there is a 4-dimensional space U containing ξ and two points x, y of $\mathcal{H}$ not on O. Then $[x, y]$ contains the line xy, which intersects ξ in some point z, which must necessarily belong to O in view of Condition (H2). But then the ovoid $X([x, y])$ contains three collinear points x, y, z, a contradiction.

It follows that, if W_4 is an $(N - 4)$-dimensional subspace of $\mathsf{PG}(N, \mathbb{K})$ skew to ξ, then the projection ρ_4 of $X \setminus O$ from ξ onto W_4 is injective. Let p be any point of O. Then $T(p) := T(p, \xi^*)$, for an arbitrary elliptic space ξ^* not containing p, is 4-dimensional and intersects ξ in the plane $T_p(O)$. Hence $T(p)^{\rho_4}$ is a line, which we denote by L_p. For any member $\xi' \in \Theta$ containing p, with $\xi' \neq \xi$, each line in ξ' through p is either contained in $T(p)$ or intersects $X(\xi')$ in a second point. This immediately implies that the projection of $X(\xi') \setminus \{p\}$ is an affine plane $\alpha_{\xi'}$ in W_4 which, completed with L_p, becomes a projective plane $\pi_{\xi'}$. Choose $q \in O$ with $q \neq p$ and let $\xi'' \in \Theta$ be arbitrary but such that $q \in \xi''$ and $\xi'' \neq \xi$. First, remark that $\alpha_{\xi'}$ and $\alpha_{\xi''}$ share exactly one point (namely, the projection of the intersection $\xi' \cap \xi''$; they cannot share more by injectivity of ρ_4). Hence, W_4 has dimension at least 4. Also, considering the varying ovoid $X([q', p])$, with $q' \in X(\xi'') \setminus \{q\}$, all points of $(X \setminus O)^{\rho_4}$ arise in all planes of W_4 that are spanned by L_p and some point of $\alpha_{\xi''}$, and no such point is contained in L_p (as L_p always corresponds with the tangent plane at p to $X([q', p])$). It follows that $\dim W_4 = 4$ (and hence $N = 8$), and that the projection of $X \setminus O$ from ξ coincides with the affine space obtained from W_4 by deleting the 3-space Σ generated by L_p and L_q (this is indeed a 3-space for, if the lines L_p and L_q would intersect in a point x, then the projection of the elliptic spaces through an arbitrary point z of $X \setminus O$ and p, respectively, q, would intersect in the set $\langle z^{\rho_4}, x\rangle \setminus \{x\}$, which contains at least one point of the image of ρ_4 coming from different points on $X([p, z])$ and $X([q, z])$, a contradiction to the injectivity of ρ_4 and Condition (H2)). It also follows from this argument that, for any point $r \in O$, the line L_r is contained in Σ. Moreover, if s is an arbitrary point of Σ, then we can choose two points s_1, s_2 in $W_4 \setminus \Sigma$ such that s, s_1, s_2 are collinear. Considering the inverse images under ρ_4 of s_1 and s_2, and the unique member ξ^* of Θ containing both these inverse images, we see that, if $\xi^* \cap \xi = \{t\}$, the point s is contained in L_t. We conclude that the lines L_r, for r ranging through O, form a spread $\mathcal{S}$ of Σ, and we obtain an André construction of the projective plane $\mathcal{P}$. Hence, each line of $\mathcal{P}$ is a translation line. So $\mathcal{P}$ is a Moufang plane. The inverse image of Σ under ρ_4 is a 7-dimensional space which we shall denote by $T(\xi)$ or $T(O)$ and refer to as the tangent space to X at ξ or at O. Clearly, it intersects X in O.

Since $T(O)$ contains $T(p)$, it follows that $T(p)$ intersects X in just p. Moreover, since L_p and L_q are skew, we see that the spaces $T(p)$ and ξ', for $\xi' \ni q$ and $\xi' \neq \xi$, generate $\mathsf{PG}(8, \mathbb{K})$, and so are complementary.

Now consider a point $x \in X$ not on O. From the previous paragraph, we know that the spaces $T(x)$ and ξ are complementary. Consider a point $a \in X \setminus \{x\}$. The space $[a, x]$ meets $T(x)$ in the plane $T_x([a, x])$ and so the projection of a from $T(x)$ onto ξ coincides with the unique intersection point $O \cap [a, x]$. This implies that the image of the projection of $X \setminus \{x\}$ from $T(x)$ onto ξ coincides with O. Denote the projection operator by ρ_3 for further reference. The previous argument now easily implies that the image under ρ_3 of any member of Θ not containing x coincides with O. By varying x we deduce that all members of Θ are projectively equivalent.

Note that the projections ρ_3 and ρ_4 are in a certain sense "opposite". Indeed, we can choose $W_4 = T(x)$ and then the kernel of one projection is the image of the other. Let ζ be a member of Θ containing x. Put $z = \zeta \cap \xi$. Then $\zeta \cap T(x)$ equals $T_x(X(\zeta))$, and hence, since $\langle z, T_x(X(\zeta))\rangle = \zeta$ and $z \in \xi$, the image under ρ_4 of ζ coincides with $\zeta \cap T_x(X(\zeta))$. But, with the above notation, the latter plane is spanned by x and L_z. It follows that the spread $\mathcal{S}$ is projectively equivalent with the set of lines arising from the planes $T_x(O^*)$, with O^* ranging through the set of members of Θ containing x, by projecting from x inside $T(x)$. Consequently, we can state the following remark.

Remark 3.3. The tangent planes at x cover the whole 4-dimensional tangent space $T(x)$.

Now we return to the André representation above. Let ξ'' be as above, and put $O'' := X(\xi'')$. Let α be a plane in ξ'' through q different from the tangent plane at q to O''. The intersection $C = \alpha \cap O''$ is an oval. The projection under ρ_4 of $C \setminus \{q\}$ is some "affine" line ℓ. Considering a plane π in W_4 containing ℓ and intersecting Σ in a line not belonging to the spread $\mathcal{S}$, we obtain a subplane $\mathcal{P}'$ of $\mathcal{P}$. Taking inverse images with respect to ρ_4, we see that $\mathcal{P}'$ contains C and lives in a 6-dimensional space $\pi^{\rho_4^{-1}}$ containing O and that the lines of $\mathcal{P}'$ not contained in ξ correspond to plane sections of members of Θ (and we call such plane sections *ovals*). Since $\mathcal{P}'$ is determined by two such ovals, and since an oval always projects from one of its points b onto a line (modulo the point b), we can obtain the same subplane

$\mathcal{P}'$ by projecting from another elliptic space containing a line of $\mathcal{P}'$. Intersecting now the two 6-spaces thus obtained, we see that $\mathcal{P}'$ is contained in a 5-space Ω, and the lines of $\mathcal{P}'$ are plane ovals. Moreover, $\mathcal{P}'$ generates Ω (indeed, the 6-space $\pi^{\rho_4^{-1}}$ is generated by $\mathcal{P}'$ and ξ, hence by $\mathcal{P}'$ and one further point of O; so $\langle \mathcal{P}' \rangle$ has codimension at most one in $\pi^{\rho_4^{-1}}$). Consequently, by Theorem 2.3, the plane $\mathcal{P}'$ has a Veronesean embedding in Ω, which implies that $\mathbb{K}$ is commutative and that all ovals above are conics.

Varying C, we now see that all ovals on members of Θ are conics. We claim that this, in turn, implies that all ovoids of X are elliptic quadrics. Indeed, consider the ovoid O and consider two plane sections C_1 and C_2 intersecting in two distinct points x_1 and x_2. Let y be some point of O not contained in $C_1 \cup C_2$.

We claim that O is the unique object S of the 3-space $\langle O \rangle$ containing C_1, C_2 and y and such that

(∗) Every plane of $\langle O \rangle$ intersects S in a possibly degenerate conic, and
(∗∗) Through each point of S, there is a unique plane (called the *tangent plane*) containing all tangents at this point to arbitrary plane sections through that point.

Note that, in particular, Condition (∗∗) implies that no point of S is incident with three lines contained in S.

We now prove the claim. We start by showing that the intersection of S with an arbitrary line through y is determined by C_1, C_2, y and the Conditions (∗) and (∗∗).

Put $\pi_i = \langle C_i \rangle$, $i = 1, 2$. Let $L_{i,j}$ be the tangent to C_i in π_i at the point x_j, $i, j \in \{1, 2\}$. Put $z_i = L_{i,1} \cap L_{i,2}$, $i = 1, 2$, and set $\alpha_j = \langle L_{1,j}, L_{2,j} \rangle$, $j = 1, 2$. Let L be any line through y and suppose first that L is not incident with either z_1 or z_2, and that it does not meet both $L_{1,1}$ and $L_{2,2}$, and neither both $L_{1,2}$ and $L_{2,1}$ (and let us call the lines through y not satisfying these conditions *exceptional*). Let y_i be the intersection of L with π_i, $i = 1, 2$. The conditions on L imply that for some $i \in \{1, 2\}$ none of the lines $y_1 x_i$ and $y_2 x_i$ is tangent to C_1 and C_2, respectively. Hence, they intersect C_1 and C_2, respectively, in the point x_i and two further points u_1 and u_2, respectively. The points u_1 and u_2 coincide if and only if L meets the intersection of π_1 and π_2, and then $u_1 = u_2 = x_{3-i}$. Hence, in the plane $\alpha := \langle L, x_i \rangle$ we already count either four points of S, or three points. Hence, $\alpha \cap S$ is a conic C. The tangent to C at x_i is the line obtained by intersecting α with $\langle L_{1,i}, L_{2,i} \rangle$, and if $u_1 = u_2 = x_{3-i}$, we have something similar at x_{3-i}. Hence C is uniquely determined and so is the intersection of S with L.

Now let L be one of the four exceptional lines through y. Since this is a finite number (and we may suppose that $\mathbb{K}$ is infinite), the tangent plane to S at y is determined by the points we already traced. So we can consider a plane distinct from that tangent plane, containing L. It contains infinitely many of the already constructed points of O, and hence the corresponding conic section is uniquely determined. Our claim is proved.

Now any elliptic or hyperbolic quadric through C_1, C_2, y satisfies Conditions $(*)$ and $(**)$, as well as our ovoid O. So if we show that there is an elliptic or hyperbolic quadric through C_1, C_2, y, then it must coincide with O, and consequently O is an elliptic quadric.

We introduce coordinates. We may assume that C_1 contains the points $x_1 := (1,0,0,0)$, $x_2 := (0,0,1,0)$ and $(1,1,1,0)$ and that z_1 has coordinates $(0,1,0,0)$. Also, C_2 can be assumed to contain, besides x_1 and x_2, the point $(1,0,1,1)$ and z_2 can be chosen as $(0,0,0,1)$. Then C_1 has equations

$$\begin{cases} X_1^2 = X_0X_2, \\ X_3 = 0, \end{cases}$$

whereas C_2 has equations

$$\begin{cases} X_3^2 = X_0X_2, \\ X_1 = 0. \end{cases}$$

Hence, the equation of a generic quadric through C_1 and C_2, distinct from the union of the two planes π_1 and π_2, is $X_1^2 + kX_1X_3 + X_3^2 = X_0X_2$, with $k \in \mathbb{K}$. Since y is not contained in $\pi_1 \cup \pi_2$, there is a unique $k \in \mathbb{K}$ such that the coordinates of y satisfy the above equation. There remains to show that this quadric cannot be an elliptic cone. If it was, then the generator through y must meet $C_1 \cup C_2$ in a unique point, and this must either be x_1 or x_2 (otherwise three points of O are collinear), say x_1. But then y is contained in α_1, which is ridiculous because α_1 is the plane tangent to O at x_1.

So we have shown that O is an elliptic quadric.

Since the rest of the proof of Theorem 4.1 in [1] holds for arbitrary fields $\mathbb{K}$ we conclude that, if $\mathbb{L}$ is the quadratic extension of $\mathbb{K}$ with the roots of $x^2 + kx + 1$, then $\mathcal{P}$ is isomorphic to $\mathsf{PG}(2, \mathbb{L})$.

3.3 The General Case

In Sect. 5 of [1] in the induction argument from index 2 to index n, the only point where the finiteness assumption is used is the starting point of the induction which we handled here in Theorem 3.2 and the fact that the tangent planes at x cover the whole 4-dimensional tangent space $T(x)$, which we proved for the infinite case in Remark 3.3. Finally, we exclude the possibility of $\mathcal{P}$ being infinite-dimensional. By Remark 3.3 we can proceed as in [1] to prove that for a Hermitian Veronesean cap of index r the tangent space at a point has dimension $2r$, yielding the general version of Proposition 3.1 of [1]. This immediately shows that $\mathcal{P}$ is finite-dimensional.

Hence, Theorem 2.4 is proved.

4 An Application of Hermitian Veronesean Caps

We now show Theorem 2.5. We start with the case of index 2. Recall that we assume that $|\mathbb{K}| > 2$. At a crucial point in the proof, it will become clear why this theorem does not hold for $|\mathbb{K}| = 2$ and a counterexample naturally pops up. We also note that, at some point, one argument needs $|\mathbb{K}| > 4$ (which is allowed since the finite case follows from [7]).

Let $\mathcal{S} = (\mathcal{P}, \mathcal{L}, \mathtt{I})$ be a projective plane. Let $\mathsf{PG}(d, \mathbb{K})$ be a d-dimensional projective space over the skew field $\mathbb{K}$, with $d \geq 8$. A projective space of dimension l in a projective space of dimension d has *codimension* $d - l$. Suppose that $\mathcal{P}$ is a spanning subset of the point set of $\mathsf{PG}(d, \mathbb{K})$ and that every element of $\mathcal{L}$ induces an ovoid in some solid of $\mathsf{PG}(d, \mathbb{K})$. In the sequel, we identify a line of $\mathcal{S}$ with the set of points incident with it. Our aim is to prove that $\mathbb{K}$ is commutative, that there is a quadratic extension $\mathbb{L}$ of $\mathbb{K}$ such that $\mathcal{S} \cong \mathsf{PG}(2, \mathbb{L})$ and that $\mathcal{P}$ is the Hermitian Veronesean cap of $\mathsf{PG}(2, \mathbb{L})$.

In order to do so, we need to show that two lines of $\mathcal{S}$ generate a six-dimensional subspace of $\mathsf{PG}(d, \mathbb{K})$, and that the tangent planes $T_x(O)$ at a fixed point x to a variable ovoid O corresponding with a line of $\mathcal{S}$ containing x, are contained in a four-space. We first need some preliminary lemma's.

Lemma 4.1. *Every triangle of lines of $\mathcal{S}$ generates a subspace of codimension at most* 1.

Proof. Let L_1, L_2, L_3 be such a triangle of lines. Let x_{ij} be the intersection of L_i with L_j (this implies $x_{ij} = x_{ji}$), for $i, j \in \{1, 2, 3\}$, $i \neq j$. Let x be a point of $\mathcal{S}$ not contained in $\langle L_1, L_2, L_3\rangle$. Let L be any line of $\mathcal{S}$ through x. If L is not incident with x_{12}, x_{23}, x_{31}, then L meets $L_1 \cup L_2 \cup L_3$ in three points, say y_1, y_2, y_3, which are non-collinear in $\mathsf{PG}(d, \mathbb{K})$. Since x is not contained in $\langle L_1, L_2, L_3\rangle$, we see that $\langle L\rangle = \langle y_1, y_2, y_3, x\rangle \subseteq \langle L_1, L_2, L_3, x\rangle$.

Since $|\mathbb{K}| > 2$, it is clear that the number of points of any line M not contained in a given subspace is either 0 or exceeds 5. But from the foregoing it follows that every line of $\mathcal{S}$ not through x has at most 3 points not in $\langle L_1, L_2, L_3, x\rangle$, namely potentially the ones lying on lines through x and x_{12}, x_{13} or x_{31}. Hence every such line is contained in $\langle L_1, L_2, L_3, x\rangle$. This implies that $\mathsf{PG}(d, \mathbb{K})$ coincides with $\langle L_1, L_2, L_3, x\rangle$ and so $\langle L_1, L_2, L_3\rangle$ has codimension 0 or 1 in $\mathsf{PG}(d, \mathbb{K})$.

Remark 4.2. If $|\mathbb{K}| = 2$ if follows from the proof of the foregoing lemma that at most six points of $\mathcal{S}$ are not contained in $\langle L_1, L_2, L_3, x\rangle$. Indeed, the only possible ones lie on the three lines through x and x_{12}, x_{13} and x_{23} respectively. Each such line contains five points and intersects $\langle L_1, L_2, L_3\rangle$ in at least two points. In case this maximum is attained these six points must form a hyperoval in $\mathcal{S}$. This situation really occurs. Indeed, let $|\mathbb{K}| = 2$. There is an example in dimension 10, and it is the largest one. To see this, we note that the points of every line add up to zero. Hence we can try to find the universal embedding by relating to each point of $\mathsf{PG}(2, 4)$ a generator of an elementary abelian 2-group, and then factor out the subgroup generated by the sums of points corresponding to the lines.

This subgroup is exactly the code of $\mathsf{PG}(2,4)$ generated by the lines, which has dimension 10, see page 722 of [2]. Hence we obtain an eleven-dimensional vector space over $\mathsf{GF}(2)$, and hence this "universal" example lives in 10-dimensional projective space. Now, with previous notation, the subspace $\langle L_1, L_2, L_3, x\rangle$ is at most 9-dimensional, and it easily follows from the arguments above that it is at most 1-codimensional. Consequently, $\langle L_1, L_2, L_3, x\rangle$ has dimension 9 and the points of $\mathcal{S}$ not in $\langle L_1, L_2, L_3, x\rangle$ form a hyperoval.

We now continue with the case $|\mathbb{K}| > 2$. The spans of two lines of $\mathcal{S}$ cannot meet in a plane, since otherwise the span of these two lines and any third line not through the intersection point would be contained in a subspace of dimension $4 + 2 = 6$, contradicting Lemma 4.1. Also, if $d = 9$, then the spans of two lines of $\mathcal{S}$ cannot meet in a line by a similar argument. Finally, d cannot be larger than 9, as a triangle of lines cannot generate a subspace of dimension > 8.

We now analyze the case where a point x_0 of $\mathcal{S}$ is contained in a subspace $\langle L_0\rangle$, where L_0 is a line of $\mathcal{S}$ not incident with x_0 in $\mathcal{S}$. By the previous paragraph this can only happen if $d = 8$. So in the next proof, we may restrict to $d = 8$.

Lemma 4.3. *If* $x_0 \notin \mathcal{P}\backslash L_0$, *then* $x_0 \notin \langle L_0\rangle$.

Proof. Let x_0 and L_0 be as above and suppose by way of contradiction that $x_0 \in \langle L_0\rangle$. Let $\tilde{L}$ be any line of $\mathcal{S}$ not through x_0. Let x be the intersection of $\tilde{L}$ with L_0. Let L' be any line of $\mathcal{S}$ through x_0 but not incident with x. Then $\langle L'\rangle$ shares at least one plane with $\langle \tilde{L}, L_0\rangle$, and so the dimension of $\langle L_0, \tilde{L}, L'\rangle$ is at most one more than the dimension of $\langle L_0, \tilde{L}\rangle$. By Lemma 4.1, it follows that the dimension of $\langle L_0, \tilde{L}\rangle$ is at least six, and hence equal to six. Suppose there is another point $y_0 \in \mathcal{P}$ contained in $\langle L_0\rangle$ not incident with L_0. Then the line L'' through x_0 and y_0 passes through x since otherwise $\langle L_0, \tilde{L}, L''\rangle$ is 6-dimensional, contradicting Lemma 4.1. But replacing L with a line not through x, we arrive again at a contradiction. Hence it follows that x_0 is the only element of $\mathcal{P}$ in $\langle L_0\rangle \setminus L_0$.

Consider the projection of $\mathcal{P} \setminus (L_0 \cup \{x_0\})$ from $\langle L_0\rangle$ onto some 4-dimensional space U skew to $\langle L_0\rangle$. The points of every line L not through x_0 not in $\langle L_0\rangle$ are mapped onto an affine plane α_L of U. The points of every line M through x_0 different from x_0 and from $x_M := M \cap L_0$ are mapped onto the points of a line λ_M, and the point fibers are plane ovals through x_0 and x_M (with x_0 and x_M themselves omitted).

Now we fix a line L not through x_0 and distinct from L_0, and a line M through x_0. We choose L and M such that L, M and L_0 are not concurrent. Then α_L and λ_M meet in a point z_M (they share the projection of the intersection $L \cap M$, but not more as otherwise L_0, L, M would be contained in a 6-dimensional space). Hence, $\langle \alpha_L, \lambda_M\rangle$ is a solid Σ. Now let M' be any line of $\mathcal{S}$ through x_0. We claim that $\lambda_{M'}$ is contained in Σ.

First, we assume that L_0, L, M' are not concurrent. Then α_L and $\lambda_{M'}$ share a unique point $z_{M'}$. We choose two arbitrary points u, v in α_L with the only restriction that the line $\langle u, v\rangle$ of U does not contain any of the points $z_M, z_{M'}$ (this is possible). Let u_L and v_L be the unique points of L mapping down to u and v, respectively.

The fiber of v_L is an oval minus two points. Now we assume that $|\mathbb{K}| > 4$, so that we can choose a point $\tilde{v} \in \mathcal{P}$ in the fiber of v (which has size at least 4) such that $\tilde{v} \neq v_L$, and such that the line $L' := u_L\tilde{v}$ of $\mathcal{S}$ contains neither x_M nor $x_{M'}$. Then $\alpha_{L'}$ contains both u and v and intersects λ_M in the projection of $L' \cap M$. Hence, $\alpha_{L'}$ belongs to Σ. Now $\lambda_{M'}$ shares the projection of $M' \cap L'$ with $\alpha_{L'}$ (which is different from $z_{M'}$ as otherwise $\alpha_{L'} = \alpha_L$, which would imply that L_0, L, L' are contained in a 6-dimensional space), and it also contains $z_{M'}$; hence, it is entirely contained in Σ.

Now we assume that M' is incident with $L_0 \cap L$. We may replace L with a line L' as introduced in the previous paragraph, and the claim is proved.

This claim immediately implies that the entire projection is contained in Σ, and so $\mathcal{P}$ is contained in a 7-dimensional space, a contradiction. This completes the proof.

The next lemma is trivial if $d = 9$ by Lemma 4.1. Hence we may assume $d = 8$.

Lemma 4.4. *The space generated by two lines of $\mathcal{S}$ is 6-dimensional, so* (H2) *holds.*

Proof. Suppose, by way of contradiction, that two lines L_0, M generate a 5-space (it is impossible that $\langle L_0, M\rangle$ is 4-dimensional, as this would imply that $\langle L_0, M, N\rangle$ is at most 6-dimensional, for every line N not incident with $L \cap M$). We again consider the projection from $\langle L_0\rangle$ of $\mathcal{P} \setminus L_0$ onto a suitable 4-dimensional subspace U skew to $\langle L_0\rangle$. Now, $M \setminus \{L_0 \cap M\}$ is projected onto an affine line λ_M and the fibers are pointed plane ovals.

Let L be any line of $\mathcal{S}$ not incident with $L_0 \cap M$. If the projection of $L \setminus \{L_0 \cap L\}$ were a line λ_L, then, since λ_L and λ_M have a point in common (namely, the projection of $L \cap M$), the lines L_0, L, M would be contained in a 6-space, a contradiction. Hence, the projection of $L \setminus \{L_0 \cap L\}$ is an affine plane α_L. So the projection is injective on lines not incident with $L_0 \cap M$.

We now claim that every line M' of $\mathcal{S}$ incident with $L_0 \cap M$ is projected onto an affine line (where we do not project $L_0 \cap M'$ of course). Indeed, let M' be such a line and let x' be the intersection of M' with L. Also, let x be the intersection of M with L. Let u and u' be the projection of x and x', respectively. We choose a point w on the line $\langle u, u'\rangle$ inside α_L (this is possible since we assume that there are at least 4 points on a line in the projective space $\mathsf{PG}(8, \mathbb{K})$). Let y be a point in the fiber of u distinct from x. Let z be the point of L projected onto w. Then the line $yz \in \mathcal{L}$ does not contain $L_0 \cap M$ and is hence projected onto an affine plane α. This plane α contains u and w and hence intersects α_L in either an affine line or an affine line minus a point.

If u' is in the intersection of α and α_L, then at least two points (one point is x'; let's denote a second one by x'') are projected onto u'. By this non-injectivity the join in $\mathcal{S}$ of x' and x'' must be a line through $L_0 \cap M$, and hence coincides with M'.

So we may assume that u' is not in α. Let p be the unique point on $\langle u, u'\rangle$ not in α_L. Since u' is not in α it follows that p belongs to α. We can now vary the point y, and by the foregoing, we may assume that the point p belongs each time to the corresponding affine plane. Hence the fiber of the point p contains at least two elements, say q and q'. The line $qq' \in \mathcal{L}$ must be incident with $L_0 \cap M$ and projects onto an affine line. But this affine line must have a point in common with α_L. This implies that this affine line is contained in $\langle \alpha_L \rangle$, which implies that L_0, L, qq' is contained in a 6-space, a contradiction by Lemma 4.1.

But now, similarly as in the previous lemma, one shows that the projection of $\mathcal{P} \setminus L_0$ is contained in the 3-space generated by λ_M and α_L. This contradiction proves that $\langle L_0, M \rangle$ is 6-dimensional.

From now on, we assume $d \in \{8, 9\}$ again.

Next we construct quadratic subveroneseans. Let C_1 and C_2 be two plane ovals contained in the lines L_1 and L_2, respectively, of $\mathcal{S}$, and suppose that $C_1 \cap C_2 = \{x\}$ is a point of $\mathcal{S}$. Let L be a line of $\mathcal{S}$ incident with a point x_i of C_i, for $i = 1, 2$, and with $x_1 \neq x \neq x_2$. We project $\mathcal{P} \setminus L$ from $\langle L \rangle$ onto a $(d-4)$-dimensional subspace U skew to $\langle L \rangle$. By Lemma 4.4, this projection is injective. The points of every line M of $\mathcal{S}$ (except for the point $M \cap L$) are mapped onto the points of an affine plane α_M (bijectively). Let p be a point of $\langle \alpha_M \rangle \setminus \alpha_M$. We claim that

(∗) p is not the projection of any element of $\mathcal{P} \setminus L$.

Indeed, assume by way of contradiction that $q \in \mathcal{P}$ is mapped onto p. Take an arbitrary point y of M. Then the projection of the line yq of $\mathcal{S}$ intersects α_M in at least two points, contradicting injectivity of the projection operator on $\mathcal{P} \setminus L$.

Now it is clear that $(C_1 \cup C_2) \setminus \{x_1, x_2\}$ is mapped onto the union of two intersecting affine lines, say $\lambda_1 \cup \lambda_2$, with λ_i the projection of $C_i \setminus \{x_i\}$, $i = 1, 2$. Let π be the plane of $\mathsf{PG}(d, \mathbb{K})$ generated by $\lambda_1 \cup \lambda_2$ and let p_i^∞ be the unique non-affine point of the projective extension of λ_i, $i = 1, 2$. Let z_i be an arbitrary point of $C_i \setminus (L \cup \{x\})$, $i = 1, 2$. Then the projection of the oval on $z_1 z_2$ determined by z_1, z_2 and $z_1 z_2 \cap L$ is projected onto an affine line $\lambda_{z_1 z_2}$ in π. Let z^∞ be the unique point on $\langle \lambda_{z_1 z_2} \rangle$ not in $\lambda_{z_1 z_2}$. If $z^\infty \notin \langle p_1^\infty, p_2^\infty \rangle$, then we choose distinct points z_i' and z_i'' on C_i, $i = 1, 2$, such that, with corresponding notation, z^∞ belongs to $\langle \lambda_{z_1' z_2'} \rangle$ and to $\langle \lambda_{z_1'' z_2''} \rangle$. By the claim (∗), this implies $z^\infty = z'^\infty = z''^\infty$. Hence $\langle p_1^\infty, p_2^\infty \rangle \cap (\lambda_{z_1' z_2'} \cup \lambda_{z_1'' z_2''})$ contains exactly two elements u', u''. Let $v', v'' \in \mathcal{P} \setminus L$, respectively, be the inverse image (under the projection) of the latter. Then the line vv' of $\mathcal{S}$ is projected onto an affine plane containing all points of $\langle u', u'' \rangle$ except for one. But (∗) implies that both points p_1^∞ and p_2^∞ are outside that affine plane, contradicting $p_i^\infty \in \langle u', u'' \rangle$, $i = 1, 2$. Hence, the points in π that are a projection of points of $\mathcal{S}$ not in L form an affine plane α. Each affine line of α corresponds with a plane oval, and all these form an affine subplane $\mathcal{A}$ of $\mathcal{S}$.

Lemma 4.5. *The abstract projective closure of $\mathcal{A}$ is a projective subplane $\mathcal{S}'$ of $\mathcal{S}$, which forms a quadratic Veronesean.*

Proof. For this, we have to show that three parallel affine lines of $\mathcal{A}$ meet in a point of $\mathcal{S}$, and that all points thus obtained are collinear in $\mathcal{S}$. We start by showing that, if D_1 and D_2 are two distinct ovals corresponding with two parallel affine lines of α, then $D_1 \cap D_2$ is nonempty and is contained in L. Indeed, let K_i be the line of $\mathcal{S}$

containing D_i, $i = 1, 2$. Let t be the intersection point of K_1 and K_2. If $t \notin L$, then the projections of K_1 and K_2 would share an affine line, a contradiction. Hence, $t \in L$, and since D_1 and D_2 both contain a point of L (as they are projected onto affine lines), we must necessarily have $t \in D_1 \cap D_2$. Since t is uniquely determined by any of D_1 or D_2 only, we automatically have that three lines of $\mathcal{S}$ which induce parallel lines in $\mathcal{A}$ meet in a point of $\mathcal{S}$. Since all these points lie on L, our claim is proved.

Hence, we see that $\mathcal{S}'$ is a subplane of $\mathcal{S}$ contained in a 6-dimensional subspace W of $\mathsf{PG}(d, \mathbb{K})$. However, if we consider a different line L, then we obtain the same subplane contained in a different 6-dimensonal space W' (indeed, different since it will now not contain all points of L but only an oval). Hence, it is now easy to see that $\mathcal{S}'$ spans a 5-dimensional subspace V. Consequently, $\mathcal{S}'$ forms a quadratic Veronesean.

Lemma 4.6. *Condition* (H3) *holds.*

Proof. Now let $x \in \mathcal{P}$ be an arbitrary point and let L_1, L_2 be two distinct lines of $\mathcal{S}$ through x. Then the tangent planes at x in $\langle L_i \rangle$, $i = 1, 2$, to L_i together span a 4-space Ξ. We show that, if L is an arbitrary line of $\mathcal{S}$ through x, then the tangent plane at x to L in $\langle L \rangle$ is contained in Ξ. This will follow if we show that an arbitrary tangent line T at x to L in $\langle L \rangle$ is contained in Ξ. Therefore, let C be an arbitrary oval on L through x with T tangent to C at x, and let L' be any line of $\mathcal{S}$ not through x but containing a point y of C. Let C' be the oval on L' containing y and a point of each of L_1 and L_2. Then C, C' are contained in a unique subplane inducing a quadratic subveronesean $\mathcal{V}$ on $\mathcal{S}$, as shown in the previous paragraphs. Note that $\mathcal{V}$ contains a conic lying on L_1 and one on L_2. The line T is contained in the plane spanned by the tangent lines at x to the conics of $\mathcal{V}$ on L_1 and L_2, and hence T is contained in Ξ.

This completes the proof of the fact that $\mathcal{S}$ is a Hermitian cap; in particular $d = 8$ and the case $d = 9$ cannot occur after all. Indeed, we can now also give a short proof for the fact that the case $d = 9$ cannot occur. Suppose $d = 9$, consider a line L and a point p not on L and project $\mathcal{P} \setminus L$ from $\langle L \rangle$ onto a 5-dimensional space skew to $\langle L \rangle$. Then, since we also assume that X has the structure of a projective plane, it follows that $\mathcal{P} \setminus L$ is projected into the 4-dimensional space which is the projection of the tangent space through p. Hence $d = 8$.

Now, concerning the case $n > 2$ of Theorem 2.5, similarly to the finite case, it suffices to show that, if we consider a subset Y of X corresponding to the point set of a plane of $\mathsf{PG}(n, \mathbb{L})$, then Y generates a subspace of dimension 8. But the proof of the finite case, see Sect. 4 of [7], applies verbatim.

5 Another Application

We now define the following object. Let $\mathbb{L}$ be a quadratic extension of the field $\mathbb{K}$. Let V be a vector space of dimension n over $\mathbb{L}$. Consider V as a vector space W of dimension $2n$ over $\mathbb{K}$, and let $\mathfrak{L}$ be the set of 2-dimensional subspaces of W arising from the vector lines of V. Then consider the line Grassmannian of

$\mathsf{PG}(W)$. The image $\mathcal{G}(\mathfrak{L})$ of $\mathfrak{L}$ is precisely the Hermitian Veronesean variety of index $n-1$.

Indeed, there exists a $d \in \mathbb{K}$ such that the equation $x^2+x+d = 0$ has no solution in $\mathbb{K}$ and two solutions in $\mathbb{L}$. Note that $\mathfrak{L}$ has the natural structure of a projective space $\mathfrak{P}$ isomorphic to $\mathsf{PG}(V)$. A line of $\mathfrak{P}$ corresponds to the set of members of $\mathfrak{L}$ arising from vector lines of V contained in a vector plane π. This vector plane over $\mathbb{L}$ becomes a 4-space over $\mathbb{K}$. We now coordinatize the situation as follows. Each element x of $\mathbb{L}$ can be written as a couple (x_1, x_2), where $x = x_1 + ix_2$, with $i^2 + i + d = 0$. Note that $i(x_1 + ix_2) = -dx_2 + i(x_1 - x_2)$. Then a vector of V with coordinates $(x^{(1)}, x^{(2)}, \ldots, x^{(n)})$ can be given the coordinates $(x_1^{(1)}, x_2^{(1)}, x_1^{(2)}, x_2^{(2)}, \ldots, x_1^{(n)}, x_2^{(n)})$ in W. For π we can now take the set of vectors with coordinates $(x, y, 0, 0, \ldots, 0)$, $x, y \in \mathbb{L}$, and the vector plane in W corresponding to the above vector $(x, y, 0, \ldots, 0)$ is spanned by the vectors $(x_1, x_2, y_1, y_2, \ldots, 0)$ and $(-dx_2, x_1-x_2, -dy_2, y_1-y_2, 0, \ldots, 0)$. The image of this vector plane under the line Grassmannian in $\mathsf{PG}(W)$ is the point with coordinates

$$\begin{aligned} p_{01} &= x_1^2 - x_1x_2 + dx_2^2, & p_{12} &= -x_1y_1 + x_2y_1 - dx_2y_2, \\ p_{02} &= -dx_1y_2 + dx_2y_1, & p_{13} &= -x_1y_2 + x_2y_1, \\ p_{03} &= x_1y_1 - x_1y_2 + dx_2y_2, & p_{23} &= y_1^2 - y_1y_2 + dy_2^2. \end{aligned}$$

We observe that $p_{02} = dp_{13}$ and $p_{13} = p_{12} + p_{03}$; hence, we may concentrate on the coordinates $p_{01}, p_{12}, p_{13}, p_{23}$. One easily verifies that these satisfy the equation $p_{01}p_{23} = p_{12}^2 - p_{12}p_{13} + dp_{13}^2$ and hence all these points are contained in an elliptic quadric Q. Conversely, suppose the four field elements $a, b, c, e \in \mathbb{K}$ satisfy $ab = c^2 - ce + de^2$ (where we think of a, b, c, e as potential values for $p_{23}, p_{01}, p_{12}, p_{13}$, respectively), then $(a, b) \neq (0, 0)$. Suppose without loss of generality that $a \neq 0$, then we may even assume $a = 1$. Put $y_1 = 1$, $y_0 = 0$, $x_1 = e - c$ and $x_2 = e$. Then $b = c^2 - ce + de^2 = x_1^2 - x_1x_2 + dx_2^2$, and so the corresponding point on Q belongs to $\mathcal{G}(\mathfrak{L})$. It follows that the lines of $\mathfrak{P}$ are elliptic quadrics in solids of $\mathsf{PG}(W)$.

As clearly every point of $\mathcal{G}(\mathfrak{L})$ contained in $\langle Q \rangle$ belongs to Q (since every other point has to involve an additional coordinate p_{ij}, with $\{i, j\} \not\subseteq \{0, 1, 2, 3\}$), left to show is the fact that the dimension of he span of $\mathcal{G}(\mathfrak{L})$ is at least $(n-1)(n+1)$. We show this by induction on n. For $n = 2$, this is proved above, as Q spans a 3-space.

Now suppose that $n > 2$. Consider a basis of size n in V and let V' be the subspace generated by the first $n-1$ basis vectors. Let W' be the corresponding subspace of W, let $\mathfrak{L}'$ be the corresponding subset of $\mathfrak{L}$ and let $\mathcal{G}(\mathfrak{L}')$ be the corresponding image on the line Grassmannian. The induction hypothesis implies that $\langle \mathcal{G}(\mathfrak{L}') \rangle$ has dimension at least $(n-2)n$. Now consider the $2n-1$ points of V with coordinates $(0, 0, \ldots, 0, 1)$, $(1, 0, 0, \ldots, 0, 1)$, $(0, 1, 0, 0, \ldots, 0, 1)$, $\ldots$, $(0, 0, \ldots, 0, 1, 1)$, $(i, 0, 0, \ldots, 0, 1)$, $(0, i, 0, 0, \ldots, 0, 1)$, $\ldots$, $(0, 0, \ldots, 0, i, 1)$. Then it is a routine exercise to verify that the corresponding points in $\mathcal{G}(\mathfrak{L})$ generate a $(2n-2)$-dimensional projective subspace skew to $\langle \mathcal{G}(\mathfrak{L}') \rangle$ (this follows from the fact that the first point gives rise to the Grassmann coordinate $p_{2n-1,2n}$, the next $n-1$

points additionally to $p_{2k-1,2n} - p_{2k,2n-1}$, $1 \leq k < n$, and the corresponding points on the Grassmannian of the last $n-1$ vectors involve $p_{2k,2n} + p_{2k,2n-1} + p_{2k-1,2n-1}$, $1 \leq k < n$).

Hence we checked (H1*), (H2*) and (H3*) and we are done.

References

1. B. Cooperstein, J. A. Thas & H. Van Maldeghem, Hermitian Veroneseans over finite fields, *Forum Math.* **16** (2004), 365–381
2. R. Graham, M. Grötschel and L. Lovász. *Handbook of Combinatorics*, Elsevier, Amsterdam, 1995
3. G. Lunardon, Normal spreads, *Geom. Dedicata* **75** (1999), 245–261
4. J. Schillewaert & H. Van Maldeghem, Quadric Veronesean caps, *Discrete Mathematics*, submitted
5. J. A. Thas & H. Van Maldeghem, Classification of finite Veronesean caps, *European J. Combin.* **25** (2004), 275–285
6. J. A. Thas & H. Van Maldeghem, Characterizations of the finite quadric Veroneseans $\mathcal{V}_n^{2^n}$, *Quart. J. Math.* **55** (2004), 99–113
7. J. A. Thas & H. Van Maldeghem, Some characterizations of finite Hermitian Veroneseans, *Des. Codes Cryptogr.* **34** (2005), 283–293
8. O. Veblen and J. Young, *Projective geometry Vol I+II*, Blaisdell Publishing Co. Ginn and Co., New York-Toronto-London, 1965

On a Class of $c.F_4$-Geometries

Antonio Pasini

Abstract This paper is a survey of the work done by a number of authors on the classification of flag-transitive geometries with diagram and orders as below:

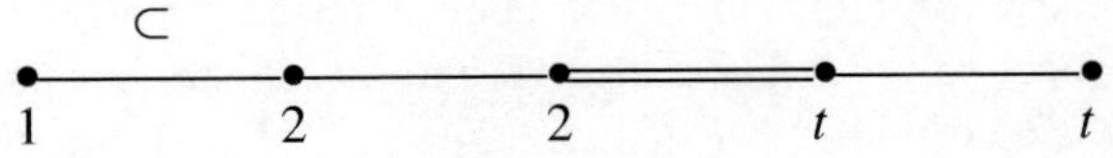

$(t \in \{1, 2, 4\})$.

MSC 2000: 51E24, 20D28, 20E42.

Keywords Diagram geometry • Circular extensions • Buildings • Finite simple groups

1 Introduction

The following diagram, which we call $c.F_4(2,t)$, is the subject of this paper.

$(c.F_4(2,t))$

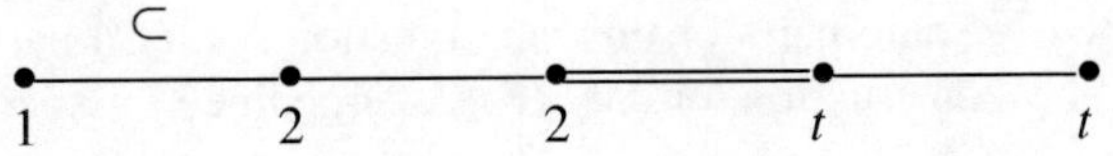

The numbers $1, 2, 2, t, t$ are orders and $t < \infty$. Hence, $t \in \{1, 2, 4\}$. The label $\subset$ on the first stroke of the diagram stands for the class of complete graphs, regarded as point-line geometries with the vertices as points and the edges as lines.

A. Pasini (✉)
Department of Mathematics, University of Siena, Pian dei Mantellini 44, 53100 Siena, Italy
e-mail: pasini@unisi.it

N.S.N. Sastry (ed.), *Buildings, Finite Geometries and Groups*, Springer Proceedings in Mathematics 10, DOI 10.1007/978-1-4614-0709-6_10,

Several flag-transitive geometries are known for this diagram. I will describe some of them later (Sects. 3 and 5.3). I only say here that the Fischer group Fi_{22} and the Baby Monster group B occur as automorphism groups of two of them, with $t = 1$ and $t = 4$, respectively.

A number of mathematicians (Ivanov and Wiedorn [15], Ivanov, Pasechnik and Shpectorov [16], Wiedorn [28], Stroth [24]) have been busy to prove that, even if we do not know all flag-transitive $c.F_4(2,t)$-geometries, nevertheless we know all of those that satisfy certain hypotheses, called (TR) and (LR) in this paper, which I will discuss in Sect. 2.

This is indeed true when $t = 1$ and $t = 4$, as proved by Ivanov, Pasechnik, Shpectorov and Wiedorn nearly ten years ago. They proved that Fi_{22}, its non-split central extension $3^{\cdot}Fi_{22}$ and the Baby Monster group B are the only groups that act flag-transitively on a $c.F_4(2,t)$-geometry satisfying (TR) and (LR) and with $t = 1$ or $t = 4$. See Ivanov and Wiedorn [15] for Fi_{22} and $3^{\cdot}Fi_{22}$ (where $t = 1$) and Ivanov, Pasechnik and Shpectorov [16] for B (where $t = 4$). On the other hand, the case of $t = 2$ stood against all efforts for long. Only four flag-transitive $c.F_4(2,2)$-geometries are known satisfying (TR) and (LR). They admit $2^{26}{:}F_4(2)$, $E_6(2)$, $3^{\cdot 2}E_6(2)$ and $3^{\cdot}E_6(2)$, respectively, as flag-transitive automorphism groups. The first three geometries, associated with $2^{26}{:}F_4(2)$, $E_6(2)$ and $3^{\cdot 2}E_6(2)$, are simply connected (Ivanov and Pasini [13]) while the fourth one is a quotient of the third. A thorough analysis of the possible amalgams in this case has been accomplished by Corinna Wiedorn [28], but her analysis fails to end up with a complete classification. Sadly, Corinna is no more among us. Some time ago, Ivanov and myself agreed to take over that job starting from the point that she had reached, but we met some difficulties which stopped us. Quite recently, Stroth has decided to try in his turn. He has succeded [Str]: no example as we want can exist besides those mentioned above. Thus, the classification is complete.

In this paper, I will expose the main steps of this classification, starting from the work by Ivanov, Pasechnik, Shpectorov and Wiedorn on the cases of $t = 1$ and $t = 4$ and finishing with Stroth's contribution on $t = 2$. I will skip most of technical details, but I will put a bit of emphasis on certain general geometrical constructions exploited in some parts of this classification, which I particularly like. I will also give a new construction for the $c.F_4(2,2)$-geometry associated to $E_6(2)$.

1.1 Organization of the Paper

In Sects. 2–4, we shall report on the work by Ivanov, Pasechnik, Shpectorov, Stroth and Wiedorn on flag-transitive $c.F_4(s,t)$-geometries. We shall firstly discuss hypotheses (TR) and (LR) (Sect. 2). Next, we describe the known examples that satisfy those hypotheses, sticking to the description that Ivanov, Pasechnik, Shpectorov and Wiedorn have chosen for them (Sect. 3). After that, we turn to the classification (Sect. 4).

In Sect. 5, we construct an infinite family of geometries associated to Chevalley groups of type E_6 and belonging to the diagram $Af.F_4$. The smallest member of that family is the $c.F_4(2,2)$-geometry for $E_6(2)$. Many $Af.F_4$-geometries can be obtained by a different construction, as affine extensions of F_4-buildings. We discuss them too in Sect. 5.

The last five sections of this paper are appendices containing notions, constructions and results essential to fully understand Sects. 2–5. Section 6 is mainly devoted to embeddings and affine extensions. Section 7 deals with affine polar spaces and their quotients. Section 8 is a concise introduction to shrinkings and geometries at infinity. A few facts and notions on F_4-geometries and F_4-buildings are recalled in Sects. 9 and 10.

We could have referred the reader to the literature for the material of Sects. 6–8, but fusing the various sources that one can find in the literature in one single picture is not so easy. We have preferred to save the reader that labor.

1.2 Notation and Conventions

We use the notation of [5] for finite groups. We follow [19] for basic notions of diagram geometry, except that we call $(n-1)$-coverings and $(n-1)$-quotients of geometries of rank n just *coverings* and *quotients* for short and we say that a geometry of rank n is *simply connected* if it is $(n-1)$-simply connected, namely it is its own universal $(n-1)$-cover.

Geometries are defined in [19] in such a way that all geometries are residually connected, by definition. We keep that convention in this paper.

Given a geometry $\mathcal{G}$ and an element x of $\mathcal{G}$, we denote the residue of x in $\mathcal{G}$ by $\mathrm{Res}_{\mathcal{G}}(x)$, also $\mathrm{Res}(x)$ for short when no ambiguity arises. We will often write $x \in \mathcal{G}$ as a shortening of the phrase "x is an element of $\mathcal{G}$."

As in [19], we denote the Intersection Property by the symbol (IP).

Conventions for Geometries with String-Shaped Diagrams Let $\mathcal{G}$ be a geometry of rank n belonging to a diagram as follows, where $X_1, X_2, \dots, X_{n-1}$ denote classes of geometries of rank 2 other than generalized digons and the integers $1, 2, \dots, n$ are the types:

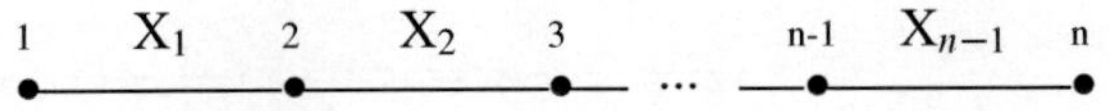

We may regard $\mathcal{G}$ as a poset by setting $x < y$ for two elements $x, y \in \mathcal{G}$ if and only if x and y are incident in $\mathcal{G}$ and $\tau(x) < \tau(y)$, where τ is the type-function of $\mathcal{G}$. The residue of an element $x \in \mathcal{G}$ of type $\tau(x) = k$ with $1 < k < n$ is a direct sum $\mathrm{Res}(x) = \mathrm{Res}^-(x) \oplus \mathrm{Res}^+(x)$ where $\mathrm{Res}^-(x)$ is the residue of any $\{k, k+1, \dots, n\}$-flag of $\mathcal{G}$ containing x and $\mathrm{Res}^+(x)$ is the residue of any $\{1, \dots, k-1, k\}$-flag containing x. We call $\mathrm{Res}^-(x)$ and $\mathrm{Res}^+(x)$ the *lower* and *upper* residue of x, respectively. We can extend this notation to 1- and n-elements: if $\tau(x) = n$ then $\mathrm{Res}^-(x) := \mathrm{Res}(x)$, and similarly for 1-elements.

The elements of $\mathcal{G}$ of type 1 and 2 are called *points* and *lines*:

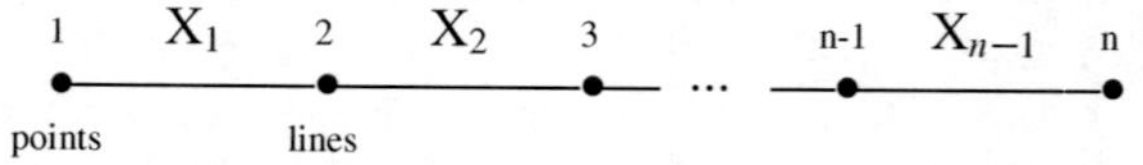

We denote the set of points and the set of lines of $\mathcal{G}$ by $P(\mathcal{G})$ and $L(\mathcal{G})$ respectively. The rank 2 geometry $\mathcal{G}_{|1,2} := (P(\mathcal{G}), L(\mathcal{G}))$ is called the *point-line system* of $\mathcal{G}$. The *collinearity graph* of $\mathcal{G}$ is the collinearity graph of $\mathcal{G}_{|1,2}$.

Suppose that $\mathcal{G}$ satisfies (IP). Then, denoted by $P(x)$ the set of points incident with an element $x \in \mathcal{G}$, if $x \neq y$ then $P(x) \neq P(y)$, we have $x < y$ in the poset $\mathcal{G}$ if and only if $P(x) \subset P(y)$ and, if $P(x) \cap P(y) \neq \emptyset$ then there exists a unique element $z \in \mathcal{G}$ such that $P(z) = P(x) \cap P(y)$. In short, the poset $(\mathcal{G}, <)$ is isomorphic to $(\{P(x)\}_{x \in \mathcal{G}}, \subset)$ and the latter, enriched with $\emptyset$ as the minimal element, is a lower semilattice with respect to $\cap$.

We denote by $\mathcal{G}^*$ the dual poset of $(\mathcal{G}, <)$ and we say that $\mathcal{G}^*$ is the geometry *dual* of $\mathcal{G}$. Clearly, the points and the lines of $\mathcal{G}^*$ are the elements of $\mathcal{G}$ of type n and $n-1$.

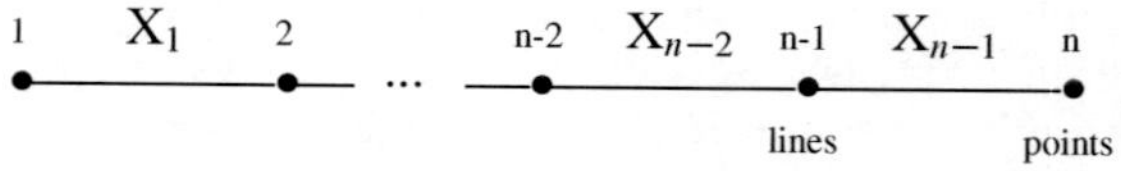

Still on Points and Lines Points and lines can be defined in a more general setting. Let $\mathcal{G}$ be a geometry of rank > 1 belonging to a connected diagram. Let Δ be the underlying graph of the diagram of $\mathcal{G}$. Chosen a type, say 0, let $\Delta(0)$ be the neighborhood of 0 in Δ. The 0-elements of $\mathcal{G}$ are taken as points while the flags of type $\Delta(0)$ are the lines. In other words, we take as points and lines the points and the lines of the 0-grassmann geometry of $\mathcal{G}$, which indeed belongs to a string-shaped diagram (see [19, Chap. 5]).

Conventions for F_4-Geometries We choose the integers $1, 2, 3$ and 4 as types for the diagram F_4, calling the elements of type 3 and 4 *planes* and *symps* respectively. Elements of type 1 and 2 are called points and lines, according to the convention stated for geometries with string-shaped diagrams.

(F_4)

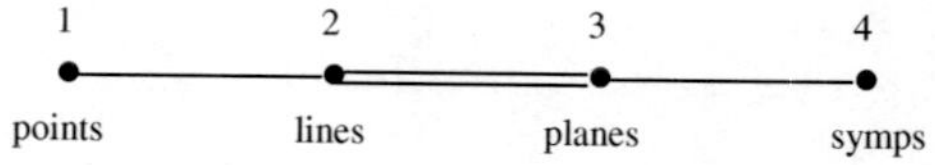

We denote by $F_4(2, t)$ the F_4-diagram with orders $2, 2, t, t$:

$(F_4(2, t))$

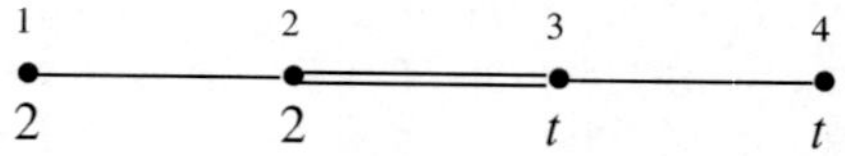

An F_4-geometry (in particular, an F_4-building) with orders $2, 2, t, t$ will be called an $F_4(2,t)$-*geometry* ($F_4(2,t)$-*building*).

Conventions for $c.F_4(2,t)$-Geometries We choose $0, 1, 2, 3$ and 4 as types for the diagram $c.F_4(2,t)$, as follows:

$(c.F_4(2,t))$

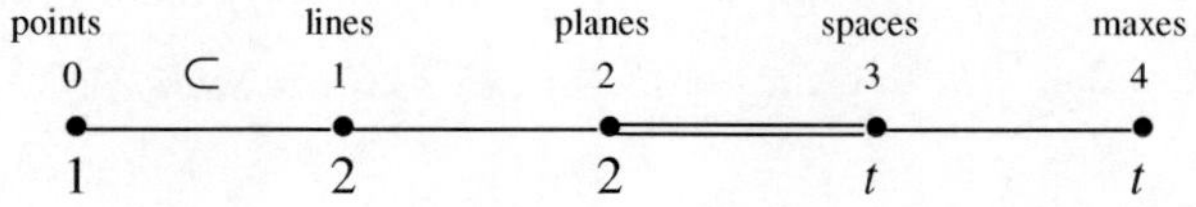

The elements of type 0 and 1 are called points and lines. The elements of type 2, 3 and 4 will be called *planes*, *spaces* and *maxes* respectively, as indicated in the above picture.

Note that, as a complete graph on 4 vertices is the same as an affine plane of order 2, the diagram $c.F_4(2,t)$ can also be regarded as a special case of the following, where Af denotes the class of affine planes.

$(Af.F_4)$

2 Looking for the Right Hypotheses

In this section, we shall discuss the hypotheses assumed on flag-transitive $c.F_4(2,t)$-geometries in [15,16,24,28]. However, before to come to those hypotheses, we must see what flag-transitivity implies.

Throughout this section $\mathcal{G}$ is a given flag-transitive $c.F_4(2,t)$-geometry. Lemma 9.1 of Sect. 9 immediately implies the following:

Lemma 2.1. *The residues of the points of $\mathcal{G}$ are buildings of type F_4.*

Lemma 2.2. *No two distinct lines of $\mathcal{G}$ are incident with the same pair of points.*

Proof. Let G be a flag-transitive subgroup of Aut($\mathcal{G}$). Given a point a of $\mathcal{G}$, let $L(a)$ be the set of lines of $\mathcal{G}$ incident with a and let Θ_a be the equivalence relation defined on $L(a)$ by declaring that two lines $l, m \in L(a)$ correspond in Θ_a if $P(l) = P(m)$. By Lemma 2.1, Res(a) is a building. Hence, the group induced on Res(a) by the stabilizer G_a of a in G contains a subgroup isomorphic to $O_8^+(2) : S_3$, $F_4(2)$ or ${}^2E_6(2)$, according to whether t is 1, 2 or 4. In any case, $G_{a,l}$ is maximal in G_a. In other words, G_a acts primitively on $L(a)$. On the other hand, G_a preserves Θ_a. Therefore, Θ_a is either the identity relation or the trivial relation. In the first case, no two lines through a have the same points, as we wanted to prove. In the latter case, there is a unique point b such that $P(l) = \{a, b\}$ for every $l \in L(a)$, but this is clearly impossible. □

Lemma 2.3. *Property* (IP) *holds in the residue of every element of* $\mathcal{G}$.

Proof. This immediately follows from [19, Lemma 7.25], Lemmas 2.1 and 2.2 and the fact that (IP) holds in every building. □

Lemma 2.3 and Proposition 7.3 of Sect. 7 imply the following:

Lemma 2.4. *The residues of the maxes of* $\mathcal{G}$ *are standard quotients of affine polar spaces.*

(See Sect. 7 for the definition of affine polar spaces and their standard quotients.) It is now time to cope with a difficulty that might sit in the shape of the diagram $c.F_4(2,t)$. The underlying minimal circuit diagram of $c.F_4(2,t)$ is the affine diagram $\widetilde{F}_4$. So, it would be no wonder if $\mathcal{G}$ or some of its covers were infinite. However we would like to keep infinite geometries out of the door. Indeed, we would like to end up with a classification, but the usual tools we can use to that goal would be ineffective in the infinite case. So, we need residues of maxes to be sufficiently tight to force finiteness. To this aim, we assume the following:

(TR) (**Tight Residues**) The collinearity graph of the residue of any max of $\mathcal{G}$ has diameter equal to 1.

This property drastically reduces the range of possibilities for residues of maxes allowed by Lemma 2.4. Explicitly,

Lemma 2.5. *Let* (TR) *hold. Then the residues of the maxes of* $\mathcal{G}$ *are minimal standard quotients of an affine polar space* $\mathcal{P} \setminus H$ *where* $\mathcal{P}$ *and* H *are as follows, according to whether* t *is* 1, 2 *or* 4.

(1) Let $t = 1$. *Then* $\mathcal{P}$ *is the* $O_8^+(2)$*-polar space and* H *is isomorphic to the* $O_7(2)$*-polar space. The minimal standard quotient of* $\mathcal{P} \setminus H$ *is a tangent geometry of the* $O_6^+(2)$*-polar space.*
(2) Let $t = 2$. *Then* $\mathcal{P}$ *is the* $S_8(2)$*-polar space and* H *is singular. The minimal standard quotient of* $\mathcal{P} \setminus H$ *is the affine extension of the* $S_6(2)$*-polar space, naturally embedded in (the projective geometry of) the vector space* $V(6,2)$.
(3) Let $t = 4$. *Then* $\mathcal{P}$ *is the* $O_{10}^-(2)$*-polar space and* H *is isomorphic to the* $O_9(2)$*-polar space. The minimal standard quotient of* $\mathcal{P} \setminus H$ *is a tangent geometry of the* $O_8^-(2)$*-polar space.*

(See Sect. 7.2 for the minimal standard quotient of an affine polar space, Sect. 7.3 for tangent geometries and Sect. 6.3 for affine extensions.) Property (TR) has an additional nice effect.

Lemma 2.6. *If* $\mathcal{G}$ *satisfies* (TR) *then* (IP) *also holds in* $\mathcal{G}$.

Proof. Given two elements x and y of $\mathcal{G}$ of type at least 1, suppose that $P(x) \cap P(y) \neq \emptyset$. Pick a point $p \in P(x) \cap P(y)$ and let x' and y' be maxes incident with x and y respectively, possibly, $x' = x$ or $y' = y$. By (TR), $P(x)$ and $P(y)$ are cliques in the collinearity graphs of $\mathrm{Res}(x')$ and $\mathrm{Res}(y')$. For every point $q \in P(x) \cap P(y)$ different from p, let pq_x and pq_y be the lines through p and q in $\mathrm{Res}(x')$ and

Res(y'), respectively. Then pq_x is incident with x and pq_y is incident with y. By Lemma 2.2, $pq_x = pq_y$. The line $pq := pq_x = pq_y$ is incident with either of x and y (possibly, $pq = x$ or $pq = y$). Hence, we can replace the set of points $P(x) \cap P(y)$ with the set of lines through p incident with both x and y, thus turning to Res(p). The latter is a building (Lemma 2.1), hence (IP) holds in it. Therefore, there exists an element $z \in$ Res(p) incident with x and y and such that a 2-element of Res(p) is incident with z if and only if it is incident with both x and y. By the above, $P(z) = P(x) \cap P(y)$. Property (IP) is proved. □

Another hypothesis can be considered besides (TR). When trying to classify a class of geometries, it can be useful to focus on their collinearity graphs, turning the original problem into a problem of local recognition of graphs. However, for we can safely go from point-residues to neighborhoods in the collinearity graph and back, without missing any piece of information along the way, we need a hypothesis that says that, given a point p, every edge that exists in the neighborhood of p in the collinearity graph is somehow recognizable in Res(p). In our case, this can be phrased as follows, where $\Gamma(\mathcal{G})$ stands for the collinearity graph of $\mathcal{G}$:

(LR) (**Local Recognizability**) Every 3-clique of $\Gamma(\mathcal{G})$ is contained in at least one max of $\mathcal{G}$.

According to (LR), only two kinds of 3-cliques exist in $\Gamma(\mathcal{G})$: a 3-clique can be either contained in a plane or not. In the first case that 3-clique belongs to exactly seven maxes while in the second case it belongs to exactly one max (at least one by (LR) and at most one by (IP)).

Properties (TR) and (LR) are indeed the hypotheses assumed in [12, 13, 15, 16, 24, 28]. Clearly, each of (TR) and (LR) is preserved when taking universal covers (but (TR) is not preserved when taking 2-covers). Property (TR) forces finiteness. Indeed:

Proposition 2.7. *If $\mathcal{G}$ satisfies* (TR) *then $\Gamma(\mathcal{G})$ has diameter at most* 4.

Proof. Assume firstly that $\mathcal{G}$ satisfies both (TR) and (LR). Then the previous statement is the main result of Ivanov and Pasini [13] when $t = 2$ and it is a by-product of the classification theorems of Ivanov and Wiedorn [15] and Ivanov, Pasechnik and Shpectorov [16] when $t = 1$ and $t = 4$ (see Theorems 4.2 and 4.3 of Sect. 4 and the information given on $\mathcal{G}(3^{\cdot}Fi_{22})$ and $\mathcal{G}(B)$ in Sect. 3).

Suppose now that $\mathcal{G}$ satisfies (TR) but not (LR). Let p be a point of $\mathcal{G}$ and let $\Gamma(p)$ be its neighborhood in $\Gamma(\mathcal{G})$. By (TR), every collinear pair and every symplectic pair of points of the F_4-building Res(p) contributes an edge of $\Gamma(p)$. (See Sect. 10 for the definition of symplectic and special pairs of points of an F_4-building.) However, since now (LR) fails to hold in $\mathcal{G}$, additional edges of $\Gamma(p)$ are contributed by special pairs of points of Res(p) or pairs of points of Res(p) at distance 3 in Res(p). It is not difficult to prove that this forces diam($\Gamma(\mathcal{G})$) ≤ 2. We leave the details of this proof for the reader. □

Problem 2.8. The situation where $\mathcal{G}$ satisfies (TR) but not (LR) has not been investigated in the literature, but it is worth of consideration. What can we say on $\mathcal{G}$ in that case? Can we prove the following conjecture?

Conjecture 2.9. If $\mathcal{G}$ satisfies (TR) and is simply connected, then it also satisfies (LR).

Problem 2.10. Let $\mathcal{G}$ be a simply connected flag-transitive $c.F_4(2,t)$-geometry satisfying (TR) and let $\widetilde{\mathcal{G}}$ be its universal 2-cover. Is it true that $\widetilde{\mathcal{G}}$ is always a proper 2-cover of $\mathcal{G}$? (Note that, as $\mathcal{G}$ is assumed to be simply connected, $\widetilde{\mathcal{G}}$ is a proper 2-cover of $\mathcal{G}$ if and only if (TR) does not hold in $\widetilde{\mathcal{G}}$.) Is $\widetilde{\mathcal{G}}$ always infinite?

For instance, the geometry $\mathcal{G}(2^{26}F_4(2))$ (to be defined in the next section) admits a proper 2-cover (see Sect. 5.3), but we do not know if it is 2-simply connected. Perhaps it isn't. More flag-transitive $c.F_4(2,t)$-geometries where (TR) fails to hold will be discussed in Sect. 5.3, but we do not know so much on their universal 2-covers.

3 The Examples

We shall now describe the known flag-transitive $c.F_4(2,t)$-geometries that satisfy (TR) and (LR), sticking to the constructions offered in [12, 15, 16]. Throughout this section $\mathcal{F}$ is an $F_4(2,t)$-building and $\Phi(\mathcal{F})$ is the graph with the points of $\mathcal{F}$ as vertices, two points being adjacent in $\Phi(\mathcal{F})$ when they are either collinear or form a symplectic pair (see Sect. 10).

3.1 Preliminaries

All examples considered in [12, 15, 16] are constructed as follows. A graph Γ is considered that is locally $\Phi(\mathcal{F})$. The graph Γ is related to a group G which acts transitively on the set of directed edges of Γ and such that the stabilizer G_a of a in G of a vertex a of Γ acts as a flag-transitive subgroup F of $\mathrm{Aut}(\mathcal{F})$ on the neighborhood $\Gamma(a)$ of a. The vertices and the edges of Γ are taken as points and lines respectively of the geometry $\mathcal{G}$ that we are going to define. For every vertex a of Γ, the subgraphs of $\Gamma(a)$ corresponding to lines, planes and symps of $\mathcal{F}$ in the isomorphism $(G_a, \Gamma(a)) \cong (F, \Phi(\mathcal{F}))$, joined with $\{a\}$, are taken as planes, spaces and maxes of $\mathcal{G}$ on a. This definition is consistent. Indeed, in each of the cases to consider the pair (G, Γ) satisfies the following:

(C) (**Consistency Condition**) For every edge $\{a, b\}$ of Γ, for every class $\mathcal{X}$ of subgraphs of $\Gamma(a)$ corresponding to the set of lines, planes or symps of $\mathcal{F}$, and every $X \in \mathcal{X}$ containing b, there is an element $g \in G$ that maps a to b and X to $(X \setminus \{b\}) \cup \{a\}$.

Planes, spaces and maxes of $\mathcal{G}$ can also be recovered from Γ as follows. The maxes of $\mathcal{G}$ are maximal cliques of Γ, but not all maximal cliques of Γ are maxes of $\mathcal{G}$. Indeed Γ admits maximal cliques of different size. The largest ones have size $4\cdot(7t+2)$. These are the maxes of $\mathcal{G}$. Having recognized maxes of $\mathcal{G}$ in this way, we can recover spaces and planes as intersections of maxes. Indeed, given two maxes X and Y of $\mathcal{G}$, either $|X \cap Y| \leq 2$ or $|X \cap Y| = 8$. If $|X \cap Y| = 8$ then $X \cap Y$ is a space. Finally, given three maxes X, Y and Z, either $|X \cap Y \cap Z| = 8$ (whence $X \cap Y \cap Z = X \cap Y$ is a space) or $|X \cap Y \cap Z| = 4$ or $|X \cap Y \cap Z| \leq 2$. If $|X \cap Y \cap Z| = 4$ then $X \cap Y \cap Z$ is a plane.

By construction, $\mathcal{G}$ belongs to $c.F_4(2,t)$ and G acts flag-transitively on it. Clearly, Γ is the collinearity graph of $\mathcal{G}$. Moreover, $\mathcal{G}$ satisfies the following:

($*$) Three distinct points of $\mathcal{G}$ form a clique in Γ if and only if there is max of $\mathcal{G}$ that contains them all.

This property is just the conjuction of properties (TR) and (SR) considered in Sect. 2.

We shall now go into details, explaining which group G and which graph Γ are to be chosen. We shall firstly consider the cases of $t = 1$ and $t = 4$, keeping the case of $t = 2$ for last. Our exposition is based on Ivanov and Wiedorn [15] for $t = 1$ and $t = 4$, Ivanov, Pasechnik and Shpectorov [16] for all cases and Ivanov and Pasechnik [12] for $t = 2$.

3.2 $\mathcal{G}(Fi_{22})$ and $\mathcal{G}(3^{\cdot}Fi_{22})$ $(t = 1)$

Let $\Gamma(Fi_{22})$ be the graph with the $2D$-involutions of $\mathrm{Aut}(Fi_{22}) = Fi_{22}{:}2$ as vertices, two such involutions being adjacent if and only if they commute. (Note that these involutions are outer automorphisms of Fi_{22}.) The graph $\Gamma(Fi_{22})$ is locally $\Phi(\mathcal{F})$, where $\mathcal{F}$ is the $F_4(2,1)$-building. Moreover, Fi_{22} is transitive on the set of directed edges of $\Gamma(Fi_{22})$ and the pair $(Fi_{22}, \Gamma(Fi_{22}))$ satisfies condition (C). Thus, a geometry $\mathcal{G}(Fi_{22})$ exists, belonging to $c.F_4(2,1)$, satisfying (TR) and (SR) and admitting Fi_{22} as a flag-transitive automorphism group. Actually, $\mathrm{Aut}(\mathcal{G}(Fi_{22})) = \mathrm{Aut}(Fi_{22}) = Fi_{22}{:}2$.

This geometry is not simply connected. Its universal cover is a 3-fold cover $\mathcal{G}(3^{\cdot}Fi_{22})$ with $\mathrm{Aut}(\mathcal{G}(3^{\cdot}Fi_{22})) = 3^{\cdot}Fi_{22}{:}2$. Accordingly, $\Gamma(Fi_{22})$ admits a 3-fold cover $\Gamma(3^{\cdot}Fi_{22})$ which lives in $3^{\cdot}Fi_{22}{:}2$. The geometry $\mathcal{G}(3^{\cdot}Fi_{22})$ arises from it.

The geometry $\mathcal{G}(3^{\cdot}Fi_{22})$ is simply connected. This is a by-product of Theorem 4.2 of Sect. 4, but it also follows from Proposition 2.7 and the fact that $\Gamma(3^{\cdot}Fi_{22})$ has diameter equal to 4.

3.3 $\mathcal{G}(B)$ $(t = 4)$

Let $\Gamma(B)$ be the Baby Monster graph, namely the graph on the set of $\{3,4\}$-transposition in the Baby Monster group B in which two such transpositions are adjacent if their product is a central involution in B, namely an involution centralized by a Sylow 2-subgroup of B. The graph $\Gamma(B)$ is locally $\Phi(\mathcal{F})$, where $\mathcal{F}$ is the $F_4(2,4)$-building. Moreover, B acts transitively on the set of directed edges of $\Gamma(B)$ and the pair $(B, \Gamma(B))$ satisfies condition (C). Thus, a geometry $\mathcal{G}(B)$ exists, belonging to $c.F_4(2,4)$, satisfying (TR) and (SR) and admitting B as a flag-transitive automorphism group. Indeed $\mathrm{Aut}(\mathcal{G}(B)) = B$.

The geometry $\mathcal{G}(B)$ is simply connected. This is a by-product of Theorem 4.3 of Sect. 4. It also follows from Proposition 2.7, recalling that $\Gamma(B)$ has diameter equal to 3.

3.4 $\mathcal{G}(2^{26}F_4(2))$, $\mathcal{G}(E_6(2))$, $\mathcal{G}({}^2E_6(2))$ and $\mathcal{G}(3^{\cdot 2}E_6(2))$ $(t = 2)$

Let G be any of the groups $2^{26}{:}F_4(2)$, $E_6(2)$, ${}^2E_6(2)$ or $3^{\cdot 2}E_6(2)$, where $2^{26}{:}F_4(2)$ is the extension of the Chevalley group $F_4(2)$ by its 26-dimensional GF(2)-module. In any case, G admits a maximal subgroup $F \cong F_4(2)$. Let V be the set of cosets of F (left cosets, to fix ideas). The transitive action of G on V by left translation admits several subdegrees but, in any case, one of them is equal to 69615, which is the number of points of the $F_4(2)$-building. Let $\overline{E}$ be the orbital of G on V corresponding to that subdegree, E the set of unordered pairs corresponding to members of $\overline{E}$ and Γ the graph with V as the set of vertices and E as the set of edges. Then Γ is locally $\Phi(\mathcal{F})$, where $\mathcal{F}$ is the $F_4(2)$-building. Moreover, (G, Γ) satisfies condition (C). Thus, we obtain geometries $\mathcal{G}(2^{26}F_4(2))$, $\mathcal{G}(E_6(2))$, $\mathcal{G}({}^2E_6(2))$ and $\mathcal{G}(3^{\cdot 2}E_6(2))$ belonging to $c.F_4(2,2)$, satisfying (TR) and (SR) and admitting $2^{26}{:}F_4(2)$, $E_6(2)$, ${}^2E_6(2)$ and $3^{\cdot 2}E_6(2)$ respectively as flag-transitive automorphism groups. Actually $\mathrm{Aut}(\mathcal{G}(2^{26}F_4(2))) = 2^{26}{:}F_4(2)$ but $\mathrm{Aut}(\mathcal{G}(E_6(2))) = E_6(2){:}2$, $\mathrm{Aut}(\mathcal{G}({}^2E_6(2))) = {}^2E_6(2){:}2$ and $\mathrm{Aut}(\mathcal{G}(3^{\cdot 2}E_6(2))) = 3^{\cdot 2}E_6(2){:}2$.

The graph Γ has diameter 4, 3, 2 or 4 according to whether G is $2^{26}F_4(2)$, $E_6(2)$, ${}^2E_6(2)$ or $3^{\cdot 2}E_6(2)$. Hence, $\mathcal{G}(2^{26}F_4(2))$, $\mathcal{G}(E_6(2))$ and $\mathcal{G}(3^{\cdot 2}E_6(2))$ are simply connected, by Proposition 2.7. The geometry $\mathcal{G}({}^2E_6(2))$ is a quotient of $\mathcal{G}(3^{\cdot 2}E_6(2))$.

The geometries $\mathcal{G}(2^{26}F_4(2))$ and $\mathcal{G}(E_6(2))$ can also be produced by constructions of more geometrical nature. Indeed, as proved in [13], $\mathcal{G}(2^{26}F_4(2))$ is nothing but the affine extension of the 26-dimensional projective embedding of the $F_4(2)$-building $\mathcal{F}$ (see Sect. 6.3 for the definition of affine extensions). In Sect. 5, we will obtain

$\mathcal{G}(E_6(2))$ as a special case of a construction which works for any building of type E_6. On the other hand, the graph-theoretic constructions given at the beginning of this subsection and in Remark 3.2 (see below) are so far the only ones available for $\mathcal{G}({}^2E_6(2))$.

Problem 3.1. Find a different, more geometric construction for $\mathcal{G}({}^2E_6(2))$. (Compare Problem 5.8.)

Remark 3.2. Both $\mathcal{G}(Fi_{22})$ and $\mathcal{G}({}^2E_6(2))$ are subgeometries of $\mathcal{G}(B)$, namely the elements of $\mathcal{G}(Fi_{22})$ and $\mathcal{G}({}^2E_6(2))$ are elements of $\mathcal{G}(B)$, with the same types as in $\mathcal{G}(B)$, and the incidence relations of $\mathcal{G}(Fi_{22})$ and $\mathcal{G}({}^2E_6(2))$ are suitable refinements of the incidence relation inherited from $\mathcal{G}(B)$. To see this, we must recall a few properties of $\Gamma(B)$. There are two types of pairs of vertices $\{a, b\}$ at distance 2 in $\Gamma(B)$, according to whether ab has order 3 or 4. Let $\{a, b\}$ be a pair of vertices of $\Gamma(B)$ at distance 2 with $(ab)^3 = 1$ and let $\Gamma_3(a, b)$ be the subgraph induced by $\Gamma(B)$ on the set of vertices at distance 3 from either of a and b. Then $\Gamma_3(a, b) \cong \Gamma(Fi_{22})$. We can now recover $\mathcal{G}(Fi_{22})$ inside $\mathcal{G}(B)$. The points and the lines of $\mathcal{G}(Fi_{22})$ are the vertices and the edges of $\Gamma_3(a, b)$. The 4-elements of $\mathcal{G}(Fi_{22})$ are provided by the 4-elements x of $\mathcal{G}(B)$ with $P(x) \cap \Gamma_3(a, b)$ of maximal size, while the 2- and 3-elements of $\mathcal{G}(Fi_{22})$ are the 2- and 3-elements y of $\mathcal{G}(B)$ incident to a 4-element as above and such that $P(y) \subset \Gamma_3(a, b)$.

Turning to $\mathcal{G}({}^2E_6(2))$, its collinearity graph is isomorphic to the graph $\Gamma_3(a)$ induced by $\Gamma(B)$ on the set of vertices at distance 3 from a given vertex a. The geometry $\mathcal{G}({}^2E_6(2))$ can be recovered from $\Gamma_3(a)$ like $\mathcal{G}(Fi_{22})$ from $\Gamma_3(a, b)$.

4 The Classification

Theorem 4.1. *Let $\mathcal{G}$ be a flag-transitive $c.F_4(2,t)$-geometry satisfying both properties* (TR) *and* (LR). *Then $\mathcal{G}$ is one of the geometries described in Sect. 3.*

As explained in the introduction, different results are collected in this theorem, one for each of the cases $t = 1$, $t = 4$ and $t = 2$. We shall discuss them one by one.

4.1 The Case of $t = 1$

Theorem 4.2 (Ivanov and Wiedorn [15]). *$\mathcal{G}(Fi_{22})$ and its cover $\mathcal{G}(3^{\cdot}Fi_{22})$ are the only flag-transitive $c.F_4(2,1)$-geometries that satisfy both properties* (TR) *and* (LR).

Sketch of the proof. Let $\mathcal{G}$ be a flag-transitive $c.F_4(2, 1)$-geometry satisfying (TR) and (LR). Ivanov and Wiedorn firstly investigate the graphs defined on the set of 1-elements and the set of 2-elements of $\mathcal{G}$ by taking "being parallel inside the same $\{0, 1, 2\}$-residue" as the adjacency relations. The connected components of these two graphs are the classes of the equivalence relations called Λ_1 and Λ_2 in Sect. 8.2. Condition (RS) of Sect. 8.1 and its type 2 analogue hold. So, we can consider the geometry at infinity $\mathcal{G}^{\infty,2}$, which is also flag-transitive. Ivanov and Wiedorn prove that $\mathcal{G}^{\infty,2}$ belongs to the following diagram:

$(C_3.c^*)$

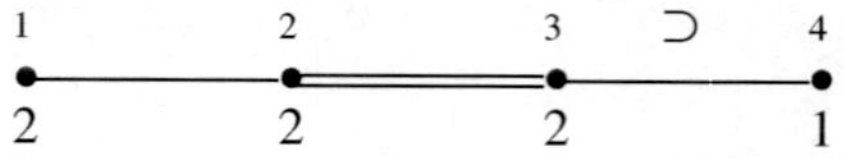

If we find out which group $\mathrm{Aut}(\mathcal{G}^{\infty,2})$ is, then we are done. Indeed, $\mathrm{Aut}(\mathcal{G}^{\infty,2}) = \mathrm{Aut}(\mathcal{G})$ by Proposition 8.4.

Thus, we turn from $\mathcal{G}$ to $\mathcal{G}^{\infty,2}$. One can prove that (IP) holds in all residues of $\mathcal{G}^{\infty,2}$ of rank 3. Hence, the $\{1,2,3\}$-residues of $\mathcal{G}^{\infty,2}$ are polar spaces, isomorphic to the $S_6(2)$-polar space. So, $\mathcal{G}^{\infty,2}$ is the dual of a c-extended dual polar space. Flag-transitive c-extended dual polar spaces have been classified by Ivanov [10], [11]. We can exploit that classification to finish (but I warn that this is not the way that Ivanov and Wiedorn choose in [15]). In view of that, we must determine the isomorphism type of the $\{2,3,4\}$-residues of $\mathcal{G}^{\infty,2}$. As (IP) holds in these residues, they are dually isomorphic to standard quotients of affine polar spaces (Proposition 7.3). Having stated this, a range of possibilities still remains for those residues. In order to select the right one from it, we ask shrinkings for help. The classes of Λ_1, which provide the 1-elements of $\mathcal{G}^{\infty,2}$, are the point-sets of the shrinkings of $\mathcal{G}$. Hence, the stabilizer in $\mathrm{Aut}(\mathcal{G})$ of a shrinking of $\mathcal{G}$ is the same as the stabilizer in $\mathrm{Aut}(\mathcal{G}^{\infty,2})$ of a 1-element of $\mathcal{G}^{\infty,2}$. Let $\mathrm{Shr}(\mathcal{G})$ be a shrinking of $\mathcal{G}$. Then $\mathrm{Shr}(\mathcal{G})$ is flag-transitive with diagram and orders as follows:

$(c.C_3^*)$

The $\{2,3,4\}$-residues of $\mathrm{Shr}(\mathcal{G})$ are isomorphic to the $\{2,3,4\}$-residues of $\mathcal{G}$. The latters are C_3-buildings. It is well known that every C_3-building with orders $1,1,2$ is isomorphic to the geometry of vertices, edges and 3-cliques of a complete 3-partite graph with all classes of size 3. Denoted this geometry by $\mathcal{K}$, let $\mathcal{K}^*$ be its dual. Thus, the $\{2,3,4\}$-residues of $\mathrm{Shr}(\mathcal{G})$ are isomorphic to $\mathcal{K}^*$. On the other hand, the $\{0,1,2,3\}$-residues of $\mathcal{G}$ are isomorphic to the minimal standard quotient of the affine polar space obtained from the $O_8^+(2)$-polar space by removing an $O_7(2)$-hyperplane (Lemma 2.5). Hence, the $\{1,2,3\}$-residues of $\mathrm{Shr}(\mathcal{G})$ are isomorphic to the minimal standard quotient of the affine polar space obtained from the $O_6^+(2)$-polar space by removing an $O_5(2)$-hyperplane (compare Example 8.2). It follows that the collinearity graph of $\mathrm{Shr}(\mathcal{G})$ is locally the distance 1-or-2 graph of the collinearity graph of $\mathcal{K}^*$. We can now apply a result of Cuypers [6] to recover $\mathrm{Shr}(\mathcal{G})$ as a subgeometry of the $F_4(2,1)$-building with $\mathrm{Aut}(\mathrm{Shr}(\mathcal{G})) \cong U_4(2){:}2$ (see also [16, Propsition 5.2]).

By the above, the stabilizer of a 1-element of $\mathcal{G}^{\infty,2}$ in $\mathrm{Aut}(\mathcal{G}^{\infty,2})$ acts as $U_4(2){:}2 \cong \mathrm{PSO}_6^-(2)$ in the residue of thatelement. This makes it clear that the $\{2,3,4\}$-residues of $\mathcal{G}^{\infty,2}$ are dually isomorphic to the affine polar space obtained by removing an $O_6^-(2)$-hyperplane from the $O_7(2)$-polar space. By Ivanov [10, Proposition 2.5], $\mathrm{Aut}(\mathcal{G}^{\infty,2})$ is either $Fi_{22}{:}2$ or $3^{\cdot}Fi_{22}{:}2$.

However, as I have said before, the argument used by Ivanov and Wiedorn in [15] is different from the above. It is worth to give a shortened exposition of it, too.

The dual $\mathcal{G}^{\infty,2,*}$ of $\mathcal{G}^{\infty,2}$ belongs to the following diagram, which is the same as that of $\mathcal{G}^{\infty,2}$, but switched. The types 1, 2, 3, 4 of $\mathcal{G}^{\infty,2,*}$ correspond to the types 4, 3, 2, 1 of $\mathcal{G}^{\infty,2}$.

$(c.C_3^*)$

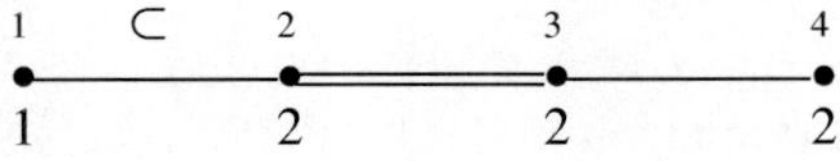

The $\{1,2,3\}$-residues of $\mathcal{G}^{\infty,2,*}$ are the duals of the $\{2,3,4\}$-residues of $\mathcal{G}^{\infty,2}$. According to what we have previously said on the latter, the $\{1,2,3\}$-residues of $\mathcal{G}^{\infty,2,*}$ are isomorphic to the affine polar space obtained by a removing an $O_6^-(2)$-hyperplane from the $O_7(2)$-polar space.

The geometry $\mathcal{G}^{\infty,2,*}$ admits shrinkings. Let $\mathrm{Shr}(\mathcal{G}^{\infty,2,*})$ be one of them. The isomorphism type of the $\{1,2,3\}$-residues of $\mathcal{G}^{\infty,2,*}$ forces $\mathrm{Shr}(\mathcal{G}^{\infty,2,*})$ to belong to the following diagram, where P stands for the Petersen graph (compare Example 8.2):

(P_3)

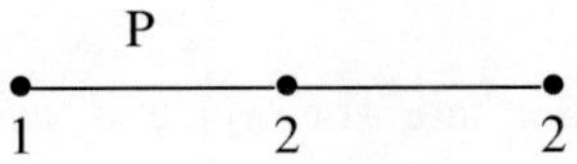

So, $\mathrm{Shr}(\mathcal{G}^{\infty,2,*})$ is a flag-transitive P_3-geometry. Flag-transitive P_n-geometries are classified (see Ivanov and Shpectorov [14]). Only two flag-transitive P_3-geometries exist. They arise from M_{22} and $3^{\cdot}M_{22}$, respectively. $\mathrm{Shr}(\mathcal{G}^{\infty,2,*})$ is isomorphic to one of these two geometries. In order to finish, Ivanov and Wiedorn must combine the previous information on shrinkings with the investigation of another geometry $\mathcal{B}$ such that $\mathrm{Aut}(\mathcal{B}) = \mathrm{Aut}(\mathcal{G}^{\infty,2,*})$. The construction of $\mathcal{B}$ by Ivanov and Wiedorn is beautiful but its details are too complicated for I can summarize them here. I only mention that they consider a graph B having the shrinkings of $\mathcal{G}^{\infty,2,*}$ as the vertices, two shrinkings being adjacent in B precisely when they contain 2-elements supported by the same 2-element of $\mathcal{G}^{\infty,2,*}$. The shrinkings of $\mathcal{G}^{\infty,2,*}$ are taken as 3-elements of $\mathcal{B}$, like if we constructed the dual of a geometry at infinity. The 2- and 1-elements of $\mathcal{B}$ are certain 3-cliques and certain subgraphs of B. Since the shrinkings of $\mathcal{G}^{\infty,2,*}$ are P_3-geometries for M_{22} or $3^{\cdot}M_{22}$, the stabilizer in $\mathrm{Aut}(\mathcal{B})$ of a 3-element x of $\mathcal{B}$ induces on $\mathrm{Res}_{\mathcal{B}}(x)$ a group isomorphic to M_{22}, $M_{22}{:}2$, $3^{\cdot}M_{22}$ or $3^{\cdot}M_{22}{:}2$. With this information at hand, Ivanov and Wiedorn can prove that $\mathcal{B}$ is a truncation of a flag-transitive $c.C_3$-geometry $\overline{\mathcal{B}}$ with diagram and orders as below:

$(c.C_3)$

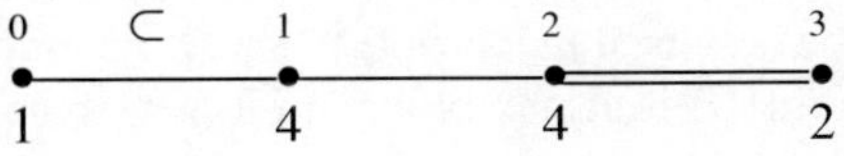

We obtain $\mathcal{B}$ from $\overline{\mathcal{B}}$ by removing all 0-elements of $\overline{\mathcal{B}}$. Flag-transitive $c.C_3$-geometries with orders as above have been characterized since long ago (Buekenhout and Hubaut [3]; see also Del Fra et al. [8]). In view of that characterization, $\mathrm{Aut}(\overline{\mathcal{B}})$ is isomorphic to either $Fi_{22}{:}2$ or $3^{\cdot}Fi_{22}{:}2$. Ivanov and Wiedorn also prove that $\mathrm{Aut}(\mathcal{B}) = \mathrm{Aut}(\mathcal{G}^{\infty,2})$. Hence, $\mathrm{Aut}(\mathcal{G})$ is either $Fi_{22}{:}2$ or $3^{\cdot}Fi_{22}{:}2$. □

4.2 The Case of $t = 4$

Theorem 4.3 (Ivanov et al. [16]). *$\mathcal{G}(B)$ is the unique flag-transitive $c.F_4(2,4)$-geometry satisfying* (TR) *and* (LR).

Sketch of the proof. Let $\mathcal{G}$ be a flag-transitive $c.F_4(2,4)$-geometry satisfying (TR) and (LR). In this case, the geometry at infinity $\mathcal{G}^{\infty,2}$ is harder to understand than $\mathcal{G}$ itself. No insight into $\mathcal{G}$ can be got from it. On the other hand, shrinkings can be of some help. Every shrinking $\mathrm{Shr}(\mathcal{G})$ of $\mathcal{G}$ is a flag-transitive c-extended dual polar space with orders as follows:

$(c.C_3)$

Its $\{2,3,4\}$-residues are isomorphic to the dual of the $U_6(2)$-polar space while its $\{1,2,3\}$-residues are isomorphic to the minimal standard quotient of the affine polar space obtained by removing an $O_7(2)$-hyperplane from the $O_8^-(2)$-polar space (see Example 8.2). Hence, the collinearity graph of $\mathrm{Shr}(\mathcal{G})$ is locally the distance 1-or-2 graph of the dual of the $U_6(2)$-polar space. One can now exploit the classification of flag-transitive c-extended dual polar spaces by Ivanov [10], [11] or a result by Cuypers [6] to determine the isomorphism type of $\mathrm{Shr}(\mathcal{G})$. It turns out that $\mathrm{Shr}(\mathcal{G})$ is isomorphic to a well known geometry admitting the Conway group Co_2 as its full automorphism group [16, Proposition 5.2].

The isomorphism type of $\mathrm{Shr}(\mathcal{G})$ is thus determined, but a lot of work still remains to do in order to finish the proof. At this stage, the authors of [16] turn to a thorough investigation of the collinearity graph Γ of $\mathcal{G}$, focusing on the common neighborhood $\Gamma(x,y)$ of two vertices x and y at distance 2 and the second neighborhood $\Gamma_2(x)$ of a vertex x. In particular, they prove that, for every vertex $y \in \Gamma_2(x)$, there is exactly one vertex $y' \in \Gamma_2(x)$ different from y and such that $\Gamma(x,y) = \Gamma(x,y')$. Let π_x be the permutation of Γ that fixes all vertices of $\Gamma \setminus \Gamma_2(x)$ and switches all pairs $\{y,y'\} \subset \Gamma_2(x)$ with $\Gamma(x,y) = \Gamma(x,y')$. The authors of [16] prove that π_x is an automorphism of Γ (hence, it also induces an automorphism of $\mathcal{G}$). Some amount of work is needed to prove this claim. We will not go into the details of that proof. We only mention that the information previously obtained on $\mathrm{Shr}(\mathcal{G})$ is also exploited in it.

If x and z are adjacent vertices of Γ then $\pi_z(x) = x$, whence π_z induces an automorphism of the neighborhood $\Gamma(x)$ of x. In fact, the family $\{\pi_z\}_{z\in\Gamma(x)}$

generates a copy of ${}^2E_6(2)$. So far, we know that the graph Γ is locally $\Phi(\mathcal{F})$, where $\mathcal{F}$ is the $F_4(2,4)$-building, and it admits an automorphism group G such that, for every vertex x, the stabilizer G_x of x in G induces on $\Gamma(x)$ an action containing ${}^2E_6(2)$. By a result of Ivanov [9], G is the Baby Monster group B and $\Gamma = \Gamma(B)$ (the Baby Monster graph). □

4.3 The Case of $t = 2$

As said in the introduction, this case turned out to be the most difficult one. Till a short time ago, we only knew the following.

Theorem 4.4 (Wiedorn [28]). *Besides $\mathcal{G}(2^{26}F_4(2))$, $\mathcal{G}(E_6(2))$ and $\mathcal{G}(3^{\cdot 2}E_6(2))$, at most four simply connected flag-transitive $c.F_4(2,2)$-geometries exist that satisfy* (TR).

Sketch of the proof. Let G be a group acting flag-transitively on a $c.F_4(2,2)$-geometry $\mathcal{G}$ satisfying (TR) and for $i = 0,1,2,3,4$ let G_i be the stabilizer in G of the i-element a_i of a given chamber $c = \{a_i\}_{i=0}^4$ of $\mathcal{G}$. Let $\mathcal{A} = (G_i)_{i=0}^4$ be the amalgam of the subgroups G_0, G_1, G_2, G_3, G_4 with intersections as in G. Then $\mathcal{G}$ is simply connected if and only if G is the universal completion of $\mathcal{A}$ (see [19, Theorem 12.28]). So, we must determine all possibilities for the amalgam $\mathcal{A}$.

Henceforth, for an element $x \in \mathcal{G}$ we denote by G_x the stabilizer of x in G and by K_x the kernel of the action of G_x on $\mathrm{Res}(x)$ and we put $\overline{G}_x := G_x/K_x$. We also put $G_{x,y} := G_x \cap G_y$. If x has type $\neq 0, 4$, we denote by K_x^+ and K_x^- the kernels of the actions of G_x on $\mathrm{Res}^+(x)$ and $\mathrm{Res}^-(x)$ respectively. In the previous paragraph we wrote G_i for G_{a_i}. Accordingly, we write K_i for K_{a_i}, $G_{i,j}$ for G_{a_i,a_j} and so on.

Clearly, $\overline{G}_0 \cong F_4(2)$, $\overline{G}_1 \cong 2 \times S_6(2)$ and $\overline{G}_4 \cong 2^6{:}S_6(2)$ (recall that, by Lemma 2.5, $\mathrm{Res}(a_4)$ is the affine extension of the $S_6(2)$-polar space embedded in $V(6,2)$). Moreover, $\overline{G}_2$ is isomorphic to either $S_4 \times L_3(2)$ or $A_4 \times L_3(2)$ and $\overline{G}_3$ is isomorphic to either $(2^3{:}L_3(2)) \times S_3$ or $(2^3{:}L_3(2)) \times 3$. However, $\overline{G}_2$ must fit with $\overline{G}_3$. Hence, $\overline{G}_2 \cong S_4 \times L_3(2)$ and $\overline{G}_3 \cong (2^3{:}L_3(2)) \times S_3$.

Wiedorn firstly proves that $|K_0| \leq 2$ and that if $|K_0| = 2$ then $G_0 = K_0 \times \overline{G}_0$. Hence, $G_{0,1} = K_0 \times 2^{1+6+8}{:}S_6(2)$. We must now describe G_1. Clearly $[G_1 : G_{0,1}] = 2$. Hence $G_1 = G_{0,1} \cdot \langle \tau \rangle$ for an element τ switching the two points of the line a_1 and such that $\tau^2 \in G_{0,1}$. Modulo multiplying τ be a suitable element of $G_{0,1}$ we can assume that $\tau \in K_{a_1}^+$. We can do more, choosing $\tau \in K_{a_1}^+$ in such a way that it moves as little as possible of $G_{0,1}$. Wiedorn proves that if we can choose $\tau \in C_G(G_{0,1})$ (which can be only if $K_0 = 1$) then $\mathcal{G} = \mathcal{G}(2^{26}F_2(4))$.

Let $\mathcal{G} \neq \mathcal{G}(2^{26}F_2(4))$. Hence, $\tau \notin C_G(G_{0,1})$. Let $G'_{0,1} \cong 2^{1+6+8}{:}S_6(2)$ be the commutator subgroup of $G_{0,1}$. Wiedorn proves that we can choose τ in such a way that it centralizes a subgroup $C_\tau < G'_{0,1}$ containing $O_2(G'_{0,1})$ and such that $C_\tau/O_2(G'_{0,1}) \cong O_6^+(2)$ (where $O_6^+(2)$ stands for the commutator subgroup of $PGO^+(6,2)$, as in [5]). With the above constraints, once C_τ has been chosen, the element τ is uniquely determined modulo multiplication by elements of $Z(G_{0,1}) = K_0 \times \langle z \rangle$, where z is the unique involution of $Z(G'_{0,1})$. Wiedorn also proves that τ has order either 2 or 4 and if τ has order 4 then $\tau^2 = z$. It is worth giving τ a name. We call it the *switch element* of G_1.

At this stage we have got at most four isomorphism types $\mathcal{A}^{(1)}_{\varepsilon,\eta}$ for the amalgam $\mathcal{A}^{(1)} := (G_0, G_1)$, where $\varepsilon \in \{1,2\}$ is the size of K_0 and $\eta \in \{2,4\}$ is the order $o(\tau)$ of the switch element τ. However, when $\varepsilon = 2$ we can replace τ with τk, where k is the unique involution of K_0. If $o(\tau) = 2$ then $o(tauk) = 4$ and if $o(\tau) = 4$ then $o(\tau k) = 2$. Thus, three possibilities exist for $\mathcal{A}^{(1)}$, namely $\mathcal{A}^{(1)}_{1,2}$, $\mathcal{A}^{(1)}_{1,4}$ and $\mathcal{A}^{(1)}_{2,*}$, where $*$ can be read as 2 or 4, as we like.

Turning to the rank 3 amalgam $\mathcal{A}^{(2)} = (G_0, G_1, G_2)$, Wiedorn proves that each of the amalgams $\mathcal{A}^{(1)}_{1,2}$ and $\mathcal{A}^{(1)}_{1,4}$ can be extended in exactly one way to a realization of $\mathcal{A}^{(2)}$, which we will denote by $\mathcal{A}^{(2)}_{1,2,*}$ and $\mathcal{A}^{(2)}_{1,4,*}$, respectively. On the other hand, $\mathcal{A}^{(1)}_{2,*}$ can be extended in at most to ways. We denote these two extensions by the symbols $\mathcal{A}^{(2)}_{2,*,\diamond}$ and $\mathcal{A}^{(2)}_{2,*,\circ}$. We warn that no explicit description of $\mathcal{A}^{(2)}_{2,*,\diamond}$ and $\mathcal{A}^{(2)}_{2,*,\circ}$ is given in [28]. Wiedorn only gives an indirect argument which implies that at most two ways exist to extend $\mathcal{A}^{(1)}_{2,*}$.

Next, Wiedorn proves that for every choice of $(\varepsilon, \eta, \theta) = (1,2,*), (1,4,*), (2,*,\diamond)$ or $(2,*,\circ)$ the amalgam $\mathcal{A}^{(2)}_{\varepsilon,\eta,\theta}$ can be extended in a unique way to a rank 4 amalgam $\mathcal{A}^{(3)}_{\varepsilon,\eta,\theta}$. Finally, again by an indirect argument, she proves that for every choice of $(\varepsilon, \eta, \theta)$ there are at most two ways to extend $\mathcal{A}^{(3)}_{\varepsilon,\eta,\theta}$ to a rank 5 amalgam $\mathcal{A} = (G_0, G_1, G_2, G_3, G_4)$. We denote those two extensions by the symbols $\mathcal{A}^{(4)}_{\varepsilon,\eta,\theta,\diamond}$ and $\mathcal{A}^{(4)}_{\varepsilon,\eta,\theta,\circ}$.

So far, at most eight possibilities exist for $\mathcal{A} = (G_0, G_1, G_2, G_3, G_4)$, corresponding to the eight quadruples $(1,2,*,\diamond)$, $(1,2,*,\circ)$, $(1,4,*,\diamond)$, $(1,4,*,\circ)$, $(2,*,\diamond,\diamond)$, $(2,*,\diamond,\circ)$, $(2,*,\circ,\diamond)$ and $(2,*,\circ,\circ)$. However, it might happen that two of these possibilities are mutually isomorphic or that one of them does not admit any completion. Moreover, if for some choice of $\theta, \zeta, \eta, \zeta'$ the amalgam $\mathcal{A}^{(4)}_{2,*,\theta,\zeta}$ embodies $\mathcal{A}^{(4)}_{1,\eta,*,\zeta'}$, then these two amalgams define the same geometry.

Each of $\mathcal{G}(E_6(2))$ and $\mathcal{G}(3{\cdot}^2E_6(2))$ gives rise to an amalgam $\mathcal{A}^{(4)}_{1,\eta,*,\zeta}$ as well as an amalgam $\mathcal{A}^{(4)}_{2,*,\theta,\zeta}$, since $E_6(2){:}2$ and $3{\cdot}^2E_6(2){:}2$ also acts on $\mathcal{G}(E_6(2))$ and $\mathcal{G}(3{\cdot}^2E_6(2))$, respectively. So, at most four possibilities remain for $\mathcal{A}^{(4)}_{\varepsilon,\eta,\theta,\zeta}$ that do not correspond to any of the known examples. □

Lately, Ivanov and I have made some attempts to improve the previous theorem, but unsuccessfully. Eventually, G. Stroth decided to try in his turn, and he has succeded.

Theorem 4.5 (Stroth [24]). *Only three simply connected flag-transitive c.$F_4(2,2)$-geometries exist that satisfy* (TR), *namely* $\mathcal{G}(2^{26}F_4(2))$, $\mathcal{G}(E_6(2))$ *and* $\mathcal{G}(3{\cdot}^2E_6(2))$.

Sketch of the proof. Stroth's proof starts from Wiedorn's analysis, discussed above. Wiedorn [28] neither gives a detailed description of $O_2(G_4)$ nor explains how the switch element τ acts in Res(a_4). Stroth [24] fills these two gaps. By elementary group theory (but very skillfully exploited) and with the help of some detailed information on generators of $F_4(2)$, he proves the following:

Claim 1. Let $K_0 = 1$. Then $O_2(G_4)$ is extraspecial of order 2^{21} and it contains the switch element τ (equivalently, τ acts as an affine translation on Res(a_4)). In fact we can say more, namely $O_2(G_4) = 2_+^{1+20}$ with $O_2(G_4)/Z(O_2(G_4)) \cong \wedge^3 V$ as an $S_6(2)$-module, $V = V(6,2)$ being the 6-dimensional natural module for $S_6(2)$.

Claim 2. Let $|K_0| = 2$. Then $O_2(G_4)$ admits a normal subgroup E of index 2 with E extraspecial of order 2^{21} as in Claim 1 and $O_2(G_4) = E \cdot K_0$. Moreover, the switch element τ can be chosen in such a way that $\tau \in E$ (hence it acts as an affine translation in Res(a_4)). We also have $K_4 = K_0 \times (K_4 \cap E)$ and $K_4 \cap E = 2^{1+6+8}$.

Let $K_0 = 1$. By Claim 1, τ is an involution. Indeed $\tau^2 \in Z(O_2(G_{0,1}))$ but $Z(O_2(G_{0,1})) \neq Z(O_2(G_4)) = Z(O_2(G_{0,4}))$. This observation is sufficient to rule out either of the amalgams $\mathcal{A}^{(4)}_{1,4,*,\zeta}$. Thus, the amalgams $\mathcal{A}^{(4)}_{1,2,*,\zeta}$ only survive, which lead to $E_6(2)$ and $3^{\cdot 2}E_6(2)$ respectively.

Turning to the case of $K_0 \neq 1$, we can now go on in two different ways. With the help of Claim 2 we can prove that each of the amalgams $\mathcal{A}^{(4)}_{2,*,\theta,\zeta}$ contains an $\mathcal{A}^{(4)}_{1,\eta,*,\zeta}$. As the latter arises from either $E_6(2)$ or $3^{\cdot 2}E_6(2)$, only two of the amalgams $\mathcal{A}^{(4)}_{2,*,\theta,\zeta}$ survive and they must arise from either $E_6(2)$:2 or $3^{\cdot 2}E_6(2)$:2.

Otherwise, one can go a bit deeper into the analysis of $\mathcal{A}^{(2)}_{2,*,\theta}$. This is the way chosen by Stroth [24]. By exploiting some information on generators of the subgroups $G_{0,i}$ ($i = 1,2,3,4$) he shows that only one amalgam $\mathcal{A}^{(2)}_{2,*,\theta}$ exists that can be extended (actually in two ways) to a rank 5 amalgam $\mathcal{A}^{(4)}_{2,*,\theta,\zeta}$ with $O_2(G_4)$ as in Claim 2. So, only two amalgams $\mathcal{A}^{(4)}_{2,*,\theta,\zeta}$ exist when $K_0 \neq 1$, and we get the same conclusion as above. □

By Theorem 4.5 we immediately obtain the following:

Theorem 4.6. *Only four flag-transitive $c.F_4(2,2)$-geometries exist that satisfy* (TR), *namely* $\mathcal{G}(2^{26}F_4(2))$, $\mathcal{G}(E_6(2))$, $\mathcal{G}(3^{\cdot 2}E_6(2))$ *and its quotient* $\mathcal{G}({}^2E_6(2))$.

Remark 4.7. In an earlier version of [24], Stroth gives a slightly different proof of Theorem 4.5, based on complete descriptions of the amalgams $\mathcal{A}^{(4)}_{\epsilon,\eta,\theta,\zeta}$ in terms of generators and relations. He considers a set of generators of G formed by a suitable set U of involutions generating a Sylow 2-subgroup S of $F_4(2)$, four involutions r_1, r_2, r_3, r_4 generating a copy W of the Weyl group of $F_4(2)$, and the switching element τ itself. The involutions r_1, r_2, r_3, r_4 form a Coxeter system of type $F_4(2)$ and are chosen in such a way that (S, W) is a BN-pair and the panel stabilized by r_i has type i. Moreover, for every $i = 1,2,3,4$ the set U contains exactly one element u_i such that $(r_i u_i)^3 = 1$. The element τ is assumed to centralize r_3, r_4 and a subset U_τ of U such that $\langle U_\tau, r_3, r_4\rangle = 2^{1+6+8}{:}O_6^+(2)$, as in Wiedorn [28]. It also centralizes r_2 modulo the unique involution z of $Z(G'_{0,1})$. Moreover, $\tau^2 = 1$ when $K_0 \neq 1$, as allowed by Wiedorn [28]. Put $s = \tau^{r_1}$. Stroth proves that one of the followings must hold, for a certain (uniquely determined) element $u \in U$:

(I.a) $K_0 = 1$, $\tau^2 = 1$, $[s, \tau] = 1$ and $s^{u_1} = s\tau$;
(I.b) $K_0 = 1$, $\tau^2 = z$, $s^2 = u$, $[s, \tau] = 1$ and $s^{u_1} = s\tau uz$;
(II.a) $|K_0| = 2$, $[s, \tau] = 1$ and $s^{u_1} = s\tau$;
(II.b) $|K_0| = 2$, $[s, \tau] = uz$, $s^{u_1} = s\tau ku$, where $K_0 = \langle k \rangle$.

Cases (I.a) and (I.b) correspond to $\mathcal{A}^{(2)}_{1,2,*}$ and $\mathcal{A}^{(2)}_{1,4,*}$ respectively, while (II.a) and (II.b) lead to $\mathcal{A}^{(2)}_{2,*,\circ}$ and $\mathcal{A}^{(2)}_{2,*,\diamond}$. Let $\mathcal{A}^{(2)}_{2,*,\circ}$ correspond to (II.a), to fix ideas. Going on with his detailed analysis, Stroth eventually proves that neither $\mathcal{A}^{(2)}_{1,4,*}$ nor $\mathcal{A}^{(2)}_{2,*,\diamond}$ can be extended to a rank 5 amalgam as we want. The conclusion of Theorem 4.5 follows.

Remark 4.8. We have mentioned shrinkings in the proofs of Theorems 4.2 and 4.3, but no use of them is made in the proofs of Theorems 4.4 and 4.5. However, shrinkings can also be considered in the case of $t = 2$. Indeed, let $\mathcal{G}$ be a flag-transitive $c.F_4(2,2)$-geometry $\mathcal{G}$ satisfying (TR) and (LR). By arguments similar to those used in the proofs of Theorems 4.2 and 4.3 one can prove that the shrinkings of $\mathcal{G}$ are isomorphic to the affine extension of the dual of the $O_7(2)$-polar space embedded in $V(8,2)$ via the spin embedding (see also Ivanov, Pasechnik and Shpectorov [16, Proposition 5.2]). One can also prove that property (RS) of Sect. 8 as well as its type 2 analogue hold in $\mathcal{G}$. Hence, a geometry at infinity can also be defined, either as in Sect. 8 or in other ways (see [22]). Perhaps shrinkings and a suitable geometry at infinity could be exploited to give another proof of Theorem 4.6, but this does not seem to be so easy to do.

Remark 4.9. Wiedorn assumes property (LR) in [28] but she never uses it in her investigation. Stroth neither uses it in [24]. Hence, I have not mentioned (LR) in Theorems 4.4–4.6. Actually, I cannot see how (LR) could be exploited when working with amalgams. This is one of the reasons that make me believe that simple connectedness implies (LR), as suggested in Conjecture 2.9. On the contrary, the proofs of Theorems 4.2 and 4.3 given in [15] and [16] do use (LR), although this may be not so evident from my shortened exposition of those proofs.

Problem 4.10. Can we drop (LR) from the hypotheses of Theorems 4.2 and 4.3?

5 A Geometric Construction of $\mathcal{G}(E_6(2))$ and More $Af.F_4$-Geometries

The main goal of this section is to produce a family of $Af.F_4$-geometries which contains $\mathcal{G}(E_6(2))$ as its smallest member. We shall firstly recall a construction of buildings of type E_6 and F_4 as geometries embedded in vector spaces of dimension 27 and 26, respectively. After that, we will construct our family of $Af.F_4$-geometries. In the last part of this section we will discuss a few more families of $Af.F_4$-geometries, which can be created as affine extensions from projective embeddings of F_4-buildings.

5.1 Preliminaries on Buildings of Type E_6 and F_4

The material of this subsection is taken from Chap. 18 of Buekenhout and Cohen [2]. We will omit many details and all proofs. The reader can find them in the above quoted chapter of [2] (also in Cohen and Cooperstein [4]).

Given a field $\mathbb{F}$, let $M = M_3(\mathbb{F})$ be the algebra of (3×3)-matrices over $\mathbb{F}$ and let $V = M \times M \times M$ (a 27-dimensional vector space over $\mathbb{F}$). For a vector $x = (X_1, X_2, X_3)$ of V, put

$$f(x) = \det(X_1) + \det(X_2) + \det(X_3) - \mathrm{tr}(X_1 X_2 X_3), \quad (1)$$

$$x^\sharp = (\mathrm{adj}(X_1) - X_2 X_3,\ \mathrm{adj}(X_3) - X_1 X_2,\ \mathrm{adj}(X_2) - X_3 X_1), \quad (2)$$

where for a matrix $X \in M$ we denote by $\det(X)$ and $\mathrm{tr}(X)$ its determninat and its trace and $\mathrm{adj}(X)$ is the adjugate matrix of X, namely the transpose of the cofactor matrix. Thus, we have defined a cubic form f on V and a mapping $(.)^\sharp : V \to V$. A symmetric bilinear form ϕ can also be defined on V by putting

$$\phi(x, y) = \mathrm{tr}(X_1 Y_1 + X_2 Y_3 + X_3 Y_2) \quad (3)$$

for any two vectors $x = (X_1, X_2, X_3)$ and $y = (Y_1, Y_2, Y_3)$ of V. We have

$$f(x + y) = f(x) + f(y) + \phi(x^\sharp, y) + \phi(x, y^\sharp) \quad (\forall x, y \in V). \quad (4)$$

Also, $x^{\sharp\sharp} = f(x)x$ for every $x \in V$. Hence, $x^\sharp = 0$ implies $f(x) = 0$ (but the converse is false in general). A commutative (but non-associative) operation $\times$ can also be defined, as follows:

$$x \times y := (x + y)^\sharp - x^\sharp - y^\sharp. \quad (5)$$

We are now ready to construct the $E_6(\mathbb{F})$-building, denoted by $\mathcal{E}$ throughout the rest of this section. We choose types as follows:

(E_6)

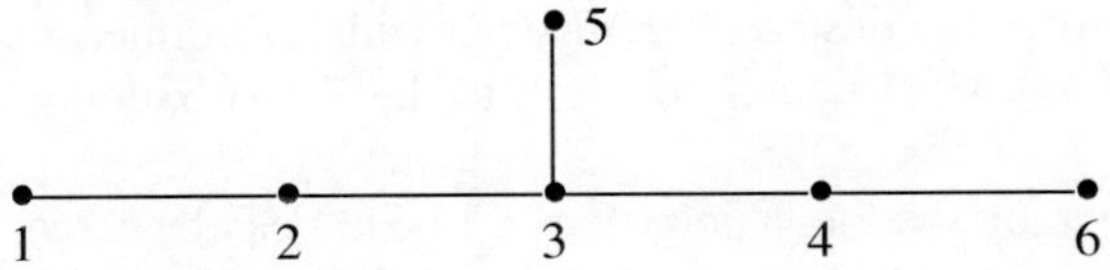

The elements of type 1, 2 and 3 will be called *points*, *lines* and *planes* respectively. The points of $\mathcal{E}$ are the points $\langle x \rangle$ of $\mathrm{PG}(V)$ such that $x^\sharp = 0$. Denoted the set of points of $\mathcal{E}$ by $P(\mathcal{E})$, the lines and the planes of $\mathcal{E}$ are the lines and the planes of $\mathrm{PG}(V)$ entirely contained in $P(\mathcal{E})$. Note that two points $\langle x\rangle$ and $\langle y\rangle$ of $\mathcal{E}$ are collinear in $\mathcal{E}$ if and only if $x \times y = 0$.

The set $P(\mathcal{E})$ spans $\mathrm{PG}(V)$. Thus, we have produced a projective embedding of the point-line-plane system $\mathcal{E}_{|1,2,3}$ of $\mathcal{E}$ in $\mathrm{PG}(V)$, which sends points, lines and planes of $\mathcal{E}$ to points, lines and planes of $\mathrm{PG}(V)$. It remains to recover the remaining elements of $\mathcal{E}$ as subspaces or subgeometries of $\mathrm{PG}(V)$.

The geometry $\mathcal{E}_{|1,2,3}$ admits two families of maximal singular subspaces (see Sect. 6.4). One of those two families is formed by 4-dimensional projective spaces. These are the 4-elements of $\mathcal{E}$. The other family consists of 5-dimensional projective spaces. These are the 5-elements of $\mathcal{E}$.

We shall now define the elements of type 6, called *symps*. For every point $\langle x\rangle \in P(\mathcal{E})$, let $\sigma(x)$ be the projective subspace of $\mathrm{PG}(V)$ corresponding to the subspace $\{x \times y\}_{y\in V}$ of V (clearly, the choice of the representative x of $\langle x\rangle$ is irrelevant in this definition). The points and the lines of $\mathcal{E}$ contained in $\sigma(x)$ form the point-line system of a $D_5(\mathbb{F})$-building, say $\mathcal{S}(x)$. The planes of $\mathcal{S}(x)$ are the planes of $\mathcal{E}$ contained in $\sigma(x)$. One of the two families of maximal singular subspaces of $\mathcal{S}(x)$ consists of the 4-elements of $\mathcal{E}$ contained in $\sigma(x)$. The other family is formed by the intersections $\sigma(x) \cap y$ for y a 5-element of $\mathcal{E}$ such that $y \cap \sigma(x)$ has projective dimension 4. (The reader should notice that we are taking the liberty not to distinguish between an element of $\mathcal{E}$ of type 4 or 5 and the subspace of $\mathrm{PG}(V)$ that corresponds to it.)

The incidence relation is defined via inclusion except when an element of type 5 is involved together with an element of type 4 or 6. Let y be an element of type 5. If z is a 4-element then y and z are incident if and only if they intersect in a 3-dimensional subspace of $\mathrm{PG}(V)$. If $z = \mathcal{S}(x)$ is an element of type 6, then y and z are incident if and only if $y \cap \sigma(x)$ is 3-dimensional.

The group $\mathrm{Aut}(f)$ of all invertible linear mappings of V that preserve f acts flag-transitively on $\mathcal{E}$ with kernel Z_0 isomorphic to the group of cubic roots of 1 in $\mathbb{F}$. The quotient $\mathrm{Aut}(f)/Z_0$ is the Chevalley group of adjoint type $E_6(\mathbb{F})$. Let Z be the group of scalar transformations of V. The product $Z \cdot \mathrm{Aut}(f)$ is a subgroup of the group of all linear transformations of V that preserve f modulo a scalar. Clearly $Z \cap \mathrm{Aut}(f) = Z_0$ and $(Z \cdot \mathrm{Aut}(f))/Z \cong \mathrm{Aut}(f)/Z_0$. Hence $\mathrm{Aut}(f)$ and $Z \cdot \mathrm{Aut}(f)$ induce the same group on $\mathcal{E}$.

Note that, in general, the elements of $\mathrm{Aut}(f)$ neither preserve ϕ nor commute with the mapping $(.)^\sharp$ or the operation $\times$. However, if $g \in \mathrm{Aut}(f)$ then $x^\sharp = 0$ if and only if $g(x)^\sharp = 0$.

So far, we have constructed $\mathcal{E}$ together with a 27-dimensional projective embedding $\varepsilon : \mathcal{E} \to V$. We shall now turn to the $F_4(\mathbb{F})$-building, denoted by $\mathcal{F}$ in the sequel.

The mapping sending a point x of $\mathcal{E}$ to $\mathcal{S}(x)$ can be extended to a polarity of $\mathcal{E}$, which we will also denote by the symbol σ. Referring the reader to [2, Chap. 18] for details, we only say that this polarity permutes a point x with $\mathcal{S}(x)$ and sends every element y of $\mathcal{E}$ of type 2, 3 or 4 to $\cap_{x\in P(y)}\sigma(x)$, where $P(y)$ is the set of points of

y. If y is a 5-element then $\sigma(y) = \cup_{x\in\pi(y)}\sigma(z)$, where $\pi(y)$ stands for the set of planes contained in y. A point x of $\mathcal{E}$ is σ-absolute if and only if $x \in \sigma(x)$, a line x is σ-absolute if and only if $x \subset \sigma(x)$, a 4- or 6-element x is σ-absolute if and only if $\sigma(x)$ is contained in x, a plane or a 5-element x is σ-absolute if and only if $x = \sigma(x)$ (if and only if all points of x are σ-absolute).

The building $\mathcal{F}$ is formed by the σ-absolute elements of $\mathcal{E}$ of type 1, 2, 3 and 5. If we like, we can also regard σ-absolute elements of type 1 or 2 as σ-absolute flags of type $\{1,6\}$ and $\{2,4\}$. We keep the types 1, 2 and 3 for σ-absolute elements of $\mathcal{E}$ of type 1, 2 or 3 when we regard them as elements of $\mathcal{F}$ but we take 4 as the type of a σ-absolute 5-element of $\mathcal{E}$ when we regard it as an element of $\mathcal{F}$.

(F_4)

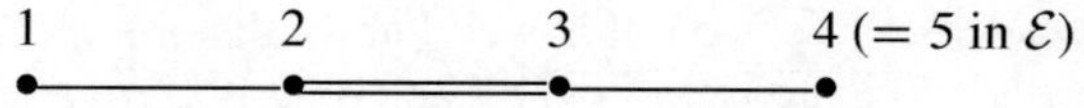

The $\{1,2,3\}$-residues of $\mathcal{F}$ are isomorphic to polar spaces of symplectic type while $\{2,3,4\}$-residues are dually isomorphic to polar spaces of orthogonal type. In Dynkin notation,

(F_4)

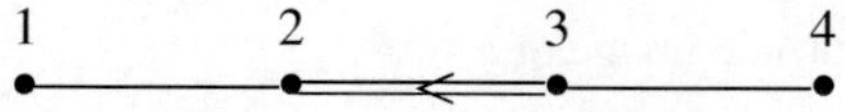

Another description of $\mathcal{F}$ is possible. Let $\iota := (I, O, O)$, where I is the identity 3×3-matrix while O stands for the null matrix. Then $f(\iota) = 1$. As proved in [2, Chap. 18], a vector $x \in V$ represents a σ-absolute point of $\mathcal{E}$ if and only if $\phi(\iota, x) = 0$. Hence, the points (1-elements) of $\mathcal{F}$ are the points of $\mathcal{E}$ contained in the hyperplane H_ι of V represented by the equation $\phi(\iota, x) = 0$. The σ-absolute lines and the σ-absolute planes of $\mathcal{E}$ are contained in H_ι, but not all lines and planes of $\mathcal{E}$ contained in H_ι are σ-absolute. The σ-absolute points, lines and planes incident to a given σ-absolute 5-element x of $\mathcal{E}$ form a polar space of symplectic type, naturally embedded in the 5-dimensional projective space x. The stabilizer $\mathrm{Aut}(f)_\iota$ of ι in $\mathrm{Aut}(f)$ induces on $\mathcal{F}$ a flag-transitive group of automorphisms, isomorphic to the Chevalley group of adjoint type $F_4(\mathbb{F})$.

Note that we have defined $\mathcal{F}$ as a poset together with a 26-dimensional projective embedding $e : \mathcal{F} \to H_\iota$, induced by the 27-dimensional projective embedding $\varepsilon : \mathcal{E} \to V$. If x is a symp of $\mathcal{F}$ (namely a 4-element of $\mathcal{F}$) and e_x is the embedding of $\mathrm{Res}_{\mathcal{F}}(x)$ induced by e, then x is a 6-dimensional subspace of H_ι and e_x embeds $\mathrm{Res}_{\mathcal{F}}(x)$ in that subspace as a polar space of symplectic type.

The building $\mathcal{E}$ admits projective embeddings different from ε. Every Weyl module provides one of them. For instance, with the 5- and 3-elements of $\mathcal{E}$ chosen as points and lines, we can embed $\mathcal{E}$ in a 78-dimensional vector space W. Let us denote this 78-dimensional embedding by ε^*. As the 4- and 3-elements of $\mathcal{F}$ arise from 5- and 3-elements of $\mathcal{E}$, the embedding ε^* induces a projective embedding e^* of the dual $\mathcal{F}^*$ of $\mathcal{F}$ in the subspace W' of W spanned by the image $\varepsilon^*(\mathcal{F}^*)$ of $\mathcal{F}^*$

via ε^*. The embedding $e^* : \mathcal{F}^* \to W'$ arises from the 52-dimensional Weyl module associated to the root corresponding to the right hand node of the Dynkin diagram:

(F_4) 

Hence $\dim(W') = 52$. Let x be a symp of $\mathcal{F}^*$ (namely a point of $\mathcal{F}$) and let e^*_x be the embedding induced by e^* on $\mathrm{Res}_{\mathcal{F}^*}(x)$. Then e^*_x embeds $\mathrm{Res}_{\mathcal{F}^*}(x)$ as a polar space of orthogonal type in a 7-dimensional subspace of W'.

When either $\mathrm{char}(\mathbb{F}) \neq 2$ or $\mathrm{char}(\mathbb{F}) = 2$ but $\mathbb{F}$ is non-perfect, the posets $\mathcal{F}$ and $\mathcal{F}^*$ are not isomorphic. Indeed the $\{1,2,3\}$-residues of $\mathcal{F}$ are symplectic polar spaces, associated to $\mathrm{PSp}(6,\mathbb{F})$, while $\{4,3,2\}$-residues are orthogonal polar spaces associated to $\mathrm{P}\Omega(7,\mathbb{F})$. It is well known that a $\mathrm{PSp}(6,\mathbb{F})$-polar space and a $\mathrm{P}\Omega(7,\mathbb{F})$-polar space are isomorphic if and only if $\mathbb{F}$ is perfect of characteristic 2. On the other hand, let $\mathbb{F}$ be perfect of characteristic 2. In this case $\mathrm{Aut}(\mathcal{F})$ admits a diagram automorphism, namely $\mathcal{F} \cong \mathcal{F}^*$. Therefore, when $\mathrm{char}(\mathbb{F}) = 2$ and $\mathbb{F}$ is perfect, the embedding e is not its own linear hull. Indeed let $\tilde{e}$ be the absolutely universal embedding of $\mathcal{F}$, which exists whatever $\mathbb{F}$ is (Kasikova and Shult [17]). Since now $\mathcal{F} \cong \mathcal{F}^*$, both e and e^* are projective embeddings of $\mathcal{F}$, hence each of them is a morphic image of $\tilde{e}$.

5.2 A Construction for $\mathcal{G}(E_6(2))$

We keep the notation of the previous subsection. For every $k \in \mathbb{F} \setminus \{0\}$, the group $\mathrm{Aut}(f)$ is transitive on the set $\mathcal{V}_k$ of vectors $x \in V$ such that $f(x) = k$ (see Buekenhout and Cohen [2, Chap. 18]).

Put $G := Z \cdot \mathrm{Aut}(f)$ and $\mathbb{F}^3 := \{t^3\}_{t \in \mathbb{F}}$. The orbits of G on $V \setminus \{0\}$ are joins $\overline{\mathcal{V}}_k := \cup(\mathcal{V}_{sk} \mid s \in \mathbb{F}^3 \setminus \{0\})$ of orbits of $\mathrm{Aut}(f)$. Clearly, $\overline{\mathcal{V}}_k = \overline{\mathcal{V}}_h$ if and only if $h^{-1}k \in \mathbb{F}^3 \setminus \{0\}$. Moreover, $\overline{\mathcal{V}}_k \cup \{0\}$ is the union of 1-dimensional linear subspaces of V (points of $\mathrm{PG}(V)$). We denote by $\mathcal{P}_k$ the set of points of $\mathrm{PG}(V)$ contained in $\overline{\mathcal{V}}_k \cup \{0\}$.

Let $\iota = (I, O, O)$, as in Sect. 5.1. Then $\langle \iota \rangle \in \mathcal{P}_1$. As in Sect. 5.1, H_ι is the hyperplane of V orthogonal to ι with respect to ϕ, but we take the liberty to use the symbol H_ι also for the corresponding hyperplane of $\mathrm{PG}(V)$.

Recalling the construction of $\mathcal{F}$ as a subgeometry of H_ι, we define $R(\langle \iota \rangle)$ as the collection of all projective subspaces of $\mathrm{PG}(V)$ that contain $\langle \iota \rangle$ and meet H_ι in an element of $\mathcal{F}$.

Lemma 5.1. *Let $X \in R(\langle \iota \rangle)$. Then:*

(i) $X \setminus (X \cap H_\iota) \subset \mathcal{P}_1$.
(ii) Let $p \in X \cap \mathcal{P}_1$. Then there exists an element $g \in G$ such that $g(X) = X$ and $g(\langle \iota \rangle) = p$. (Compare condition (C) of Sect. 3.)

Proof. Let x be a non-zero vector of H_ι with $x^\sharp = 0$. Then $f(x) = 0$. Note also that $\iota^\sharp = \iota$. Hence, (4) of Sect. 5.1 forces $f(\iota + tx) = f(\iota) = 1$ for all $t \in \mathbb{F}$. Therefore, all projective lines, planes and 3-spaces of $R(\langle\iota\rangle)$ are contained in $\mathcal{P}_1 \cup H_\iota$. Moreover, all points of $\mathrm{PG}(V)$ contained in a 4-element of $\mathcal{F}$ are points of $\mathcal{F}$. Hence, the 4-elements of $R(\langle\iota\rangle)$ are joins of lines of $R(\langle\iota\rangle)$. By the above, they are contained in $\mathcal{P}_1 \cup H_\iota$. Claim (i) is proved.

Turning to (ii), let $X \in R(\langle\iota\rangle)$ and $p \in X \cap \mathcal{P}_1$. Let $L = \langle\iota, p\rangle$. Then $L \in R(\langle\iota\rangle)$, by (i). We shall firstly prove that there exists an element $g_L \in G$ such that $g_L(L) = L$ and $g_L(\langle\iota\rangle) = p$. Let $\upsilon := (E_{1,2}, O, O) \in V$, where

$$E_{1,2} := \begin{bmatrix} 0 & 1 & 0 \\ 0 & 0 & 0 \\ 0 & 0 & 0 \end{bmatrix}$$

Then $\upsilon \in H_\iota$ and $\upsilon^\sharp = 0$, namely υ represents a point of $\mathcal{F}$. Without loss, we may assume that $L = \langle\iota, \upsilon\rangle$ and $p = \langle\iota + \upsilon\rangle$. It is easy to see that, for every choice of three non-singular matrices $A, B, C \in M$ with the same determinant, the mapping $g_{A,B,C}$ sending every vector $x = (X_1, X_2, X_3) \in V$ to $(AX_1B^{-1}, BX_2C^{-1}, CX_3A^{-1})$ belongs to $\mathrm{Aut}(f)$. Put

$$A := \begin{bmatrix} a & r & s \\ 0 & a & 0 \\ 0 & t & b \end{bmatrix} \quad \text{for } a, b \neq 0, \qquad B := A - aE_{1,2}$$

and let C be any (3×3)-matrix with $\det(C) = a^2 b$. Then $g_{A,B,C}$ sends ι to $\iota + \upsilon$ and fixes υ. If $X = L$ then (ii) is proved. Let $L \subset X$. As $R(\langle\iota\rangle) \cong \mathcal{F}$ and the stabilizer G_ι of $\langle\iota\rangle$ in G induces a flag-transitive group on $\mathcal{F}$, G_ι also acts flag-transitively on $R(\langle\iota\rangle)$. Hence, there exists an element $g_X \in G_\iota$ such that $g_X(L) = L$ and $g_X(X) = g_{A,B,C}^{-1}(X)$. The element $g := g_{A,B,C}\, g_X$ has the properties required in (ii).

Let now $p \in \mathcal{P}_1$ and let $g \in G$ map $\langle\iota\rangle$ onto p. Define $R(p) := g(R(\langle\iota\rangle))$. Since the stabilizer of ι in $\mathrm{Aut}(f)$ yields an automorphism group of $\mathcal{F}$, this definition does not depend on the choice of g.

We now define an incidence structure $\mathcal{G}$ having $\mathcal{P}_1 \cup \bigcup_{p \in \mathcal{P}_1} R(p)$ as the set of elements and inclusion as the incidence relation. Types are defined as follows: the points of $\mathcal{P}_1$ are the 0-elements of $\mathcal{G}$. If $Y \in R(p)$ for a point $p \in \mathcal{P}_1$ and $Y = g(X)$ for an element $X \in R(\langle\iota\rangle)$ and an element $g \in G$ sending $\langle\iota\rangle$ to p, then the type of the element $X \cap H_\iota$ of $\mathcal{F}$ is taken as the type $\tau(Y)$ of Y.

Recall that, according to the conventions stated in the introduction of this paper, only residually connected structures deserve to be called geometries.

Lemma 5.2. *The structure $\mathcal{G}$ is residually connected (whence it is a geometry). Moreover:*

(1) $\mathrm{Res}_{\mathcal{G}}(p) = R(p) \cong \mathcal{F}$ *for every point* $p \in \mathcal{P}_1$.
(2) If X is an element of $\mathcal{G}$ with type $0 < \tau(X) < 4$ *then* $\mathrm{Res}^{-}_{\mathcal{G}}(X)$ *is the set of points and subspaces of the affine space* $X \setminus (X \cap H_\iota)$.

(3) *If X is a 4-element of $\mathcal{G}$ then* $\mathrm{Res}_{\mathcal{G}}(X)$ *is isomorphic to the affine extension of the* $\mathrm{Sp}(6, \mathbb{F})$*-symplectic polar space, embedded in* $V(6, \mathbb{F})$.

Proof. We firstly prove that $\mathcal{G}$ is connected. As the group G acts transitively on $\mathcal{P}_1$, it transitively permutes the connected component of $\mathcal{G}$. Let $\mathcal{X}$ be the component of $\mathcal{G}$ containing $\langle \iota \rangle$ and $G_{\mathcal{X}}$ the stabilizer of $\mathcal{X}$ in G. The stabilizer G_ι of $\langle \iota \rangle$ is contained in $G_{\mathcal{X}}$. However, $G_{\mathcal{X}}$ is larger than G_ι, by (ii) of Lemma 5.1. On the other hand, G_ι is maximal in G, since the Chevalley group $F_4(\mathbb{F})$ is a maximal subgroup of $E_6(\mathbb{F})$. It follows that $G_{\mathcal{X}} = G$, namely $\mathcal{X} = \mathcal{G}$, since G acts transitively on the set of components of $\mathcal{G}$. The connectedness of $\mathcal{G}$ is proved. The residual connectedness of the residues of the elements of $\mathcal{G}$ follows from claims (1)–(3). So, it only remains to prove these claims.

Let $Y = g_1(X)$ for $X \in R(\langle \iota \rangle)$ and $g_1 \in G$. Let $p \in \mathcal{P}_1 \cap Y$. Then $g_1^{-1}(p) \in X$. By (ii) of Lemma 5.1, an element $g_2 \in G$ exists such that $g_2(X) = X$ and $g_2(\langle \iota \rangle) = g_1^{-1}(p)$. Hence $Y = g_1 g_2(X)$ and $p = g_1 g_2(\langle \iota \rangle)$. Claim (1) is proved. Claim (2) follows from (i) of Lemma 5.1 and claim (1). Finally, (3) follows from (2) and the fact that, if X is a 4-element of $R(\langle \iota \rangle)$, all points of $X \cap H_\iota$ belong to $\mathcal{F}$. □

Theorem 5.3. *The geometry $\mathcal{G}$ belongs to the following diagram:*

$(Af.F_4)$

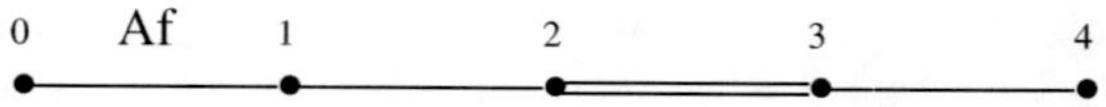

(Easy, by Lemma 5.2.)

Theorem 5.4. *The group G acts flag-transitively on $\mathcal{G}$ with Z as the kernel of that action.*

Proof. Let c be a given chamber of $\mathcal{G}$ containing $\langle \iota \rangle$ as the 0-element and let d be any other chamber of $\mathcal{G}$. We must prove that $g(d) = c$ for a suitable element $g \in G/Z$. As G/Z is transitive on the point-set $\mathcal{P}_1$ of $\mathcal{G}$, there are elements of G/Z that map the 0-element of d onto $\langle \iota \rangle$. So, we may assume that $\langle \iota \rangle$ is the 0-element of d too. The stabilizer G_ι of $\langle \iota \rangle$ in G acts flag-transitively on $\mathcal{F}$ and $\mathrm{Res}_{\mathcal{G}}(\langle \iota \rangle) \cong \mathcal{F}$. Hence, there exists an element $g \in G_\iota$ such that $g(d) = c$. □

Corollary 5.5. *Let* $\mathbb{F} = \mathrm{GF}(2)$. *Then* $\mathcal{G} = \mathcal{G}(E_6(2))$.

Proof. In this case $Z = 1$. Hence, $G = G/Z = E_6(2)$. The group $E_6(2)$ contains just one conjugacy class Φ of subgroups isomorphic to $F_4(2)$. Note that the polarity σ of $\mathcal{E}$ exploited to define the F_4-building $\mathcal{F}$ centralizes the member of Φ that stabilizes $\mathcal{F}$. The $\{0, 1, 2, 3\}$-residues of $\mathcal{G}$ are just as in $\mathcal{G}(E_6(2))$, by (3) of Lemma 5.2 (see also Lemma 2.5). The conclusion is now obvious. □

Remark 5.6. Let $\widehat{G}$ be the group of linear transformations of V that preserve f modulo a scalar. In general, $\widehat{G}$ is larger than G. If so, then $\widehat{G}$ fuses the orbits of G on $\mathrm{PG}(V)$ in larger orbits. Let $\widehat{\mathcal{P}}_1$ be the $\widehat{G}$-orbit containing $\mathcal{P}_1$ and define an incidence structure $\widehat{\mathcal{G}}$ on it in the same way as $\mathcal{G}$ on $\mathcal{P}_1$. Then $\widehat{\mathcal{G}}$ is the disjoint union of copies of $\mathcal{G}$, one copy for each of the orbits of G fused in $\widehat{\mathcal{P}}_1$.

Problem 5.7. As noticed in Sect. 3.4, the geometry $\mathcal{G}(E_6(2))$ is simply connected. Is $\mathcal{G}$ simply connected whatever $\mathbb{F}$ is?

Problem 5.8. Suppose that $\mathbb{F}$ is a quadratic extension of a field $\mathbb{F}_0$. Is it possible to modify the construction of $\mathcal{G}$ in such a way as to obtain an $Af.F_4$-geometry for the twisted group ${}^2E_6(\mathbb{F}_0)$ with F_4-residues isomorphic to the $F_4(\mathbb{F}_0)$-building?

5.3 More $Af.F_4$-Geometries

Affine Extensions of Buildings of Type $F_4(\mathbb{F})$ Let $\mathcal{F}$ be the $F_4(\mathbb{F})$-building, with points and lines chosen as in Sect. 5.1, let $\mathcal{F}^*$ be its dual and $e : \mathcal{F} \to H_\iota$ and $e^* : \mathcal{F}^* \to W'$ the 26- and 52-dimensional embeddings considered in Sect. 5.1. The affine extensions $\mathrm{Ex}_e(\mathcal{F})$ and $\mathrm{Ex}_{e^*}(\mathcal{F}^*)$ are flag-transitive $Af.F_4$-geometries. Take $\{0, 1, 2, 3, 4\}$ as the type-set for both $\mathrm{Ex}_e(\mathcal{F})$ and $\mathrm{Ex}_{e^*}(\mathcal{F}^*)$, with 0 standing for points and the types 4, 3, 2 and 1 of $\mathcal{F}^*$ replaced by 1, 2, 3 and 4:

$(Af.F_4)$

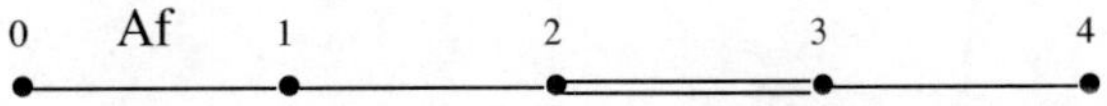

The $\{1, 2, 3, 4\}$-residues of $\mathrm{Ex}_e(\mathcal{F})$ are isomorphic to the affine extension of the $\mathrm{Sp}(6, \mathbb{F})$-polar space, naturally embedded in $V(6, \mathbb{F})$. This extension is isomorphic to the minimal standard quotient of the affine polar space obtained by removing a singular hyperplane from the $\mathrm{Sp}(8, \mathbb{F})$-polar space. On the other hand, the $\{1, 2, 3, 4\}$-residues of $\mathrm{Ex}_{e^*}(\mathcal{F}^*)$ are isomorphic to the affine extension of the $O(7, \mathbb{F})$-polar space. This extension is isomorphic to the affine polar space obtained by removing a singular hyperplane from the $O(9, \mathbb{F})$-polar space.

If $\mathrm{char}(\mathbb{F}) = 2$ and $\mathbb{F}$ is perfect then $\mathcal{F} \cong \mathcal{F}^*$ and the embedding e is a morphic image of e^*. Consequently, $\mathrm{Ex}_{e^*}(\mathcal{F}^*)$ is a 2-cover of $\mathrm{Ex}_e(\mathcal{F})$ (see the comments at the end of Sect. 5.1).

In particular, let $\mathbb{F} = \mathrm{GF}(2)$. Then $\mathrm{Ex}_e(\mathcal{F}) = \mathcal{G}(2^{26}F_4(2))$. The latter is simply connected (Sect. 3.4) but it is not 2-simply connected, since $\mathrm{Ex}_{e^*}(\mathcal{F}^*)$ is a proper 2-cover of it. Clearly, property (TR) of Sect. 2 fails to hold in $\mathrm{Ex}_{e^*}(\mathcal{F}^*)$.

Problem 5.9. Determine the universal 2-cover of $\mathcal{G}(2^{26}F_4(2))$. This is the same as determining the universal representation group of the point-line geometry of the $F_4(2)$-building (see Proposition 6.2).

Problem 5.10. Are $\mathrm{Ex}_e(\mathcal{F})$ and $\mathrm{Ex}_{e^*}(\mathcal{F}^*)$ simply connected for any choice of the field $\mathbb{F}$? What about their universal 2-covers? (Compare Propositions 6.1 and 6.2.)

More Affine Extensions Let $\mathcal{F}$ be the metasymplectic space of the $D_4(\mathbb{F})$-building. It is well known that $\mathcal{F}$ admits a projective embedding e of vector dimension 27 or 28, according to whether $\mathrm{char}(\mathbb{F})$ is 2 or different from 2.In any case, $\mathrm{Ex}_e(\mathcal{F})$ is a flag-transitive $Af.F_4$-geometry. If $\mathbb{F} = \mathrm{GF}(2)$ then $\mathrm{Ex}_e(\mathcal{F})$ has the same orders as $\mathcal{G}(Fi_{22})$ and $\mathcal{G}(3^{\cdot}Fi_{22})$, but it has nothing to do with either of them. Needless to say, (TR) fails to hold in $\mathrm{Ex}_e(\mathcal{F})$.

Suppose now that $\mathbb{F}$ is a quadratic extension of a subfield $\mathbb{F}_0$ and let $\mathcal{E}$ be the $E_6(\mathbb{F})$-building. It is well known that $\mathcal{E}$ contains a subgeometry $\mathcal{F}_{\mathrm{tw}}$ isomorphic to the F_4-building of twisted type ${}^2E_6(\mathbb{F}_0)$. Regarding $\mathcal{F}_{\mathrm{tw}}$ as a poset with set of types $\{1,2,3,4\}$ as we have done for $\mathcal{F}$, the $\{1,2,3\}$-residues of $\mathcal{F}_{\mathrm{tw}}$ are isomorphic to the polar space associated to $\mathrm{PSU}(6,\mathbb{F})$ while $\{4,3,2\}$-residues are isomorphic to the $O^-(8,\mathbb{F}_0)$-polar space. The 27-dimensional embedding ε of $\mathcal{E}$ induces a projective embedding e_{tw} of $\mathcal{F}_{\mathrm{tw}}$. The extension $\mathrm{Ex}_{e_{\mathrm{tw}}}(\mathcal{F}_{\mathrm{tw}})$ is a flag-transitive $Af.F_4$-geometry.

Problem 5.11. The 78-dimensional embedding $e^* : \mathcal{E} \to W$ induces a representation of the dual $\mathcal{F}^*_{\mathrm{tw}}$ of $\mathcal{F}_{\mathrm{tw}}$ in a subspace S of W. The lines of $\mathcal{F}_{\mathrm{tw}}$ are mapped by e^* into lines of $\mathrm{PG}(S)$, but not onto them. Can we find a Baer subgeometry $\mathcal{S}_0$ of $\mathrm{PG}(S)$ such that e^*_{tw} induces a projective embedding $e^*_0 : \mathcal{F}^*_{\mathrm{tw}} \to \mathcal{S}_0$?

If $\mathcal{S}_0$ existed, then $\mathrm{Ex}_{e^*_0}(\mathcal{F}^*_{\mathrm{tw}})$ would be an $Af.F_4$-geometry with point-residues isomorphic to $\mathcal{F}^*_{\mathrm{tw}}$. In particular, for $\mathbb{F} = \mathrm{GF}(4)$ the extension $\mathrm{Ex}_{e^*_0}(\mathcal{F}^*_{\mathrm{tw}})$ would have the same orders as $\mathcal{G}(B)$, but (TR) would not hold in it.

6 Representations, Projective Embeddings, Affine Extensions and Hyperplane Complements

As said in the introduction, this and the subsequent sections are appendices where we concisely expose various notions, constructions and results to which we have referred in Sects. 2–5. We start with representations and extensions.

Throughout this section $\mathcal{G}$ is a geometry with a string-shaped diagram of rank $n \geq 2$ over the set of types $\{1,2,\ldots,n\}$. We stick to all conventions stated in Sect. 1.2 for geometries of this kind. We do not assume (IP) on $\mathcal{G}$, but we assume that (IP) holds in the point-line system $\mathcal{G}_{|1,2}$ of $\mathcal{G}$, as it is customary in the literature on embeddings and hyperplanes.

Leaving geometries with non-string diagrams out of our exposition is not really restrictive. Indeed, when considering embeddings or hyperplanes of a geometry belonging to a non-string diagram, we actually deal with its Grassmann geometry with respect to a given type [19, Chap. 5]. The diagram of a Grassmann geometry is indeed string-shaped.

6.1 Representations and Extensions

Following [20], we choose a very general setting for our definitions. Later we shall turn to the definitions usually given in the literature.

Given a group R, let $S(R)$ be its subgroup lattice. A *complete representation* of $\mathcal{G}$ in R (a *representation* of $\mathcal{G}$ in R, for short) is a mapping $\rho : \mathcal{G} \to S(R)$ satisfying all the following:

(R1) $\langle \rho(p) \rangle_{p \in P(\mathcal{G})} = R$;
(R2) $\langle \rho(p) \rangle_{p \in P(x)} = \rho(x)$ for every $x \in \mathcal{G}$ and $\rho(p) \neq 1$ for every point $p \in P(\mathcal{G})$;
(R3) $\rho(p) \cap \rho(q) = 1$ for any two distinct points $p, q \in P(\mathcal{G})$.

Note that, regarded $\mathcal{G}$ and $S(R)$ as posets, (R2) implies that ρ is a morphism from $\mathcal{G}$ to $S(R)$. Henceforth we write $\rho : \mathcal{G} \to R$ as a shortening of the phrase "ρ is a representation of $\mathcal{G}$ in R".

Given a representation $\rho : \mathcal{G} \to R$, the *extension* $\mathrm{Ex}_\rho(\mathcal{G})$ of $\mathcal{G}$ via ρ (called "expansion" in [19–21]) is the geometry over the set of types $\{0, 1, \ldots, n\}$ defined as follows: the elements of R are the 0-elements of $\mathrm{Ex}_\rho(\mathcal{G})$ while for $i = 1, 2, \ldots, n$ the elements of $\mathrm{Ex}_\rho(\mathcal{G})$ of type i are the pairs $(a\rho(x), x)$ where x is an i-element of $\mathcal{G}$, $a \in R$ and $a\rho(x)$, called the *support* of $(a\rho(x), x)$, is a coset of the subgroup $\rho(x)$ of R. Incidence is defined as follows: a 0-element a and an i-element $(b\rho(x), x)$ are incident if and only if $a \in b\rho(x)$; two elements $(a\rho(x), x)$ and $(b\rho(y), y)$ with $1 \leq \tau(x) < \tau(y) \leq n$ are incident if and only if $x < y$ in $\mathcal{G}$ and $a\rho(x) \subseteq b\rho(y)$. The residual connectedness of $\mathrm{Ex}_\rho(\mathcal{G})$ follows from (R1) and (R2). The diagram of $\mathrm{Ex}_\rho(\mathcal{G})$ is still string-shaped. The residues of the 0-elements of $\mathrm{Ex}_\rho(\mathcal{G})$ are isomorphic to $\mathcal{G}$.

The $\{0, 1\}$-residues of $\mathrm{Ex}_\rho(\mathcal{G})$ are partial linear spaces, by (R3). In order to say more on them we need to know more on the mappings induced by ρ on the lines of $\mathcal{G}$. For instance, given $l \in L(\mathcal{G})$, suppose that $\rho(l)$ is abelian and $\langle \rho(p), \rho(q) \rangle = \rho(l)$ for any two points $p, q \in P(l)$. Then the lower residue of $\rho(l)$ in $\mathrm{Ex}_\rho(\mathcal{G})$ is a net. If, moreover, $\rho(l) = \cup_{p \in P(l)} \rho(p)$, then that residue is an affine plane.

When (IP) holds in $\mathcal{G}$ and ρ satisfies the following:

(R4) for every choice of $p \in P(\mathcal{G})$ and $x \in \mathcal{G}$, if $\rho(p) \leq \rho(x)$ then $p \in P(x)$,

then ρ induces an isomorphism from the poset $\mathcal{G}$ to the poset induced by $S(R)$ on $\rho(\mathcal{G}) = \{\rho(x)\}_{x \in \mathcal{G}}$. In this case, the elements of $\mathrm{Ex}_\rho(\mathcal{G})$ of type > 0 bijectively correspond to their supports, whence we can simplify the definition of $\mathrm{Ex}_\rho(\mathcal{G})$ by taking cosets $a\rho(x)$ instead of pairs $(a\rho(x), x)$ as elements and inclusion as the incidence relation.

Given two representations $\rho_1 : \mathcal{G} \to R_1$ and $\rho_2 : \mathcal{G} \to R_2$, a *morphism* from ρ_1 to ρ_2 is a homomorphism $f : R_1 \to R_2$ such that, for every $x \in \mathcal{G}$, f induces an isomorphism from $\rho_1(x)$ to $\rho_2(x)$. Note that f is surjective, by (R1). Every morphism from ρ_1 to ρ_2 induces a covering from $\mathrm{Ex}_{\rho_1}(\mathcal{G})$ to $\mathrm{Ex}_{\rho_2}(\mathcal{G})$. Moreover, let $\widetilde{R}$ be the universal completion of the amalgam $\mathcal{A}_\rho(\mathcal{G}) := \{\rho(x)\}_{x \in \mathcal{G}}$, where we only take a record of the inclusions $\rho(x) < \rho(y)$ for $x < y$. Let $\tilde{\rho}$ be the representation of $\mathcal{G}$ in $\widetilde{R}$ naturally induced by ρ, namely $\tilde{\rho}(x) = \rho(x)$ for every $x \in \mathcal{G}$, with $\rho(x)$ being now regarded as a subgroup of $\widetilde{R}$. The canonical projection $\tilde{f} : \widetilde{R} \to R$ is a morphism from $\tilde{\rho}$ to ρ and, for every morphism of representations $f : \rho' \to \rho$

there is unique morphism $g : \tilde{\rho} \to \rho'$ such that $\tilde{f} = f \circ g$. This property uniquely determines $\tilde{\rho}$ up to isomorphism. The following also holds (see [20]):

Proposition 6.1. *The extension* $\mathrm{Ex}_{\tilde{\rho}}(\mathcal{G})$ *is the universal cover of* $\mathrm{Ex}_{\rho}(\mathcal{G})$.

We call $\tilde{\rho}$ the *hull* of ρ, also the *abstract hull* of ρ when we need to distinguish it from linear hulls, to be defined later.

A representation $\rho : \mathcal{G} \to R$ is said to be *abelian* if R is abelian. If ρ is abelian then we can also consider its *abelian hull*, defining it as the composition of $\tilde{\rho}$ with the canonical projection of $\widetilde{R}$ onto the abelianization $\widetilde{R}/\widetilde{R}'$ of $\widetilde{R}$.

With $\mathcal{G}$ and R as above, a *point-line representation* of $\mathcal{G}$ in R is a (complete) representation $e : \mathcal{G}_{|1,2} \to R$. Point-line representations may also be called representations for short, when there is no danger of confusion with complete representations. If e is a point-line representation of $\mathcal{G}$, we take the liberty of writing $e : \mathcal{G} \to R$ instead of $e : \mathcal{G}_{|1,2} \to R$, provided that no ambiguity arises.

Let $e : \mathcal{G} \to R$ be a point-line representation of $\mathcal{G}$. We can extend e to a complete representation $\rho_e : \mathcal{G} \to R$ by setting $\rho_e(x) := \langle e(p) \rangle_{p \in P(x)}$, according to (R2). We call ρ_e the *completion* of e. Clearly, if $f : \bar{e} \to e$ is a morphism of point-line representations then f induces a morphism from the completion $\rho_{\bar{e}}$ of $\bar{e}$ to the completion ρ_e of e. So, if e is its own hull then ρ_e is its own hull as well, but the converse fails to hold in general. Consequently, if $\widetilde{\rho_e}$ and $\rho_{\tilde{e}}$ are the hull of ρ_e and the completion of the hull $\tilde{e}$ of e respectively, then $\mathrm{Ex}_{\rho_{\tilde{e}}}(\mathcal{G})$ is a 2-cover of $\mathrm{Ext}_{\widetilde{\rho_e}}(\mathcal{G})$. Nevertheless, $\mathrm{Ext}_{\widetilde{\rho_e}}(\mathcal{G})$ is simply connected by Proposition 6.1. Note that $\mathrm{Ex}_{\rho_{\tilde{e}}}(\mathcal{G})$ is also simply connected (but possibly not 2-simply connected), because $\rho_{\tilde{e}}$ is its own hull. By combining Proposition 6.1 with [18, Theorem 1] we can say something more.

Proposition 6.2. *Let e be a point-line representation of $\mathcal{G}$. Suppose that $\mathcal{G}$ has rank $n \geq 3$ and that $\mathcal{G}$ as well as all of its residues of rank at least* 3 *are simply connected. Let $\tilde{e}$ be the hull of e. Then* $\mathrm{Ex}_{\tilde{e}}(\mathcal{G}_{|1,2})$ *is a truncation of the universal* 2*-cover of* $\mathrm{Ex}_e(\mathcal{G})$. *If moreover $\mathcal{G}$ satisfies* (IP) *then* $\mathrm{Ex}_{\rho_{\tilde{e}}}(\mathcal{G})$ *is the universal* 2*-cover of* $\mathrm{Ex}_e(\mathcal{G})$.

6.2 *GF*(2)-*Representations*

Suppose that all lines of $\mathcal{G}$ have exactly three points. A GF(2)-*representation* of $\mathcal{G}$ is a point-line representation $e : \mathcal{G} \to R$ such that

(R5) $e(p)$ has order 2 for every point $p \in P(\mathcal{G})$ and $e(l)$ is elementary abelian of order 2^2 for every line $l \in L(\mathcal{G})$.

We refer the reader to Ivanov and Shpectorov [14, Chap. 2] for properties of these representations. We only make a few remarks here.

Let $e : \mathcal{G} \to R$ be a GF(2)-representation and let ρ_e be its completion. Then the $\{0, 1\}$-residues of $\mathrm{Ex}_{\rho_e}(\mathcal{G})$ are affine planes of order 2.

In view of (R5), the amalgam $\mathcal{A}_{1,2}(\mathcal{G}) := \mathcal{A}_e(\mathcal{G}_{|1,2}) = \{e(x)\}_{x\in\mathcal{G}_{|1,2}}$ is completely determined by $\mathcal{G}_{|1,2}$ itself, with no need of any further information on e. Hence, all GF(2)-representations of $\mathcal{G}$ have the same hull, which is called the *universal representation* of $\mathcal{G}$. The universal completion of $\mathcal{A}_{1,2}(\mathcal{G})$ is called the (*universal*) *representation group* of $\mathcal{G}$. When R is abelian, the abelian hull of e is called the (*universal*) *representation module* of $\mathcal{G}$.

Remark 6.3. We warn the reader that, in the literature, it is customary to consider the amalgam $\mathcal{A}_{1,2}(\mathcal{G})$ and its universal completion even if $\mathcal{G}$ admits no GF(2)-representation, as when the universal completion of $\mathcal{A}_{1,2}(\mathcal{G})$ collapses to 1, for instance. However, in this paper, we never meet situations like these.

6.3 Projective Embeddings and Affine Extensions

Let V be a vector space over a field $\mathbb{F}$ and let $e : \mathcal{G} \to V$ be a point-line representation of $\mathcal{G}$ in the additive group of V. We say that e is a *projective embedding* (defined over $\mathbb{F}$) if it satisfies the followings:

(E1) $e(p)$ is 1-dimensional vector subspace of V for every point $p \in P(\mathcal{G})$;
(E2) for every line $l \in L(\mathcal{G})$, $e(l)$ is a 2-dimensional subspace of V and $e(l) = \cup_{p\in P(l)} e(p)$.

In view of the above, a projective embedding $e : \mathcal{G} \to V$ can also be regarded as a mapping from $\mathcal{G}$ to the set of points and lines of PG(V). Accordingly, we also write $e : \mathcal{G} \to \text{PG}(V)$ instead of $e : \mathcal{G} \to V$.

Let ρ_e be the completion of e. The extension $\text{Ex}_{\rho_e}(\mathcal{G})$ (also denoted by $\text{Ext}_e(\mathcal{G})$ for short) is called the *affine extension* of $\mathcal{G}$ via e. The $\{0,1\}$-residues of $\text{Ex}_{\rho_e}(\mathcal{G})$ are affine planes.

Given two projective embeddings $e_1 : \mathcal{G} \to \text{PG}(V_1)$ and $e_2 : \mathcal{G} \to \text{PG}(V_2)$ defined over the same field $\mathbb{F}$, a (*projective*) *morphism* from e_1 to e_2 is a semilinear mapping $f : V_1 \to V_2$ such that $f \circ e_1 = e_2$. Clearly, morphisms of projective embeddings are also morphisms of representations as defined in Sect. 6.1.

A linear analogue of the abstract hull $\tilde{e}$ of e can also be defined. The elements of the amalgam $\mathcal{A}_e(\mathcal{G}_{|1,2})$ are $\mathbb{F}$-vector spaces. Let $\widehat{V}$ be the largest $\mathbb{F}$-vector space in which $\mathcal{A}_e(\mathcal{G}_{|1,2})$ can be embedded. Then the mapping $\hat{e} : \mathcal{G} \to \widehat{V}$ defined by the clause $\hat{e}(x) = e(x)$ for $x \in \mathcal{G}_{|1,2}$ but with $e(x)$ regarded as a subspace of $\widehat{V}$, is a projective embedding of $\mathcal{G}$ in $\widehat{V}$ and the canonical projection $\hat{f} : \widehat{V} \to V$ is a morphism from $\hat{e}$ to e. The pair $(\hat{e}, \hat{f})$ satisfies the following universal property: for every morphism of projective embeddings $f : e' \to e$ there is unique projective morphism $g : \hat{e} \to e'$ such that $\hat{f} = f \circ g$. We call $\hat{e}$ the (*linear*) *hull* of e. Clearly, $\hat{e}$ is also a point-line representation of $\mathcal{G}$. Hence a morphism of representations exists from the abstract hull $\tilde{e}$ of e to $\hat{e}$.

The category of projective embeddings of a geometry $\mathcal{G}$ often admits an initial object, called the *absolutely universal* embedding of $\mathcal{G}$ (see Kasikova and Shult [17]. In general, that object is not initial in the category of point-line representations of $\mathcal{G}$.

Remark 6.4. Suppose that all lines of $\mathcal{G}$ have precisely three points and that $\mathcal{G}$ admits a projective embedding. Then the projective embeddings of $\mathcal{G}$ are just the abelian GF(2)-representations of $\mathcal{G}$. The universal representation module of $\mathcal{G}$ affords the absolutely universal embedding of $\mathcal{G}$.

6.4 Subspaces, Hyperplanes and Their Complements

A (proper) *subspace* of $\mathcal{G}$ is a (proper) subset S of $P(\mathcal{G})$ such that every line of $\mathcal{G}$ either is fully contained in S or it meets S in at most one point. A subspace S is said to be *totally singular* (also just *singular*, for short) if all of its points are pairwise collinear.

A proper subspace of $\mathcal{G}$ meeting every line of $\mathcal{G}$ non-trivially is a *hyperplane* of $\mathcal{G}$. Given a hyperplane H of $\mathcal{G}$, the *complement* $\mathcal{G} \setminus H$ of H in $\mathcal{G}$ is the substructure of $\mathcal{G}$ formed by the points exterior to H and the elements $x \in \mathcal{G}$ such that $P(x) \not\subseteq H$, with the incidence relation and the type function inherited from $\mathcal{G}$.

In general, $\mathcal{G} \setminus H$ is not residually connected (not even connected). However, in many interesting cases $\mathcal{G} \setminus H$ is residually connected. Let this be the case. Then $\mathcal{G} \setminus H$ is a geometry. The residues of the points of $\mathcal{G} \setminus H$ are the same as in $\mathcal{G}$. If x is an element of $\mathcal{G} \setminus H$ of type $\tau(x) > 2$, then $H \cap P(x)$ is a hyperplane of $\mathrm{Res}^-_{\mathcal{G}}(x)$ and $\mathrm{Res}^-_{\mathcal{G}\setminus H}(x)$ is the complement of $P(x) \cap H$ in $\mathrm{Res}^-_{\mathcal{G}}(x)$. For instance, if $\mathrm{Res}^-_{\mathcal{G}}(x)$ is a projective geometry then $\mathrm{Res}^-_{\mathcal{G}\setminus H}(x)$ is an affine geometry.

If $\mathcal{G}$ admits a projective embedding $e : \mathcal{G} \to \mathrm{PG}(V)$, then for every projective hyperplane U of $\mathrm{PG}(V)$ the preimage $e^{-1}(U \cap e(P(\mathcal{G})))$ is a hyperplane of $\mathcal{G}$. In many embeddable geometries (as classical polar spaces, for instance) all hyperplanes arise in this way, provided that e is its own linear hull.

7 Affine Polar Spaces and Their Quotients

7.1 Affine Polar Spaces

Let $\mathcal{P}$ be a non-degenerate polar space of finite rank $n \geq 2$. We assume that all lines of $\mathcal{P}$ have at least three points. Let H be a hyperplane of $\mathcal{P}$. Its complement $\mathcal{P} \setminus H$ is residually connected. When $n \geq 3$ it belongs to the following diagram:

(Af.C_{n-1}) 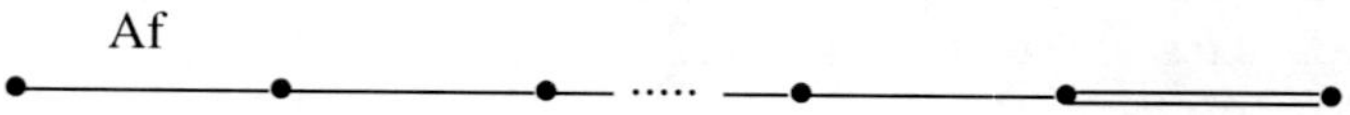

When all lines of $\mathcal{P}$ have exactly 3 points, the diagram of $\mathcal{P} \setminus H$ can also be drawn as follows:

(c.C$_{n-1}$)

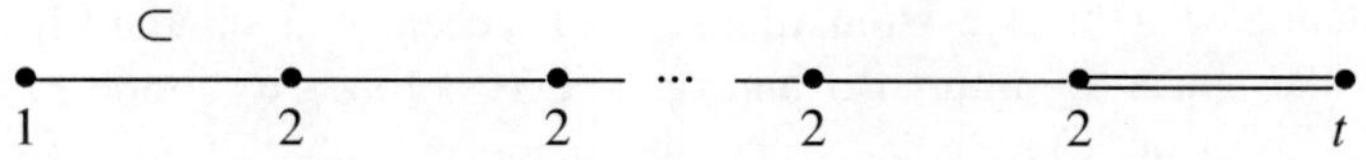

We call $\mathcal{P} \setminus H$ an *affine polar space* of rank n (*affine generalized quadrangle* when $n = 2$). The next proposition is well known (and easy to prove):

Proposition 7.1. *Property* (IP) *holds in every affine polar space.*

The following is also well known (see [19, Proposition 12.50]).

Proposition 7.2. *Affine polar spaces of rank $n > 2$ are simply connected.*

Let H be a hyperplane of $\mathcal{P}$. The polar space $\mathcal{P}$ induces a possibly degenerate polar space $\mathcal{H}$ on H, of rank n or $n-1$. The polar space $\mathcal{H}$ is degenerate if and only if $H = p^{\perp}$ for a given point p of $\mathcal{P}$ (where $\perp$ stands for the collinearity relation of $\mathcal{P}$, as usual). In this case, the hyperplane H is said to be *singular* with p as its *deepest point* (but we warn the reader that a singular hyperplane is not a singular subspace in the sense of Sect. 6.4).

In many cases (in all cases when $\mathcal{P}$ is finite), the isomorphism type of $\mathcal{H}$ uniquely determines H up to automorphisms of $\mathcal{P}$. Hence, it also determines $\mathcal{P} \setminus H$ up to isomorphisms. In these cases, a symbol denoting the isomorphism type of $\mathcal{H}$ can be used to recall what H is. For instance, if $\mathcal{H}$ is isomorphic to the polar space associated to the symplectic group $S_6(2)$ then we say that H is of $S_6(2)$-type or that it is an $S_6(2)$-hyperplane.

7.2 *Standard Quotients of Affine Polar Spaces*

In general, an affine polar space admits several quotients. Those that satisfy (IP) are called *standard quotients* in [7]. They can be described as follows. Given $\mathcal{P}$ and H as above, let Θ_H be the equivalence relation on the set of points of $\mathcal{P} \setminus H$ defined as follows: two points a and b of $\mathcal{P} \setminus H$ correspond in Θ_H if and only if $a^{\perp} \cap H = b^{\perp} \cap H$. Then Θ_H can be extended in a natural way to an equivalence relation on the set of elements of $\mathcal{P} \setminus H$ of any type and we can consider the quotient $(\mathcal{P} \setminus H)/\Theta_H$ of $\mathcal{P} \setminus H$ by the relation Θ_H thus extended. This quotient is the minimal standard quotient of $\mathcal{P} \setminus H$. Any other standard quotient of $\mathcal{P} \setminus H$ can be obtained as $(\mathcal{P} \setminus H)/\Theta$ for a refinement Θ of Θ_H (see [7]). The following is the main result of Cuypers and Pasini [7]:

Proposition 7.3. *Every geometry belonging to the diagram* Af.C$_{n-1}$ *($n \geq 3$) and satisfying* (IP) *is a standard quotient of an affine polar space.*

In all cases we are aware of, the stabilizer of H in Aut($\mathcal{P}$) acts flag-transitively on $\mathcal{P} \setminus H$. It also induces a flag-transitive group on the minimal standard quotient $(\mathcal{P} \setminus H)/\Theta_H$. Perhaps this is true in any case. Anyway, it is always true when $\mathcal{P}$ is finite.

The collinearity graph of an affine polar space has diameter at most 3 (sometimes 2) while minimal standard quotients of affine polar spaces have diameter at most 2 (sometimes just 1). Clearly, a standard quotient of an affine polar space is minimal if and only if its collinearity graph has diameter at most 2. So, if an affine polar space has diameter 2, then it does not admit any proper standard quotient.

Example 7.4. Let $\mathcal{Q}$ be the $O_5(2)$-generalized quadrangle and H a hyperplane of $\mathcal{Q}$. Then H can be either singular or an ovoid or a (3×3)-grid. If H is singular then $\mathcal{Q} \setminus H$ is the vertex-edge system of the cube. If H is an ovoid, then $\mathcal{Q} \setminus H$ is the Petersen graph. If H is a grid then $\mathcal{P} \setminus H$ is a dual grid. If H is non-singular then $\mathcal{Q} \setminus H$ admits no proper quotient. If H is singular then $\mathcal{Q} \setminus H$ admits a 2-fold quotient, isomorphic to the affine plane of order 2.

7.3 Minimal Standard Quotients as Affine Extensions or Tangent Geometries

Given a projective embedding $e : \mathcal{P}_0 \to \mathrm{PG}(V)$ of a polar space $\mathcal{P}_0$, suppose that for any two non-collinear points x and y of $\mathcal{P}_0$ the set $e(\{x, y\}^{\perp\perp})$ spans a line of $\mathrm{PG}(V)$. Then the affine extension $\mathrm{Ex}_e(\mathcal{P}_0)$ is a minimal standard quotient of an affine polar space. Indeed, let $\mathcal{P}$ be a polar space with point-residues isomorphic to $\mathcal{P}_0$ and let $H = p^{\perp}$ for a point p of $\mathcal{P}$. Then $\mathrm{Ex}_e(\mathcal{P}_0) \cong (\mathcal{P} \setminus H)/\Theta_H$.

We now turn to tangent geometries. An abstract definition of tangent geometries is stated in [7], according to which tangent geometries of polar spaces and minimal standard quotients of affine polar spaces are ultimately the same thing, but we are not interest in that definition here. A more concrete although slightly less general definition is also given in [7], which we shall now recall.

Let H be a hyperplane of $\mathcal{P}$ and let $\mathcal{H}$ be the polar space induced by $\mathcal{P}$ on H. Let $e : \mathcal{H} \to V$ be a projective embedding of $\mathcal{H}$. For instance, e can be induced by an embedding of $\mathcal{P}$, but this is not essential for the sequel. Assume that all non-singular hyperplanes of $\mathcal{H}$ arise from e as explained in the final paragraph of Sect. 6.4. According to this assumption, for every non-singular hyperplane K of $\mathcal{H}$, $\langle e(K)\rangle^{\perp_e}$ is a non-singular point of $\mathrm{PG}(V)$ and $\langle e(K)\rangle^{\perp_e\perp_e} = \langle e(K)\rangle$, where $\perp_e$ is the orthogonality relation associated to $e(\mathcal{H})$ in V. We call $\langle e(K)\rangle^{\perp_e}$ the *pole* of K. Thus, the non-singular hyperplanes of $\mathcal{H}$ bijectively correspond to their poles. Let $\mathcal{T}_{e,0}(\mathcal{H})$ be the set of the poles of the non-singular hyperplanes of $\mathcal{H}$. For $i > 0$ let $\mathcal{T}_{e,i}(\mathcal{H})$ be the set of the i-dimensional subspaces X of $\mathrm{PG}(V)$ such that X meets the image $e(H)$ of H in an $(i-1)$-dimensional projective subspace contained in $X^{\perp_e}$ and $X \setminus e(H) \subseteq \mathcal{T}_{e,0}(\mathcal{H})$. Clearly, $\mathcal{T}_{e,i}(\mathcal{H}) = \emptyset$ if i is too large. Let m be the largest i such that $\mathcal{T}_{e,i}(\mathcal{H}) \neq \emptyset$. Then $\mathcal{T}_e(\mathcal{H}) := \cup_{i=0}^{m} \mathcal{T}_{e,i}(\mathcal{H})$ equipped with the natural incidence relation (namely inclusion) is an incidence structure over the set of types $\{0, 1, \dots, m\}$. In general, $\mathcal{T}_e(\mathcal{H})$ is not connected. However, its connected components are residually connected. We call them *projective tangent geometries* of $\mathcal{H}$ at e, also *tangent geometries* of $\mathcal{H}$ for short. As proved in [7], the quotient $(\mathcal{P} \setminus H)/\Theta_H$ is isomorphic to one of these tangent geometries. The other tangent geometries of $\mathcal{H}$ at e, if any, are also isomorphic to minimal standard quotients of affine polar spaces, but they might arise from polar spaces quite different from $\mathcal{P}$.

8 Shrinkings and Geometries at Infinity

8.1 Shrinkings

The earliest explicit mention of shrinkings can be found in a paper by Stroth and Wiedorn [25], but that idea is implicit in other papers too (e.g., Ivanov [10], Ivanov and Wiedorn [15], Ivanov, Pasechnik and Shpectorov [16]). That construction has been later generalized by Pasini and Wiedorn [22]. Less ambitious expositions, closer to the setting of [25], are given in [21, 26]. In the sequel we place ourselves at a level of generality intermediate between [25] and [22].

Let $\mathcal{G}$ be a geometry of rank $n+1$ with string-shaped diagram and $0, 1, \dots, n$ as types, elements of type 0, 1 and 2 being called points, lines and planes respectively. We assume that the lower residues of the planes of $\mathcal{G}$ are affine planes.

Let $\widehat{\Lambda}$ be the binary relation on the line-set $L(\mathcal{G})$ of $\mathcal{G}$ defined as follows: two lines $l, m \in L(\mathcal{G})$ correspond in $\widehat{\Lambda}$ if and only if they are parallel in the lower residue of a plane of $\mathcal{G}$. Let Λ be the transitive closure of $\widehat{\Lambda}$. Similarly, given an element $x \in \mathcal{G}$ of type $\tau(x) > 1$, let $L(x)$ be the set of lines of $\mathcal{G}$ incident with x and let $\widehat{\Lambda}_x$ be defined on $L(x)$ by declaring that $(l, m) \in \widehat{\Lambda}_x$ precisely when l and m are parallel in a $\{0, 1\}$-residue of $\mathrm{Res}^-(x)$. Let Λ_x be the transitive closure of $\widehat{\Lambda}_x$. The relation Λ_x is a possibly proper refinement of the relation induced by Λ on $L(x)$. In particular, when $\tau(x) = 2$ then Λ_x is the usual parallelism of the affine plane $\mathrm{Res}^-(x)$, but it can happen that Λ induces the trivial relation on $L(x)$.

Let C be one of the equivalence classes of Λ. We define a geometry $\mathrm{Shr}(C)$ with $\{1, 2, \dots, n\}$ as the set of types and C as the set of 1-elements. For every $i = 2, 3, \dots, n$, the i-elements of $\mathrm{Shr}(C)$ are the pairs (C_x, x) where x is an i-element of $\mathcal{G}$, which we call the *support* of (C_x, x), and C_x is a class of Λ_x such that $C_x \subseteq C$. A 1-element $l \in C$ is incident with (C_x, x) if and only if $l \in C_x$. Two elements (C_x, x) and (C_y, y) of type i and j respectively, with $2 \leq i < j$, are incident in $\mathrm{Shr}(C)$ if and only if their supports x and y are incident in $\mathcal{G}$ and $C_x \subseteq C_y$. We call $\mathrm{Shr}(C)$ a *shrinking* of $\mathcal{G}$.

For every 1-element $l \in C$ of $\mathrm{Shr}(C)$, the residue of l in $\mathrm{Shr}(C)$ is isomorphic to the upper residue of l in $\mathcal{G}$. For every type $k = 2, 3, \dots, n$, the $\{1, 2, \dots, k\}$-residues of $\mathrm{Shr}(C)$ are shrinkings of $\{0, 1, 2, \dots, k\}$-residues of $\mathcal{G}$. In particular, the $\{1, 2\}$-residues of $\mathrm{Shr}(C)$ are shrinkings of the $\{0, 1, 2\}$-residues of $\mathcal{G}$.

If $\mathcal{G}$ satisfies (IP) then (IP) also holds in $\mathrm{Shr}(C)$. In many cases the following also holds:

(RS) (**Residual Stability of Parallelism.**) For every element $x \in \mathcal{G}$ with type $\tau(x) > 2$, the relation Λ induces Λ_x on $L(x)$.

If (RS) holds, then the class C_x is uniquely determined by x and C. So, we can replace the elements of $\mathrm{Shr}(C)$ with their supports (with the convention that a line $l \in C$ is its own support), thus regarding $\mathrm{Shr}(C)$ as a subgeometry of $\mathcal{G}$.

In general, the isomorphism type of a shrinking $\mathrm{Shr}(C)$ depends on the choice of the class C, but when $\mathrm{Aut}(\mathcal{G})$ acts transitively on $L(\mathcal{G})$ then all shrinkings of $\mathcal{G}$ are mutually isomorphic. In this case, it is customary to call $\mathrm{Shr}(C)$ *the* shrinking of $\mathcal{G}$, denoting it by the symbol $\mathrm{Shr}(\mathcal{G})$. Clearly, if $\mathcal{G}$ is flag-transitive then $\mathrm{Shr}(\mathcal{G})$ is flag-transitive.

Example 8.1. The shrinkings of a $(k+1)$-dimensional affine geometry are k-dimensional affine geometries. So, if the $\{0, 1, \ldots, k\}$-residues of $\mathcal{G}$ are affine geometries then the $\{1, 2, \ldots, k\}$-residues of $\mathrm{Shr}(C)$ are still affine geometries.

Example 8.2. The shrinkings of an affine polar space of rank $n+1 \geq 3$ are affine polar spaces of rank n (Pasini and Wiedorn [22, Proposition 8.1]). More generally, the shrinkings of a (minimal) standard quotient of an affine polar space of rank $n+1$ are (minimal) standard quotients of affine polar spaces of rank n.

For instance, let $\mathcal{P}$ be the $O_7(2)$-polar space and put $\mathcal{G} := \mathcal{P} \backslash H$, for a hyperplane H of $\mathcal{P}$. The hyperplane H can be singular or isomorphic to either the $O_6^-(2)$-quadrangle or the $O_6^+(2)$-polar space. If H is singular then $\mathrm{Shr}(\mathcal{G})$ is the cube-graph, if H is the $O_6^-(2)$-quadrangle then $\mathrm{Shr}(\mathcal{G})$ is the Petersen graph and if H is the $O_6^+(2)$-polar space then $\mathrm{Shr}(\mathcal{G})$ is a dual grid (compare Example 7.4). In the first case $\mathcal{G}$ admits a 2-fold quotient, the shrinkings of which are affine planes of order 2.

Example 8.3. Let e be a projective embedding of a geometry $\mathcal{G}_0$ with string-shaped diagram. Suppose that e admits completion and let $\mathcal{G} = \mathrm{Ex}_e(\mathcal{G}_0)$ be its affine extension. Then $\mathcal{G}$ admits shrinkings and its shrinkings are extensions of the representations induced by e on the point-residues of $\mathcal{G}_0$.

8.2 Geometries at Infinity

Different definitions of geometries at infinity can be given, suited to different situations (see Pasini and Wiedorn [22]). We will consider only one of them here.

With $\mathcal{G}$ as in the previous subsection, suppose that for a type $k > 1$ the $\{0, 1, \ldots, k\}$-residues of $\mathcal{G}$ are $(k+1)$-dimensional affine geometries. Suppose that $\mathcal{G}$ satisfies condition (RS) of the previous subsection, so that the shrinkings of $\mathcal{G}$ are subgeometries of $\mathcal{G}$. With Λ defined as in the previous subsection, put $\Lambda_1 := \Lambda$ and, for $i = 2, \ldots, k$, let Λ_i be the transitive closure of the relation "being parallel inside the same $\{0, 1, \ldots, i\}$-residue," defined on the set of i-elements of $\mathcal{G}$. Suppose that Λ_i satisfies a condition analogous to (RS). (We leave the precise statement of this condition for the reader.)

Then we can define a geometry $\mathcal{G}^{\infty,k}$ over the set of types $\{1, 2, \ldots, n\}$, which we call the *geometry at infinity* of $\mathcal{G}$ of *level* k. For $i = 1, 2, \ldots, k$, the i-elements of $\mathcal{G}^{\infty,k}$ are the classes of Λ_i while, for $i > k$, the i-elements of $\mathcal{G}^{\infty,k}$ are just the same as in $\mathcal{G}$. In particular, the 1-elements of $\mathcal{G}^{\infty,k}$ bijectively correspond to the shrinkings of $\mathcal{G}$. The incidence relation is defined as follows. Two elements X and Y of $\mathcal{G}^{\infty,k}$ of type $i, j \leq k$ are declared to be incident if a member of X is incident in $\mathcal{G}$ with a member of Y. Two elements x and y of $\mathcal{G}^{\infty,k}$ of type $i, j > k$ are incident in $\mathcal{G}^{\infty,k}$ if and only if they are incident as elements of $\mathcal{G}$. An element X of type $i \leq k$ is incident with an element y of type $j > k$ if and only if y is incident in $\mathcal{G}$ with at least one member of X.

If $k = n$ then $\mathcal{G}$ is an affine geometry and $\mathcal{G}^{\infty,k}$ is its projective geometry at infinity. When $k < n$ the $\{1, 2, \ldots, k\}$-residues of $\mathcal{G}^{\infty,k}$ are projective geometries. If, moreover, $k < n - 1$ then the $\{k + 2, \ldots, n\}$-residues of $\mathcal{G}^{\infty,k}$ are the same as in $\mathcal{G}$ and, for every element x of type $j > k$, $\mathrm{Res}^-_{\mathcal{G}^{\infty,k}}(x)$ is the geometry at infinity of $\mathrm{Res}^-_{\mathcal{G}}(x)$ of level k. (Note that property (RS) as well as its analogues for type $2, 3, \ldots, k$ are preserved when taking lower residues.)

Proposition 8.4. *Let $k < n$ and suppose that no two distinct elements of $\mathcal{G}$ are incident with the same set of n-elements (as when* (IP) *holds, for instance). Then* $\mathrm{Aut}(\mathcal{G}^{\infty,k}) = \mathrm{Aut}(\mathcal{G})$. *Moreover, if* $\mathrm{Aut}(\mathcal{G})$ *acts flag-transitively on* $\mathcal{G}$ *then* $\mathrm{Aut}(\mathcal{G}^{\infty,k})$ *is also flag-transitive.*

Proof. The latter claim is obvious. The first claim follows from the fact that, since $k < n$, $\mathcal{G}$ and $\mathcal{G}^{\infty,k}$ have the same n-elements and, since no two elements of $\mathcal{G}$ are incident with the same set of n-elements, the action of $\mathrm{Aut}(\mathcal{G})$ on $\mathcal{G}$ is uniquely determined by its action on the set of n-elements. □

Unfortunately, Proposition 8.4 is seldom of great help. Indeed, in general, $\mathcal{G}^{\infty,k}$ is far more difficult to investigate than $\mathcal{G}$ itself, due to the fact that rather odd geometries can occur as residues of $\mathcal{G}^{\infty,k}$ of type $\{k, k + 1\}$ or $\{k + 1, k + 2\}$. However, in a few lucky cases $\mathcal{G}^{\infty,k}$ turns out to be a well known object. In those lucky cases we $\mathcal{G}^{\infty,k}$ can help us to better understand $\mathcal{G}$.

9 A Lemma on Flag-Transitive F_4-Geometries

Let $\mathcal{F}$ be an F_4-geometry with finite orders s, s, t, t, where $s > 1$.

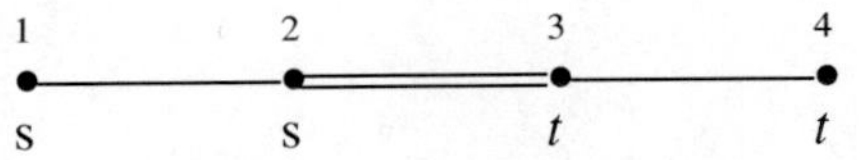

$(1 < s < \infty,\ 1 \leq t < \infty)$

The next lemma has been exploited in Sect. 2.

Lemma 9.1. *If $\mathcal{F}$ is flag-transitive then $\mathcal{F}$ is a building, except possibly when $t = 1$ and $s > 2$.*

Proof. Suppose $t > 1$. Then, by Yoshiara [29], either all C_3-residues of $\mathcal{F}$ are buildings or $s = t = 2$ and one of these residues is isomorphic to the flat C_3-geometry for the alternating group A_7. The latter case is impossible by Aschbacher [1]. Hence, all C_3-residues of $\mathcal{F}$ are buildings. By Tits [27], $\mathcal{F}$ is covered by a building $\widetilde{\mathcal{F}}$. However, $\mathrm{Aut}(\widetilde{\mathcal{F}})$ does not admit any normal subgroup that can define a proper quotient of $\widetilde{\mathcal{F}}$. Hence $\mathcal{F} = \widetilde{\mathcal{F}}$.

Let $t = 1$. Then the $\{1,2,3\}$-residues of $\mathcal{F}$ are buildings [19, Theorem 14.13]. On the other hand, $\{2,3,4\}$-residues arise from suitable sets of latin squares as explained by Rees [23]. It follows from Rees's description that when $s = 2$ these residues are either buildings or flat quotients of buildings. In this case all C_3-residues of $\mathcal{F}$ are covered by buildings and we obtain the conclusion as in the previous paragraph. □

Remark 9.2. As far as I know, the case of $t = 1$ and $s > 2$ is still open.

10 The Graph $\Phi(\mathcal{F})$ of an F_4-Building $\mathcal{F}$

Let $\mathcal{F}$ be a building of type F_4 and let $\Gamma(\mathcal{F})$ be its collinearity graph. We recall that $\Gamma(\mathcal{F})$ has diameter equal to 3 and there are two kinds of pairs of points at distance 2 in it, namely symplectic pairs and special pairs. A pair $\{x, y\}$ of points at distance two in $\Gamma(\mathcal{F})$ is said to be *symplectic* if there are at least two points collinear with both x and y. If only one point exists collinear with both x and y then $\{x, y\}$ is a *special* pair. Equivalently, $\{x, y\}$ is symplectic or special according to whether it is contained in a symp or not.

The following graph, which we denote by $\Phi(\mathcal{F})$, has been exploited in Sect. 3 to define $c.F_4(2,t)$-geometries. The points of $\mathcal{F}$ are the vertices of $\Phi(\mathcal{F})$, two points being adjacent in $\Phi(\mathcal{F})$ when they either are collinear or form a symplectic pair. It is not so difficult to prove that $\Phi(\mathcal{F})$ has diameter equal to 2.

Clearly, the symps of $\mathcal{F}$ are maximal cliques of $\Phi(\mathcal{F})$, but not all maximal cliques of $\Phi(\mathcal{F})$ arise from symps. Nevertheless, when $\mathcal{F}$ admits finite orders then we can recover $\Gamma(\mathcal{F})$ from $\Phi(\mathcal{F})$. Indeed, let $\mathcal{F}$ be an $F_4(s,t)$-building, $s, t < \infty$. For a vertex x of $\Phi(\mathcal{F})$, let $\Phi(x)$ be the set of vertices adjacent to x in $\Phi(\mathcal{F})$, with the convention that $x \notin \Phi(x)$. Let $\{x, y\}$ be an edge of $\Phi(\mathcal{F})$. If x and y are collinear points of $\mathcal{F}$ then

$$|\Phi(x) \cap \Phi(y)| = s^2(t^2 + t + 1)(1 + st + s^2t) + s - 1.$$

If $\{x, y\}$ is a symplectic pair then

$$|\Phi(x) \cap \Phi(y)| = s^2t^2(st + 1)(t + 1) + (s^2t + 1)(s^2 + s + 1) - 2.$$

The previous two numbers are never equal. Hence, $\Gamma(\mathcal{F})$ can be recovered from $\Phi(\mathcal{F})$. As the whole of $\mathcal{F}$ can be recovered from $\Gamma(\mathcal{F})$, we can also recover $\mathcal{F}$ from $\Phi(\mathcal{F})$.

References

1. M. Aschbacher. Flag structures in Tits geometries, *Geom. Dedicata* **14** (1983), 21–32
2. F. Buekenhout and A. M. Cohen. *Diagram Geometry*, unpublished book available at http://www.win.tue.nl/~amc/buek/
3. F. Buekenhout and X. Hubaut. Locally polar spaces and related rank 3-groups, J. Algebra **45** (1977), 391–434
4. A. M. Cohen e B. N. Cooperstein. The 2-spaces of the standard E_6-module, *Geom. Dedicata* **25** (1988), 467–480
5. J. H. Conway, R. T. Curtis, S. P. Norton, R. A. Parker and R. A. Wilson. *Atlas of Finite Groups*, Clarendon, Oxford 1985
6. H. Cuypers. Extended near hexagons and line systems, *Adv. Geometry* **4** (2004), 181–214
7. H. Cuypers and A. Pasini. Locally polar geometries with affine planes, *European J. Combin.* **13** (1992), 39–57
8. A. Del Fra, D. Ghinelli, T. Meixner and A. Pasini. Flag-transitive extensions of C_n geometries, *Geom. Dedicata* **37** (1990), 253–273
9. A. A. Ivanov. A geometric characterization of Fischer's Baby Monster, *J. Algebraic Combin.* **1** (1992), 43–65
10. A. A. Ivanov. Exceptional extended dual polar spaces, *European J. Combin.* **18** (1997), 859–885
11. A. A. Ivanov. Affine extended dual polar spaces, in *Groups and Geometries* (L. Di Martino et al. ed.), Brikhäuser, Basel (1998), 107–121
12. A. A. Ivanov and D. V. Pasechnik. c-extendions of the $F_4(2)$-building, *Discr. Math.* **264** (2003), 91–110
13. A. A. Ivanov and A. Pasini. A diameter dound for extensions of the $F_4(2)$-buildings, *European J. Combin.* **24** (2003), 685–707
14. A. A. Ivanov and S. V. Shpectorov. *Geometry of Sporadic Groups II. Representations and Amalgams*, Cambridge Univ. Press, Cambridge 2002
15. A. A. Ivanov and C. Wiedorn. More geometries of the Fischer group Fi_{22}, *J. Algebraic Combin.* **16** (2002), 111–150
16. A. A. Ivanov, D. V. Pasechnik and S. V. Shpectorov. Extended F_4-buildings and the Baby Monster, *Invent. Math.* **144** (2001), 399–433
17. A. Kasikova and E. E. Shult. Absolute embeddings of point-line geometries, *J. Algebra* **238** (2001), 265–291
18. A. Pasini. Covers of finite geometries with non-spherical minimal circuit diagram, in *Buildings and the Geometry of Diagrams* (L. A. Rosati ed.), Springer L. N. **1181**, Springer, Berlin (1986), 218–241
19. A. Pasini. *Diagram Geometries*, Oxford Univ. Press, Oxford 2004
20. A. Pasini. Embeddings and expansions, *Bull. Belgian Math. Soc. – Simon Stevin* **10** (2003), 585–626
21. A. Pasini. Shrinkings, structures at infinity and affine expansions, with an application to c-extended P- and T-geometries, *European J. Combin.* **28** (2007), 592–612
22. A. Pasini and C. Wiedorn. Local parallelisms, shrinkings and geometries at infinity, in *Topics in Diagram Geometry*, Quaderni di Matematica **12**, II Univ. of Naples (2003), 127–195
23. S. E. Rees. Finite C_3-geometries in which all lines are thin, *Math. Z.* **189** (1985), 263–171
24. G. Stroth. Flag transitive $c.F_4(2)$-geometries, *Adv. Geom.*, to appear

25. G. Stroth and C. Wiedorn. c-extensions of P- and T-geometries, *J. Cmbin. Th. A* **93** (2001), 261–280
26. G. Stroth and C. Wiedorn. c-extensions of P- and T-geometries – a survey of known examples, in *Topics in Diagram Geometry*, Quaderni di Matematica **12**, II Univ. of Naples (2003), 197–226
27. J. Tits. A local approach to buildings, in *The Geometric Vein* (C. Davis et al. ed.), Springer, Berlin (1981), 519–547
28. C. Wiedorn. Flag-transitive c-extensions of the $F_4(2)$-building – the amalgams, *Euroepan J. Combin.* **28** (2007), 641–652
29. S. Yoshiara. The flag-transitive C_3-geometries of finite order, *J. Algebraic Combin.* **5** (1996), 251–284

Buildings and Kac-Moody Groups

Bertrand Rémy

Abstract This survey paper provides an overview of some aspects of the theory of buildings in connection with geometric and analytic group theory.

Subject Classifications: AMS classification (2000): 20E42, 51E24, 20F32, 20F67, 20F69, 22F, 22F10, 22F50

Keywords Buildings • Geometric and analytic group theory • Kac-Moody groups

1 Introduction

1. The general goal of this survey paper is to introduce a class of metric spaces with remarkable symmetry properties (*buildings*), and a class of finitely generated groups acting on some of them (*Kac-Moody groups*). Then – and mostly – we mention what the viewpoint of geometric group theory enabled one to prove in the very recent years. More precisely, we deal with the following topics – see the *structure of the paper* at the end of the introduction to find the exact places.

- *Simplicity*: Kac-Moody groups provide a wide class of infinite finitely generated, often finitely presented and Kazhdan, simple groups (Caprace–Rémy).
- *Rigidity*: these groups enjoy strong rigidity properties, e.g., of the type "higher-rank vs hyperbolic spaces" (Caprace–Rémy).

B. Rémy (✉)
Université de Lyon, CNRS; Université Lyon 1, Institut Camille Jordan,
43 boulevard du 11 november 1918, 69622 Villeurbanne cedex, France
e-mail: remy@math.univ-lyon1.fr

N.S.N. Sastry (ed.), *Buildings, Finite Geometries and Groups*, Springer Proceedings in Mathematics 10, DOI 10.1007/978-1-4614-0709-6_11,

- *Amenability*: though these groups are not themselves amenable, they admit amenable actions on explicit compact spaces provided by boundaries of buildings (Caprace–Lécureux, Lécureux).
- *Quasi-morphisms*: the existence of some non-standard quasi-homomorphisms is understood in terms of the geometry of the buildings and the transitivity of the action (Caprace–Fujiwara).
- *Quasi-isometry*: Kac-Moody lattices provide infinitely many quasi-isometry classes of finitely presented simple groups (Caprace–Rémy).

This subject still has some motivating and fast developments; it follows a general trend according to which more and more analytic topics turn out to be relevant to geometric group theory. Therefore, in the core of the text, we mostly focus on analytic statements. Other very interesting results are alluded to below, in the second and third parts of the introduction; some of them are also described in more detail in [26].

2. We see the present paper as a kind of continuation of the previous survey paper on the topic, written slightly less than 10 years ago [58]. This is why this part of the introduction is dedicated to outline recent progress on some of the questions mentioned there (in the 5th section).

We will provide later some details on the way Kac-Moody theory provides finitely generated groups having a discrete action on the product of two (twinned) buildings. Let Λ be such a group, with associated buildings X_- and X_+. The buildings $X_\pm$ are locally finite, so the compact-open topology on the groups $\mathrm{Aut}(X_\pm)$ is locally compact. The Λ-action on a single factor $X_\pm$ has infinite stabilizers (therefore it is not discrete), and its kernel is the finite center $\mathrm{Z}(\Lambda)$. We denote by $\overline{\Lambda}^\pm$ the closure of the image of Λ in $\mathrm{Aut}(X_\pm)$. It was proved in [60] that the locally compact group $\overline{\Lambda}^\pm$ is locally pro-p, for p a well-defined prime number (equal to the characteristic of the finite ground field of Λ). By analogy with Lie groups over local fields of positive characteristic, some questions about $\overline{\Lambda}^\pm$ were addressed in [58, 5.5.2].

- *Decomposition into abstractly simple factors.* One of these questions is the decomposition of these groups into direct products of abstractly simple groups [58, Question 5.5.7] (the weaker result of a decomposition into topologically simple factors had been proved in [56]). It turns out that thanks to a clever combination of Tits' simplicity criterion for BN-pairs [11] and arguments from pro-p groups, Carbone et al. [21] could prove this result for a large class of Kac-Moody groups (those for which one, or equivalently any, cell-stabilizer is topologically finitely generated). This proves the decomposition for instance when the generalized Cartan matrix defining Λ is 2-*spherical* (i.e., any two canonical reflections in the Weyl group generate a finite group).
- *Non-linearity for compact open subgroups.* One other question was to decide whether some compact open subgroups in $\overline{\Lambda}^\pm$ are not linear (over any field) under suitable conditions on the geometry of the buildings [58, Question

5.5.6]. Strictly speaking, this question is not answered but it is related to a more interesting result on the Golod-Shafarevich property due to Ershov [31]; this result involves the pro-p completions of cell-stabilizers in Kac-Moody lattices Λ – see below.

- *Generalized arithmeticity.* Given the inclusion of a lattice Γ in a locally compact group G, one says that Γ is *arithmetic* in G if its commensurator in G

  $\mathrm{Comm}_G(\Gamma) = \{g \in G : \Gamma \cap g\Gamma g^{-1}$ has finite index both in Γ and in $g\Gamma g^{-1}\}$ is dense in G (recall that classically, according to a well-known criterion due to G. Margulis, a lattice in a semisimple Lie group is arithmetic if and only if its commensurator is dense in the ambient group [73, 6.2]). Facet stabilizers in Kac-Moody lattices provide interesting exotic examples of arithmetic *non-uniform* lattices in full tree automorphism groups [4], but it seems that the more general question asked in [58, Question 5.5.4] is still open.
- *Non-linearity for discrete subgroups.* Let Γ denote the stabilizer in Λ of some cell in X_ϵ (where $\epsilon = \pm$). Then Γ gives rise to a lattice for the building $X_{-\epsilon}$ of opposite sign. The subgroup Γ is much smaller than Λ; its closure in $\mathrm{Aut}(X_\epsilon)$ is an open virtually pro-p subgroup, contained as a finite index subgroup of a suitable maximal compact subgroup of $\overline{\Lambda}^\epsilon$. The question of deciding under which conditions Γ cannot be a group of matrices over any field [58, Question 5.5.5] is still open. Note that the only linearity to be disproved is over fields of characteristic p [53], but it may happen that Γ is linear since for suitable choices of generalized Cartan matrices and ground fields defining Λ, we can have $\Gamma \simeq \mathrm{SL}_n(\mathbf{F}_q[t])$ (for arbitrary integer $n \geqslant 2$ and prime power q). Note that according to Caprace, Gramlich and Mühlherr there is another class of very interesting lattices (for a single building) arising in this context, namely centralizers of suitable involutions of Λ; see for instance [33, 35] for an introduction to these lattices.

Note also that the (non-)linearity question for Λ is now solved by the answer to the simplicity question: according to Mal'cev, a linear finitely generated group – like Λ – is always residually finite [43], i.e. the intersection of its finite index (normal) subgroups is trivial; in particular, it cannot be simple if it is infinite. One difficulty, among others, for the linearity problem of the subgroups Γ is the fact that these groups are usually *not* finitely generated (see [3] for sufficient conditions, though).

3. Of course, most nice recent results on Kac-Moody groups (discrete and completed versions) were not proved after the questions mentioned in [58]! Here are some of them.

- *Harmonic analysis.* First, the analogy between the groups $\overline{\Lambda}^\pm$ and reductive Lie groups over local fields suggests the possibility of generalizing some results about non-commutative harmonic analysis on Lie groups. This leads to the question of proving the existence of a Gelfand pair with respect to a suitable maximal compact subgroup [62], the next step being the explicit computation of the corresponding spherical functions [42]. One striking result in this vein again reveals that the geometry of the buildings is crucial: Lécureux proved that whenever a building is *not* Euclidean, sufficiently transitive automorphism (i.e., sufficiently interesting) groups cannot have any Gelfand pair with respect to any maximal compact subgroup [41]. This means that convolution, restricted to any (Hecke) algebra of compactly supported functions bi-invariant under any compact subgroup, is never a commutative law. Actually, since maximal compact subgroups in this context are open, the Hecke algebras here are well-understood algebraic objects defined by relations closely related to infinite root systems; the resulting combinatorial problem is eventually solved by geometric arguments similar to those shortly described in Sect. 4.1.

- *The Golod–Shafarevich condition and property (T).* Roughly speaking, a pro-p group is said to have the *Golod–Shafarevich* property if it (admits a presentation which) has few relations with respect to the number of generators. This condition can be made technically very precise, and was initially designed to prove that some (Galois) groups are infinite. It turns out that this condition implies the largeness of the pro-p group under consideration in many more ways – we refer to [31, Introduction] for a nice and very efficient introduction to the subject. By passing to pro-p completions, the Golod–Shafarevich condition is relevant to the study of discrete groups. As already mentioned, to data needed to define a finitely generated Kac-Moody group are a generalized Cartan matrix and a finite field (see Sect. 2.3). If p denotes the characteristic of the latter ground field, then the rich combinatorial structure of Λ enables one to study the pro-p completion $\Gamma_{\widehat{p}}$ of a facet stabilizer Γ as above (note that such a Γ is residually finite: sufficiently many finite index subgroups are given for instance by pointwise stabilizers of combinatorial balls around the facet). After works by Lubotzky and Sarnak, some conjectures were made by Lubotzky and Zelmanov about the incompatibility between property (T) or (τ) and being Golod–Shafarevich [72]. The main result of [31] disproves some of these intuitions and says that for every sufficiently large prime number p, there exists a finitely generated group having property (T) and which is Golod–Shafarevich with respect to p. This is proved by using Kac-Moody theory, and in particular the pro-p completions $\Gamma_{\widehat{p}}$ of facet stabilizers.
- *Connection with the general theory of totally disconnected groups.* In a somewhat different direction, Willis started to develop a thorough study of arbitrary locally compact totally disconnected groups [70]. The main tool he uses to perform this study is the space of all compact open subgroups of the given group, which is non-empty (and big) by the assumption of being totally disconnected [12, III, Cor. 1, p.36]. The latter space is endowed with a natural metric which allows one to use some arguments of dynamical nature. This theory applies of course to groups defined as closed automorphism groups of locally finite cell complexes, and to algebraic groups over non-Archimedean local fields. One of its outcomes is to attach some invariants of the topological

group structure to any totally disconnected locally compact group; one of them is a notion of rank, namely the *flat rank*, which coincides with the usual notion of rational rank when the group is obtained as the rational points of a reductive algebraic group over a non-archimedean local field. When the building is not Euclidean, which excludes any algebraic group consideration, the flat rank can still be related to another notion of rank in connection with maximal flat (i.e., Euclidean) subspaces in the building; see [16,24]. Another connection between Kac-Moody groups and this theory is the study of contraction groups, which are so to speak dynamical generalizations of unipotent radicals of parabolic subgroups in the algebraic group case; it is proved in [15] that whenever the buildings are *not* Euclidean, contraction groups are *not* closed, which is another avatar, in addition to the dichotomy "simplicity vs linearity", of the dichotomy between Euclidean and non-Euclidean buildings.

– *Decomposition of abstract homomorphisms.* The last question that was solved and which can be mentioned in this introduction is purely of algebraic nature. The intuition leading to it is not the analogy with non-Archimedean Lie groups, but is the very starting point of the construction of Kac-Moody groups by Tits, namely to introduce infinite-dimensional analogues of Chevalley group schemes [67]. Having this (initial) motivation in mind, one can ask whether there is, as in the finite-dimensional case, a well understood factorization for abstract group homomorphisms between groups of rational points over fields of Kac-Moody functors. The strongest results in this direction were obtained by Caprace [20].

1.1 Structure of the Paper

In Sect. 1, we recall some basic definitions in building theory and explain how the algebraic machinery of Kac-Moody theory provides examples endowed with interesting group actions. In Sect. 2, we present some recent results in connection with quotients and actions, namely we study the problems of simplicity and rigidity for Kac-Moody lattices. In Sect. 3, we investigate some more topics from analytic and metric group theory, namely amenability (for actions), existence of (exotic) quasi-characters and distortion.

1.2 Conventions

In this paper, letters like Λ and Γ denote discrete groups, letters like G and H denote non-discrete topological groups (assumed to be locally compact) and letters like X and Y denote metric spaces (most of the time assumed to be complete).

2 Buildings and Kac-Moody Groups

In this section, we briefly provide the material in Kac-Moody and building theory needed to understand the problems investigated in the sequel. The general reference for building theory is [1]; for Kac-Moody groups it is [54].

2.1 (Simplicial) Buildings

In what follows, we only deal with the simplicial point of view on buildings. This is because we are mostly interested here in new phenomena (with respect to the classical theory of Lie groups) which occur mainly while studying groups acting on *non-affine* buildings. Non-simplicial buildings are mostly interesting (so far) when they are of Euclidean type (via the Bruhat-Tits theory of reductive groups over non-discretely valued fields [59], or appearing as asymptotic cones of symmetric spaces or Bruhat–Tits buildings [39]).

First, let us recall the following preliminary notions [11].

- A *Coxeter group*, say W, is a group admitting a presentation: $W=\langle s\in S \mid (st)^{M_{s,t}}=1\rangle$ where $M = [M_{s,t}]s,t\in S$ is a *Coxeter matrix* (i.e., symmetric with 1's on the diagonal and other entries in $\mathbf{N}_{\geqslant 2}\cup\{\infty\}$).
- For any Coxeter system (W,S) there is a natural simplicial complex Σ on the maximal simplices of which W acts simply transitively: Σ is called the *Coxeter complex* of (W,S).

We can go the other way round; namely, let us use a theorem of Poincaré's [45, IV.H.11] starting with a suitable tesselation and providing a Coxeter group (the initial tesselation is eventually a geometric realization of the alluded to above Coxeter complex):

Example 2.1. Start with a Euclidean or hyperbolic periodic tiling whose dihedral angles are integral submultiples of π. Then the group generated by the reflections in the codimension 1 faces of the fundamental tile is a discrete subgroup of the full isometry group; in fact, it is a Coxeter group and the initial tiling realizes its Coxeter complex.

The reason why we introduced Coxeter complexes is that they are so to speak "slices" in buildings, as the following definition shows. We freely use the above notation W and Σ (Coxeter complex).

Definition 2.2. A building of type (W,S) is a cellular complex, covered by subcomplexes all isomorphic to Σ, called the apartments, such that:

(i) Any two simplices, called the facets, are contained in a suitable apartment;
(ii) Given any two apartments A and A', there is a cellular isomorphism $A\simeq A'$ fixing $A\cap A'$.

The group W is called the Weyl group of the building.

When W is a Euclidean reflection group [11, V Sect. 3], one says that the building is *affine* or, equivalently, *Euclidean*.

Example 2.3. A tree with all vertices of valency $\geqslant 2$ (resp. a product of such trees) is a building with W equal to the infinite dihedral group D_∞ (resp. with W equal to a product of D_∞'s).

The above examples of trees are elementary, but they are the only ones with infinite Weyl group which can be reasonably drawn. Note that it is enough to consider these examples to see a difficulty in producing interesting group-theoretic situations from the theory of buildings: *it may very well happen that the automorphism group of a building be trivial* (take a tree in which any two distinct vertices have distinct valencies). It is precisely one feature of Kac-Moody theory to give rise to buildings automatically endowed with a highly transitive group action.

2.2 Analogies with Lie Groups and Exotic Examples

What are buildings good for? They were first designed to provide a uniform approach to simple algebraic groups whatever the ground field is, but it turned out that Chevalley's scheme-theoretic approach became more popular. Still, thanks to buildings, the idea to attach a suitable geometry to simple groups of Lie type was pushed far beyond the classical example of the interplay between simple real Lie groups and symmetric spaces – we refer to [26] for more details on these historical points.

Eventually, as the examples below show, there is no rule in general on what comes first, the building or the group: spherical (resp. Euclidean) buildings are used to better understand (isotropic) algebraic groups over arbitrary (resp. non-Archimedean local) fields, but in some recent approaches, buildings were designed first in order to provide automorphism groups for which the analogy with reductive Lie group makes sense and is often very fruitful.

Example 2.4. Pick (the rational points of) a simple algebraic group over a non-Archimedean local field, e.g. $\mathrm{SL}_n(\mathbf{Q}_p)$. Then the group acts on a Euclidean building in a suitable way: the action is strongly transitive, that is transitive on the inclusions of a chamber (= maximal facet) in an apartment; this is one of the main results of Bruhat–Tits theory – see [18] and [19].

Which other buildings (with an interesting group action) can be exhibited? This question is natural. Actually, there exist lots of interesting examples which are not relevant to Borel–Tits or Bruhat–Tits theory. We will see that some of these buildings have apartments which are periodic tilings of real hyperbolic spaces; here is a list of selected results concerning such hyperbolic buildings.

- According to Haglund and Paulin, for some of these hyperbolic buildings the full isometry group is a locally compact, totally disconnected, uncountable *abstractly simple* group [38].
- According to Bourdon, some of them are characterized by any group acting discretely and cocompactly: this is a strong rigidity result "à la Mostow" [13].
- The latter result can be pushed further: according to Bourdon and Pajot, any quasi-isometry from a right-angled Fuchsian building to itself is at finite distance from a true isometry of the latter space – see [14, 71].

All of these statements strongly support the analogy between exotic buildings and symmetric spaces, between Lie groups and full automorphism groups of these new geometries.

2.3 Kac-Moody Groups and Kac-Moody Buildings

The previously mentioned examples were studied from a geometric (more precisely: metric) viewpoint. We turn now to Kac-Moody theory, which can be seen as an algebraic machinery to produce (as a first step) groups with good combinatorial properties – refinements of Tits systems –, and (as a second step) – by formal and now standard arguments – buildings automatically endowed with strongly transitive group actions. Let us briefly sum up the situation and refer to [54] for details and constructions of twisted variants.

- Kac-Moody groups (in the "minimal" version we are interested in, see [68]) were constructed by Tits [67] in order to generalize, as group functors, (split) reductive algebraic group schemes initally due to Chevalley and Demazure.
- They are defined by a presentation generalizing the generators and relations of SL_n (using elementary unipotent matrices) [64].
- The defining datum for a Kac-Moody group is: a field $\mathbf{K}$ and a *generalized Cartan matrix*, i.e., an integral matrix $A = [A_{s,t}]_{s,t \in S}$ such that $A_{s,s} = 2$ for all $s \in S$ and $A_{s,t} \leqslant 0$ for $s \neq t$, with $A_{s,t} = 0$ if and only if $A_{t,s} = 0$.
- Unfortunately, they have not yet been endowed with any structure from algebraic geometry, which would/will be, by the way, infinite-dimensional.
- Nevertheless, they share many combinatorial properties with groups of points of Chevalley–Demazure group schemes: they have a (twin) BN-pair structure [67, Sects. 5 and 6].

Example 2.5. The standard example of such a group is $\Lambda = \mathbf{G}\big(\mathbf{K}[t, t^{-1}]\big)$ for $\mathbf{G}$ a simple (isotropic) matrix group over a field $\mathbf{K}$.

Still, we will see that the groups from this example are not the most interesting ones for the purposes mentioned in the first part of the introduction: they have an obvious matrix interpretation, and as such are residually finite groups. Note that the associated buildings are Bruhat–Tits buildings, hence are Euclidean.

The geometric counterpart to the above BN-pair combinatorics is the following.

Fact 2.6 *(i) Any Kac-Moody group Λ naturally acts on the product $X_- \times X_+$ of two isomorphic buildings $X_\pm$.*

(ii) The explicit rule for W deduces $[M_{s,t}]_{s,t\in S}$ from $A = [A_{s,t}]_{s,t\in S}$. More precisely, we have $M_{s,t} = 2, 3, 4, 6$ or ∞ according to whether $A_{s,t} \cdot A_{t,s}$ is $0, 1, 2, 3$ or is $\geqslant 4$, respectively.

Reference. This is [67, Sects. 5 and 6] for the group-theoretic part, and the geometric side is explained in [54, Sects. 1 and 2]. □

Reading backwards $[A_{s,t}]_{s,t\in S} \mapsto [M_{s,t}]_{s,t\in S}$, we can produce buildings (with nice group actions) provided the Weyl group has its Coxeter exponents in $\{2;3;4;6;\infty\}$. This is the only condition on the shape of apartments – see [55, Sect. 2] for the application to hyperbolic buildings

- The case of affine buildings corresponds exactly to the previous examples $\Lambda = \mathbf{G}(\mathbf{K}[t,t^{-1}])$, with a concrete matrix interpretation. We say then that the generalized Cartan matrix A and the corresponding Kac-Moody group Λ are of *affine* type.
- When W is a Fuchsian group, e.g. when W is generated by a right-angled hyperbolic polygon or by a regular triangle of angle $\frac{\pi}{4}$ or $\frac{\pi}{6}$, then $X_\pm$ carries a negatively curved metric.

These are only examples; the general case is a mixture.

Let us finally mention the basic results saying that over a finite ground field, Kac-Moody groups are relevant to geometric group theory.

Fact 2.7 *(i) Any Kac-Moody group Λ over any finite field is finitely generated.*

(ii) The associated buildings $X_\pm$, for a suitable non-positively curved realization, are locally finite.

Reference. More technically: according to Davis, any building admits a realization carrying a CAT(0) metric [29], and this realization is locally finite if and only if the ground field if finite. □

3 Simplicity and Rigidity

In this section, we are interested in the algebraic question of simplicity for the finitely generated groups obtained by taking Kac-Moody groups over finite fields. It turns out that in this case – where we have a nice geometric action – there is a simple answer to this basic question in group theory (it can be formulated in terms of the geometry of the buildings). The question of rigidity of group actions is less algebraic, but we review it here in order to explain the general idea to combine simplicity, non-positive curvature and representation-theoretic properties.

3.1 Covolume

From now on, we denote by Λ a Kac-Moody group defined by a generalized Cartan matrix, say $A = [A_{s,t}]_{s,t \in S}$, and a *finite* field $\mathbf{F}_q$; the associated buildings are denoted by $X_\pm$. The full automorphism groups $\mathrm{Aut}(X_\pm)$ are thus locally compact for the compact open topology; therefore $\mathrm{Aut}(X_\pm)$ admits Haar measures [10]. Using the Λ-actions on $X_\pm$ arising from the combinatorial structure of a twin BN-pair, we can see Λ as a (diagonally embedded) subgroup of $\mathrm{Aut}(X_-) \times \mathrm{Aut}(X_+)$. The starting point to combine Kac-Moody groups and geometric group theory is the following result, which establishes an analogy between Kac-Moody groups over finite fields and arithmetic groups in positive characteristics [53].

Theorem 3.1. *Assume the Weyl group W of Λ is infinite and denote by $W(t) = \sum_{w \in W} t^{\ell(w)}$ its growth series. If $W(\frac{1}{q}) < \infty$, then Λ is a lattice of $X_+ \times X_-$; it is never cocompact.*

Reference. This is the result settled in [23] or in the note [52]. □

Remark 3.2. 1. When the Kac-Moody group is affine, i.e. when $\Lambda = \mathbf{G}\big(\mathbf{F}_q[t, t^{-1}]\big)$, the condition $W(\frac{1}{q}) < \infty$ is empty since the virtually abelian group W has polynomial growth: it is a Euclidean reflection group. In this case, this is in fact a well-known consequence of a deeper result: Harder's reduction theory in positive characteristic [37].
2. The diagonal Λ-action on $X_+ \times X_-$ is always discrete and the real number $W(\frac{1}{q})$ is merely the covolume $\mu\big(\frac{\mathrm{Aut}(X_-) \times \mathrm{Aut}(X_+)}{\Lambda}\big)$ for a natural normalization μ of the Haar measure on $\mathrm{Aut}(X_-) \times \mathrm{Aut}(X_+)$.

Concerning computation of covolume, some natural questions are in order. For instance, one consists in fixing a building and checking whether the infimum of the normalized covolumes is > 0; one can also try to decide whether this infimum is reached. This is an adaptation, in a new situation of locally compact groups admitting lattices, of very classical questions and results going back to Siegel, one of the most important achievements being strong finiteness results due to Prasad [50]. In the new situation allowed by considering suitable non-affine buildings and their automorphism groups, some partial results are already available. For instance, Thomas exhibited, for right-angled Fuchsian buildings, an infinite increasing sequence of lattices [66]; she also showed that in some cases the sets of covolumes are unrestricted with respect to the classical "Lie" situation (in particular when super-rigidity holds) [65].

3.2 Simplicity

We keep the previous notation. The fact that a finitely generated Kac-Moody group such as Λ can be seen as a lattice of some reasonable geometry is the starting point to prove the following simplicity result.

Theorem 3.3. *Let Λ be a Kac-Moody group defined over the finite field $\mathbf{F}_q$. Assume that the Weyl group W is infinite and irreducible, and that $W(\frac{1}{q}) < \infty$. Then Λ is simple (modulo its finite center) whenever the buildings $X_\pm$ are not Euclidean and Λ is generated by its root subgroups.*

Reference. This statement is contained in [27] – see in particular Sect. 4 of [loc. cit.]. □

What follows in this subsection is dedicated to explaining the general two-step strategy of the proof, but let us first make some comments on the statement itself and some extensions of it.

Remark 3.4. 1. Here is a rough but striking geometric reformulation of the above statement: whenever Λ (with irreducible Weyl group) has no obvious matrix interpretation – because it is not affine – it is a simple finitely generated group; moreover there is a geometric (Coxeter-theoretic) explanation for this.
2. There exist infinitely many generalized Cartan matrices A such that Λ is a *finitely presented, Kazhdan, simple group* for any $q \gg 1$.

Point 2 above is proved by combining the previous simplicity theorem together with:

(i) (Cohomological) finiteness results due to Abramenko and Mühlherr, e.g. [3] and [2].
(ii) Dymara and Januszkiewicz's criterion for property (T) for automorphism groups of buildings [30].

We can now turn to roughly sketching the proof of the simplicity theorem. It owes a lot to Burger–Mozes' recent construction of finitely presented torsion-free simple groups. The latter groups appear as cocompact lattices in products of two trees (here, it is good to have in mind that trees are 1-dimensional buildings!).

The general idea in [8] is first to see the discrete groups under consideration as analogues of lattices in Lie groups in order to rule out infinite quotients, and finally to exploit decisive differences with the classical Lie group case in order to rule out finite quotients too.

- The *analogy part* is motivated by Margulis' normal subgroup theorem, which says that a normal subgroup in a higher-rank lattice must have finite index [73, Sect. 8]. The point is to obtain a generalization of this result without relying on any algebraic group structure on the ambient topological group given by the full automorphism group of the geometry (product of trees or, more generally, of buildings).
- The *difference part* is more specific to the situation. In the case of products of trees, it relies on the possibility of proving some *non*-residual finiteness criteria involving transitivity conditions on the local actions (around each vertex) for the projection of the lattice on each of the two trees. This part was eventually improved by the possibility to embed explicitly well-known non-residually finite

groups into suitable cocompact lattices of products of trees. This enabled Rattaggi to reduce a lot the size of the presentation of some simple groups as constructed by Burger and Mozes [51].

Remark 3.5. A group all of whose quotients are finite is called just infinite; according to Wilson, such a finitely generated group either is residually finite (examples are given by linear groups) or contains a finite index subgroup which is a direct product of finitely many isomorphic simple groups.

We refer to [6, Th. 5.6 p. 59] for a more refined version of Wilson's alternative involving the so-called branch groups ("Grigorchuk's trichotomy").

Here is a slightly more precise summary of the proof of simplicity in the Kac-Moody case, with relevant references.

1. The analogy part follows Margulis' strategy for the normal subgroup property: any normal subgroup of a irreducible Kac-Moody lattice has finite index or is finite and central (i.e. an irreducible center-free Kac-Moody lattice is just infinite). Actually, the crucial point to prove this is to use a criterion due to Shalom [63] (resp. Bader and Shalom [17]) to prove property (T) (resp. amenability) for the quotient group Λ/N, endowed with the discrete topology, where $N \lhd \Lambda$ is the normal subgroup under consideration (recall that a group which is both Kazhdan and amenable is compact). The paper [63] considers *cocompact* irreducible lattices in direct products, but Shalom notes himself that the cocompactness assumption can relaxed to a weaker integrability condition involving the induction cocycle; the latter integrability condition is checked in [57].
2. What can go wrong from being just infinite to being simple? Consider the affine (linear) example $\Lambda = \mathbf{G}(\mathbf{F}_q[t, t^{-1}])$; it has a lot of finite quotients (given for instance by the congruence kernels). This is where non-affineness of the Weyl group has to be used crucially. Indeed, a strengthening of Tits' alternative for Coxeter groups (due to Margulis–Noskov–Vinberg, see [48] and [49]) implies that non-affine Coxeter groups are "weakly hyperbolic" in the sense that there exist lots of triples of roots with empty pairwise intersections. This is what has to be combined together with a trick on infinite root systems and some defining relations for Kac-Moody groups, in order to rule out finite quotients for Λ [27, Sect. 4].

3.3 Rigidity

Why Care About Kazhdan's Property (T) for Simple Groups? A well-known general principle is that there is a deep connection between the representation theory of a locally compact group and its intrinsic topological group structure. Property (T) illustrates this perfectly: one of its features is that it has many equivalent formulations among which are definitions by representation-theoretic means. What we explain below may not be the strongest way to combine Kazhdan's property (T), simplicity and non-positive curvature for metric spaces acted upon, but it provides a reasonable motivation to investigate further rigidity questions in such situations.

Here are the main steps of the argument.

Residual Finiteness and Compactness First let us recall that a finitely generated group is *residually finite* (i.e. the intersection of its finite index subgroups is trivial) if, and only if, it embeds abstractly in a compact group. One implication is easy since residual finiteness amounts to saying that the natural homomorphism from the group to its profinite completion is injective. For the other direction, the trick is to use the Peter–Weyl theorem decomposing representations of compact groups – this is where the previous general principle is used (it is a variant of the way to use Peter–Weyl's theorem to understand general compact groups [69, Chap. V]). In our situation, where we are most interested in finitely generated groups, we can conclude that a finitely generated simple group Γ has trivial homomorphic image to any compact group.

Geometric Consequence for Cell Complexes Therefore, if such a Γ acts on a locally finite cell complex with a global fixed point, the action is actually trivial; indeed, by definition of the compact open topology on the automorphism group of a complex, the stabilizer of a point is (open and) compact.

Non-Positive Curvature The situation is better if we make some assumption on the curvature properties (in a singular sense) of the metric space acted upon. Indeed, Bruhat–Tits fixed point theorem says that a bounded subset in a complete non-positively curved space has a barycenter which is uniquely characterized by purely metric properties [18, Sect. 3.2]; this can be applied to bounded orbits of isometric group actions. In our situation, this implies that if Γ (simple) acts non-trivially on a CAT(0) locally finite space, then any of its orbits is unbounded.

Negative Curvature Recall that the CAT(0) *property* for a geodesic metric space is that all geodesic triangles be at least as thin as in the Euclidean plane [7, II.1]. The similar comparison with the real hyperbolic plane $\mathbb{H}^2_{\mathbf{R}}$ instead of $\mathbf{R}^2$ leads to the notion of a CAT(-1)-*space*. Now let Y be a proper CAT(-1)-space with Isom(Y) acting cocompactly (the properness assumption here means that closed metric balls are compact and the cocompactness assumption is a variant of requiring bounded geometry). Then according to Burger–Mozes, the stabilizer of any $\xi \in \partial_\infty Y$ is amenable, so any non-trivial action of a finitely generated *Kazhdan* simple group Γ on Y has no global fixed point in the compactification $Y \cup \partial_\infty Y$ (recall that "Kazhdan + amenable" implies "compact").

Super-Rigidity By convention in this paper, the terminology *super-rigid* applies to group actions on non-positively curved spaces (or spaces derived from them). It means that if a discrete group has a "nice" action on a specific geometry, then it cannot have a non-degenerate action on a space which doesn't look like the initial geometry ("nice" means for instance that the action of the discrete group enables one to see it as a lattice of the full isometry group of the geometry). A well-known result by Margulis [44, VII.5] says that higher-rank lattices in semisimple Lie groups are super-rigid; this means that the lattice actions on the symmetric spaces associated to the ambient Lie groups are super-rigid with respect to actions on symmetric spaces (or Bruhat–Tits buildings) of other type.

Example 3.6. There are many results disproving the existence of actions of higher-rank lattices, e.g. $\mathrm{SL}_n(\mathbf{Z})$ for $n \geqslant 3$, on the circle (which can be seen as the boundary at infinity of $\mathbb{H}^2_{\mathbf{R}} = \mathrm{SL}_2(\mathbf{R})/\mathrm{SO}(2)$, the Poincaré plane).

We refer to Ghys' paper [34] for a general report about group actions on the circle.

Let us go back to our specific topic: what makes a Kac-Moody lattice be of higher-rank? For non-affine buildings, Kazhdan's property (T) and the existence of flats of dimension $\geqslant 2$ in the buildings are independent conditions (this is a difference with the classical case of non-Archimedean simple Lie groups). Once these two conditions are fulfilled, we can draw strong consequences on actions on some hyperbolic spaces in the sense of Gromov [7, Sect. III.H.1].

Theorem 3.7. *Let Λ be a simple Kac-Moody lattice and let Y be a proper* CAT(-1)*-space with cocompact isometry group. If the buildings $X_\pm$ of Λ contain flat subspaces of dimension $\geqslant 2$ and if Λ is Kazhdan, then the group Λ has no nontrivial action by isometries on Y.*

Reference. This statement is contained in [27, Sect. 7]. □

The intuition behind this result is of course that the abstract group structure of Λ encodes a substantial part of the geometry of the buildings, so that there is not "enough room" in a hyperbolic space to be compatible with existence of higher-dimensional flats in the buildings of Λ. One crucial ingredient for the proof is a general super-rigidity result due to Monod and Shalom [47].

4 Amenability, Quasi-Homomorphisms and Quasi-Isometry

This section is finally dedicated to reviewing some more advanced topics in geometric group theory, such as the existence of non-standard quasi-characters, or the beginnings of the (still widely open) quasi-isometric classification of Kac-Moody lattices. We start be mentioning another vein of very interesting results, relevant to an analytic approach to geometric group theory, namely the study of amenability properties for actions of building lattices.

4.1 Amenability

We first report on Caprace and Lécureux's work on the classification of amenable subgroups in automorphism groups of buildings, and then mention Lécureux's work on amenability of group actions on suitably compactified buildings.

Compactifications and Amenable Subgroups Once again, the beginning of the story is the situation of semisimple real Lie groups and their associated symmetric spaces. Given a simple real Lie group G, like $\mathrm{SL}_n(\mathbf{R})$, families of compactifications of the symmetric space $X = G/K$ (where K is a maximal compact subgroup) were defined in different ways – and for different puposes – by Furstenberg [32] and Satake [61] in the 1960s. These families depend on the choice of a conjugacy class of parabolic subgroups, and for the same choice of conjugacy class, they were eventually shown to be isomorphic. We are here interested in the fact that, given G, the maximal Satake–Furstenberg compactification of X provides a geometric (virtual) parametrization of maximal amenable subgroups in G: according to Moore, any point stabilizer in G is an amenable subgroup; conversely, any amenable subgroup of G has a finite index subgroup stabilizing a point in the maximal compactification $\overline{X}$ [46]. This result was extended later to the case of non-Archimedean simple Lie groups acting on their Bruhat–Tits buildings [36].

In view of the analogy described in Sect. 2.2, it is then natural (though not obvious at all!) to try to prove the following statement.

Theorem 4.1. *Any locally finite building X admits a compactification providing the same "classification" for amenable subgroups in $G = \mathrm{Isom}(X)$ as above.*

Reference. This is one of the main results of [25]. □

The main difficulty is actually to define suitably the compactification, which is done by a clever use of projections onto residues in buildings; this idea enables Caprace–Lécureux to obtain an embedding of the set of residues of a given building to a space of maps between sets of residues, the latter space being compact whenever the building is locally compact. Note that this compactification is not the one given by asymptotic classes of geodesic rays [7, Sects. II.8 and III.H.3]; it is related to the combinatorics of infinite root systems. Caprace and Lécureux also define a compactification by using the (compact) Chabauty topology [10] on the space of closed subgroups of a given locally compact group, and relate it to the previous one.

From Amenable Groups to Amenable Actions Roughly speaking, a G-action on a space S is called *amenable* if there is a sequence of maps $\{\mu_n : S \to \mathcal{M}^1(G)\}_{n \geqslant 0}$ (where $\mathcal{M}^1(G)$ denotes the space of the probability measures on G) such that

$$\lim_{n\to\infty} \parallel \mu_n(g.x) - g_*\mu_n(x) \parallel = 0$$

uniformly on compact subsets of $G \times S$. In other words, there exists a sequence of maps $S \to \mathcal{M}^1(G)$ which is "asymptotically" equivariant.

In the situation of the previous theorem, the group $G = \mathrm{Isom}(X)$ itself is not amenable in general, but we have the following.

Theorem 4.2. *For any building X, any proper action by a locally compact group on the above compactification $\overline{X}$ is amenable.*

Reference. This is the main result of [40]. □

The main idea is to use the non-positive curvature techniques already shown to be useful in the study of the strong Tits alternative and the weak hyperbolicity of Coxeter groups as mentioned at the end of Sect. 3.2. This enables Lécureux to embed an apartement (Coxeter complex) into a product of trees and then to define the desired family of measures.

Remark 4.3. Admitting an amenable action on a compact space is an important property in analytic group theory: for instance, it is related to the Novikov conjecture; it also provides theoretical resolutions in bounded cohomology and boundary maps in rigidity theory (which, by the way, might be useful in the study of the linearity of certain building lattices).

4.2 Quasi-Homomorphisms

In this subsection, we mention quickly some results by Caprace and Fujiwara about quasi-characters of automorphism groups of buildings (again seen as analogues of semisimple Lie groups).

(i) A *quasi-character* for a group G is by definition a map $f : G \to \mathbf{R}$ such that

$$\sup_{g,h \in G} | f(gh) - f(g) - f(h) | < \infty.$$

(ii) The set of all quasi-characters of G is denoted by $\mathrm{QH}(G)$.
(iii) The set of non-trivial quasi-characters is by definition

$$\widetilde{\mathrm{QH}(G)} = \frac{\mathrm{QH}(G)}{\mathrm{Hom}(G, \mathbf{R}) \oplus \ell^\infty(G)}.$$

Quasi-homomorphisms are related to rigidity questions; according to Burger and Monod, higher-rank lattices in Lie groups don't have non-trivial quasi-characters [9]. As we saw for rigidity questions in Sect. 3.3, it is not clear what to require to consider that a building is of higher rank. The next result shows that many buildings are *not* of higher-rank with respect to quasi-homomorphims.

Theorem 4.4. *Let (W, S) be an infinite, irreducible, non-affine Coxeter system and let X be a building of type (W, S). Let G be a group acting on X by automorphisms so that at least one of the following conditions is satisfied:*

(i) *The G-action on X is Weyl-transitive.*
(ii) *For some apartment $\mathbb{A} \subset X$, the stabilizer $\mathrm{Stab}_G(\mathbb{A})$ acts cocompactly on $\mathbb{A}$.*

Then $\widetilde{\mathrm{QH}(G)}$ is infinite-dimensional.

Reference. This is the main result of [22]. □

The main idea is to combine Coxeter-theoretic ideas together with a criterion due to Bestvina and Fujiwara [5]; the key notion is that of a rank-one isometry in a CAT(0)-space (that is, an isometry with certain contraction properties).

Combined with Kac-Moody theory, this implies that, *up to isomorphism, there exist infinitely many finitely presented simple groups of strictly positive stable commutator length.*

4.3 Quasi-Isometry

Let G be a locally compact group admitting a finitely generated lattice Γ. This implies that G admits a compact generating subset, say $\widehat{\Sigma}$; we denote by $d_{\widehat{\Sigma}}$ the word metric associated with $\widehat{\Sigma}$. Similarly, we fix a finite generating set Σ for Γ and denote by d_Σ the associated word metric. The lattice Γ is called *undistorted* in G if d_Σ is quasi-isometric to the restriction of $d_{\widehat{\Sigma}}$ to Γ. This amounts to saying that the inclusion of Γ in G is a quasi-isometric embedding from (Γ, d_Σ) to $(G, d_{\widehat{\Sigma}})$.

Theorem 4.5. *Any Kac-Moody lattice* $\Lambda < \mathrm{Aut}(X_+) \times \mathrm{Aut}(X_-)$ *is undistorted.*

Reference. This is the main result of [28]. □

Again, combined with simplicity results from Kac-Moody theory, this implies that *there exist infinitely many pairwise non-quasi-isometric finitely presented simple groups.*

The above result uses the metric and combinatorial shape of the apartments in a rough way; in particular, it is far from solving the question of quasi-isometric classification of Kac-Moody lattices.

Acknowledgements I thank Narasimha Sastry for the organization of the conference "Buildings, Finite Geometries and Groups" (Indian Statistical Institute, Bangalore) in August 2010, and for having quickly edited their proceedings. I thank warmly the referees for their quick and efficient reading of the manuscript.

References

1. Peter Abramenko and Kenneth S. Brown, *Buildings*, Graduate Texts in Mathematics, vol. 248, Springer, New York, 2008, Theory and applications. MR2439729 (2009g:20055)
2. Peter Abramenko, *Finiteness properties of groups acting on twin buildings*, Groups: topological, combinatorial and arithmetic aspects, London Math. Soc. Lecture Note Ser., vol. 311, Cambridge Univ. Press, Cambridge, 2004, pp. 21–26. MR2073344
3. Peter Abramenko and Bernhard Mühlherr, *Présentations de certaines BN-paires jumelées comme sommes amalgamées*, C. R. Acad. Sci. Paris Sér. I Math. **325** (1997), no. 7, 701–706. MR1483702 (98h:20043)

4. Peter Abramenko and Bertrand Rémy, *Commensurators of some non-uniform tree lattices and Moufang twin trees*, Essays in geometric group theory, Ramanujan Math. Soc. Lect. Notes Ser., vol. 9, Ramanujan Math. Soc., Mysore, 2009, pp. 79–104. MR2605356
5. Mladen Bestvina and Koji Fujiwara, *A characterization of higher rank symmetric spaces via bounded cohomology*, Geom. Funct. Anal. **19** (2009), no. 1, 11–40. MR2507218 (2010m:53060)
6. Laurent Bartholdi, Rostislav I. Grigorchuk, and Zoran Šunić, *Branch groups*, Handbook of algebra, Vol. 3, North-Holland, Amsterdam, 2003, pp. 989–1112. MR2035113 (2005f:20046)
7. Martin R. Bridson and André Haefliger, *Metric spaces of non-positive curvature*, Grundlehren der Mathematischen Wissenschaften, vol. 319, Springer, Berlin, 1999. MR1744486 (2000k:53038)
8. Marc Burger and Shahar Mozes, *Lattices in product of trees*, Inst. Hautes Études Sci. Publ. Math. (2000), no. 92, 151–194 (2001). MR1839489 (2002i:20042)
9. Marc Burger and Nicolas Monod, *Continuous bounded cohomology and applications to rigidity theory*, Geom. Funct. Anal. **12** (2002), no. 2, 219–280. MR1911660 (2003d:53065a)
10. Nicolas Bourbaki, *Éléments de mathématique. Fascicule XXIX. Livre VI: Intégration. Chapitre 7: Mesure de Haar. Chapitre 8: Convolution et représentations*, Actualités Scientifiques et Industrielles, No. 1306, Hermann, Paris, 1963. MR0179291 (31 #3539)
11. Nicolas Bourbaki, *Éléments de mathématique. Fasc. XXXIV. Groupes et algèbres de Lie. Chapitre IV: Groupes de Coxeter et systèmes de Tits. Chapitre V: Groupes engendrés par des réflexions. Chapitre VI: systèmes de racines*, Actualités Scientifiques et Industrielles, No. 1337, Hermann, Paris, 1968. MR0240238 (39 #1590)
12. Nicolas Bourbaki, *Éléments de mathématique. Topologie générale. Chapitres 1 à 4*, Hermann, Paris, 1971. MR0358652 (50 #11111)
13. Marc Bourdon, *Immeubles hyperboliques, dimension conforme et rigidité de Mostow*, Geom. Funct. Anal. **7** (1997), no. 2, 245–268. MR1445387 (98c:20056)
14. Marc Bourdon and Hervé Pajot, *Rigidity of quasi-isometries for some hyperbolic buildings*, Comment. Math. Helv. **75** (2000), no. 4, 701–736. MR1789183 (2003a:30027)
15. Udo Baumgartner, Jacqui Ramagge, and Bertrand Rémy, *Contraction groups in complete Kac-Moody groups*, Groups Geom. Dyn. **2** (2008), no. 3, 337–352. MR2415303 (2009g:22014)
16. Udo Baumgartner, Bertrand Rémy, and George A. Willis, *Flat rank of automorphism groups of buildings*, Transform. Groups **12** (2007), no. 3, 413–436. MR2356316 (2008j:22035)
17. Uri Bader and Yehuda Shalom, *Factor and normal subgroup theorems for lattices in products of groups*, Invent. Math. **163** (2006), no. 2, 415–454. MR2207022 (2006m:22017)
18. François Bruhat and Jacques Tits, *Groupes réductifs sur un corps local*, Inst. Hautes Études Sci. Publ. Math. (1972), no. 41, 5–251. MR0327923 (48 #6265)
19. François Bruhat and Jacques Tits, *Groupes réductifs sur un corps local. II. Schémas en groupes. Existence d'une donnée radicielle valuée*, Inst. Hautes Études Sci. Publ. Math. (1984), no. 60, 197–376. MR756316 (86c:20042)
20. Pierre-Emmanuel Caprace, *"Abstract" homomorphisms of split Kac-Moody groups*, Mem. Amer. Math. Soc. **198** (2009), no. 924, xvi+84. MR2499773 (2010d:20057)
21. Lisa Carbone, Mikhail Ershov, and Gordon Ritter, *Abstract simplicity of complete Kac-Moody groups over finite fields*, J. Pure Appl. Algebra **212** (2008), no. 10, 2147–2162. MR2418160 (2009d:20067)
22. Pierre-Emmanuel Caprace and Koji Fujiwara, *Rank-one isometries of buildings and quasi-morphisms of Kac-Moody groups*, Geom. Funct. Anal. **19** (2010), no. 5, 1296–1319. MR2585575
23. Lisa Carbone and Howard Garland, *Existence of lattices in Kac-Moody groups over finite fields*, Commun. Contemp. Math. **5** (2003), no. 5, 813–867. MR2017720 (2004m:17031)
24. Pierre-Emmanuel Caprace and Frédéric Haglund, *On geometric flats in the CAT(0) realization of Coxeter groups and Tits buildings*, Canad. J. Math. **61** (2009), no. 4, 740–761. MR2541383 (2010k:20051)
25. Pierre-Emmanuel Caprace and Jean Lécureux, *Combinatorial and group-theoretic compactifications of buildings*, to appear in the Annales de l'Institut Fourier

26. Pierre-Emmanuel Caprace and Bertrand Rémy, *Groups with a root group datum*, Innov. Incidence Geom. **9** (2009), 5–77. MR2658894
27. Pierre-Emmanuel Caprace and Bertrand Rémy, *Simplicity and superrigidity of twin building lattices*, Invent. Math. **176** (2009), no. 1, 169–221. MR2485882 (2010d:20056)
28. Pierre-Emmanuel Caprace and Bertrand Rémy, *Non-distortion of twin building lattices*, Geom. Dedicata **147** (2010), 397–408. MR2660586
29. Michael W. Davis, *Buildings are* CAT(0), Geometry and cohomology in group theory (Durham, 1994), London Math. Soc. Lecture Note Ser., vol. 252, Cambridge Univ. Press, Cambridge, 1998, pp. 108–123. MR1709955 (2000i:20068)
30. Jan Dymara and Tadeusz Januszkiewicz, *Cohomology of buildings and their automorphism groups*, Invent. Math. **150** (2002), no. 3, 579–627. MR1946553 (2003j:20052)
31. Mikhail Ershov, *Golod-Shafarevich groups with property (T) and Kac-Moody groups*, Duke Math. J. **145** (2008), no. 2, 309–339. MR2449949 (2009i:20060)
32. Hillel Furstenberg, *A Poisson formula for semi-simple Lie groups*, Ann. of Math. (2) **77** (1963), 335–386. MR0146298 (26 #3820)
33. Ralf Gramlich, Max Horn, and Bernhard Mühlherr, *Abstract involutions of algebraic groups and of Kac-Moody groups*, to appear in J. Group Theory.
34. Étienne Ghys, *Groups acting on the circle*, Enseign. Math. (2) **47** (2001), no. 3-4, 329–407. MR1876932 (2003a:37032)
35. Ralf Gramlich and Bernhard Mühlherr, *Lattices from involutions of Kac-Moody groups*, in Abstracts from the mini-workshop held November 16–22, 2008, organized by S. Goodwin and R. Gramlich, Oberwolfach Reports. Vol. 5, no. 4.
36. Yves Guivarc'h and Bertrand Rémy, *Group-theoretic compactification of Bruhat-Tits buildings*, Ann. Sci. École Norm. Sup. (4) **39** (2006), no. 6, 871–920. MR2316977 (2008f:20056)
37. Günter Harder, *Minkowskische Reduktionstheorie über Funktionenkörpern*, Invent. Math. **7** (1969), 33–54. MR0284441 (44 #1667)
38. Frédéric Haglund and Frédéric Paulin, *Simplicité de groupes d'automorphismes d'espaces à courbure négative*, The Epstein birthday schrift, Geom. Topol. Monogr., vol. 1, Geom. Topol. Publ., Coventry, 1998, pp. 181–248 (electronic). MR1668359 (2000b:20034)
39. Bruce Kleiner and Bernhard Leeb, *Rigidity of quasi-isometries for symmetric spaces and Euclidean buildings*, Inst. Hautes Études Sci. Publ. Math. (1997), no. 86, 115–197 (1998). MR1608566 (98m:53068)
40. Jean Lécureux, *Amenability of actions on the boundary of a building*, Int. Math. Res. Not. IMRN (2010), no. 17, 3265–3302. MR2680274
41. Jean Lécureux, *Hyperbolic configurations of roots and Hecke algebras*, J. Algebra **323** (2010), no. 5, 1454–1467. MR2584964
42. Ian G. Macdonald, *Spherical functions on a group of p-adic type*, Ramanujan Institute, Centre for Advanced Study in Mathematics, University of Madras, Madras, 1971, Publications of the Ramanujan Institute, No. 2. MR0435301 (55 #8261)
43. Wilhelm Magnus, *Residually finite groups*, Bull. Amer. Math. Soc. **75** (1969), 305–316. MR0241525 (39 #2865)
44. Gregori A. Margulis, *Discrete subgroups of semisimple Lie groups*, Ergebnisse der Mathematik und ihrer Grenzgebiete (3), vol. 17, Springer, Berlin, 1991. MR1090825 (92h:22021)
45. Bernard Maskit, *Kleinian groups*, Grundlehren der Mathematischen Wissenschaften, vol. 287, Springer, Berlin, 1988. MR959135 (90a:30132)
46. Calvin C. Moore, *Compactifications of symmetric spaces*, Amer. J. Math. **86** (1964), 201–218. MR0161942 (28 #5146)
47. Nicolas Monod and Yehuda Shalom, *Cocycle superrigidity and bounded cohomology for negatively curved spaces*, J. Differential Geom. **67** (2004), no. 3, 395455. MR2153026 (2006g:53051)
48. Gregori A. Margulis and Èrnest B. Vinberg, *Some linear groups virtually having a free quotient*, J. Lie Theory **10** (2000), no. 1, 171–180. MR1748082 (2001h:22016)
49. Guennadi A. Noskov and Èrnest B. Vinberg, *Strong Tits alternative for subgroups of Coxeter groups*, J. Lie Theory **12** (2002), no. 1, 259–264. MR1885045 (2002k:20072)

50. Gopal Prasad, *Volumes of S-arithmetic quotients of semi-simple groups*, Inst. Hautes Études Sci. Publ. Math. (1989), no. 69, 91–117, With an appendix by Moshe Jarden and the author. MR1019962 (91c:22023)
51. Diego Rattaggi, *A finitely presented torsion-free simple group*, J. Group Theory **10** (2007), no. 3, 363–371. MR2320973 (2008c:20055)
52. Bertrand Rémy, *Construction de réseaux en théorie de Kac-Moody*, C. R. Acad. Sci. Paris Sér. I Math. **329** (1999), no. 6, 475–478. MR1715140 (2001d:20028)
53. Bertrand Rémy, *Classical and non-linearity properties of Kac-Moody lattices*, Rigidity in dynamics and geometry (Cambridge, 2000), Springer, Berlin, 2002, pp. 391–406. MR1919413 (2003g:20071)
54. Bertrand Rémy, *Groupes de Kac-Moody déployés et presque déployés*, Astérisque (2002), no. 277, viii+348. MR1909671 (2003d:20036)
55. Bertrand Rémy, *Immeubles de Kac-Moody hyperboliques, groupes non isomorphes de même immeuble*, Geom. Dedicata **90** (2002), 29–44. MR1898149 (2003c:20033)
56. Bertrand Rémy, *Topological simplicity, commensurator super-rigidity and non-linearities of Kac-Moody groups*, Geom. Funct. Anal. **14** (2004), no. 4, 810–852, With an appendix by P. Bonvin. MR2084981 (2005g:22024)
57. Bertrand Rémy, *Integrability of induction cocycles for Kac-Moody groups*, Math. Ann. **333** (2005), no. 1, 29–43. MR2169827 (2006k:22018)
58. Bertrand Rémy, *Kac-Moody groups as discrete groups*, Essays in geometric group theory, Ramanujan Math. Soc. Lect. Notes Ser., vol. 9, Ramanujan Math. Soc., Mysore, 2009, pp.105–124. MR2605357
59. Guy Rousseau, *Euclidean buildings*, Géométries à courbure négative ou nulle, groupes discrets et rigidités (Paris), Sémin. Congr., vol. 18, Soc. Math. France, 2009, pp. 77–116. MR2655310
60. Bertrand Rémy and Mark Ronan, *Topological groups of Kac-Moody type, right-angled twinnings and their lattices*, Comment. Math. Helv. **81** (2006), no. 1, 191–219. MR2208804 (2007b:20063)
61. Ichirô Satake, *On representations and compactifications of symmetric Riemannian spaces*, Ann. of Math. (2) **71** (1960), 77–110. MR0118775 (22 #9546)
62. Ichirô Satake, *Theory of spherical functions on reductive algebraic groups over $\mathfrak{p}$-adic fields*, Inst. Hautes Études Sci. Publ. Math. (1963), no. 18, 5–69. MR0195863 (33 #4059)
63. Yehuda Shalom, *Rigidity of commensurators and irreducible lattices*, Invent. Math. **141** (2000), no. 1, 1–54. MR1767270 (2001k:22022)
64. Robert Steinberg, *Lectures on Chevalley groups*, Yale University, New Haven, Conn., 1968, Notes prepared by John Faulkner and Robert Wilson. MR0466335 (57 #6215)
65. Anne Thomas, *Lattices acting on right-angled buildings*, Algebr. Geom. Topol. **6** (2006), 1215–1238. MR2253444 (2008b:22003)
66. Anne Thomas, *On the set of covolumes of lattices for Fuchsian buildings*, C. R. Math. Acad. Sci. Paris **344** (2007), no. 4, 215–218. MR2292989 (2008d:20053)
67. Jacques Tits, *Uniqueness and presentation of Kac-Moody groups over fields*, J. Algebra **105** (1987), no. 2, 542–573. MR873684 (89b:17020)
68. Jacques Tits, *Groupes associés aux algèbres de Kac-Moody*, Astérisque (1989), no. 177-178, Exp. No. 700, 7–31, Séminaire Bourbaki, Vol. 1988/89. MR1040566 (91c:22034)
69. André Weil, *L'intégration dans les groupes topologiques et ses applications*, Actual. Sci. Ind., no. 869, Hermann, 1940. MR0005741 (3,198b)
70. George Willis, *The structure of totally disconnected, locally compact groups*, Math. Ann. **300** (1994), no. 2, 341–363. MR1299067 (95j:22010)
71. Xiangdong Xie, *Quasi-isometric rigidity of Fuchsian buildings*, Topology **45** (2006), no. 1, 101–169. MR2170496 (2006m:53064)
72. Efim Zelmanov, *Infinite algebras and pro-p groups*, Infinite groups: geometric, combinatorial and dynamical aspects, Progr. Math., vol. 248, Birkhäuser, Basel, 2005, pp. 403–413. MR2195460 (2006k:20053)
73. Robert J. Zimmer, *Ergodic theory and semisimple groups*, Monographs in Mathematics, vol. 81, Birkhäuser, Basel, 1984. MR776417 (86j:22014)

Some Equations Over Finite Fields Related to Simple Groups of Suzuki and Ree Types

N.S. Narasimha Sastry

In the memory of Shri. M.N.Shamiahgaru and
Shri. M.K.Subbaiahgaru

Abstract We point out some common features (many valid for all admissible infinite fields) of Suzuki and Ree groups (Propositions 1.3, 1.4 and Sect. 3.C,E,F), leading to a proof of the uniqueness, up to conjugacy, of ${}^2B_2(2^{2n+1})$ in $B_2(2^{2n+1})$, ${}^2G_2(3^{2n+1})$ in $G_2(3^{2n+1})$ and the possibilities for the structure of the normalizer in $F_4(2^{2n+1})$ of the centre of a Sylow 2-subgroup of ${}^2F_4(2^{2m+1})$.

Subject Classifications: 20E, 20G

Keywords Suzuki groups • Ree groups

1 Introduction

A. Let X denote one of the finite Chevalley groups $B_2(q)$, $G_2(q)$ and $F_4(q)$, where $q = p^m, m \geq 3$. Throughout, p is understood to be 3 if $X = G_2(q)$ and 2 otherwise. When m is odd, X admits an automorphism σ of order two (composition of the graph and the field automorphisms of X) such that the set X^+ of the σ-fixed points in X is a simple group: the Suzuki group ${}^2B_2(q)$ if $X = B_2(q)$, the Ree group ${}^2G_2(q)$ if $X = G_2(q)$ and the Ree group ${}^2F_4(q)$ if $X = F_4(q)$. Our notation, taken mostly from [5, 6, 8], is described in Sects. 1.C and 3.A. For $A \subseteq X$, we denote by A^+ the set $\{a \in A : \sigma(a) = a\}$. The equations and the automorphisms of $\mathbb{F}_q$ discussed in this paper and stated below occur in an effort to characterize

N.S.N. Sastry (✉)
Division of Theoretical Statistics and Mathematics, Indian Statistical Institute, Bangalore, Karnataka, India
e-mail: nsastry@isibang.ac.in

N.S.N. Sastry (ed.), *Buildings, Finite Geometries and Groups*, Springer Proceedings in Mathematics 10, DOI 10.1007/978-1-4614-0709-6_12,

the normalizer $N_{X^+}(Z(S))$ in X^+ of the centre $Z(S)$ of a Sylow p-subgroup S of X^+ to prove the uniqueness of X^+ as a subgroup of X up to conjugacy in X. Conjugacy is well-known if $X = B_2(q)$, for $X = G_2(q)$ it is proved by Kleidmann and Magliore (see [4]); and it is perhaps well-known for $X = F_4(q)$. A proof for the case $X = F_4(q)$ was in fact outlined to me independently by Professors Peter Sin and Kay Magaard. However, our approach here is to use some common features of the groups $B_2(q)$, $G_2(q)$ and $F_4(q)$, where q in each case throughout is as specified above: the structure of S (Proposition 1.3) and the action by conjugation of L^+ on the unipotent radical $R = O_p(P)$ of the parabolic subgroup $P = N_X(Z(U))$ of X. Here, $U \in Syl_p(X)$ and L is a σ-invariant complement of R in P. See Lemma 2.2 for a description of the σ-invariant subgroups P, S, R and L of X. A main step here is the determination of: (1) the automorphisms θ mentioned in (B) below (Propositions 1.1 and 1.2, and [2]); and (2) the L^+-invariant subgroups of $O_p(P)$ (see Proposition 3.8 for the case $X = F_4(q)$). A consequence of (ii) useful for the case $X = F_4(q)$ is that if M is an L^+-invariant subgroup of $O_p(P)$ such that $Z(M) \neq Z(U)$ and the exponent of M and that of $O_p(P)^+$ are equal, then M is conjugate to $O_p(P)^+$. Richardson, et al. [7] have studied the structure of the unipotent radical of parabolic subgroups as a module over its Levi-factor for some groups of Lie type with abelian unipotent radical.

B. We now state the equations and the field automorphisms mentioned in the title. For a field K, $K^\times$ denotes $K\backslash\{0\}$.

($^2B_2$) θ is an automorphism of $\mathbb{F}_{2^m}$ such that for all $\lambda, \mu \in \mathbb{F}_{2^m}^\times$ with $\lambda + \mu \neq 0 \neq \lambda\mu$,

$$\left(\lambda^{1-\theta} + \mu^{1-\theta}\right)\left(\lambda^{2\theta-1} + \mu^{2\theta-1}\right) \neq 0 \tag{1}$$

and

$$\left(\lambda^{1-\theta} + \mu^{1-\theta}\right)^{2\theta-1} = \left(\lambda^{2\theta-1} + \mu^{2\theta-1}\right)^{1-\theta} \tag{2}$$

holds.

($^2G_2$) θ is an automorphism of $\mathbb{F}_{3^m}$ such that for all $\lambda, \mu \in \mathbb{F}_{3^m}^\times$ with $\lambda + \mu \neq 0 \neq \lambda\mu$,

$$\left(\lambda^{2\theta-1} + \mu^{2\theta-1}\right)\left(\lambda^{2-3\theta} + \mu^{2-3\theta}\right) \neq 0 \tag{3}$$

and

$$\left(\lambda^{2\theta-1} + \mu^{2\theta-1}\right)^{2-3\theta} = \left(\lambda^{2-3\theta} + \mu^{2-3\theta}\right)^{2\theta-1} \tag{4}$$

holds.

($^2F_4$) θ is a field automorphism of $\mathbb{F}_{2^m}$ such that there exists $z_i \in \mathbb{F}_{2^m}^\times$, $i = 1, 2, 3, 4$, such that, $z_1z_2z_3z_4 \neq 0$ and, for all $k = 1, 2, 3, \ldots$, the following hold:

$$z_1 = (z_1z_2)^\theta \tag{5}$$

$$z_1z_2^{-1}z_3z_4 = (z_2z_3)^{2\theta} \quad \text{or} \quad z_1z_2z_3^{-1}z_4^{-1} = (z_2z_3)^{2\theta} \tag{6}$$

$$(1+(z_1z_2^{-1}z_3^{-1}z_4)^k)\cdot(1+(z_2z_4^{-1})^k)\cdot(1+z_3^k)$$
$$=(1+(z_1z_2^{-1}z_3^{-1}z_4)^k)^\theta\cdot(1+(z_2z_4^{-1})^k)^{2\theta}\cdot(1+(z_3z_4)^k)^\theta \tag{7}$$

$$(1+z_4^k)^3\cdot(1+(z_2z_3^{-1})^k)\cdot(1+(z_3z_4^{-1})^{2k})\cdot(1+(z_1z_2^{-1}z_3^{-1}z_4^{-1})^k)$$
$$=(1+z_4^{4k})^\theta\cdot(1+(z_3z_4^{-1})^k)^{3\theta}\cdot(1+(z_2^2z_3^{-2})^k)^\theta\cdot(1+(z_1z_2^{-1}z_3^{-1}z_4^{-1})^k)^\theta \tag{8}$$

We prove in Sects. 4.1 and 4.2 that the only automorphisms satisfying the conditions in $(^2B_2)$ and $(^2G_2)$ are the Tits automorphisms:

Proposition 1.1. *Let K be a field of characteristic 2 and θ be an automorphism of K such that (1) and (2) of $(^2B_2)$ hold. Then, $2\theta^2 =$ Identity $\notin \{\theta, 2\theta\}$.*

Proposition 1.2. *Let K be a field of characteristic 3 and θ be an automorphism of K such that (3) and (4) of $(^2G_2)$ hold. Then, $3\theta^2 =$ Identity $\notin \{\theta, 3\theta\}$.*

Professor Bombieri has determined the set of all 5-tuples $(\theta; z_1, z_2, z_3, z_4)$ satisfying the conditions in $(^2F_4)$ for a field of characteristic 2 [2]. I am indebted to him for his interest in this problem.

C. Notation: Keeping the notation used in Sect. 1.A, we describe the rest of the notation used in this paper [5,6]. See also Sect. 3.A. I suggest that, instead of reading this section, the reader look up the notation as and when required.

(i) Let Σ denote a root system for X, Δ a system of simple roots in Σ and Σ^+ the set of all positive roots in Σ with respect to Δ. Let $\{e_1, \cdots, e_n\}$ be an orthonormal basis of the Euclidean space $\mathbb{R}^n$. We write the inner product in $\mathbb{R}^n$ as $(-,-)$ and choose the root system Σ for X as : $\{\pm e_1, \pm e_2, \pm e_1 \pm e_2\} \subseteq \mathbb{R}^2$ if $X = B_2(q)$; $\{\pm(e_i - e_j), \pm(2e_i - e_j - e_k) : i < j, \{i, j, k\} = \{1, 2, 3\}\} \subseteq (e_1 + e_2 + e_3)^\perp \subseteq \mathbb{R}^3$ if $X = G_2(q)$; and $\{\pm e_i, (\pm e_1 \pm e_2 \pm e_3 \pm e_4)/2, \pm e_i \pm e_j : 1 \le i < j \le 4\} \subseteq \mathbb{R}^4$ if $X = F_4(q)$. We choose the fundamental system of roots Δ in Σ to be: $\{e_2, e_1 - e_2\}$ if $X = B_2(q)$, $\{a = e_1 - e_2, b = -2e_1 + e_2 + e_3\}$ if $X = G_2(q)$ and $\{\alpha_1 = e_2 - e_3, \alpha_2 = e_3 - e_4, \alpha_3 = e_4, \alpha_4 = (e_1 - e_2 - e_3 - e_4)/2\}$ if $X = F_4(q)$. When confusion is unlikely, we abbreviate $\pm e_i, \pm e_i \pm e_j$ and $(\pm e_1 \pm e_2 \pm e_3 \pm e_4)/2$ as $\pm i, \pm i \pm j$ and $\pm 1 \pm 2 \pm 3 \pm 4$ respectively. Let $s = \Sigma_{\alpha\in\Delta} n_\alpha \alpha$ (respectively $\ell = \Sigma_{\alpha\in\Delta} N_\alpha \alpha$), $n_\alpha, N_\alpha \in \mathbb{N}$, be the root of highest weight among the short (respectively, long) roots in Σ. Thus, if $X = B_2(q)$, then $(s, \ell) = (e_1, e_1 + e_2)$, $(N_2, N_{1-2}) = (2, 1)$ and $(n_2, n_{1-2}) = (1, 1)$; if $X = G_2(q)$, then $(s, \ell) = (2a+b, 3a+2b)$, $(N_a, N_b) = (3, 2)$ and $(n_a, n_b) = (2, 1)$; and if $X = F_4(q)$, then $(s, \ell) = (e_1, e_1 + e_2)$, $(N_{2-3}, N_{3-4}, N_4, N_{1-2-3-4}) = (2, 3, 4, 2)$ and $(n_{2-3}, n_{3-4}, n_4, n_{1-2-3-4}) = (1, 2, 3, 2)$.

(ii) For $r, s \in \Sigma$, define $r(s) = 2(r, s)/(s, s)$. Extend this linearly in r and s to define $\alpha(\beta)$ for $\alpha, \beta \in \mathbb{Z}\Sigma$. Let $\tilde{w}_r$ be the permutation of Σ of order 2 taking $s \in \Sigma$ to $s - s(r)r$. For $r \in \Sigma$ and $z \in \mathbb{F}_q^\times := \mathbb{F}_q \backslash \{0\}$, let $\chi_{r,z}$

$\in \mathrm{Hom}_{\mathbb{Z}}(\mathbb{Z}\Sigma, \mathbb{F}_q^\times)$ denote the map taking $u \in \mathbb{Z}\Sigma$ to $z^{u(r)}$, $\mathcal{H}_r = \langle \chi_{r,z} : z \in \mathbb{F}_q^\times \rangle \simeq \mathbb{F}_q^\times$ and $\mathcal{H} = \Sigma_{r \in \Sigma} \mathcal{H}_r \subseteq \mathrm{Hom}_{\mathbb{Z}}(\mathbb{Z}\Sigma, \mathbb{F}_q^\times)$. Clearly, $\mathcal{H} \simeq (\mathbb{F}_q^\times)^{|\Delta|}$. Let $\tilde{W} = \langle \tilde{w}_r : r \in \Sigma \rangle$, the Weyl group of Σ, and $\widetilde{w}_0$ denote the longest element of $\tilde{W}$. Then, $\widetilde{w}_0(r) = -r$ for each $r \in \Sigma$.

(iii) Let $\varphi_r : SL(2, \mathbb{F}_q) \to X$ be the group homomorphism taking $\left(\begin{smallmatrix} 1 & t \\ 0 & 1 \end{smallmatrix}\right)$ to $x_r(t)$ and $\left(\begin{smallmatrix} 1 & 0 \\ t & 1 \end{smallmatrix}\right)$ to $x_{-r}(t)$, $t \in \mathbb{F}_q$. Let $w_r := \varphi_r\left(\left(\begin{smallmatrix} 0 & 1 \\ -1 & 0 \end{smallmatrix}\right)\right)$. Then, $X_r = \{x_r(t) : t \in \mathbb{F}_q\} \simeq (\mathbb{F}_q, +)$, $U := \Pi_{r \in \Sigma^+} X_r \in Syl_p(X)$ and $W = \langle w_r : r \in \Sigma \rangle$ is the Weyl group of X.

In a product $\Pi_{\alpha \in \Lambda} y_\alpha$ of a set of elements or subgroups y_α indexed by a subset Λ of Σ^+, the elements of Λ are always written in the increasing order with respect to the partial order '$\prec$' on Σ defined by Δ. With this convention, for $u \in U$ and $\alpha \in \Sigma^+$, we define $u_\alpha \in \mathbb{F}_q$ by $u = \Pi_{\alpha \in \Sigma^+} x_\alpha(u_\alpha)$. Further, for any function f from Λ to $\mathbb{F}_q$, we denote by $f * \Lambda$ the element u of U with $u_\alpha = f(\alpha)$.

Let $h : \mathcal{H}_r \to X$ be the homomorphism such that $h(\chi_{r,z}) = \varphi_r\left(\left(\begin{smallmatrix} z & 0 \\ 0 & z^{-1} \end{smallmatrix}\right)\right)$ for all $z \in \mathbb{F}_q^\times$. Then, $H_r := h(\mathcal{H}_r) \simeq \mathbb{F}_q^\times$, $((\mathbb{F}_q^\times)^{|\Delta|} \simeq) H := \Pi_{r \in \Sigma} H_r < X$ and $X = UHWU$. For $r, s \in \Sigma, \chi \in \mathrm{Hom}_{\mathbb{Z}}(\mathbb{Z}\Sigma, \mathbb{F}_q^\times), t \in \mathbb{F}_q$ and $w \in W$, the following hold: $h(\chi) x_r(t)\, h(\chi)^{-1} = x_r(\chi(r)t)$, $wh(\chi)w^{-1} = h(\chi^1)$ where $\chi^1(s) = \chi(w^{-1}(s))$, and $w_r\, x_s(t)\, w_r^{-1} = x_{w_r(s)}(\eta_{r,s} t)$ where $\eta_{r,s} \in \{1, -1\}$, $\eta_{r,s}$ depends only on Σ and is as in [6], p. 442.

(iv) **The Ree-Tits Automorphism** σ of X**:** Let $\theta \in \mathrm{Aut}(\mathbb{F}_q)$ be such that $p\theta^2 = Identity$ on $\mathbb{F}_q$. Let $\Psi \in \mathrm{End}_{\mathbb{Z}}(\mathbb{Z}\Sigma)$ be the map taking e_i to $e_i + e_{i+1}$ for $i \in \{1, 3\}$ and e_i to $e_{i-1} - e_i$ for $i \in \{2, 4\}$ if $X = B_2(q)$ or $F_4(q)$; and taking a to b and b to $3a$ if $X = G_2(q)$. Let $^-$ denote the permutation of Σ of order 2 defined by $\Psi(r) = \lambda(r)\, \bar{r}$, where $\lambda(r) = 1$ if $r \in \Sigma$ is a short root and $\lambda(r) = p$ if r is a long root. Then, $\lambda(r)\, \lambda(\bar{r}) = p$. The involution σ in $\mathrm{Aut}(X)$ of interest in this paper is defined by setting $x_r(t)^\sigma = x_{\bar{r}}(t^{\lambda(\bar{r})\theta})$ for all $r \in \Sigma$ and $t \in \mathbb{F}_q$. It satisfies $w_r^\sigma = w_{\bar{r}}$ and $h(\chi)^\sigma = h(\bar{\chi})$, where $\bar{\chi} \in \mathcal{H}$ is defined by $\bar{\chi}(r) = \chi(\Psi(r))^\theta = \chi(\bar{r})^{\lambda(r)\theta}$.

(v) **Additional notation for** $F_4(q)$: For distinct $i, j \in \{1, 2, 3, 4\}$, define $\Sigma_{i,j} = (\mathbb{Z}e_i + \mathbb{Z}e_j) \cap \Sigma$ and $\Sigma_{i,j}^+ = \Sigma^+ \cap \Sigma_{i,j}$; the subgroup $F_{i,j}(q) = \langle X_r : r \in \Sigma_{i,j} \rangle$ of X and its subgroups $U_{i,j} = \langle X_r : r \in \Sigma_{i,j}^+ \rangle$, $W_{i,j} = \langle w_r : r \in \Sigma_{i,j} \rangle$ and $H_{i,j} = \langle h(\chi) \in H : \chi(e_t) = 1 \text{ for } t \in \{1, 2, 3, 4\} \backslash \{i, j\} \rangle$. Note that $F_{1,2}(q) \simeq B_2(q) \simeq F_{3,4}(q)$ and $F_{1,3}(q) \simeq A_1(q) \times A_1(q) \simeq F_{2,4}(q)$. We write $\tilde{W}_{i,j} = \langle \tilde{w}_i, \tilde{w}_j \rangle < \tilde{W}$. Let $\tilde{w}_{i,j}$ denote the element of $\tilde{W}_{i,j}$ which takes e_t, $t \in \{1, 2, 3, 4\}$, to $-e_t$ or e_t according as $t \in \{i, j\}$ or not. Let $w_{i,j}$ denote the element of $W_{i,j}$ corresponding to $\tilde{w}_{i,j}$.

For $\chi \in \mathcal{H}$, we write $h(\chi(e_1), \chi(e_2), \chi(e_3), \chi(e_4))$ for $h(\chi)$; for $z \in \mathbb{F}_q^\times$, we write $h\left(z, z^{2\theta-1}, 1, 1\right)$ as $h_{1,2}^+(z)$; $h\left(z, z^{-2\theta-1}, 1, 1\right)$ as $h_{1,2}^-(z)$; $h\left(1, 1, z, z^{2\theta-1}\right)$ as $h_{3,4}^+(z)$; and $h\left(1, 1, z, z^{-2\theta-1}\right)$ as $h_{3,4}^-(z)$. For $i \in \{1, 3\}$, let $H_{i,i+1}^+ = \{h_{i,i+1}^+(z) : z \in \mathbb{F}_q^\times\}$ and $H_{i,i+1}^- = \{h_{i,i+1}^-(z) : z \in \mathbb{F}_q^\times\}$. Note that $H = H_{1,2} H_{3,4}$ and $H_{i,i+1} = H_{i,i+1}^+ H_{i,i+1}^-$. Our notation in group theory is standard. In particular, C_n denotes the cyclic group of order n. For $x, y \in X$ and $A \subseteq X$, we write $x^y = y^{-1}xy$, $A^x = x^{-1}Ax$ and $[x, y] = x^{-1}y^{-1}xy$. For an abelian group A, $A^\times$ denotes the set of its nontrivial elements.

D. This paper is a part of an effort to identify minimum information about the structure and the embedding of a group of Lie type (here , X^+) which is sufficient to determine the possibilities of its embedding, up to conjugacy, as a maximal subgroup in a group of Lie type of larger dimension (here, X). The next two propositions motivate the Hypothesis $\mathcal{A}$ in Theorem 1.5 and the hypothesis of Proposition 5.2 which characterize $N_{X^+}(Z(S))$ as a subgroup of X, the main step in proving the uniqueness of the embedding of X^+ in X.

Proposition 1.3. *(a) The centre $Z(U)$ of U is X_sX_ℓ and that of U^+ is $(X_sX_\ell)^+$.*
(b) For $u = \Pi_{\alpha\in\Sigma^+}x_\alpha(u_\alpha) \in U$ with $u_\alpha \in \mathbb{F}_q$,

$$u^{p^{|\Delta|-1}} = x_s\left(\Pi_{\alpha\in\Delta}(-u_\alpha)^{n_\alpha}\right)\, x_\ell\left(\Pi_{\alpha\in\Delta}(-u_\alpha)^{N_\alpha}\right).$$

(c) Let $u\in U$ be as in (b) and $\lambda, \mu \in \mathbb{F}_q^\times$. Then, $u^{p^{|\Delta|-1}} = x_s(\lambda)\, x_\ell(\mu)$ if, and only if, $(u_2, u_{1-2}) = (\lambda^{-1}\mu, \lambda^2\mu^{-1})$ if $X = B_2(q)$, $(u_a, u_b) = (-\lambda^2\mu^{-1}, -\lambda^{-3}\mu^2)$ if $X = G_2(q)$ and $(u_{2-3}, u_{3-4}, u_4, u_{1-2-3-4}) = (\mu\lambda^{-1}x^{-1}, y^{-1}x^2, yx^{-1}, (\lambda^2\mu^{-1}y^{-1})^{1/2})$ with arbitrary $x, y \in \mathbb{F}_q$ if $X = F_4(q)$. In particular, $\{u^{p^{|\Delta|-1}} : u\in U\} = \{1\} \cup X_\ell^\times X_s^\times$ and $\{u^{p^{|\Delta|-1}} : u\in U^+\} = Z(U^+)$. Further, the exponents of U and U^+ are equal and to $p^{|\Delta|}$.
(d) The subgroup $H_\ell H_s$ (respectively $(H_\ell H_s)^+$) of H (respectively H^+) normalizes U (respectively U^+) and acts regularly on $X_\ell^\times X_s^\times$ (respectively $Z(U^+)\backslash\{1\}$).

A long, but a routine, calculations using Lemma 2.1 yield (a) and (b). (c) is a direct consequence of (b). Relations at the end of Sect. 1.C.iii yield (d).

Proposition 1.4. *Let D be a subgroup of X isomorphic to X^+ and $S \in Syl_p(D)$. Then: (a) $|S| = q^{|\Sigma^+|/2}$; (b) $Z(S) = \{x^{p^{|\Delta|-1}} : x \in S\} \simeq (\mathbb{F}_q, +)$; (c) S contains a subgroup of order $|R^+|$ which is normalized by a subgroup $K \simeq L^+$ of D.*

For $D = X^+$, (a) is well-known (see [5, 6, 8]); (b) follows from Proposition 1.3 (b). For (c), see Lemma 2.2. In (c), $|R^+| = q^2, q^3$or q^{10} according as $X = B_2(q), G_2(q)$ or $F_4(q)$. The properties (a), (b), (c) above determine S up to conjugacy in X (Proposition 5.2).

We collect in Hypothesis ($\mathcal{A}$) below information about the structure of X^+ and its embedding in X to prove the uniqueness of X^+ in X up to conjugacy.

Let D be a subgroup of X.

Hypothesis ($\mathcal{A}$) on D: D contains subgroups S and T such that ($\mathcal{A}_1$)–($\mathcal{A}_7$) stated below hold:

($\mathcal{A}_1$) $|S| = q^{|\Sigma^+|/2}$, $T = M \times N \leq N_D(S)$, $M \simeq C_{q-1}$, and $N \simeq C_{q-1}$ if $X = F_4(q)$ and 1 if $X \in \{B_2(q), G_2(q)\}$

($\mathcal{A}_2$) $|S : \Phi(S)| = q^{|\Delta|/2}$ ($\mathcal{A}_3$) $Z(S) = \{x^{p^{|\Delta|-1}} : x \in S\}$

($\mathcal{A}_4$) The action of M on $Z(S)^{\times}$ by conjugation is regular (in particular, $|Z(S)| = q$)

($\mathcal{A}_5$) $N = C_T(z)$ for some $z \in Z(S)^{\times}$. Further, the action of N by conjugation on the set $\{s\Phi(S) : s \in S \text{ and } s^{p^{|\Delta|-1}} = z\}$ is regular.

($\mathcal{A}_6$) In the case $X = F_4(q)$, $C_D(M)$ contains a subgroup Q of order q^2 and an involution $y \in N_D(N)$ such that: (i) $Z(Q) = \{a^2 : a \in Q\}$ and is of order q, (ii) N acts transitively on $Z(Q)\backslash\{1\}$ and (iii) Q is a Sylow 2– subgroup of $K = \langle Q, N, y\rangle$.

Set $\tilde{K} = \langle K, T\rangle$ if $X = F_4(q)$ and $\tilde{K} = T$ if $X \in \{B_2(q),\ G_2(q)\}$.

($\mathcal{A}_7$) $\tilde{K}$ normalizes a subgroup of S of order $|R^+|$ and $S\tilde{K}$ is a subgroup of X.

Remark. $X^+ \leq X$ satisfies ($\mathcal{A}$): take $S = U^+$ for all X; $M = H^+$ and $N = \langle 1\rangle$ if $X \in \{B_2(q), G_2(q)\}$ and $M = H^+_{1,2}$, $N = H^+_{3,4}$, $Q = U^+_{3,4}$, $y = w_{3,4}$, and so $K = F^+_{3,4}(q)$ and $\widetilde{K} = F^+_{3,4}(q)\, H^+_{1,2}$, if $X = F_4(q)$.

We now state the main theorem of this paper and prove it in Sect. 4.3. It is a step towards a characterization of X^+ in X, up to conjugacy, using the structure of X^+ and its embedding in X.

Theorem 1.5. *Let D, S and T be subgroups of X such that ($\mathcal{A}_1$)–($\mathcal{A}_5$) of hypothesis ($\mathcal{A}$) hold.*

(a) There exists $g \in X$ and an automorphism θ of $\mathbb{F}_q$ such that $S^g \subseteq U$ and $T^g \subseteq H$.

(b) Assume further that $S \subseteq U$ and $T \subseteq H$. Then,

(i) $Z(S)^g = \{x_s(t^\theta)\, x_l(t) : t \in \mathbb{F}_q\} = \{\left(u^{p^{|\Delta|-1}}\right)^h : h \in T\}$ for some $u \in S$.

(ii) $Z(S)^g = \Phi(S)^g$ and $S^g = \cup_{t\in T} u^t \Phi(S^g)$

(c) Let $X = F_4(q)$, $h(\chi) \in T^g$ and $\chi(e_i) = z_i, i = 1, 2, 3, 4$. Then the 5-tuple $(\theta; z_1, z_2, z_3, z_4)$ satisfies the conditions stated in ($^2F_4$) of Sect. 1.A.

(d) Let $X \in \{B_2(q),\ G_2(q)\}$. Then, $p\theta^2 =$ Identity and $S^g = U^+$, $T^g = H^+$ and so $(ST)^g = P^+$. Further, if $D \simeq X^+$, then $D^g = X^+$.

See part (**D**) of the proof of Theorem 1.5 in Sect. 4.3 for the reasons to consider the automorphisms of $\mathbb{F}_q$ satisfying the conditions stated in ($^2B_2$) and ($^2G_2$). In Sect. 3, we point out the similarity between the group $U = R$ with operators H^+ in the case $X = B_2(2^{2n+1})$ and the subgroup R of U with operators $F^+_{3,4}(2^{2n+1})$ in the case $X = F_4(2^{2n+1})$, with conjugation action in both the cases. Our effort in this section is largely to explain the comment: the $F^+_{3,4}(q)$-module structure of R is 'obtained' by replacing each root subgroup X_α, $\alpha \in \Sigma^+_{1,2}$, of U in $B_2(q)$, a H^+-module, by the unique indecomposable 5-dimensional $F^+_{3,4}$-module. We end with some remarks in Sect. 5.

2 Preliminaries

Lemma 2.1. *(Chevalley's commutator relations for X [6, p. 443], [5, p. 404]) Let* $X \in \{B_2(q), G_2(q), F_4(q)\}$*;* $r, s \in \Sigma^+$*;* $t, u \in \mathbb{F}_q$ *and the characteristic of* $\mathbb{F}_q$ *as specified in Sect. 1.A. Then,* $[x_r(t), x_s(u)] = 1$ *with the following exceptions:*

(a) If $X = G_2(q)$*, then*

(i) $[x_a(t), x_b(u)] = x_{a+b}(-tu)\, x_{2a+b}(-t^2u)\, x_{3a+b}(t^3u)\, x_{3a+2b}(t^3u^2)$*;*
(ii) $[x_a(t), x_{a+b}(u)] = x_{2a+b}(tu)$*; and*
(iii) $[x_b(t), x_{3a+b}(u)] = x_{3a+2b}(tu)$

(b) If $X \in \{B_2(q), F_4(q)\}$*, then* $[x_r(t), x_s(u)]$ *equals* $X_{r+s}(tu)$ *if* $r + s \in \Sigma^+$ *and* $r, s, r+s$ *are all of equal length; and* $x_{r+s}(tu)x_{2r+s}(t^2u)$ *if* $\lambda(r) = 1, \lambda(s) = 2$ *and* $r + s \in \Sigma^+$*.*

Lemma 2.2. *(a) The subgroup* $P = N_X(X_l X_s)$ *of X is parabolic and* σ*-invariant. Further, P admits a* σ*-invariant complement L of* $R = O_p(P)$*. If* $X = B_2(q)$ *or* $G_2(q)$*, then* $P = UH$*,* $R = U$ *and* $L = H$*. If* $X = F_4(q)$*, then* $P = UHW_{3,4}U = C_X(X_1X_{1+2})\,H_{1,2}$*,* $R = \Pi_{r\in\Sigma^+\setminus\Delta}X_r$ *and* $L = F_{3,4}H_{1,2}$*.*
(b) For each $z \in X_l^{\times}X_s^{\times}$*,* $C_X(z) = C_X(X_lX_s) = U$ *if* $X \in \{B_2(q), G_2(q)\}$ *and* $UH_{3,4}W_{3,4}U = O_2(P) \rtimes F_{3,4}$ *if* $X = F_4(q)$*. For* $1 \neq z \in (X_\ell X_s)^+$*,* $C_{X^+}(z) = C_{X^+}((X_\ell X_s)^+) = (C_X(z))^+$*.*
(c) $N_X((X_lX_s)^+) = C_X(X_lX_s)\,H$ *and* $N_{X^+}((X_lX_s)^+) = C_{X^+}((X_\ell X_s)^+)H^+$*.*
(d) (i) If $X = B_2(2^{2n+1})$*, then* $U^+ = \alpha(\mathbb{F}_q)\,\beta(\mathbb{F}_q)$*, where* $\alpha(\mathbb{F}_q) = \{\alpha(t) : t \in \mathbb{F}_q\}$ *and* $\beta(\mathbb{F}_q) = \{\beta(t) : t \in \mathbb{F}_q\}$*; and, for* $t \in \mathbb{F}_q$*,* $\alpha(t) = x_2(t^\theta)x_{1-2}(t)x_1(t^{\theta+1})$ *and* $\beta(t) = x_1(t^\theta)x_{1+2}(t)$*.*
(ii) If $X = G_2(3^{2n+1})$*, then* $U^+ = \alpha(\mathbb{F}_q)\beta(\mathbb{F}_q)\gamma(\mathbb{F}_q)$*, where* $\alpha(\mathbb{F}_q) = \{\alpha(t) : t \in \mathbb{F}_q\}$*, etc. and, for* $t \in \mathbb{F}_q$*,* $\alpha(t) = x_a(t^\theta)x_b(t)x_{a+b}(t^{\theta+1})x_{2a+b}(t^{2\theta+1})$*,* $\beta(t) = x_{a+b}(t^\theta)x_{3a+b}(t)$ *and* $\gamma(t) = x_{2a+b}(t^\theta)x_{3a+2b}(t)$*.*

In both the cases (i) and (ii), $H^+ = \{h(\chi) \in H : \chi(l)^\theta = \chi(s)\}$*.*

(See [1,8].) The next lemma is verified by explicit calculations using Lemma 2.1.

Lemma 2.3. *Let X and* Σ *be of type* G_2*. Let* $u, v \in U$ *and* $u' \in U$ *be defined by* $u = u' \cdot x_{2a+b}(-u_{2a+b}) \cdot x_{3a+2b}(-u_{3a+2b})$*. Then*

(a) $u'v' = x_a(u_a + v_a) \cdot x_b(u_b + v_b) \cdot x_{a+b}(u_{a+b} + v_{a+b} + u_bv_a) \cdot x_{3a+b}(u_{3a+b} + v_{3a+b} - u_bv_a^3) \cdot x_{2a+b}(v_a^2u_b - u_{a+b}v_a) \cdot x_{3a+2b}\{-v_a^3u_b^2 - v_b(u_{3a+b} - v_a^3u_b)\}$*.*
(b) $(u')^2 = x_a(-u_a) \cdot x_b(-u_b) \cdot x_{a+b}(-u_{a+b} + u_au_b) \cdot x_{3a+b}(-u_{3a+b} - u_bu_a^3) \cdot x_{2a+b}(u_a^2u_b - u_au_{a+b}) \cdot x_{3a+2b}(-u_{3a+b}u_b)$*,*
(c) $u^3 = (u')^3 = x_{2a+b}(-u_a^2u_b) \cdot x_{3a+2b}(-u_a^3u_b^2)$*,*
(d) $(u')^{-1} = x_a(-u_a) \cdot x_b(-u_b) \cdot x_{a+b}(-u_{a+b} + u_au_b) \cdot x_{3a+b}(-u_{3a+b} - u_a^3u_b) \cdot x_{2a+b}(-u_au_{a+b} - u_a^2u_b)\, x_{3a+2b}(-u_{3a+b}u_b + u_a^3u_b^2)$*, and*
(e) $[u, v] = [u', v'] = x_{a+b}(v_au_b - u_av_b) \cdot x_{3a+b}(u_a^3v_b - u_bv_a^3) \cdot x_{2a+b}(u_av_{a+b} - v_au_{a+b} + v_a^2u_b - u_a^2v_b) \cdot x_{3a+2b}(u_bv_{3a+b} - v_bu_{3a+b} + u_av_b(v_a^3 - u_a^3) + u_a^3v_b^2 - v_a^3u_b^2)$*.*

□

3 Structure of R When $X = F_4(q)$

A. Decomposition of R: We partition Σ^+ by its subsets $\{A(\alpha) : \alpha \in \{e_0\} \cup \Sigma^+_{1,2}\}$, where $A(e_0) = \Sigma^+_{3,4}$ and, for $\alpha \in \Sigma^+_{1,2}$, $A(\alpha) = \{\alpha\} \cup \{\lambda(\alpha)^{-1}(\alpha + r) : r \in \Sigma_{3,4}$ with $\lambda(r) = \lambda(\alpha)\}$. So, $A(e_i) = \{e_i\} \cup \{e_i \pm e_j : j \in \{3,4\}\}$ for $i \in \{1,2\}$ and $A(e_1 + \gamma e_2) = \{e_1 + \gamma e_2\} \cup \{(e_1 + \gamma e_2 + \epsilon e_3 + \tau e_4)/2\}$ with $\gamma, \epsilon, \tau \in \{+1, -1\}$. We set $A(e_\infty) = \Sigma^+_{1,2}$. Note that $\overline{A(\alpha)} = A(\overline{\alpha})$ for each $\alpha \in \Sigma^+_{1,2}$. Following [5, p. 406–407], we define the subsets $R_i, i = 1, \cdots, 12$, of Σ^+ by their minimal elements r as follows : R_1, R_4, R_5, R_6 are of the form $\{r, \overline{r}, r + \overline{r}\}$ with $r = 4, 2-4, 2, 2+4$, respectively; and $R_2, R_3, R_i, 7 \le i \le 12$, are of the form $\{r, \overline{r}\}$ with $r = 3+4, 2-3, 2+3, 1-3, 1-4, 1+4, 1+3, 1$, respectively. Note that $\min(R_i) \prec \min(R_j)$ for $1 \le i < j \le 12$.

Consider the subgroups $U(\alpha) = \Pi_{r \in A(\alpha)} X_r$, $\alpha \in \Sigma^+_{1,2} \cup \{e_0, e_\infty\}$, and $R = \Pi_{\alpha \in \Sigma^+_{1,2}} U(\alpha)$ of U; and the subgroups $V_i = \langle \alpha_i(t) : t \in \mathbb{F}_q \rangle$ and $V^{(i)} = V_i V_{i+1} \cdots V_{12}$ $(i = 1, \cdots, 12)$ of U^+, where $\alpha_i(t) = x_s(t^\theta) x_{\overline{s}}(t) x_{s+\overline{s}}(t^{\theta+1})$ if $R_i = \{s, \overline{s}, s + \overline{s}\}$ and $\alpha_i(t) = x_s(t^\theta) x_{\overline{s}}(t)$ if $R_i = \{s, \overline{s}\}$ with $s \in R_i$ being the smallest element in R_i of length 1 (see [5, p. 407]). The following facts about these subgroups are immediate consequences of Chevalley commutator relations (Lemma 2.1(b)).

Lemma 3.1. *(a)* $U = R \rtimes U(e_0)$, $Z(U) = X_1 X_{1+2}$, $R = O_2(C_{F_4(q)}(Z(U))) = \Pi_{\alpha \in \Sigma^+_{1,2}} U(\alpha)$, $U(e_0) \in Syl_2(F_{3,4}(q))$ *and* $U(e_\infty) \in Syl_2(F_{1,2}(q))$. *Further,* $U(\alpha)$ $(\alpha \in \Sigma^+_{1,2})$, $R' = U(e_1)U(e_1 + e_2)$ *and* $R/R' \simeq U(e_2) \times U(e_1 - e_2)$ *are elementary abelian subgroups of orders* q^5, q^{10} *and* q^{10}, *respectively.*

(b) $U^+ = V^{(1)} = R^+ \rtimes U^+_{3,4}$, $R^+ = V^{(3)}$, $(R^+)' = V^{(8)}$, $Z(U^+) = V_{12}$ *and* $V_i' = Z(V_i) = V_k$ *with* $(i,k) \in \{(1,2), (4,8), (6,11), (5,12)\}$. *Further,* V_i *is isomorphic to* $(\mathbb{F}_q, +)$ *if* $|R_i| = 2$ *and to a Sylow* 2*–subgroup of* ${}^2B_2(q)$ *if* $|R_i| = 3$. □

B. $U(\alpha)$ *as a* $O(5, q)$ - Module: Throughout Sect. 3.B, the letter α (respectively, β, f, f', g, r, s) denotes an arbitrary element of $\Sigma^+_{1,2}$ (respectively, of $\{e_2, e_1 - e_2\}$, $\mathbb{F}_q^{A(\alpha)}, \mathbb{F}_q^{A(\beta)}, F_{3,4}(q), A(\alpha), A(\beta)$).

(i) Define $\widetilde{\beta} = \beta + \lambda(\beta)\overline{\beta}$ and $s^* = s + n(s, \beta)\overline{\beta}$ with $n(s, \beta) = 2$ if $s = \beta = e_1 - e_2$ and $n(s, \beta) = 1$ otherwise. Note that the map $x \mapsto \widetilde{x}$ takes e_2 to e_1 and $e_1 - e_2$ to $e_1 + e_2$. The map $y \mapsto y^*$ takes e_2 to e_1, $e_1 - e_2$ to $e_1 + e_2$, $e_2 \pm e_i$ to $e_1 \pm e_i$, $i \in \{3,4\}$, and $(e_1 - e_2 \pm e_3 \pm e_4)/2$ to $(e_1 + e_2 \pm e_3 \pm e_4)/2$. Further, $A(\beta)^* = A(\beta^*)$.

(ii) Define the following abelian group isomorphisms: $^-$ from $\mathbb{F}_q^{A(\alpha)}$ to $\mathbb{F}_q^{A(\overline{\alpha})}$ by $\overline{f}(\overline{r}) = f(r)$; τ_α from $\mathbb{F}_q^{A(\alpha)}$ to $U(\alpha)$ by setting $\tau_\alpha(f)$ to be $\Pi_{r \in A(\alpha)} x_r(f(r))$ if $\lambda(\alpha) = 1$ and as $\Pi_{r \in A(\alpha)} x_r(f(r)^{\lambda(r)\theta})$ if $\lambda(\alpha) = 2$; $\widetilde{\pi}_\beta$ from $\mathbb{F}_q^{A(\beta)}$ to $\mathbb{F}_q^{A(\widetilde{\beta})}$ by $(\widetilde{\pi}_\beta(f'))(\overline{\beta} + s) = f'(s)$; π_β from $U(\beta)$ to $U(\widetilde{\beta})$ by $\pi_\beta(x_s(t_s)) = x_{s^*}(t_s)$. Note that $\pi_{e_2} =: \pi_2$ is the restriction to $U(e_2)$ of the conjugation by $w_{1-2}w_2$ and

$\pi_{e_1-e_2} =: \pi_{1-2}$ is the restriction to $U(e_1 - e_2)$ of the conjugation by $w_2 w_{1-2}$. Then, the diagram

$$\begin{array}{ccccccc}
U(e_2) & \xleftarrow{\ \tau_2\ } & \mathbb{F}_q^{A(e_2)} & \xleftrightarrow{\ -\ } & \mathbb{F}_q^{A(e_1-e_2)} & \xrightarrow{\ \tau_{1-2}\ } & U(e_1-e_2) \\
\downarrow{\scriptstyle \pi_2} & & \downarrow{\scriptstyle \widetilde{\pi}_2} & & \downarrow{\scriptstyle \widetilde{\pi}_{1-2}} & & \downarrow{\scriptstyle \pi_{1-2}} \\
U(e_1) & \xleftarrow{\ \tau_1\ } & \mathbb{F}_q^{A(e_1)} & \xleftrightarrow{\ -\ } & \mathbb{F}_q^{A(e_1+e_2)} & \xrightarrow{\ \tau_{1+2}\ } & U(e_1+e_2)
\end{array}$$

commutes; that is, $\widetilde{\pi}_{\overline{\beta}} \circ {}^- = {}^- \circ \widetilde{\pi}_\beta$ and $\pi_\beta \circ \tau_\beta = \tau_{\overline{\beta}} \circ \widetilde{\pi}_\beta$ on $\mathbb{F}_q^{A(\beta)}$. Note that $\tau_{\overline{\alpha}} \circ {}^- \circ \tau_\alpha^{-1} = \sigma$ on $U(\alpha)$ and $F_{3,4}^\sigma(q) = F_{3,4}(q)$.

Lemma 3.2. *For* $\alpha \in \Sigma_{1,2}^+$, $[X_\alpha, F_{3,4}(q)] = 1$ *and* $[x, F_{3,4}(q)] = [x, F_{3,4}^+(q)] = U(\alpha)$ *for each* $x \in U(\alpha)\backslash X_\alpha$.

This lemma is true because it holds for $\alpha = e_1$ and the diagram above is commutative. Conjugation in $F_4(q)$ by the elements of $H_{1,2}^+ F_{3,4}(q)$ imparts a 5-dimensional indecomposable $\mathbb{F}_q B_2(q)$-module structure on $U(e_1)$ which is unique (and is in fact the natural $\mathbb{F}_q O(5,q)$-module). We now transport this $\mathbb{F}_q B_2(q)$-module structure to $U(\alpha)$ for each $\alpha \in \Sigma_{1,2}^+$.

(iii) Consider $\mathbb{F}_q^{A(\alpha)}$ as a vector space over $\mathbb{F}_q$. We equip $U(\alpha)$ with a $\mathbb{F}_q$- vector space structure with the scalar multiplication '$\circ$' defined, for $z \in \mathbb{F}_q^\times$ and $r \in A(\alpha)$, as

$$z \circ x_r(t) = {}^\alpha h(z)^{-1} \cdot x_r(t) \cdot {}^\alpha h(z) = x_r(({}^\alpha\chi(r))^{-1} t) = x_r(zt) \text{ or } x_r(z^{\lambda(r)\theta} t)$$

according as $\lambda(\alpha) = 1$ or 2. Here, ${}^\alpha h(z) \in H$ where

$$\begin{aligned}
{}^2h(z) &:= h_{1,2}^-(z^{2\theta-1}), \quad {}^{1-2}h(z) := ({}^2h(z))^\sigma = h_{1,2}^-(z^{1-2\theta}) \\
{}^1h(z) &:= {}^{1+2}h(z) := ({}^2h(z))^{w_{1-2}w_2} = h_{1,2}^+(z^{-1}).
\end{aligned}$$

For $\alpha \in \Sigma_{1,2}^+$ and $z \in \mathbb{F}_q^\times$, $({}^\alpha h(z))^\sigma = {}^{\overline{\alpha}}h(z)$. So, $\sigma \in \mathrm{Hom}_{\mathbb{F}_q}(U(\alpha), U(\overline{\alpha}))$ and $\tau_{\overline{\alpha}} \circ {}^- \circ \tau_\alpha^{-1}$ is the restriction of σ to $U(\alpha)$. Also, $\widetilde{\pi}_\beta, \tau_\alpha$ and ${}^-$ are $\mathbb{F}_q$-linear maps. Finally, π_β is $\mathbb{F}_q$- linear because: π_{e_2} (respectively $\pi_{e_1-e_2}$) is the restriction of the conjugation by $w_{1-2}w_2$ to $U(e_2)$ (respectively, by $w_2 w_{1-2}$ to $U(e_{1-2})$); and $({}^2h(z))^{w_{1-2}w_2} = {}^1h(z)$ and $({}^{1-2}h(z))^{w_2 w_{1-2}} = {}^{1+2}h(z)$.

(iv) We define a group homomorphism I_α from $F_{3,4}(q)$ to $\mathrm{Aut}(U(\alpha))$ by setting, for $g \in F_{3,4}(q)$ and $r \in A(\alpha)$, $I_\alpha(g)(x_r(t)) = x_r(t)^g$ or $x_r(t)^{g^\sigma}$ according as $\lambda(\alpha) = 1$ or 2; and a group homomorphism φ_α from $F_{3,4}(q)$ to $\mathrm{Aut}\left(\mathbb{F}_q^{A(\alpha)}\right)$ by setting $\varphi_\alpha(g) = \tau_\alpha^{-1} \circ I_\alpha(g) \circ \tau_\alpha$. Since scalar multiplication on $U(\alpha)$ is conjugation by an element of $H_{1,2}$, $F_{3,4}^\sigma(q) = F_{3,4}(q)$ and $[H_{1,2}, F_{3,4}(q)] = 1$, it follows that τ_α and $I_\alpha(g)$ are $\mathbb{F}_q$-linear. So, $\varphi_\alpha(g)$ is also $\mathbb{F}_q$-linear. Thus, I_α and φ_α are (in fact indecomposable, but not irreducible) $\mathbb{F}_q F_{3,4}(q)$-representations.

(v) The isomorphisms $\tau_\alpha, {}^-, \pi_\beta$ and $\tilde{\pi}_\beta$ in the diagram above are $\mathbb{F}_q F_{3,4}(q)$-module isomorphisms: in fact, for each $g \in F_{3,4}(q)$, $\tau_\alpha \circ \varphi_\alpha(g) = \tau_\alpha \circ \left(\tau_\alpha^{-1} \circ I_\alpha(g) \circ \tau_\alpha\right) = I_\alpha(g) \circ \tau_\alpha$, so $\tau_\alpha \in \mathrm{Hom}_{\mathbb{F}_q F_{3,4}(q)}\left(\mathbb{F}_q^{A(\alpha)}, U(\alpha)\right)$; ${}^- \circ \varphi_\alpha(g) = \varphi_{\overline{\alpha}}(g) \circ {}^-$ on $\mathbb{F}_q^{A(\alpha)}$ (because this is equivalent to $\sigma \circ I_\alpha(g) = I_{\overline{\alpha}}(g) \circ \sigma$ on $U(\alpha)$), so ${}^- \in \mathrm{Hom}_{\mathbb{F}_q F_{3,4}(q)}\left(\mathbb{F}_q^{A(\alpha)}, \mathbb{F}_q^{A(\overline{\alpha})}\right)$; π_{e_2} and $\pi_{e_1-e_2}$ are restrictions of conjugations by $w_{1-2}w_2$ and $w_2 w_{1-2}$ respectively and $[w_2 w_{1-2}, F_{3,4}(q)] = 1$, so $\pi_\beta \circ I_\beta(g) = I_{\widetilde{\beta}}(g) \circ \pi_\beta$ and π_β and $\pi_{\widetilde{\beta}}$ are $\mathbb{F}_q F_{3,4}(q)$-module isomorphisms; and, finally, $\widetilde{\pi}_\beta \circ \varphi_\beta(g) = \widetilde{\pi}_\beta \circ (\tau_\beta^{-1} \circ I_\beta(g) \circ \tau_\beta) = (\tau_{\widetilde{\beta}}^{-1} \circ \pi_\beta) \circ I_\beta(g) \circ \tau_\beta = \tau_{\widetilde{\beta}}^{-1} \circ (I_{\widetilde{\beta}}(g) \circ \pi_\beta) \circ \tau_\beta = \tau_{\widetilde{\beta}}^{-1} \circ I_{\widetilde{\beta}}(g) \circ (\tau_{\widetilde{\beta}} \circ \widetilde{\pi}_\beta) = \varphi_{\widetilde{\beta}}(g) \circ \widetilde{\pi}_\beta$ on $U(\beta)$, so $\widetilde{\pi}_\beta \in \mathrm{Hom}_{\mathbb{F}_q F_{3,4}(q)}\left(\mathbb{F}_q^{A(\beta)}, \mathbb{F}_q^{A(\widetilde{\beta})}\right)$.

(vi) For a given $\alpha \in \Sigma_{1,2}^+$, let $\{s_i, s_i'\}, i \in \{1,2\}$, be the two disjoint 2-subsets of $A(\alpha)$ with $s_i + s_i' = \lambda(\overline{\alpha})\alpha$. The $\mathbb{F}_q$-valued functions q_α' on $\mathbb{F}_q^{A(\alpha)}$ and q_α on $U(\alpha)$ defined for $\underline{t} \in \mathbb{F}_q^{A(\alpha)}$ by

$$q_\alpha'(\underline{t}) = t_\alpha^2 + t_{s_1}t_{s_1'} + t_{s_2}t_{s_2'}; \quad q_\alpha(\underline{t} * A(\alpha)) = q_\alpha'(\underline{t}) \text{ if } \lambda(\alpha) = 1$$

$$q_\alpha(\underline{t} * A(\alpha)) = (t_\alpha + t_{s_1}t_{s_1'} + t_{s_2}t_{s_2'})^{2\theta} \text{ if } \lambda(\alpha) = 2$$

are nondegenerate quadratic forms. For each $\alpha \in \Sigma_{1,2}^+$ and $\beta \in \{e_2, e_1 - e_2\}$, $q_\alpha' = q_{\overline{\alpha}}' \circ {}^- = q_\alpha \circ \tau_\alpha$ on $\mathbb{F}_q^{A(\alpha)}$, $q_\beta' = q_{\widetilde{\beta}}' \circ \widetilde{\pi}_\beta$ on $\mathbb{F}_q^{A(\beta)}$, $q_\alpha = q_{\overline{\alpha}} \circ \sigma$ on $U(\alpha)$ and $q_\beta = q_{\widetilde{\beta}} \circ \pi_\beta$ on $U(\beta)$. Thus, the maps in the previous paragraph are isomorphisms of quadratic spaces.

(vii) Finally, for each $\alpha \in \Sigma_{1,2}^+$, the actions I_α and φ_α leave the quadratic forms invariant; that is, for each $g \in F_{3,4}(q)$, $q_\alpha \circ I_\alpha(g) = q_\alpha$ on $U(\alpha)$ and $q_\alpha' \circ \varphi_\alpha(g) = q_\alpha' \circ \left(\tau_\alpha^{-1} \circ I_\alpha(g) \circ \tau_\alpha\right) = (q_\alpha \circ I_\alpha(g)) \circ \tau_\alpha = q_\alpha'$ on $\mathbb{F}_q^{A(\alpha)}$. Thus, if $O(f)$ denotes the orthogonal group defined by the quadratic form f, then $I_\alpha(g) \in O(q_\alpha)$ and $\varphi_\alpha(g) \in O(q_\alpha')$. Further, if $\hat{\tau}_\alpha : O(q_\alpha') \to O(q_\alpha)$ is the group isomorphism induced by τ_α and $\hat{\pi}_2, \hat{\pi}_{1-2}$ are similarly defined, then the 'triangles' in the diagram commute.

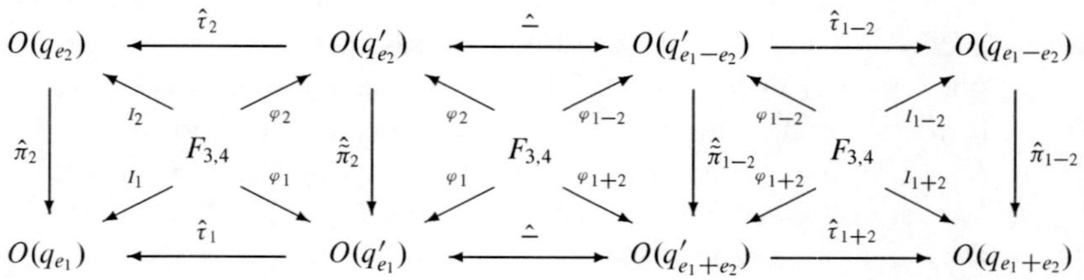

C. Multiplication in $\boldsymbol{R}$**:** From Sect. 3.A, $U = U(e_0)\,U(e_2)\,U(e_1 - e_2)\,U(e_1)\,U(e_1 + e_2)$. So, each $x \in U$ can be written uniquely as $x(e_0)\,x(e_2)\,x(e_1 - e_2)\,x(e_1)\,x(e_1 + e_2)$ with $x(\alpha) \in U(\alpha)$. We call $x(\alpha)$, the *α-component* of x. Consider the maps

$$\lambda : U(e_1+e_2) \times U(e_1-e_2) \to \mathbb{F}_q,\ \eta : U(e_1) \times U(e_2) \to \mathbb{F}_q,$$
$$\mu : U(e_2) \times U(e_1-e_2) \times U(e_1-e_2) \to \mathbb{F}_q$$

defined as follows: we take $i, j \in \{3,4\}, i \neq j$ and $\epsilon, \tau \in \{+,-\}$ in the rest of Sect. 3.C:

$$\lambda(x(e_1+e_2), y(e_1-e_2)) = \Sigma_{\epsilon,\tau}\, x_{1+2+\epsilon 3+\tau 4} y_{1-2-\epsilon 3-\tau 4};$$
$$\eta(x(e_1), y(e_2)) = \Sigma_{i,\epsilon}\, x_{1+\epsilon i} y_{2-\epsilon i}$$
$$\mu z(e_2), x(e_1-e_2), y(e_1-e_2)) = \Sigma z_a \left(x_{a'} y_{b'} + y_{a'} x_{b'}\right),$$

where the last summation is over all ordered pairs $(a, \{a', b'\})$ with $a \in A(e_2)\backslash\{e_2\}$ and $a', b' \in A(e_1-e_2)\backslash\{e_1-e_2\}$ such that $a+a'+b' = e_1$.

If $x, y \in R$ and $z = xy$, then by direct computation using Lemma 2.1.(b), we see that, for $r \in \Sigma^+\backslash\Sigma^+_{3,4}$,

$$z_r = x_r + y_r + C_r + A\delta_{e_1,r} + B\delta_{e_1+e_2,r},$$

where $C = [y(e_2), x(e_1-e_2)]$, $A = \lambda(x(e_1+e_2), y(e_1-e_2)) + \mu(y(e_2), x(e_1-e_2), y(e_1-e_2))$, $B = \eta(x(e_1), y(e_2))$ and δ is the Kronecker's delta. Further, C_r is zero if $r \in U(1-2) \cup U(2)$,

$$y_{2+\epsilon i} x_{1-2} + y_{2-j} x^2_{1-2+\epsilon i+j} + y_{2+j} x^2_{1-2+\epsilon i-j} \text{ if } r = 1+\epsilon i,$$
$$y_2 x_{1-2+\epsilon 3+\tau 4} + y_{2+\epsilon 3} x_{1-2-\epsilon 3+\tau 4} + y_{2+\tau 4} x_{1-2+\epsilon 3-\tau 4} \text{if } r = 1+2+\epsilon 3+\tau 4,$$
$$y_2 q_{e_1-e_2}(x(e_1-e_2))^{2\theta} + \Sigma_{i,\epsilon} y_{2+\epsilon i} x_{1-2-\epsilon i+j} x_{1-2-\epsilon i-j} \text{ if } r = 1 \text{ and}$$
$$q_{e_2}(y(e_2)) x_{1-2} + \Sigma_{\epsilon,\tau} y_{2+\epsilon 3} y_{2+\tau 4} x^2_{1-2-\epsilon 3-\tau 4} \text{ if } r = 1+2.$$

Thus, for each $\alpha \in \Sigma^+_{1,2}$, the α-component of xy is $x(\alpha)\cdot y(\alpha)\cdot C(\alpha)\cdot x_1(\delta_{\alpha,e_1} A)\cdot x_{1+2}(\delta_{\alpha,e_1+e_2} B)$. Compare this with the multiplication of elements in $U_{1,2}$.

D. Exponents of U and U_i For $i = 0, 1, 2, \cdots$, define $\Delta_i = \{s \in \Sigma^+ : (x^{2^i})_s = 0$ for each $x \in U\}$ and $U_i = \langle X_s : s \in \Sigma^+\backslash\Delta_i\rangle \leq U$. Explicit calculations using Lemma 2.1 yield Proposition 3.3 and Corollary 3.4 below.

Proposition 3.3. *(a)* $\Delta_0 = \phi, \Delta_1 = \Delta, \Delta_2 = A(e_0) \cup A(e_2) \cup A(e_1-e_2) \setminus \{r, \overline{r}\}$, *where* $r = (e_1-e_2+e_3+e_4)/2$, $\Delta_3 = \Sigma^+\backslash\{e_1, e_1+e_2\}$ *and* $\Delta_n = \Sigma^+$ *for all integers* $n \geq 4$.

(b) Let $i \in \{0,1,2\}$ *and* $u \in U$ *be such that* $u_s = 0$ *for each* $s \in \Delta_i$.

(i) If $i = 2$, *then* $u^2 = x_1(u_r u_{1-r}) x_{1+2}(u_{\overline{r}} u_{(1+2)-\overline{r}})$ *with* r *as in (a).*
(ii) If $i = 1$, *then* $u^4 = x_1(u_3 u_{2-4} u^2_{1-2-3+4}) x_{1+2}(u_{3+4} u^2_{2-4} u^2_{1-2-3+4})$.
(iii) If $i = 0$, *then* $u^8 = x_1(u_{2-3} u^2_{3-4} u^3_4 u^2_{1-2-3-4}) x_{1+2}(u^2_{2-3} u^3_{3-4} u^4_4 u^2_{1-2-3-4})$.
In particular, $\{u^8 : u \in U\} = \{1\} \cup X_1^\times X_{1+2}^\times$ *and* $\{u^8 : u \in U^+\} = V_{12}$.

(c) *Let* $y, z \in U$. *Assume that* $y_s = z_s = 0$ *for each* $s \in \Delta_i$. *Then,* $(yz)^{2^{3-i}} \in X_1 X_{1+2}$ *and is obtained by replacing* u_β *by* $y_\beta + z_\beta$ *for the roots* β *appearing in the expressions in (b) above.* □

Corollary 3.4. *Let* $u, v \in U, t \in \{0, 1, 2, 3\}$ *and* $z_t = u^{2^t} v^{2^t}$. *Then,* $z_t^{2^{3-t}} = x_1(r_1^{(t)}) x_{1+2}(r_{1+2}^{(t)})$, *where* $r_1^{(t)}, r_{1+2}^{(t)} \in \mathbb{F}_q$ *are obtained as follows: for* $\alpha \in \Sigma^+$, *let* $\xi(\alpha) = u_\alpha + v_\alpha$ *and, for a formal sum* $\nu = \Sigma e_\beta[\beta]$ *with* $e_\beta \in \mathbb{Z}$ *and* $\beta \in \Sigma^+$, *let* $\xi(\nu)$ *denote* $\prod \xi(\beta)^{e_\beta}$. *Then,* $r_1^{(t)} = \xi(\nu_1^{(t)})$ *and* $r_{1+2}^{(t)} = \xi(\nu_{1+2}^{(t)})$, *where*

$$\nu_1^{(3)} = [e_1]\,,\ \nu_{1+2}^{(3)} = [e_1 + e_2];\ \nu_1^{(2)} = [r] + [e_1 - r],$$

$$\nu_{1+2}^{(2)} = [\bar{r}] + [e_1 + e_2 - \bar{r}];\ \nu_1^{(1)} = [e_3] + [e_2 - e_4] + 2[r],$$

$$\nu_{1+2}^{(1)} = [e_3 + e_4] + 2[e_2 - e_4] + 2[(e_1 - e_2 - e_3 + e_4)/2];$$

$$\nu_1^{(0)} = 3[e_4] + 2[e_3 - e_4] + [e_2 - e_3] + 2[(e_1 - e_2 - e_3 - e_4)/2],$$

$$\nu_{1+2}^{(0)} = 4[e_4] + 3[e_3 - e_4] + 2[e_2 - e_3] + 2[(e_1 - e_2 - e_3 - e_4)/2]$$

and r *is as in Proposition 3.3 above.*

E. $\boldsymbol{H_{1,2}^+}$**- Invariant Subgroups of** $\boldsymbol{U_{1,2}}$**:** For convenience of notation, in the case $X = B_2(q)$, we take X as the subgroup $F_{1,2}(q)$ of $F_4(q)$. Thus, the subgroups $H, H^+, W, U, U^+, Z(U), Z(U^+)$ of $B_2(q)$ correspond to $H_{1,2}$, $H_{1,2}^+, W_{1,2}$, $U_{1,2}, V_5$, X_1X_{1+2}, V_{12}, respectively, of $F_{1,2}(q)$. For $z \in \mathbb{F}_q^\times$, let $h_z = h_{1,2}^-(z)$. Consider the following subgroups of $U_{1,2}$ with $z \in \mathbb{F}_q^\times$:

$$M_0^\flat = X_{1+2}, M_\infty^\flat = X_1, M_z^\flat = V_{12}^{h_z^{-1}}, M_z^0 = (V_5)^{h_z^{-1}}$$

$$M_0^\top = X_{1-2}U'_{1,2}, M_\infty^\top = X_2 U'_{1,2}, M_z^\top = M_z^0 X_1 X_{1+2},$$

where V_{12} and V_5 are as defined in Sect. 3.A. Note that $M_1^\flat = V_{12} = (X_1X_{1+2})^+$, $M_1^0 = V_5 = U_{1,2}^+$ and, for $z \in \mathbb{F}_q^\times, \{x^2 : x \in M_z^0\} = M_z^\flat$ and $\{x \in U_{1,2} : x^2 \in M_z^\flat\} = M_z^\top$.

Let $1 \neq a \in U_{1,2}$ and $z \in \mathbb{F}_q^\times$. Clearly, $a \in X_r$ for some $r \in \Sigma_{1,2}^+$ if, and only if, $a_r \neq 0$ and $a_s = 0$ for all $r, s \in \Sigma_{1,2}^+, r \neq s$. Using Lemma 2.1, we verify that $a \in M_z^\flat$ if, and only, if $a_2 = a_{1-2} = 0$ and $a_1 = z^2 a_{1+2}^\theta \neq 0$; and $a \in M_z^\top$ if, and only if, $a_2 a_{1-2} \neq 0$ and $a_2 = z^{-4\theta-2} a_{1-2}^\theta$. Further, if $a \in M_z^\top$, then $a \in M_z^0$ if, and only if, $a_2 a_{1-2} + z^2 a_{1+2}^\theta = a_1$. For distinct $z, z' \in \mathbb{F}_q \cup \{\infty\}$, $M_z^\flat M_{z'}^\flat = X_1 X_{1+2}$ and $M_z^\flat \cap M_{z'}^\flat = 1$. Further, if $z, z' \in \mathbb{F}_q^\times, z \neq z', \{\tau, \eta\} = \{0, \infty\}$ and $\varepsilon, \delta \in \{0, \top\}$, then

$$M_z^0 \subset M_z^\top,\ M_z^\epsilon M_{z'}^\flat = M_z^\epsilon M_\tau^\flat = M_z^\top \text{ and } M_z^\epsilon M_{z'}^\delta = M_z^\epsilon M_\tau^\top = M_\tau^\top M_\eta^\top = U_{1,2}.$$

These observations yield

Proposition 3.5. *Any nonzero $H_{1,2}^{+}$ - invariant subgroup of $U_{1,2}$, under conjugation, is equal to one of: $U_{1,2}$, $X_1, X_{1+2}, M_z^{\flat}, M_z^{\top}$, X_1X_{1+2} or is contained in $M_0^{\top}$ or in $M_{\infty}^{\top}$.* □

F. $F_{3,4}^{+}(q)$ - Invariant Subgroups of R: Consider the following invariant subgroups of R, invariant under the action of $\mathbb{F}_q F_{3,4}^{+}(q)$ described in Sect. 3 (B iii, iv):

$$\mathcal{M}_0^{\flat} = U(e_1+e_2), \mathcal{M}_{\infty}^{\flat} = U(e_1), \mathcal{M}_0^{\top} = U(e_1-e_2)R', \mathcal{M}_{\infty}^{\top} = U(e_2)R'$$

$$\mathcal{M}_z^{\flat} = ((R')^{+})^{h_z^{-1}}, \mathcal{M}_z^{0} = (R^{+})^{h_z^{-1}} \text{ and } \mathcal{M}_z^{\top} = \mathcal{M}_z^{0}R', z \in \mathbb{F}_q^{\times}$$

Note that $\mathcal{M}_1^{\flat} = (R^{+})' = V^{(8)}, \mathcal{M}_1^{0} = R^{+} = V^{(3)}$ and $\mathcal{M}_1^{\top} = V^{(3)}R'$. Further, for $\mu \in \mathbb{F}_q \cup \{\infty\}$, $[x, F_{3,4}(q)] = \mathcal{M}_{\mu}^{\flat}$ for each $x \in \mathcal{M}_{\mu}^{\flat} \backslash X_1X_{1+2}$. See Lemma 3.2.B for $\mu = 0$ and ∞. It holds for $\mu \in \mathbb{F}_q^{\times}$ because it does for $\mu = 1$ and $[F_{3,4}^{+}(q), H_{1,2}] = 1$. Straight forward calculations using Lemma 2.1 yield

Lemma 3.6. *Let $x \in R$ and $z \in \mathbb{F}_q^{\times}$. Then:*

(a) $x \in \mathcal{M}_z^{\flat}$ if, and only if, $x_r = x_{\bar{r}} = 0$ for each $r \in A(e_2)$ and $x_r = z^2 x_{\bar{r}}^{\lambda(r)\theta}$ for each $r \in A(e_1)$;
(b) $x \in \mathcal{M}_z^{\top}$ if, and only if, $x_r = z^{-4\theta-2} x_{\bar{r}}^{\lambda(r)\theta}$ for each $r \in A(e_2)$; and
(c) If $x \in \mathcal{M}_z^{\top}$, then $x \in \mathcal{M}_z^{0}$ if, and only if, $x_r + z^2 x_{\bar{r}}^{\lambda(r)\theta} = [x(e_2), x(e_1-e_2)]_r = z^2([x(e_2), x(e_1-e_2)]_r)^{\lambda(r)\theta}$ if $r \in A(e_1)$.

The following reformulation of Lemma 3.6 is also a pointer to the analogy between the subgroup $U_{1,2}$ of $B_2(q)$ and the subgroup R of $F_4(q)$.

Lemma 3.7. *Let $x \in R$ and $z \in \mathbb{F}_q^{\times}$. Then:*

(a′) $x \in \mathcal{M}_z^{\flat}$ if, and only if, $x(e_2) = x(e_1-e_2) = 1$ and $x(e_1) = z^2 \circ (x(e_1+e_2))^{\sigma}$
(b′) $x \in \mathcal{M}_z^{\top}$ if, and only if, $x(e_2) = \gamma^{-4\theta-2} \circ (x(e_1-e_2))^{\sigma}$.
(c′) If $x \in \mathcal{M}_z^{\top}$, then $x \in \mathcal{M}_z^{0}$ if, and only if,

$$x(e_1) = [x(e_2), x(e_1-e_2)](e_1) + z_0^2([x(e_2), x(e_1-e_2)](e_1+e_2))^{\sigma}.$$

Proposition 3.8. *(i) Any nontrivial $F_{3,4}^{+}(q)$-invariant submodule of R' is either $M_z^{\flat}$ for some $z \in \mathbb{F}_q \cup \{\infty\}$ or is contained in X_1X_{1+2} such that it projects surjectively onto X_1 and X_{1+2}.*
(ii) Any nontrivial $F_{3,4}^{+}$-invariant subgroup of R not contained in R' is either one of R, $\mathcal{M}_z^{0}$, $\mathcal{M}_z^{\top}$ for some $z \in \mathbb{F}_q^{\times}$ or is contained in $\mathcal{M}_0^{\top}$ or $\mathcal{M}_{\infty}^{\top}$. □

Proposition 3.9. *For any $z \in \mathbb{F}_q^{\times}$, $\{x^2 : x \in \mathcal{M}_z^{\top}\} = \mathcal{M}_z^{\flat}X_1X_{1+2}$ and $\{x^2 : x \in \mathcal{M}_z^{0}\} = \mathcal{M}_z^{\flat}$.* □

4 Proofs

1. Proof of Proposition 1.1: By (1), neither θ nor 2θ is Identity. We show that, if $k = \{\lambda \in K : \lambda^{2\theta^2} = \lambda\} \subsetneq K$, then either θ or 2θ is the identity function on $K \setminus k$, a contradiction to (1). This would prove the proposition.

(i) $\lambda^{2\theta^2} = \lambda^{3\theta-1}$ *for each* $\lambda \in K \setminus k$.

Proof. By (2),

$$(\lambda^{1-\theta} + \mu^{1-\theta})^{2\theta} (\lambda^{2\theta-1} + \mu^{2\theta-1})^{\theta} = (\lambda^{2\theta-1} + \mu^{2\theta-1}) (\lambda^{1-\theta} + \mu^{1-\theta}) .\text{So,}$$
$$\lambda^{2\theta-2\theta^2}\mu^{2\theta^2-\theta} + \lambda^{2\theta^2-\theta}\mu^{2\theta-2\theta^2} = \lambda^{2\theta-1}\mu^{1-\theta} + \lambda^{1-\theta}\mu^{2\theta-1}.$$

Taking $\mu = 1$, we see $\lambda^2 \left(\lambda^{2\theta^2-1} + 1\right) \left(\lambda^{2\theta^2} + \lambda^{3\theta-1}\right) = 0$. So, (i) follows. □

(ii) *If* $\lambda \in K \setminus k$, *then* $\lambda + \lambda^{-1} \notin K \setminus k$.

Proof. Suppose not. Then, using (i), we get

$$\lambda + \lambda^{-1} = (\lambda + \lambda^{-1})^{2\theta^2} = \lambda^{2\theta^2} + \lambda^{-2\theta^2} = \lambda^{3\theta-1} + \lambda^{1-3\theta}.$$

So, $(\lambda^{3\theta} + \lambda^2)(\lambda^{3\theta} + 1) = 0$ and either $\lambda^{3\theta} = \lambda^2$ or $\lambda^{3\theta} = 1$. Since $\lambda \notin k$, (i) and the definition of k imply that $\lambda^{3\theta} \neq \lambda^2$. So, $\lambda^3 = 1 \neq \lambda$ and $\lambda \in \mathbb{F}_4 \subseteq K$. Since $\lambda^{2\theta^2} \neq \lambda$, it follows that $\lambda^{2\theta^2} = \lambda^{-1} = \lambda^2$ and so, $\lambda^{\theta^2} = \lambda$ and $\lambda^{\theta} = \lambda$ or $\lambda^{\theta} = \lambda^{-1}$. In either case, (1) is contradicted. So, (ii) holds. □

(iii) *If* $\lambda \in K \setminus k$, *then* $\lambda^{\theta} = \lambda$ *or* $\lambda^{2\theta} = \lambda$.

Proof. By (ii), $\lambda + \lambda^{-1} \notin k$. Applying (i) to $\lambda + \lambda^{-1}$, we get $(\lambda^{\theta} + \lambda)(\lambda^{2\theta} + \lambda) = 0$ and (iii) follows. □

(iv) *Either* θ *or* 2θ *is the identity map on* $K \setminus k$.

Proof. In view of (iii), it is enough to show that $K \setminus k$ does not contain distinct elements λ and μ such that $\lambda^{\theta} = \lambda$ and $\mu^{2\theta} = \mu$. Deny. Then, if $\lambda + \mu \notin k$, then, (i) applied to $\lambda + \mu$ yields

$$(\lambda + \lambda^2)^2(\mu + \mu^2) = 0.$$

So either λ or $\mu \in \mathbb{F}_2$, a contradiction. Now, suppose that $\lambda + \mu \in k$. Since $\lambda^2 + \mu^2 \in k$, we get $\lambda^4 + \lambda^2 = \mu + \mu^2$. Applying 2θ, we get $\lambda^8 + \lambda^4 = \mu + \mu^2$. So, $\lambda^2 + \lambda \in \mathbb{F}_2$ and $\lambda^2 + \lambda = 1$. Thus, $\lambda^{-1} = \lambda + 1$ and $\lambda + \lambda^{-1} = 1 \in k$, a contradiction to (ii). Thus, (iv) and Proposition 1.1 follow. □

2. Proof of Proposition 1.2: By (3), neither θ nor 3θ is Identity. Let $\lambda \in K$ and $\lambda \neq \pm 1$. Adding and subtracting the equations obtained by taking $\mu = 1$ and $\mu = -1$ in (4), we get

$$\lambda^{6\theta^2-1}(\lambda^2 - \lambda^{5\theta-1}) = \lambda^{9\theta^2} - \lambda^{5\theta} \quad \text{and} \quad \lambda^{3\theta^2}(\lambda^2 - \lambda^{5\theta-1}) = \lambda^{9\theta^2} - \lambda^{5\theta}.$$

If $\lambda^2 \neq \lambda^{5\theta-1}$, then clearly $\lambda^{3\theta^2} = \lambda$. And $\lambda^2 = \lambda^{5\theta-1}$ if, and only if, $\lambda^{9\theta^2} = \lambda^{5\theta}$. This holds if, and only if, $\lambda^{9\theta^2} = \lambda^{5\theta} = \lambda^3$. Thus, $\lambda^{3\theta^2} = \lambda$ if $\lambda \notin \mathbb{F}_3$. This clearly holds for $\lambda \in \mathbb{F}_3$ also. Thus $\lambda^{3\theta^2} = \lambda$ for all $\lambda \in K$ and the Proposition 1.2 follows. □

3. Proof of Theorem 1.5: By Sylow's Theorem, we can assume that $S < U$. Since $Z(S) \neq 1$, by $(\mathcal{A}_3)$, S contains an element u of order $p^{|\Delta|}$. By Proposition 1.3.(b), $u_\alpha \neq 0$ for each $\alpha \in \Delta$ and $1 \neq i = u^{p^{|\Delta|-1}} \in Z(S) \subseteq \{1\} \cup X_s^\times X_\ell^\times$.

(A) *Replacing* S,T *and* u *by* S^x, T^x *and* u^x *for an appropriate* x *in* X, *we may assume that : (i)* $T \leq H$, *(ii)* $u_\alpha = 1$ *for each* $\alpha \in \Delta$ *and so,* $i = x_s(-1)\, x_\ell(-1) \in Z(S)$, *and* (iii) $S \subseteq U$ *if* $X \in \{B_2(q), G_2(q)\}$.

Proof. By Lemma 2.2.(b), Proposition 1.3.(d) and $(\mathcal{A}_4)$, it follows that $\langle U, T\rangle \leq N_X(Z(S)) = C_X(i)\,\widehat{H}$, where $\widehat{H} = \{h \in H_l H_s : i^h \in Z(S)\}$. Since H is an abelian Hall π- subgroup of X for the set π of prime divisors of $q-1$, by Wielandt's Theorem ([3], Satz 5.8, p.285), there exists $b \in C_X(i)$ such that $T^b \subseteq \widehat{H}$ if $X \in \{B_2(q), G_2(q)\}$; and $T^b \subseteq \widehat{H} H_{3,4}$ if $X = F_4(q)$. If $X = B_2(q)$ or $G_2(q)$, then $C_X(i) = U$ and so $S^b \subseteq U$. In all cases, by a further conjugation by an element of H (see Lemma 2.2.(b)), we can assume that $i \in x_s(-1)x_\ell(-1) \in Z(S)$. □

Let $M = \langle h_1\rangle$ for some $h_1 = h(\chi_1) \in H$. If $X = B_2(q)$ or $G_2(q)$, then $M = T$ and $N = \langle 1\rangle$. If $X = F_4(q)$, let $N = \langle h_2\rangle$ with $h_2 = h(\chi_2) \in H$.

(B) $\mathbb{F}_q$ *admits a field automorphism* θ *such that: (i)* $\chi_1(\ell)^\theta = \chi_1(s)$ *and (ii)* $Z(S)^\times = i^M = \{x_s(t^\theta)x_\ell(t) : t \in F_q^\times\}$.

Proof. By (**A.(ii)**) and $(\mathcal{A}_4)$,

$$Z(S)^\times = \{h_1^{-k} i h_1^k = x_s(-\chi_1^k(s))x_\ell(-\chi_1^k(\ell)) : k = 1, \cdots, q-1\}. \tag{9}$$

We note that

$$x_s(t_s)x_\ell(t_\ell)\cdot x_s(t_s')x_\ell(t_\ell') = x_s(t_s+t_s')\cdot x_\ell(t_\ell+t_\ell'), \tag{10}$$

a crucial, albeit trivial, step in our argument. By (9) and (10), the subsets

$$\{\chi_1^k(\alpha) : k = 1, \cdots, q-1\} \cup \{0\}, \alpha \in \{\ell, s\},$$

of $\mathbb{F}_q$ are additively closed. Since $\chi_1^k \in \mathrm{Hom}_{\mathbb{Z}}\left(\mathbb{Z}\Sigma, \mathbb{F}_q^\times\right)$, they are multiplicatively closed also. So they are subfields of $\mathbb{F}_q$. Their equality with $\mathbb{F}_q$ follows from putting together: (9), the assumption that $|Z(S)| = q$ and the fact that, if $x_s(\lambda)x_\ell(\mu) \in Z(S)$, then λ and μ both are zero or both are nonzero. Thus,

$$\langle \chi_1(s)\rangle = \langle \chi_1(\ell)\rangle = \mathbb{F}_q^\times. \tag{11}$$

Now, the map θ from $\mathbb{F}_q$ to itself fixing zero and taking $\chi_1^i(s)$ to $\chi_1^i(\ell)$ respects addition (by (10)) and, clearly, multiplication also. Thus, θ is a field automorphism. Now, (i) and (ii) of (**B**) are clear from (9). □

(C) *For $z \in Z(S)$, let $S_z = \{x \in S : x^{p^{|\Delta|-1}} = z\}$. Then, (i) $Z(S) \subseteq \Phi(S) \subseteq S_{id}$ and (ii) $S \setminus S_{id}$ is the union of the cosets $u^{mn}\Phi(S)$ with u* as in (**A**), $m \in M$ *and $n \in \mathbb{N}$*. Further, $S_{id} = \Phi(S)$.

Proof. Since the hypothesis of Theorem 1.5 holds for D^x, S^x and T^x for $x \in X$, we may assume that $S \subseteq U$. For any finite p - group P, $\Phi(P) = P' \langle x^p : x \in P\rangle$ ([3], Satz 3.14(a), p.272). So, by $(\mathcal{A}_3)$, $Z(S) \subseteq \Phi(S) \subseteq \Phi(U)$. From the commutator relations in Lemma 2.1, $\Phi(U) = \Pi_{\alpha \in \Sigma^+ \setminus \Delta} X_\alpha$. So $y_\alpha = 0$ for each $y \in \Phi(U)$ and $\alpha \in \Delta$. Now, by Proposition 1.3.(b), $x^{p^{|\Delta|-1}} = 1$ for $x \in \Phi(S)$ and (i) holds. Now, (ii) follows because, from **A**.ii, $(\mathcal{A}_2)$, $(\mathcal{A}_3)$ and $(\mathcal{A}_4)$, for each $y \in S \setminus S_{id}$, then $y^{p^{|\Delta|-1}} = i^m$ for some $m \in M$. □

(D) *(a)* Let $X = B_2(q)$. *Then, for all $\lambda, \mu \in F_q^\times$ with $\lambda + \mu \neq 0$,*

$$(\lambda^{1-\theta} + \mu^{1-\theta})(\lambda^{2\theta-1} + \mu^{2\theta-1}) \neq 0 \tag{12}$$

and

$$(\lambda^{1-\theta} + \mu^{1-\theta})^{2\theta-1} = (\lambda^{2\theta-1} + \mu^{2\theta-1})^{1-\theta}. \tag{13}$$

(b) Let $X = G_2(q)$. *Then, for all $\lambda, \mu \in F_q^\times$ with $\lambda + \mu \neq 0$,*

$$(\lambda^{2\theta-1} + \mu^{2\theta-1})(\lambda^{2-3\theta} + \mu^{2-3\theta}) \neq 0 \tag{14}$$

and

$$(\lambda^{2\theta-1} + \mu^{2\theta-1})^{2-3\theta} = (\lambda^{2-3\theta} + \mu^{2-3\theta})^{2\theta-1}. \tag{15}$$

Proof. Let y be the product of a pth-root in S of $x_s(\lambda^\theta)x_\ell(\lambda)$ and a pth-root in S of $x_s(\mu^\theta)x_\ell(\mu)$. If $X = B_2(q)$, then $p = 2$, $y_2 = \lambda^{1-\theta} + \mu^{1-\theta}$ and $y_{1-2} = \lambda^{2\theta-1} + \mu^{2\theta-1}$; and if $X = G_2(q)$, then $p = 3$, $y_a = -\lambda^{2\theta-1} - \mu^{2\theta-1}$ and $y_b = -\lambda^{2-3\theta} - \mu^{2-3\theta}$ (by Proposition 1.3.(c)). Now $y^p = X_s(\lambda^\theta + \mu^\theta)X_\ell(\lambda + \mu)$ by ([3], Satz 9.4, p.317) and $y^p \in Z(S)$ by $(\mathcal{A}_3)$. Clearly, $y^p = 1$ if, and only if, $\lambda + \mu = 0$. Further, $y^p \neq 1$ if, and only if, $y_\alpha \neq 0$ for each $\alpha \in \Delta$ (Proposition 1.3 (b)). So, (12) and (14) hold. If $X = B_2(q)$, then $y_1 = y_2 y_{1-2}$ and $y_{1+2} = y_2^2 y_{1-2}$ and, by (**B**), $y_2^{2\theta-1} = y_{1-2}^{1-\theta}$. Similarly, $y_a^{3\theta-2} = y_b^{1-2\theta}$ if $X = G_2(q)$. So, (13) and (15) hold. □

Now, (**D**) and Propositions 1.1 and 1.2 imply

(E) *If $X = B_2(q)$, then $2\theta^2 =$ Identity $\notin \{\theta, 2\theta\}$. If $X = G_2(q)$, then $3\theta^2 =$ identity $\notin \{\theta, 3\theta\}$. In both these cases, $q = p^{2n+1}$ for some natural number n.* □

Recall the definition of $\alpha(\mathbb{F}_q, \beta(\mathbb{F}_q) \leq B_2(q)$ and $\alpha(\mathbb{F}_q, \beta(\mathbb{F}_q), \gamma(\mathbb{F}_q) \leq G_2(q)$ in Lemma 2.2.(d) and that of $V_{12} \leq F_4(q)$ in Sect. 3.A.

(F) *(i) If $X = B_2(q)$ or $G_2(q)$ then $M = T = H^+$.*
(ii) $Z(S) = \beta(\mathbb{F}_q), \gamma(\mathbb{F}_q), V_{12}$ according as $X = B_2(q), G_2(q), F_4(q)$.

Proof. (i) follows from $(\mathcal{A}_4)$, (**B**.ii) and (**E**). (ii) follows now by definition. □

(G) *If* $X = B_2\left(2^{2n+1}\right)$, *then* $S = \alpha\left(\mathbb{F}_q\right)\beta\left(\mathbb{F}_q\right) = U^+$.

Proof. By (**C**.ii), S is the union of the cosets $\Phi(S), u^{h_1^k}\Phi(S), k = 1,\cdots,q-1$. We show that, for each k,

$$u^{h_1^{-k}}\Phi(S) \cap X_2X_{1-2}X_1 = \{\alpha(\chi_1^k(1-2))\}, \tag{16}$$

where $\alpha(t), \beta(t)$ are as defined in Lemma 2.2.d(i). Since $\Phi(S) = Z(S) = \beta(\mathbb{F}_q)$ (by (**C**) and (**F**.i)), (16) would then imply that $u^{h_1^{-k}}\Phi(S) = \alpha(\chi_1^k(1-2))\beta(\mathbb{F}_q)$. Since $\mathbb{F}_q^\times = \langle\chi_1(1-2)\rangle$, (**G**) would follow.

Since $\Phi(S) = \beta(\mathbb{F}_q)$, the left-side of (16) is a singleton. We write this as $\tau(u^{h_1^{-k}})$. Let $u = x_2(1)x_{1-2}(1)x_1(t_1)\,x_{1+2}(t_{1+2})$. Then,

$$\tau(u) = x_2(1)x_{1-2}(1)x_1(t_1 + t_{1+2}^\theta).$$

We now show that $t_1 + t_{1+2}^\theta = 1$; equivalently, that $\tau(u) = \alpha(1)$. Since $\chi_1(1) = \chi_1^\theta(1+2)$,

$$(\tau(u^{h_1^{-k}}))_1 = \chi_1^k(1)t_1 + \chi_1^{k\theta}(1+2)t_{1+2}^\theta = \chi_1^k(1)(t_1 + t_{1+2}^\theta), \text{ and} \tag{17}$$

$$(\tau(uu^{h_1^{-k}}))_1 = [(1+\chi_1^k(1))t_1 + \chi_1^k(2)] + [(1+\chi_1^k(1+2))t_{1+2} + \chi_1^{2k}(2)]^\theta \tag{18}$$

for $k \in \{1,\cdots,q-1\}$. Since $(uu^{h_1^{-k}}) \in u^{h_1^l}\Phi(S)$ for some l, $(uu^{h_1^{-k}})_2 = 1+\chi_1^k(2)$, $(uu^{h_1^{-k}})_{1-2} = 1+\chi_1^k(1-2)$ and $\chi \in \mathrm{Hom}_{\mathbb{Z}}\left(\mathbb{Z}\Sigma, \mathbb{F}_q^\times\right)$, we also have, from (17), that

$$(\tau(uu^{h_1^{-k}}))_1 = (1+\chi_1^k(2))(1+\chi_1^k(1-2))t_1 + [(1+\chi_1^{2k}(2))(1+\chi_1^k(1-2))t_{1+2}]^\theta. \tag{19}$$

Comparing (18) and (19), we get

$$(\chi_1^k(2) + \chi_1^k(1-2))t_1 + [(\chi_1^{2k}(2) + \chi_1^k(1-2))t_{1+2}]^\theta = \chi_1^k(2) + \chi_1(2)^{2k\theta}. \tag{20}$$

Now, if $\chi_1(1+2) = \lambda$, then $\chi_1(1) = \lambda^\theta$, $\chi_1(2) = \lambda^{1-\theta}$ and $\chi_1(1-2) = \lambda^{2\theta-1}$. So, from (20), it follows that $(t_1 + t_{1+2}^\theta - 1)(\lambda^{k(2\theta-1)} + \lambda^{k(1-\theta)}) = 0$ for each $k \in \{1,\cdots,q-1\}$. Since $2\theta^2 =$ Identity $\notin \{\theta, 2\theta\}$ (by (**E**)), $\lambda^{k(2\theta-1)} \neq \lambda^{k(1-\theta)}$ for some k. So, $\lambda(u) = t_1 + t_{1+2}^\theta = 1$. Now, by (17), $(\tau(u^{h_1^{-k}}))_1 = \chi_1^k(1)$ and $\tau(u^{h_1^{-k}}) = \alpha(\chi_1^k(1-2))$. □

We now show that (**A**.iii) holds for $X = F_4(q)$ also:

(H) *Let* $X = F_4(q)$. *Replacing* S,T *and* u *by* S^x, T^x *and* u^x *for an appropriate* x *in* X, *we may assume that : (i)* $T \leq H$, *(ii)* $u_\alpha = 1$ *for each* $\alpha \in \Delta$ *and so,* $i = x_s(-1)\,x_\ell(-1) \in Z(S)$, *and* (iii) $S \subseteq U$.

Proof. We continue with the statement of (**A**) and its notation and proof. If $S^b \not\subseteq U$, then the images $\bar{S}^b$ and $\bar{T}^b$ of S and T in $Y = N_X(X_1X_{1+2})/R \simeq F_{3,4}(q)H_{1,2}$ satisfy the conditions $(\mathcal{A}_1)$ to $(\mathcal{A}_5)$ of the hypothesis $(\mathcal{A})$. So, (**A**) applied to the group Y yields an element $b_1 \in F_{3,4}(q)$ such that $\bar{S}^{bb_1} \subseteq U_{3,4}$ and $\bar{T} \subseteq H$. Also, $[b_1, i] = 1$. So, we can assume that $S \subseteq U$, $T \subseteq RH$ and $i \in Z(S)$. Now, conjugating, if necessary, by an element of R, we can ensure that $S \subseteq U, i \in Z(S)$ and $T \subseteq H$ hold simultaneously. □

(I) *If* $X = G_2\left(3^{2n+1}\right)$, *then* $S' = \beta(\mathbb{F}_q)\gamma(\mathbb{F}_q) = \Phi(S)$.

Proof. Since $M = H^+$ (by (**F**.i)) and M normalizes the nontrivial subgroup $Z(S) \cap S'$ containing i, $(\mathcal{A}_4)$ and the definition of $\gamma(\mathbb{F}_q)$ in Lemma 2.2.d(ii) imply that $Z(S) = \gamma(\mathbb{F}_q) \subseteq S'$. We show that $\beta(\mathbb{F}_q) \subseteq S'$. Then, $q^2 = |\Phi(S)| \geq |S'| \geq |\beta(\mathbb{F}_q)\nu(\mathbb{F}_q)| = q^2$ (see $(\mathcal{A}_1)$ and $(\mathcal{A}_2)$) and (**I**) would follow.

We first show that $\beta(\mathbb{F}_q) \cap S' \neq 1$. Let $x \in S\backslash\Phi(S)$. If $m \in S$ is a cube root of $x^3 = \gamma(t) \in Z(S)^\times$, then $m_a = -t^{2\theta-1}$, $m_b = -t^{2-3\theta}$ (Proposition 1.3.(c)). So $m_a = m_b^\theta$ (by (**E**)). If m and n are distinct cube roots in S of x^3, then $m_a = n_a$ and $m_b = n_b$. So, by Lemma 2.3.(e) and $(\mathcal{A}_4)$,

$$[m, n] = x_{2a+b}\left[m_a\left(n_{a+b} - m_{a+b}\right)\right] x_{3a+2b}\left[m_b\left(n_{3a+b} - m_{3a+b}\right)\right] \in Z(U) \cap S = Z(S).$$

Now, since $m_a = m_b^\theta \neq 0$, (**B**.i) implies that

$$n_{a+b} - m_{a+b} = n_{3a+b}^\theta - m_{3a+b}^\theta. \tag{21}$$

Note that, if k is an integer such that $\chi_1^k(a) = \chi_1^k(b) = -1$, then m^{-1} and $m^{h_1^k}$ are distinct cube roots of x^{-3} (see Lemma 2.3.(d)). So, by (21) and Lemma 2.3.(d),

$$m_{a+b} + m_a m_b = m_{3a+b}^\theta - (m_a^3 m_b)^\theta.$$

Since $m_a = m_b^\theta$ and $3\theta^2 =$ Identity, it follows that $\left(m_a^3 m_b\right)^\theta = m_a m_b$. So,

$$m_{a+b} = m_{3a+b}^\theta + m_a m_b. \tag{22}$$

Since each element of $S\backslash\Phi(S)$ is a cube root of some element of $Z(S)$, (22) holds for all elements of $S\backslash\Phi(S)$.

Now, choose $m \in S$ such that $0 \neq m_b \neq m_a (= m_b^\theta)$. This is possible because: $u \in S$, $u_a = u_b = 1$, $\mathbb{F}_q^\times = \langle \chi_1(b) \rangle$, θ is not the identity map on $\mathbb{F}_q^\times$ and so, we can choose m as $u^{h_1^k}$ for an appropriate integer k. Using Lemma 2.3.(e), we have

$$\begin{aligned} &1 \neq [u, m] = xy, \text{ where } x = x_{a+b}(m_a - m_b)x_{3a+b}(m_b - m_a^3) \text{ and} \\ &y = x_{2a+b}\left(m_{a+b} - m_a u_{a+b} + m_a^2 - m_b\right) \\ &\quad x_{3a+2b}\left(m_{3a+b} - m_b u_{3a+b} + m_b\left(m_a^3 - 1\right) + m_b^2 - m_a^3\right) \end{aligned}$$

Since $m_a = m_b^\theta$, $x \in \beta(\mathbb{F}_q)$. Since (22) holds for both m and n, it follows that $y \in \gamma(\mathbb{F}_q)$. Since $Z(S) = \gamma(\mathbb{F}_q) \subseteq S'$, $1 \neq x \in \beta(\mathbb{F}_q) \cap S'$. Let $x = \beta(t_0)$.

By (**B**.i), $\chi_1(3a+2b)^\theta = \chi_1(2a+b)$. So, $\chi_1(a)^{3\theta-2} = (\chi_1(b)^\theta)^{3\theta-2}$. Since $t^\theta = t^{3^n}$ for $t \in \mathbb{F}_q$ and $\left(3^{n+1}-2, 3^{2n+1}-1\right) = 1$, it follows that $\chi_1(a) = \chi_1(b)^\theta$. Further, since $\left(1+3^n, 3^{2n+1}-1\right) = 2$, $\langle \chi_1(a+b) \rangle = \{t^2 : t \in \mathbb{F}_q\}$. Now, since $M = \langle h(\chi_1) \rangle$ normalizes $\beta(\mathbb{F}_q) \cap S' \neq \langle 1 \rangle$, $\{\beta(t^2 t_0) : t \in \mathbb{F}_q\} \subseteq \beta(\mathbb{F}_q) \cap S'$. Since $\beta(\mathbb{F}_q) \cap S'$ is abelian and any element of $\mathbb{F}_q$ is a sum of two squares in $\mathbb{F}_q$, it follows that $\beta(\mathbb{F}_q) \subseteq S'$, completing the proof of (**I**). □

(**J**) $S = \alpha(F_q)\beta(F_q)\gamma(F_q) = U^+$ *if* $X = G_2(q)$.

Proof. The proof of (**J**) is similar to the proof of (**G**). By (**C**.ii), S is the union of the cosets $\Phi(S)$, $u^{h_1^k}\Phi(S)$, $k = 1, \cdots, q-1$. For $x \in S \backslash \Phi(S)$, $x\Phi(S) \cap X_a X_b X_{a+b} X_{2a+b}$ is a singleton (by **I**). We write it as $\{\tau(x)\}$ and show that, if $k \in \{1, \cdots, q-1\}$ is given by $x^3 = i^{h_1^k}$ (see (**B**.i)), then $\tau(x) = \alpha(\chi_1(b)^k)$ and so the set $x\Phi(S)$ of the cube roots of x^3 in S is $\alpha(\chi_1(b)^k)\Phi(S)$. Since $\chi_1(b) = \chi_1(3a+2b)^{2-3\theta}$, $\mathbb{F}_q^\times = \langle \chi_1(3a+2b) \rangle$, $t \to t^{2-3\theta}$ is a bijection of $\mathbb{F}_q^\times$ and $\mathbb{F}_q^\times = \langle \chi_1(b) \rangle$, it follows that S contains $\alpha(\mathbb{F}_q)\beta(\mathbb{F}_q)\gamma(\mathbb{F}_q)$ (by (**I**)). Since $|S| = q^3$, (**J**) would then follow. Thus, to complete the proof, we now show that $u\Phi(S) = \alpha(1)\Phi(S)$.
By (**A**.ii), for each k,

$$\tau(u^{h_1^{-k}}) = x_a(\chi_1^k(a)) \cdot x_b(\chi_1^k(b)) \cdot x_{a+b}[\chi_1^k(a+b)u_{a+b} - \chi_1^{k\theta}(3a+b)u_{3a+b}^\theta] \cdot$$
$$x_{2a+b}[\chi_1^k(2a+b)u_{2a+b} - \chi_1^{k\theta}(3a+2b)u_{3a+2b}^\theta]. \tag{23}$$

Calculating $\tau(uu^{h_1^{-k}})$ in two ways as in the proof of (**G**) and comparing $(\tau(uu^{h_1^{-k}}))_{a+b}$, we get

$$(\chi_1^k(a) + \chi_1^k(b))(u_{a+b} - u_{3a+b}^\theta - 1) = 0$$

for each k. This implies that $u_{a+b} - u_{3a+b}^\theta = 1$. Using this to compare the two expressions for $(\tau(uu^{h_1^{-k}}))_{2a+b}$, we get

$$(\chi_1^k(a) - 1)(\chi_1^k(a) - \chi_1^k(b))(u_{2a+b} - u_{3a+2b}^\theta - 1) = 0$$

for each k. This implies that $u_{2a+b} - u_{3a+2b}^\theta = 1$. Thus, $u = \alpha(1)\beta(1)\gamma(1)$ and $u\Phi(S) = \alpha(1)\Phi(S)$ by (**I**), as claimed. □

(**K**) *Let* $X = F_4(q)$, $h = h(\chi) \in T$, $\chi(e_i) = z_i$, $i = 1, 2, 3, 4$ *and* $\theta \in \mathrm{Aut}(\mathbb{F}_q)$ *be as in (**B**). Then,* z_i *and* θ *satisfy the equations* $({}^2F_4)$ *stated in Sect. 1.B.*

Proof. Consider $u \in S$ (as in (**A**)). From (**H**) and Proposition 1.3.(b), $(u^8)^{-h} = x_1(z_1)x_{1+2}(z_1 z_2) \in Z(S)$. Now (5) follows from (**B**.i).

For $k \in \mathbb{Z}$ and $i = 0, 1, 2, 3, \cdots$, define $A_{i,k} = u^{2^i}(u^{2^i})^{h^k} \in S$. By Proposition 3.3.(d), $A_{i,k}^{2^{3-i}} \in Z(S)$. If $A_{i,k}^{2^{3-i}} = x_1\left(\lambda_{i,k}^{e_1}\right) x_{1+2}\left(\lambda_{i,k}^{e_1+e_2}\right)$, then, by (**B**.i),

$$\lambda_{i,k}^{e_1} = (\lambda_{i,k}^{e_1+e_2})^{\theta}. \tag{24}$$

Equation (8) is the case $i = 0$ of (24) in view of Proposition 3.3.(biii) and (d). We now prove (6). Since $u^4 \in U_2$, (see Sect. 3.D) by Proposition 3.3.(bi), $u^8 = x_1\left((u^4)_r(u^4)_{e_1-r}\right) x_{1+2}\left((u^4)_{\bar{r}}(u^4)_{e_1+e_2-\bar{r}}\right)$, where $r = (e_1 - e_2 + e_3 + e_4)/2$. Also $u^8 = i = x_1(1)\, x_{1+2}(1)$, by (**H**). So, $(u^4)_r\ (u^4)_{e_1-r} = 1 = (u^4)_{\bar{r}}(u^4)_{e_1+e_2-\bar{r}}$ by (**B**.i). This together with Proposition 3.3.(d) applied to $A_{2,k}$ implies that, for each k,

$$\lambda_{2,k}^{e_1} = (1+\chi^k(r))(1+\chi^k(e_1-r)) \text{ and } \lambda_{2,k}^{e_1+e_2} = (1+\chi^k(\bar{r}))(1+\chi^k(e_1+e_2-\bar{r})).$$

Using (5) and (24), we deduce that $(z_2^{-1}z_3z_4)^k + (z_2^{-1}z_3z_4)^{-k} = z_1^{(1-2\theta)k}{z_3}^{2k\theta} + z_1^{(2\theta-1)k}{z_3}^{-2k\theta}$ for each $k \in \mathbb{Z}$. Now, $(z_2^{-1}z_3z_4)^k = (z_1^{-1}z_2z_3^2)^{k\theta}$ or $(z_1^{-1}z_2z_3^2)^{-k\theta}$ for each k, in view of the remark: for elements λ and μ of a field, $\lambda + \lambda^{-k} = \mu^k + \mu^{-k}$ holds for each $k \in \mathbb{Z}$ if, and only if, $\lambda = \mu$ or $\lambda = \mu^{-1}$. Equation (6) now follows from (5).

Finally, (7) in Sect. 1.B follows from (24) by calculating both sides of (24) with $i = 1$ using Proposition 3.3.(b(ii), d) and Corollary 3.4. Note that, $(u^2)_\alpha = 1$ for $\alpha \in \{3, \bar{3}, 2-4, \overline{2-4}\}$ appearing in Proposition 3.3.b(ii), and so $(A_{1,k})_\alpha = 1+\chi^k(\alpha)$. □

Now, (a) of Theorem 1.5 is proved in (**A**) and (**H**), (9) in (**B**), (10) in (**C**) and (c) in (**K**). By (**G**), (**J**) and (**F**.i), $U^+H^+ \subseteq X^+ \cap D$. Since both X^+ and D are generated by U^+ and $N_X(H^+)$, they are equal and (d) of Theorem 1.5 follows.

5 Remarks

In the case $X = F_4(q)$, the following holds in addition to Theorem 1.5.(a) and (b):

Theorem 5.1. *Let D, S and T be subgroups of $F_4(q)$ such that the conditions $(\mathcal{A}_1)$ to $(\mathcal{A}_7)$ of hypothesis $(\mathcal{A})$ hold. Then, m is odd and $F_4(q)$ contains an element x such that $S^x = U^+$, $M^x = H_{1,2}^+$, $N^x = H_{3,4}^+$ and $\tilde{K}^x = F_{3,4}^+H_{1,2}^+$. In particular, $\langle S, \tilde{K}\rangle^x = N_{X^+}(Z(U^+))$.*

We state another characterization of S and $N_{X^+}(Z(S))$:

Proposition 5.2. *Let $X \in \{B_2(q), G_2(q), F_4(q)\}$ and m be odd. If S and K are subgroups of X such that (a),(b),(c) of Proposition 1.4 hold, then, there exists an element $x \in X$ such that $S^x \in Syl_p(X^+)$ and $(SK)^x = N_{X^+}(Z(S))$.*

The following result is crucial to our proofs of Theorem 5.1 and Proposition 5.2 for $X = F_4(q)$ and is perhaps of independent interest.

Theorem 5.3. *Let $X = F_4(q)$ and $D \in \{X, P^+ = N_X(Z(U))^+, X^+\}$. Let $K \le D$ be isomorphic to ${}^2B_2(q)$. Then, K is conjugate to $F_{1,2}^+(q)$ in D.*

The proofs of the above statements will appear elsewhere. We remark that many crucial results and arguments in this paper are true for an arbitrary field of the characteristic specified in Sect. 1.A (for example, Propositions 1.1–1.3, Sect. 3, Bombieri's solution [2]).

Acknowledgement This work was partially supported by a DST grant No.SR/54/MS : 583/07.

References

1. Aschbacher, M. and Seitz, G., Involutions in Chevalley groups over fields of even order, Nagoya Math. J. **63** (1976) 1–91
2. Bombieri, E., Sastry automorphisms, J. Algebra, **257** (2002) 222–243
3. Huppert, B., *Endliche gruppen, I*, Springer-Verlag, Berlin, New York, 1967
4. Kleidmann, P. B., The maximal subgroups of the Chevalley group $G_2(q)$ with q odd, the Ree groups ${}^2G_2(q)$, and their automorphism groups, J. Algebra 117 (1988) 30–71
5. Ree R., A family of simple groups associated with simple Lie algebras of type F_4, Amer. J. Math. **83** (1961) 401–420
6. Ree. R., A family of simple groups associated with the simple Lie algebra G_2, Amer. J. Math. **83** (1961) 432–462
7. Richardson, R. Röhrle, G. and Steinberg, R., Parabolic subgroups with abelian unipotent radical, Inv. Maths. **110** (1992), 649–671
8. Suzuki, M., On a class of doubly transitive groups, Annals of Math. **75** (1962) 105–145

Oppositeness in Buildings and Simple Modules for Finite Groups of Lie Type

Peter Sin

Abstract In the building of a finite group of Lie type, we consider the incidence relations defined by oppositeness of flags. Such a relation gives rise to a homomorphism of permutation modules (in the defining characteristic) whose image is a simple module for the group. The p-rank of the incidence relation is then the dimension of this simple module. We give some general reductions towards the determination of the character of the simple module. Its highest weight is identified and the problem is reduced to the case of a prime field. The reduced problem can be approached through the representation theory of algebraic groups and the methods are illustrated for some examples.

Subject Classifications: 20E42, 20C33, 51E24

Keywords Buildings • Oppositeness • Simple modules • Modular representations

1 Introduction

Let G be a finite group with a split BN-pair of characteristic p and rank ℓ, and let $I = \{1, \ldots, \ell\}$. The subgroup B of the BN-pair is the semidirect product of a normal p-subgroup U with an abelian group T of order prime to p. The Weyl group W is a euclidean reflection group in a real vector space V, containing a root system

P. Sin (✉)
Department of Mathematics, University of Florida, P. O. Box 118105, Gainesville, FL 32611, USA
e-mail: sin@ufl.edu

N.S.N. Sastry (ed.), *Buildings, Finite Geometries and Groups*, Springer Proceedings in Mathematics 10, DOI 10.1007/978-1-4614-0709-6_13,

R and if $S = \{\alpha_i \mid i \in I\}$ is a set of simple roots then W is a Coxeter group with generators the fundamental reflections w_i, $i \in I$, where w_i is the reflection in the hyperplane perpendicular to the simple root α_i.

For $J \subset I$, let $W_J := \langle w_i \mid i \in J \rangle$ be the associated standard parabolic subgroup of W and $P_J = BW_JB$ the standard parabolic subgroup of G. By a *type* we simply mean a nonempty subset of I. An *object of type* $I \setminus J$, or *of cotype* J is, by definition, a right coset of P_J in G.

The associated building [17] is the simplicial complex in which the simplices are the cosets of the standard parabolic subgroups of G, and the face relation is the reverse of inclusion. An object of type $I \setminus J$ is a $(\mid I \setminus J \mid -1)$-simplex.

An example which has an elementary description is the building for $G = SL(V)$ of type A_ℓ, where V is an $(\ell + 1)$-dimensional vector space. A k-simplex is a flag of $k + 1$ nontrivial, proper subspaces of V, and its type is the set of dimensions of the subspaces in the flag.

Given two fixed types there are various incidence relations between objects of the respective types which are preserved by the action of G. In the above example, if we consider in V the complete flags (objects of type I) and the i-dimensional subspaces (objects of type $\{i\}$), then an incidence relation can be specified by prescribing a sequence of ℓ dimensions for the intersections of an i-dimensional subspace with the subspaces in a complete flag.

An incidence relation $\mathcal{R}$ between the sets of objects of cotypes J and K can be encoded by an incidence matrix or, more conveniently for our purposes, by an *incidence map of permutation modules*. Let k be a commutative ring with 1. Let $\mathcal{F}_J$ denote the space of functions from the set $P_J \backslash G$ of objects of cotype J to k. Then $\mathcal{F}_J$ is a left kG-module by the rule

$$(xf)(P_Jg) = f(P_Jgx), \quad f \in \mathcal{F}_J, \quad g, x \in G. \tag{1}$$

Let δ_{P_Jg} denote the characteristic function of the object $P_Jg \in P_J \backslash G$. These characteristic functions are permuted transitively by G and form a basis for $\mathcal{F}_J$.

A G-invariant relation $\mathcal{R}$ defines a kG-homomorphism $\eta : \mathcal{F}_J \to \mathcal{F}_K$ given by

$$\eta(f)(P_Kh) = \sum_{P_Jg\mathcal{R}P_Kh} f(P_Jg). \tag{2}$$

The characteristic function of an object of cotype J is sent to the sum of the characteristic functions of all objects of cotype K which are incident with it.

Naturally, we would like to compute invariants of η. We can ask for its Smith Normal Form when $k = \mathbb{Z}$, its rank when k is a field, or its eigenvalues if $J = K$. For greater detail we can also consider the kG-module structure of the image of η. We have described the problem in great generality and, as might be expected, the nature and difficulty of a specific instance will depend very much on G, $\mathcal{R}$ and k. For example, if k is a field then whether or not its characteristic is the same as that of G is crucial, because the representation theory of G in its defining characteristic is closely related to the representation theory of reductive algebraic groups, while the cross-characteristic representation theory of G has a closer connection to its complex character theory and to Brauer's theory of blocks. This dichotomy can be observed in the case of generalized quadrangles by comparing the results in [3,8,13]. In the first two papers, k has characteristic 2, while G has odd characteristic in the first and even characteristic in the second. In the third, the group G and k have the same odd characteristic.

In this paper, we restrict ourselves to the case where k is a field of the same characteristic as G and turn next to the choice of $\mathcal{R}$. We know from examples (e.g., [4]) that the permutation kG-module $\mathcal{F}_J$ can have many composition factors, growing with respect to the Lie rank of G, and also growing for fixed rank and fixed characteristic as the size of the field increases. The same may hold for the image of η. Even for rank 2, examples (e.g., [13]) demonstrate that the number of composition factors of $\mathrm{Im}\eta$ can grow without bound. In many cases, one knows neither the number of composition factors nor their dimensions, and in these cases we have no formula for $\mathrm{rank}\eta$. There are also examples of two problems which seem very similar on the surface, but one turns out to be much harder than the other. For example, the problem of computing the rank of the inclusion relation of 2-dimensional subspaces in $(n-2)$-dimensional subspaces in a vector space of dimension $n \geq 5$ is unsolved. In contrast, a simple formula is known [14] for the rank of the relation of nonzero intersection. It is therefore desirable, for a given $\mathcal{R}$, to have some idea *a priori* of the complexity of the kG-submodule lattice of $\mathrm{Im}\eta$. From this point of view, the oppositeness relations (defined below), which will be shown to give simple modules, can be considered to form the starting point of the theory.

2 Oppositeness in Buildings

Let R^+ denote the set of positive roots. The length of an element $w \in W$, denoted $\ell(w)$, is the length of the shortest expression for w as a word in the generators w_i. This is also equal to the number of positive roots which w transforms to negative roots. There is a unique element of maximal length, denoted w_0, which sends all positive roots to negative roots.

Notions of oppositeness exist at the level of types and at the level of objects. Two types J and K are *opposite* if $\{-w_0(\alpha_i) \mid i \in J\} = \{\alpha_j \mid j \in K\}$, or, equivalently, if $\{w_0 w_i w_0 \mid i \in J\} = \{w_i \mid i \in K\}$. In other words, two types are opposite when they are mapped to each other by the symmetry of the Dynkin diagram induced by w_0. If $w_0 = -1$, then every type is opposite itself; this holds for all connected root systems except for those of type A_ℓ, D_ℓ (ℓ odd) and E_6. Let J and K be fixed opposite cotypes. An object $P_J g$ of cotype J and an object $P_K h$ of cotype K are *opposite* each other if $P_K h g^{-1} P_J = P_K w_0 P_J$ ($\Longleftrightarrow$ $P_K h \subseteq P_J w_0 P_K g$ $\Longleftrightarrow$ $P_J g \subseteq P_K w_0 P_J h$).

The definition of oppositeness is a precise way to formulate the intuitive idea of two objects being in "general position." From now on we shall let J and K be fixed opposite cotypes.

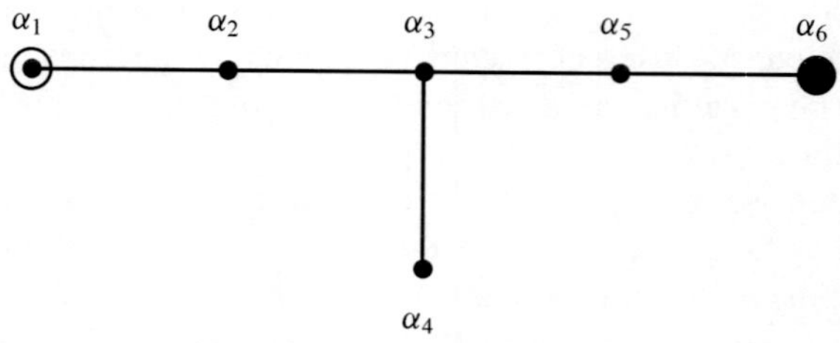

Fig. 1 E_6 diagram showing two opposite types

Example 2.1. Consider a universal Chevalley group G type A_ℓ and $J = I \setminus \{i\}$. Then $G \cong SL(V)$ for an $(\ell + 1)$-dimensional vector space V. The objects of type $\{i\}$ can be identified with i-dimensional subspaces of V and objects of the opposite type $\{\ell+1-i\}$ can be identified with $\ell+1-i$-dimensional subspaces. A subspace of type $\{i\}$ is opposite one of type $\{\ell+1-i\}$ if their intersection is the zero subspace. A familar special case is when $\ell = 3$ and $i = \ell+1-i = 2$. If we think projectively, the objects are lines in space and the oppositeness relation is skewness.

More generally, an object of cotype $J = \{j_1, \dots, j_m\}$ is a flag

$$V_{j_1} \subset V_{j_2} \subset \cdots \subset V_{j_m}$$

of subspaces of V with $\dim V_{i_j} = i_j$. If $V'_{k_1} \supset V'_{k_2} \supset \cdots \supset V'_{k_m}$ is an object of the opposite type, then the two flags are opposite iff $V_{i_j} \cap V'_{k_j} = \{0\}$, for $j = 1, \dots, m$.

Example 2.2. Let G be of type B_ℓ or C_ℓ with $\ell \geq 2$ or D_ℓ with $\ell \geq 3$ and let $J = I \setminus \{1\}$. Then J is opposite to itself. In the B_ℓ case, objects of cotype J can be identified with singular points (one-dimensional subspaces) with respect to a nondegenerate quadratic form in a finite vector space of dimension $2\ell + 1$. Two singular points are opposite if and only if they are not orthogonal. Similarly, objects of cotype J can be viewed as singular points of a 2ℓ-dimensional vector space with respect to a symplectic symplectic form for type C_ℓ or a quadratic form for type D_ℓ, with oppositeness interpreted as non-orthogonality. Two points are opposite if and only if they do not lie on a singular line. Thus, the concept of oppositeness generalizes the concept of collinearity of singular points in these classical geometries.

Example 2.3. Consider a universal Chevalley group G of type $E_6(q)$. This group has a concrete description as the group of linear transformations which preserve a certain cubic form on a 27-dimensional vector space V over $\mathbf{F}_q$. The geometry of this space has been studied in great detail. (See [1, 9, 10].) Consider the objects of type 1 and the opposite type 6. (See Fig. 1.) We can view these, respectively, as the singular points and singular (in a dual sense) hyperplanes of V. A singular point $\langle v \rangle$ is opposite a singular hyperplane H if and only $v \notin H$.

For further examples of oppositeness, we refer the reader to [5].

Let $A = A_{J,K}$ be the incidence matrix of the oppositeness relation, with rows indexed by objects of cotype J (in some order) and columns indexed by objects of cotype K. Suppose G is defined over $\mathbf{F}_q$, where q is a power of p. Brouwer [5] has shown that each eigenvalue of AA' is a power of q, where A' is the transpose of A.

In the following sections, we show that the p-rank of the incidence matrix A is the dimension of an irreducible p-modular representation of G. This fact is derived as a corollary of a general theorem of Carter and Lusztig [7]. Then we describe the simple module in terms of its highest weight and show, using Steinberg's Tensor Product Theorem that, given a root system and choice of opposite types, the p-ranks can be computed for all q once they are known in the case $q = p$. We then discuss methods for computing the character of the simple module in some examples.

3 Some Lemmas on Double Cosets

Let V_J be the subspace of V spanned by $S_J = \{\alpha_i \mid i \in J\}$. Then $R_J = R \cap V_J$ is a root system in V_J with simple system S_J and Weyl group W_J. For $w \in W_J$, its length as an element of W_J is the same as for W. Let w_J be the longest element in W_J.

The following is immediate from the definition of opposite types.

Lemma 3.1. *If $w_0 = w^* w_J = w_K w'$ then $w^* = w'$.* □

For $w \in W$, we recall that $U_w^- = U \cap w^{-1} w_0 U w_0 w$. For $w \neq 1$ this group is a nontrivial p-group. We also have that given a choice of preimage $n_w \in N$ of w, each element of BwB can be written as a unique product $g = bn_w u$, with $b \in B$ and $u \in U_w^-$. Thus, $\mid U_w^- \mid$ equals the number of cosets Bg' in BwB, which is equal to $q^{\ell(w)}$ in the untwisted case. In the case of twisted groups, the equation $\mid U_w^- \mid = q^{\ell(w)}$ is also valid if q is taken to be the *level* [11, Definition 2.1.9] of the Frobenius endomorphism defining G (which is a prime power or, for Suzuki and Ree groups, the square root of an odd power of 2 or 3) and $\ell(w)$ means the length in the Weyl group of the untwisted root system. We refer to [11, Theorems 2.3.8 and 2.4.1] and [6]. We shall use only the obvious fact that U_w^- is a nontrivial p-group when $w \neq 1$.)

Lemma 3.2 ([5], Proposition 3.1). *We have $P_K w_0 P_J = P_K w^* P_J = P_K w^* B$. The number of cosets $P_K g$ in $P_K w_0 P_J$ is $q^{\ell(w^*)}$.* □

Lemma 3.3 (cf. [5], Corollary 3.2). *Let $w \in W_J$. Then $B(w^* w)B \subseteq P_K w_0 P_J$. Let $x \in G$. For a given coset $P_K h \subseteq P_K w_0 P_J x$, the number of cosets $Bg \subseteq Bw^* wBx$ such that $Bg \subseteq P_K h$ is $q^{\ell(w)}$.*

Proof. The first assertion is true because $P_K w^* w P_J = P_K w^* P_J = P_K w_0 P_J$. The rest follows by counting. The number of cosets $P_K h \subseteq P_K w_0 P_J g$ is $q^{\ell(w^*)}$ and the number of cosets $Bg \subseteq Bw^* wBx$ is $q^{\ell(w^* w)} = q^{\ell(w^*)+\ell(w)}$.

4 Permutation Modules on Flags and Their Oppositeness Homomorphisms

Let k be an algebraically closed field of characteristic p. Recall from the Introduction that $\mathcal{F}_J$ denotes the space of functions from the set $P_J\backslash G$ of objects of cotype J to k, with left kG-module structure given by (1). Also, $\delta_{P_J g}$ denotes the characteristic function of the object $P_J g \in P_J\backslash G$ and these functions form a permutation basis of $\mathcal{F}_J$. (The module $\mathcal{F}_J$ is isomorphic to the kG-permutation module on the *l*eft cosets of P_J in G by the map sending $\delta_{P_J g}$ to $g^{-1}P_J$.) The relation of oppositeness defines the kG-homomorphism $\eta : \mathcal{F}_J \to \mathcal{F}_K$ given by

$$\eta(f)(P_K h) = \sum_{P_J g \subseteq P_J w_0 P_K h} f(P_J g) \tag{3}$$

and we have

$$\eta(\delta_{P_J g}) = \sum_{P_K h \subseteq P_K w_0 PJg} \delta_{P_K h}.$$

The following result is essentially a corollary of more general results in [7].

Theorem 4.1. *The image of η is a simple module, uniquely characterized by the property that its one-dimensional U-invariant subspace has full stablizer equal to P_J, which acts trivially on it.*

The next subsections describe some results in [7] and explain how Theorem 4.1 follows from them.

4.1 Fundamental Endomorphisms of $\mathcal{F}_\emptyset$

Let $\mathcal{F} = \mathcal{F}_\emptyset$. For each $w \in W$ we define $T_w \in \mathrm{End}_k(\mathcal{F})$ by the formula

$$T_w(f)(Bg) = \sum_{Bg' \subseteq Bw^{-1}Bg} f(Bg').$$

Then

$$T_w \in \mathrm{End}_{kG}(\mathcal{F}), \quad \text{for all } w \in W.$$

One can show (see [7]) that

$$T_{ww'} = T_w T_{w'} \quad \text{if } \ell(ww') = \ell(w) + \ell(w').$$

Let $w \in W$ have reduced expression

$$w_{j_n} \cdots w_{j_1}.$$

We consider the partial products $w_{j_1}, w_{j_2}w_{j_1}, \ldots w_{j_n} \cdots w_{j_1}$. The positive roots sent to negative roots by each partial product are also sent to negative roots by each of its successors and each partial product sends exactly one more positive root to a negative root than its predecessors. Explicitly, the new positive root sent by $w_{j_i} \cdots w_{j_1}$ to a negative root is $w_{j_1} \cdots w_{j_{i-1}}(r_{j_i})$. With this in mind, we can define the endomorphsim $\Theta^J_{w_0}$ of [7, p. 363] for any J subset of I.

For any reduced expression

$$w_0 = w_{j_k} \cdots w_{j_1}$$

define

$$\Theta_{j_i} = \begin{cases} T_{w_{j_i}} & \text{if } w_{j_1} \cdots w_{j_{i-1}}(r_{j_i}) \notin V_J \\ I + T_{w_{j_i}} & \text{if } w_{j_1} \cdots w_{j_{i-1}}(r_{j_i}) \in V_J \end{cases} \tag{4}$$

and set

$$\Theta^J_{w_0} = \Theta_{j_k}\Theta_{j_{k-1}} \cdots \Theta_{j_1}.$$

The definition depends on the choice of reduced expression but it is shown in [7] that different expressions give the same endomorphism up to a nonzero scalar multiple.

We can now state the result from which Theorem 4.1 can be deduced. We recall that in every simple kG-module the subspace fixed by the subgroup U is one-dimensional. Therefore, the full stabilizer in G of this subspace must contain $N_G(U) = B$, so must be equal to some standard parabolic subgroup P_Q, $Q \subseteq I$. Thus this line is a one-dimensional kP_Q-module.

Theorem 4.2 ([7], Theorem 7.1). *The image $\Theta^J_{w_0}(\mathcal{F})$ is a simple kG-module. The full stablizer of the one-dimensional subspace of U-fixed points in this module is P_J and the action of P_J on this one-dimensional subspace is trivial.*

(Our space $\mathcal{F}$ is the space denoted $\mathcal{F}_\chi$ in [7] when χ is the trivial character.)

4.2 *Proof of Theorem 4.1*

Let

$$w_J = w_{i_m} \cdots w_{i_2} w_{i_1} \tag{5}$$

be a reduced expression for w_J. The above expression can be extended to a reduced expresion

$$w_0 = w_{i_k} \cdots w_{i_{m+1}} w_{i_m} \cdots w_{i_1} \tag{6}$$

of w_0. Thus $m = | R_J^+ |$ and $k = | R^+ |$.

Then

$$w^* = w_{i_k} \cdots w_{i_{m+1}} \tag{7}$$

is a reduced expression for w^*. We choose the special expression (6) for w_0 to define $\Theta_{w_0}^J$. Since w_J sends all positive roots in V_J to negative roots and w_0 sends all negative roots to positive roots, it is clear that for the first m partial products the new positive root sent to a negative root belongs to V_J, and that the new positive roots for the remaining partial products are the elements of $R^+ \setminus R_J^+$, so do not belong to V_J. Thus we have

$$\Theta_{w_0}^J = T_{w^*}(1 + T_{i_m}) \cdots (1 + T_{i_1}), \tag{8}$$

Since $\ell(w^* w) = \ell(w^*) + \ell(w)$ for all $w \in W_J$, we see that $\Theta_{w_0}^J$ is a sum of endomorphisms of the form $T_{w^* w}$, for certain elements $w \in W_J$, with exactly one term of this sum equal to T_{w^*}.

Let $\pi_J : \mathcal{F} \to \mathcal{F}_J$ be defined by

$$(\pi_J(f))(P_J g) = \sum_{Bh \subseteq P_J g} f(Bh)$$

and π_K defined similarly. It is easily checked that π_J and π_K are kG-module homomorphisms and they are surjective since $\pi_J(\delta_B) = \delta_{P_J}$.

We can now complete the proof of Theorem 4.1. The main step is to compare $\eta\pi_J$ with $\pi_K T_{w^* w}$ for $w \in W_J$. For $f \in \mathcal{F}$, we compute

$$\begin{aligned}
[\eta(\pi_J(f))]\,(P_K g) &= \sum_{P_J h \subseteq P_J w^{*-1} P_K g} \; \sum_{Bx \subseteq P_J h} f(Bx) \\
&= \sum_{Bx \subseteq P_J w^{*-1} P_K g} f(Bx)
\end{aligned} \tag{9}$$

and

$$\begin{aligned}
[\pi_K(T_{w^* w}(f))]\,(P_K h) &= \sum_{Bg \subseteq P_K h} (T_{w^* w} f)(Bg) \\
&= \sum_{Bg \subseteq P_K h} \; \sum_{Bx \subseteq B(w^* w)^{-1} Bg} f(Bx) \\
&= \sum_{Bg \subseteq P_K h} \; \sum_{Bg \subseteq B(w^* w) Bx} f(Bx) \\
&= q^{\ell(w)} \sum_{Bx \subseteq P_J w^{*-1} P_K g} f(Bx),
\end{aligned} \tag{10}$$

where the last equality follows from Lemma 3.3, since $Bx \subseteq B(w^*w)^{-1}Bg$ if and only if $Bg \subseteq Bw^*wBx$. Thus, we have for each $w \in W_J$ a commutative diagram

$$\begin{array}{ccc} \mathcal{F}_J & \xrightarrow{\;q^{\ell(w)}\eta\;} & \mathcal{F}_K \\ {\scriptstyle \pi_J}\Big\uparrow & & \Big\uparrow{\scriptstyle \pi_K} \\ \mathcal{F} & \xrightarrow{\;T_{ww^*}\;} & \mathcal{F}, \end{array} \tag{11}$$

which means that for $w \neq 1$ we have $\pi_K T_{ww^*} = 0$. It now follows from (8) that $\pi_K \Theta^J_{w_0} = \pi_K T_{w^*} = \eta\pi_J$. Therefore, since $\Theta^J_{w_0}(\mathcal{F})$ is simple and $\eta\pi_J(\mathcal{F}) \neq 0$, we see that $\eta\pi_J(\mathcal{F}) \cong \Theta^J_{w_0}(\mathcal{F})$. Finally, since π_J is surjective, we have $\eta(\mathcal{F}_J) \cong \Theta^J_{w_0}(\mathcal{F})$. This completes the proof of Theorem 4.1. □

5 Highest Weights

If G is a universal Chevalley group or a twisted subgroup of such a group then there is an algebraic group $G(k)$, the Chevalley group over k, and a Frobenius endomorphism σ of $G(k)$ such that G is the group of fixed points of σ. References for this theory are the original paper [16, 12.4] or the very complete account in [11, Chap. 2].

The simple kG-modules are restrictions of certain simple rational modules $L(\lambda)$ of $G(k)$ [16, 13.1], where λ is the highest weight of the module. This connection is well known; for definitions and details we refer to the original sources [15, 16]. The simple module $L(\lambda)$ has a unique B-stable line, and T acts on this line by the character λ.

Assume that G is a universal Chevalley group over $\mathbf{F}_q$. By Steinberg's theorem [16, 13.3], the simple $G(k)$ modules with highest weights in the set of *q-restricted weights*

$$X_q = \left\{ \lambda = \sum_{i=1}^{\ell} a_i \omega_i \in X_+ \mid 0 \leq a_i \leq q-1 \ (\forall i) \right\}$$

remain irreducible upon restriction to G and this gives a complete set of mutually non-isomorphic simple kG-modules.

It is useful to identify the highest weight of the module in Theorem 4.1. We may assume that R is indecomposable, since simple modules for direct products are easily described in terms of the factors.

Let λ_{opp} denote the q-restricted highest weight such that the restriction of the simple $G(k)$-module $L(\lambda_{\mathrm{opp}})$ to G is the simple module in Theorem 4.1. The condition in Theorem 4.1 that P_J is the full stabilizer in G of the one-dimensional highest weight space of $L(\lambda_{\mathrm{opp}})$ is equivalent to the condition that for $i = 1, \ldots,$

ℓ, $\langle \lambda_{\text{opp}}, \alpha_i^\vee \rangle = 0$ if and only if $i \in J$. The condition in Theorem 4.1 that P_J acts trivially on the highest weight space means that the restriction of λ_{opp} to T is the trivial character, which in turn means that, when λ_{opp} is written as a linear combination of fundamental weights, all of the coefficients must be 0 or $q-1$. These conditions allow us to identify λ_{opp}. The fundamental weights ω_i for the ambient algebraic group are indexed by I, and $\lambda_{\text{opp}} = \sum_{i \in I \setminus J}(q-1)\omega_i$.

Consider next a twisted group G, constructed from an overlying universal Chevalley group G^* over $\mathbf{F}_q$, as the group of fixed points of an automorphism induced by an isometry ρ of order e of the Dynkin diagram of G^*. Then $q = q_0^e$ for some prime power q_0. Let $I^* = \{1, \dots, \ell^*\}$ index the fundamental roots for G^*. Then the index set I for G labels the ρ-orbits on I^*. Let ω_i, $i \in I^*$ be the fundamental weights of the ambient algebraic group. For $J \subseteq I$, let $J^* \subset I^*$ be the union of the orbits in J. Then $\lambda_{\text{opp}} = \sum_{i \in I^* \setminus J^*}(q_0 - 1)\omega_i$.

Finally, we consider the case of Suzuki and Ree groups. Here G is the subgroup of fixed points in $G(k)$ of a Steinberg endomorphism τ which induces a length-changing permutation of the Dynkin diagram of $G(k)$ and such that $\tau^2 = \sigma$ is a Frobenius endomorphism of $G(k)$ with respect to a rational structure over a finite field $\mathbf{F}_q$. Then the set I for G indexes the subset of fundamental weights of the ambient algebraic group which are orthogonal to the long simple roots, and for $J \subset I$, we have $\lambda_{\text{opp}} = \sum_{i \in I \setminus J}(q-1)\omega_i$.

Example 5.1. As examples, we can consider the extreme cases. If $J = K = I$, then $L(\lambda_{\text{opp}}) \cong k$. If $J = K = \emptyset$, $L(\lambda_{\text{opp}})$ is the *Steinberg module* of the group G, which has dimension equal to the the order of U, a p-Sylow subgroup of G.

From the discussion above, we see that λ_{opp} has the form $(q-1)\tilde{\omega}$, where $\tilde{\omega}$ is a sum of fundamental weights. If $q = p^t$, then by Steinberg's Tensor Product Theorem, we have an isomorphism

$$L((q-1)\tilde{\omega}) \cong L((p-1)\tilde{\omega}) \otimes L((p-1)\tilde{\omega})^{(p)} \otimes \cdots \otimes L((p-1)\tilde{\omega})^{(p^{t-1})} \quad (12)$$

as modules for the algebraic group. Here, the superscripts indicate twisting by powers of the Frobenius morphism. This twisting changes the isomorphism type of a module, but does not change its dimension. In particular, we have the following numerical result.

Proposition 5.2. *Let the root system R and opposite cotypes J and K be given and let $A(q) = A(q)_{J,K}$ denote the oppositeness incidence matrix for objects of cotypes J and K in the building over $\mathbf{F}_q$, where $q = p^t$. Then* $\operatorname{rank}_p A(q) = (\operatorname{rank}_p A(p))^t$.

Remark 5.3. The proposition states that once $\operatorname{rank}_p A(p)$ is known then we know $\operatorname{rank}_p A(q)$ for all powers q of p. This reduction of the computation to the prime case is significant from the viewpoint of representation theory of algebraic groups, where the Weyl modules with modules highest weight $(p-1)\tilde{\omega}$ are much less complex in structure than those of highest weight $(p^2-1)\tilde{\omega}$, say.

A very useful tool for analyzing Weyl modules is the *Jantzen Sum Formula* [12, II.8.19]: The Weyl module $V(\lambda)$ has a descending filtration, of submodules $V(\lambda)^i$, $i > 0$, such that

$$V(\lambda)^1 = \mathrm{rad}V(\lambda), \quad \text{(so that } V(\lambda)/V(\lambda)^1 \cong L(\lambda))$$

and

$$\sum_{i>0} \mathrm{Ch}(V(\lambda)^i) = -\sum_{\alpha>0} \sum_{\{m:0<mp<\langle\lambda+\rho,\alpha^\vee\rangle\}} \nu_p(mp)\chi(\lambda - mp\alpha) \qquad (13)$$

First, we recall the notation in this formula (which is standard, and follows from [12]). The module $V(\lambda)$ is the Weyl module and the module $L(\lambda)$ is its simple quotient. The weight ρ is the half-sum of the positive roots and $\nu_p(m)$ denotes the exponent of p in the prime factorization of m. Finally, the character $\chi(\mu)$ is the so-called Weyl character; there is a unique weight of the form $\mu' = w(\mu + \rho) - \rho$ in the region $\{\nu : \langle\nu + \rho, \alpha^\vee\rangle \geq 0, \forall\alpha \in R^+\}$, where $w \in W$. Then $\chi(\mu)$ is the sign(w)Ch$V(\mu')$ if μ' is dominant, and zero otherwise. The usefulness of the sum formula is based on the fact that the characters of the Weyl modules themselves are given by Weyl's Character Formula, so that the right hand side can be computed from p, R and λ. In [2] there is a detailed description of a procedure for performing this computation. We shall refer to the quantity in (13) as the Jantzen sum for $V(\lambda)$. As can be seen from the left hand side of (13), the Jantzen sum gives an upper estimate on the composition multiplicities in the radical of the Weyl module $V(\lambda)$ in terms of the composition factors of Weyl modules which have lower highest weights. Sometimes, for weights of a special form, more information can be obtained, using induction and other facts from representation theory. For example when we start with a weight of the form $\lambda = (p-1)\tilde{\omega}$, it may be that the highest weights of the Weyl characters $\chi(\mu)$ in the Jantzen sum are very few in number or all have a similar form, such as $r\tilde{\omega}$ for $r < p - 1$. In such cases, it is possible to deduce the character of $L((p-1)\tilde{\omega})$. Numerous examples of this method were worked out in detail in [2].

6 Examples

We consider again some of the examples from the Introduction.

Example 6.1. For type A_ℓ, $J = I\backslash\{i\}$, the p-ranks have been computed in [14]. (In fact it is the p-rank of the complementary relation of nonzero intersection which is computed, which equals one plus the p-rank for zero intersection.) In this case, the simple modules $L((p-1)\omega_i)$ can be found without reference to Weyl modules. Let $S(i(p-1))$ denote the degree $i(p-1)$ homogeneous component of the truncated polynomial ring $k[x_0,\ldots,x_\ell]/(x_i^p; 0 \leq i \leq \ell)$, with $G \cong \mathrm{SL}((\ell+1,k)$ acting through linear substitutions. Then it is well known and elementary to show that

$S(i(p-1))$ is a simple kG-module. By inspecting the highest weights, we see that $S(i(p-1)) \cong L((p-1)\omega_{\ell+1-i})$, for $i = 1,\dots,\ell$.

Example 6.2. For the examples of types B_ℓ, C_ℓ and D_ℓ, with $J = I \setminus \{1\}$ concerning singular points in classical spaces, the p-ranks have been computed in [2] by analysis of the Weyl modules, as outlined above. The Weyl modules in question are $V((p-1)\omega_1)$ and for type C_ℓ they are simple, so work is only needed for types B_ℓ and D_ℓ. The method can also be extended to the classical modules of the classical groups of twisted type, namely the non-split orthogonal groups (type ${}^2D_\ell$) and the unitary groups (type ${}^2A_\ell$). In the latter case, one must study the Weyl module $V((p-1)(\omega_1 + \omega_\ell))$, in accordance with our discussion in Sect. 5.

Example 6.3. We will sketch the computation of the p-rank for Example 2.3. The Jantzen sum for $V((p-1)\omega_1)$ is equal to the following:

$$\chi((p-7)\omega_1 + 3\omega_6) - \chi((p-8)\omega_1 + \omega_4 + 2\omega_6) + \chi((p-9)\omega_1 + \omega_3 + \omega_6)$$
$$-\chi((p-10)\omega_1 + \omega_2 + \omega_5) + \chi((p-11)\omega_1 + 2\omega_2) \tag{14}$$

The next step is to study the structure of the Weyl modules appearing in this character, again by using the sum formula.

We observe first that the Jantzen sum for $V((p-11)\omega_1 + 2\omega_2)$ is zero, so this Weyl module is simple.

Next, the Jantzen sum for $V((p-10)\omega_1 + \omega_2 + \omega_5)$ is $\chi((p-11)\omega_1 + 2\omega_2)$. It follows that the radical of $V((p-10)\omega_1 + \omega_2 + \omega_5)$ is isomorphic to the simple module $V((p-11)\omega_1 + 2\omega_2)$.

The Jantzen sum for $V((p-9)\omega_1 + \omega_3 + \omega_6)$ is equal to

$$\chi((p-10)\omega_1 + \omega_2 + \omega_5) - \chi((p-11)\omega_1 + 2\omega_2),$$

which by the previous paragraph is equal to $\mathrm{Ch}L((p-10)\omega_1 + \omega_2 + \omega_5)$. Therefore, $\mathrm{rad}V((p-9)\omega_1 + \omega_3 + \omega_6) \cong L((p-10)\omega_1 + \omega_2 + \omega_5)$.

The Jantzen sum for $V((p-8)\omega_1 + \omega_4 + 2\omega_6)$ is

$$\chi((p-9)\omega_1 + \omega_3 + \omega_6) - \chi((p-10)\omega_1 + \omega_2 + \omega_5) + \chi((p-11)\omega_1 + 2\omega_2)$$
$$= \mathrm{Ch}L((p-9)\omega_1 + \omega_3 + \omega_6),$$

which leads us to conclude that $\mathrm{rad}V((p-8)\omega_1 + \omega_4 + 2\omega_6) \cong L((p-9)\omega_1 + \omega_3 + \omega_6$.

The Jantzen sum for $V((p-7)\omega_1 + 3\omega_6)$ is

$$\chi((p-8)\omega_1 + \omega_4 + 2\omega_6) - \chi((p-9)\omega_1 + \omega_3 + \omega_6) + \chi((p-10)\omega_1 + \omega_2 + \omega_5)$$
$$-\chi((p-11)\omega_1 + 2\omega_2) \cong \mathrm{Ch}L((p-8)\omega_1 + \omega_4 + 2\omega_6). \tag{15}$$

Hence, $\mathrm{rad}V((p-7)\omega_1 + 3\omega_6) \cong L((p-8)\omega_1 + \omega_4 + 2\omega_6)$.

Finally, by (14), we have $\mathrm{rad}V((p-1)\omega_1) \cong L((p-7)\omega_1 + 3\omega_6)$. Therefore, there is an exact sequence

$$\begin{aligned} 0 &\to V((p-11)\omega_1 + 2\omega_2) \to V((p-10)\omega_1 + \omega_2 + \omega_5) \\ &\to V((p-9)\omega_1 + \omega_3 + \omega_6) \to V((p-8)\omega_1 + \omega_4 + 2\omega_6) \\ &\to V((p-7)\omega_1 + 3\omega_6) \to V((p-1)\omega_1) \to L((p-1)\omega_1) \to 0 \end{aligned}$$

The dimensions of the $V(\mu)$ are given by Weyl's formula. Hence,

$$\begin{aligned} \dim L((p-1)\omega_1) &= \frac{1}{2^7.3.5.11} p(p+1)(p+3) \\ &\times \big(3p^8 - 12p^7 + 39p^6 + 320p^5 \\ &-550p^4 + 1240p^3 + 2080p^2 - 1920p + 1440\big) \end{aligned}$$

Acknowledgement Thanks are due to the anonymous referees for their careful reading and helpful suggestions.

References

1. Aschbacher, M. The 27-dimensional module for E_6, I, Inventiones Mathematicae **89** (1987), 159–195
2. Arslan, O. and Sin, P. Some simple modules for classical groups and p-ranks of orthogonal and Hermitian geometries, *J. Algebra* **327** (2011), 141–169
3. Bagchi, B. Brouwer, A.E. and Wilbrink, H.A. Notes on binary codes related to the $O(5, q)$ generalized quadrangle for odd q, Geometriae Dedicata **39** (1991), 339–355
4. Bardoe, M. and Sin, P. The permutation modules for $\mathrm{GL}(n+1, \mathbb{F}_q)$ acting on $\mathbb{P}^n(\mathbb{F}_q)$ and $\mathbb{F}_q^{n+1}$, *J. London Math. Soc.* **61** (2000), 58–80
5. Brouwer, A. E. The eigenvalues of oppositeness graphs in spherical buildings, *Combinatorics and Graphs*, R.A. Brualdi, S. Hedayat, H. Kharaghani, G.B. Khosrovshahi, S. Shahriari (eds.), AMS Contemporary Mathematics Series 531, 2010
6. Carter, R.W. *Simple groups of Lie type*, Pure and Applied Mathematics, Vol. 28, John Wiley & Sons, London-New York-Sydney, 1972
7. Carter, R.W. and Lusztig, G. Modular representations of finite groups of Lie type, Proc. London Math. Soc. (3), **32** (1976), 347–384
8. Chandler, D.B. Sin, P. Xiang, Q. The permutation action of finite symplectic groups of odd characteristic on their Standard Modules, J. Algebra **318** (2007), 871–892
9. Cohen, A, Cooperstein, The 2-spaces of the standard E_6-module, Geometriae Dedicata 25 (1988) 467–480
10. Cooperstein, B. On a Connection Between Hyperbolic Ovoids on the Hyperbolic Quadric Q (10,q) and the Lie Incidence Geometry $E_{6,1}(q)$, 55–64 in Groups and Geometries, Birkhauser, 1998
11. Gorenstein, D. Lyons, R. Solomon, R. *The Classification of the Finite Simple Groups. Number 3. Part I. Chapter A, Almost Simple K-Groups*, Mathematical Surveys and Monographs, vol. 40.3, American Mathematical Society, Providence, RI, 1998
12. Jantzen, J.C. *Representations of Algebraic Groups*, Academic Press, London, 1987

13. Sastry, N.S.N. and Sin, P. The code of a regular generalized quadrangle of even order, in *Group Representations: Cohomology, Group Actions and Topology*, Proc. Symposia in Pure Mathematics **63** (1998), 485–496
14. Sin, P. The p-rank of the incidence matrix of intersecting linear subspaces, Designs, Codes and Cryptography **31** (2004), 213–220
15. Steinberg, R. Representations of algebraic groups, Nagoya Math. J. **22** (1963), 33–56
16. Steinberg, R. Endomorphisms of linear algebraic groups, Memoirs Amer. Math. Soc **80** (1968)
17. Tits, J. *Buildings of spherical type and finite BN-pairs*, Lecture Notes in Mathematics, Vol. 386, Springer, Berlin 1974

Modular Representations, Old and New

Bhama Srinivasan

To the memory of Harish-Chandra

Abstract The modular representation theory of finite groups has its origins in the work of Richard Brauer. In this survey article, we first discuss the work being done on some outstanding conjectures in the theory. We then describe work done in the eighties and nineties on modular representations in non-defining characteristic of finite groups of Lie type. In the second part of the article, we discuss some recent developments in the theory for symmetric groups and Hecke algebras, where remarkable connections with Lie theory and graded representation theory have been made.

Subject Classifications: Primary 20C20, Secondary: 20C33, 20C30, 20C08

Keywords Modular representations

1 Introduction

The art of story telling is very old in India, as is evidenced by the great epics Ramayana and Mahabharata. Even now lectures by scholars who tell contemporary versions of these epics are very popular, as was a TV series on the Ramayana, where it is said that a train stopped at a station long enough for the passengers to get out and see the latest episode.

In this paper, I would like to tell a story of modular representations of symmetric groups, Hecke algebras and related objects, starting from Brauer's introduction of

---

B. Srinivasan (✉)
Department of Mathematics, Statistics, and Computer Science (MC 249),
University of Illinois at Chicago, 851 South Morgan Street, Chicago,
IL 60607-7045, USA
e-mail: srinivas@uic.edu

N.S.N. Sastry (ed.), *Buildings, Finite Geometries and Groups*, Springer Proceedings in Mathematics 10, DOI 10.1007/978-1-4614-0709-6_14,

the concept and describing some recent developments connecting this theory with Lie theory.

We begin with stating some conjectures in the representation theory of finite groups, some of them being long-standing, and discuss recent progress in them. We then discuss the ordinary, or characteristic 0, representation theory of finite reductive groups, including Harish–Chandra theory and Deligne–Lusztig theory, which also play a role in the modular representation theory. The main theorem in the classification of blocks in the ℓ-modular representation theory is stated.

In recent years, there has been a new direction in the modular representation theory of symmetric groups and Hecke algebras via connections with Lie theory, leading to the concept of graded representation theory. A detailed, definitive account of these developments has been given by Kleshchev [32]. In the second part of the paper we give an introduction to these ideas, and hope that the reader will then be encouraged to read [32] and other papers in the literature.

1.1 Notation

Given an algebra A, $A - mod$ (resp. $A - pmod$) is the category of finitely generated A-modules (resp. projective A-modules). For a category $\mathcal{A}$, $K_0(\mathcal{A})$ is the Grothiendieck group. For an algebra A, $K_0(A)$ is the Grothiendieck group of the category $A - mod$.

For a finite group G, $\mathrm{Irr}(G)$ is the set of ordinary irreducible characters of G. The usual inner product on the space of class functions on a finite group will be denoted by $\langle -, - \rangle$.

2 Finite Groups

References for this section are [14, Chap. 12]. All groups in this section will be finite.

Let G be a finite group. The theory of group characters goes back to the work of Frobenius. If $\rho : G \to GL(n, K)$ is a representation of degree n over a field K, the character χ of ρ is the class function on G defined by $\chi(g) = \mathrm{Tr}(\rho(g))$. If $K = \mathbf{C}$ (or a sufficiently large field of characteristic 0) the (ordinary) irreducible characters form an orthonormal basis of the space of K-valued class functions on G. The number of ordinary characters is the number of conjugacy classes of the group, and this leads to a "character table" for the group which yields a lot of information about the group. We can now state

Problem 1.

(1) Classify and describe the (ordinary) irreducible characters of G over K as above, e.g., give their degrees.
(2) Describe the character table of G.

For some well-known groups such as the symmetric group S_n or the finite general linear group $GL(n, q)$ we can answer these questions. In a fundamental paper in 1955, Green constructed the characters of the groups $GL(n, q)$ (see e.g., Springer, Characters of special groups, in [7]).

Frobenius computed the first character table, that of $PSL(2, p)$. He also defined the important notion of induced characters, which we still use today even though there are generalizations for groups of Lie type, for example.

The theory of modular representations, i.e., the representations of G over a field k of characteristic p where p divides the order of G was developed by Richard Brauer starting in the 1930s. In this case, Brauer defined characters of the irreducible representations as complex-valued functions on the p-regular classes of the group, and they are now called Brauer characters. Brauer then divided the ordinary characters into subsets called "blocks" as follows.

Let K be a sufficiently large field of characteristic 0, $\mathcal{O}$ a complete discrete valuation ring with quotient field K, and k a residue field of $\mathcal{O}$ such that the characteristic of k is p. Consider the algebras KG, $\mathcal{O}G$, kG. We have $\mathcal{O}G = B_1 \oplus B_2 \oplus \ldots \oplus B_n$ where the B_i are "block algebras," indecomposable ideals of $\mathcal{O}G$. We have a corresponding decomposition of kG. The principal block is the one which contains the trivial character of G.

An invariant of a block B of G is the defect group, a p-subgroup P of G, unique up to G-conjugacy. One definition of P is that P is minimal with respect to: Every B-module is a direct summand of an induced module from P. The "Brauer correspondence" then gives a bijection between blocks of G of defect group P and blocks of $N_G(P)$ of defect group P.

An ordinary representation of G, i.e., a representation over K, is equivalent to a representation over $\mathcal{O}$, and can then be reduced mod p to get a modular representation of G over k. Then one can define the "decomposition matrix," the transition matrix between ordinary characters and Brauer characters. This leads to:

– A partition of the ordinary characters, or KG-modules, into blocks
– A partition of the Brauer characters, or kG-modules, into blocks
– A partition of the decomposition matrix into blocks

Example: A_5, $p = 2$: Ordinary characters in the principal block:

Order of element	1	2	3	3	5
Class size	1	15	20	12	12
χ_1	1	1	1	1	1
χ_2	5	1	−1	0	0
χ_3	3	−1	0	$\frac{1-\sqrt{5}}{2} - 1$	$\frac{1+\sqrt{5}}{2} - 1$
χ_4	3	−1	0	$\frac{1+\sqrt{5}}{2} - 1$	$\frac{1-\sqrt{5}}{2} - 1$

Example: A_5, $p = 2$: Brauer characters in the principal block:

Order of element	1	3	3	5
Class size	1	20	12	12
ψ_1	1	1	1	1
ψ_2	2	-1	$\frac{1+\sqrt{5}}{2} - 1$	$\frac{1-\sqrt{5}}{2} - 1$
ψ_3	2	-1	$\frac{1-\sqrt{5}}{2} - 1$	$\frac{1+\sqrt{5}}{2} - 1$

Decomposition matrix for Principal Block of A_5: $\begin{pmatrix} 1 & 1 & 1 & 1 \\ 0 & 0 & 1 & 1 \\ 0 & 1 & 0 & 1 \end{pmatrix}$

Some of the main problems in the modular theory are:

Problem 2.

1. Describe the Brauer characters of G over a (sufficiently large) field k of characteristic p as above, e.g., give their degrees.
2. Describe the blocks.
3. Find the decomposition matrix D.
4. Global to local: Describe information on the block B by "local information," i.e., from blocks of subgroups of the form $N_G(P)$, P a p-group.

Problem 2(1) is hard, and still open even for S_n and $GL(n, q)$. We will say more about (2) and (3) later. We discuss (4) below.

The Sylow theorems are among the first facts we learn about finite groups. One of Brauer's insights was to realize the importance of what we now call local subgroups of a group G, i.e., subgroups of the form $N_G(P)$ where P is a p-subgroup of G, for p- modular representation theory. His idea is that "global information" about a p-block of G must be obtainable from "local information," i.e., the characters of $N_G(P)$ where P is the defect group of the block. A striking example of this is when the defect group of the block is cyclic (see [1, p. 126]).

There have been several conjectures put forward in modular representation theory relating the characters in a block of G with the characters in blocks of local subgroups of G. Of these, the simplest to state is the McKay conjecture, which asserts that for every finite group G and every prime p, the number of irreducible characters of G having degree prime to p is equal to the number of such characters of the normalizer of a Sylow p-subgroup of G. There also exists a block-wise version, the Alperin–Mckay conjecture, comparing characters in blocks of G and in the normalizer of their defect groups. The Alperin weight conjecture counts the number of Brauer characters [13, p. 96], [42]. There are various further deep conjectures in the representation theory of finite groups, the most elaborate ones being the Isaacs–Navarro conjectures, Dade's conjectures and Broué's conjectures [2, 42].

The McKay-conjecture was proved for solvable groups and more generally for p-solvable groups as well as for various classes of non-solvable groups, like the symmetric groups and the general linear groups over finite fields, but it remained open in general. Recently, Isaacs, Malle and Navarro [22] reduced the McKay conjecture to a question about simple groups and gave a list of conditions that they hoped all simple groups will satisfy. They showed that the McKay conjecture will hold for a group G if every simple group involved in G satisfied these conditions. For more on this problem, see [39, 40]. These are conjectures at the level of characters.

Let $\mathcal{O}$ be as above. Given a block B of a group G and a block b of a group H (e.g., $H = N_G(D)$, D defect group of B), a perfect isometry is a bijection between $K_0(B)$ and $K_0(b)$ preserving certain invariants of B and b, and an isotypy is a collection of compatible perfect isometries [9]. Here, B and b are regarded as $\mathcal{O}$-algebras. At the level of characters, Broué's conjecture is that there is a perfect isometry between $K_0(B)$ and $K_0(b)$ where $H = N_G(D)$. At the module level we have Broué's abelian defect group conjecture [9], which we state below.

If A is an $\mathcal{O}$-algebra, $\mathcal{D}^b(A)$ is the bounded derived category of the category $A - mod$. It is a triangulated category.

2.1 Broué's Abelian Defect Group Conjecture

Conjecture 2.1. Let B be a block of G with the abelian defect group D, b the Brauer correspondent of B in $N_G(D)$. Then $\mathcal{D}^b(B)$ and $\mathcal{D}^b(b)$ are equivalent as triangulated categories.

A discussion of the Abelian Defect Group Conjecture is in the 1998 ICM address of Rickard [44]. Marcus [42] gives the current status of the conjecture and describes some of the methods used to prove it.

3 Finite Reductive Groups

References for this section are [10, 11, 13, 15, 19].

Until 1955 the known finite simple groups which included the classical groups and the alternating groups were all studied separately. A fundamental paper of Chevalley [12] changed this by giving a unified treatment of finite simple groups, arising from simple Lie algebras. A treatment of this theory can be found in [11].

The modern view is as follows. Let $\mathbf{G}$ be a connected reductive group defined over $\mathbf{F}_q$, F a Frobenius endomorphism, $F : \mathbf{G} \to \mathbf{G}$. Then $G = \mathbf{G}^F$, the group of F-fixed points of F, is called a finite reductive group (or finite group of Lie type). Using the structure of reductive algebraic groups we get subgroups of $\mathbf{G}$ as follows. A torus is a closed subgroup $\mathbf{T} \simeq \mathbf{F}^\times \times \mathbf{F}^\times \times \cdots \times \mathbf{F}^\times$, where $\mathbf{F}$ is an algebraic closure of $\mathbf{F}_q$. A Levi subgroup $\mathbf{L}$ is the centralizer $\mathbf{C_G(T)}$ of a torus $\mathbf{T}$. A Borel subgroup $\mathbf{B}$ is a maximal connected solvable subgroup. By taking F-fixed points we get tori T, Borel subgroups B and Levi subgroups L in G. In G there may be several conjugacy classes of maximal tori, but there is, up to conjugacy, a distinguished pair $T \subset B$ of a maximal torus and a Borel subgroup in G. A subgroup containing a Borel subgroup is a parabolic subgroup of G. A parabolic subgroup P has a Levi decomposition $P = LV$ where L is a Levi subgroup and V is the unipotent radical of P.

Examples. $G = GL(n,q), U(n,q), Sp(2n,q)$, or $SO^{\pm}(2n,q)$. In $GL(n,q)$ Levi subgroups L are isomorphic to $\prod_i GL(m_i, q^{e_i})$.

The modular representation theory of finite reductive groups over a field of characteristic p where p divides q (the "defining characteristic") was in fact developed earlier than the ordinary representation theory. The work of Curtis on this in the setting of "BN-pairs" is described in [7]. This theory is modeled on the highest weight theory for representations of semisimple Lie algebras and Lie groups. Later work in this theory, again viewing the finite groups as coming from algebraic groups via a Frobenius morphism, is due to Andersen, Humphreys, Jantzen, Lusztig, Soergel and others. The reader is referred to [21] for this rich theory which we will not describe in this paper.

In the ordinary representation theory of finite reductive groups, Curtis, Iwahori and Kilmoyer constructed the "principal series" for G by inducing characters of a Borel subgroup and showing that the endomorphism algebra of the induced representation is a Hecke algebra (these days sometimes called Iwahori-Hecke algebras) (see [15, Sect. 67]). This then led to Harish–Chandra theory, described below.

Let $\mathbf{P}$ be an F-stable parabolic subgroup of $\mathbf{G}$ and $\mathbf{L}$ an F-stable Levi subgroup of $\mathbf{P}$ so that $L \leq P \leq G$. Then Harish-Chandra induction is the following map: $R_L^G : K_0(KL) \to K_0(KG)$, where K is a sufficiently large field of characteristic 0 as before, such that if $\psi \in \mathrm{Irr}(L)$ then $R_L^G(\psi) = \mathrm{Ind}_P^G(\tilde{\psi})$ where $\tilde{\psi}$ is the character of P obtained by inflating ψ to P, using the Levi decomposition $P = LV$ of P.

We say $\chi \in \mathrm{Irr}(G)$ is cuspidal if $\langle \chi, R_L^G(\psi) \rangle = 0$ for any $L \leq P < G$ where P is a proper parabolic subgroup of G. The pair (L, θ) is a cuspidal pair if $\theta \in \mathrm{Irr}(L)$ is cuspidal. We then have the main theorem of Harish-Chandra induction:

Theorem 3.1. *(i) Let (L,θ), (L',θ') be cuspidal pairs. Then $\langle R_L^G(\theta), R_{L'}^G(\theta') \rangle = 0$ unless the pairs (L,θ), (L',θ') are G-conjugate.*

(ii) If χ is a character of G, then $\langle \chi, R_L^G(\theta) \rangle \neq 0$ for a cuspidal pair (L,θ) which is unique up to G-conjugacy. Thus $Irr(G)$ is partitioned into Harish-Chandra families: A family is the set of constituents of $R_L^G(\theta)$ where (L,θ) is cuspidal.

The endomorphism algebras of Harish-Chandra-induced cuspidal representations from parabolic subgroups were then investigated and described by Howlett and Lehrer (see [10, Chap. 10]). However, not all irreducible representations, in particular the cuspidal ones, were obtained this way. There were some character tables as well as the work of Green on the characters of $GL(n,q)$ which led to the idea (attributed to Macdonald by Springer in [Cusp forms in finite groups [7]]), that there should be families of characters of G corresponding to characters of maximal tori. This was in fact what was proved in the spectacular paper of Deligne and Lusztig in 1976 using ℓ-adic cohomlogy, where ℓ is a prime not dividing q. They introduced a map $R_{\mathbf{T}}^{\mathbf{G}} : K_0(\overline{\mathbf{Q}}_l T) \rightarrow K_0(\overline{\mathbf{Q}}_l G)$ where $\mathbf{T}$ is a maximal torus of $\mathbf{G}$. Lusztig later generalized this map, replacing maximal tori by Levi subgroups, and we will describe this below.

Suppose $\mathbf{L}$ is an F-stable Levi subgroup, not necessarily in an F-stable parabolic $\mathbf{P}$ of $\mathbf{G}$. Let ℓ be a prime not dividing q. The Lusztig linear operator is a map $R_{\mathbf{L}}^{\mathbf{G}}$: $K_0(\overline{\mathbf{Q}}_l L) \rightarrow K_0(\overline{\mathbf{Q}}_l G)$, which has the property that every χ in $\mathrm{Irr}(G)$ is in $R_{\mathbf{T}}^{\mathbf{G}}(\theta)$ for some $(\mathbf{T}, \theta)$, where $\mathbf{T}$ is an F-stable maximal torus and $\theta \in \mathrm{Irr}(T)$.

The unipotent characters of G are the irreducible characters χ in $R_{\mathbf{T}}^{\mathbf{G}}(1)$ as $\mathbf{T}$ runs over F-stable maximal tori of $\mathbf{G}$. Here 1 is the trivial character of T. If $L \leq P \leq G$, where $\mathbf{P}$ is a F-stable parabolic subgroup, $R_{\mathbf{L}}^{\mathbf{G}}$ is just Harish-Chandra induction.

To see how the map $R_{\mathbf{L}}^{\mathbf{G}}$ is constructed using an algebraic variety on which G and L act, see [19, Chap. 11].

In a series of papers and in his book [37], Lusztig classified all the irreducible characters of G, provided $\mathbf{G}$ has a connected center. (This restriction was removed by him later.) This classification leads to two new orthonormal bases of the space $\mathcal{C}(G)$ of $\overline{\mathbf{Q}}_l$-valued class functions of G: the basis of "almost characters" and the basis of characteristic functions of F-stable character sheaves on $\mathbf{G}$. Character sheaves are certain perverse sheaves in the bounded derived category $\mathcal{D}G$ of constructible $\overline{\mathbf{Q}}_l$-sheaves on $\mathbf{G}$. Lusztig then conjectured that the almost characters coincide with the characteristic functions of F-stable character sheaves on $\mathbf{G}$, up to a scalar multiple, if the characteristic p of $\mathbf{F_q}$ is "almost good". This conjecture has now been proved by Shoji and others in many cases, including groups $\mathbf{G}$ with a connected center (see [46]). If the conjecture is true, including the precise values of certain scalars, the character table of G is determined in principle. This shows the power of geometrical methods in representation theory. Indeed, the theory of character sheaves is an aspect of "geometric representation theory", a flourishing area of research.

4 Symmetric Groups, General Linear Groups, Finite Reductive Groups

References for this section are [13, 23, 26–29].

Many expositions of the ordinary representation theory of the symmetric group S_n over $\mathbf{Q}$ are available (see e.g., [28, 29]). If $\chi \in \mathrm{Irr}(G)$ then we can write $\chi = \chi_\lambda$ where λ is a partition of n. Then there is a Young diagram corresponding to λ and p-hooks, p-cores are defined, so that we can talk of the p-core of a partition λ for a prime p.

In the modular representation theory, the p-blocks were classified in the famous

Theorem 4.1 (Brauer–Nakayama [29, p. 245]). *The characters χ_λ, χ_μ of S_n are in the same p-block (p prime) if and only if λ and μ have the same p-core.*

However, the p-modular decomposition numbers are not known for S_n. The modular theory has taken surprising new directions which will be described later.

We now look at the ℓ-modular theory for $GL(n,q)$ where ℓ is a prime not dividing q (the "non-defining characteristic") which was started in [23] and led to the development of the theory by several authors for finite reductive groups.

Let q be odd. The ℓ-blocks were classified in [23] for ℓ odd and extended to $\ell = 2$ by M. Broué. Let e be the order of q mod ℓ. The unipotent characters of $GL(n,q)$ are parametrized by partitions of n, and we denote the character corresponding to a partition λ by χ_λ.

Theorem 4.2 (Fong-Srinivasan). *χ_λ, χ_μ are in the same ℓ-block of $GL(n,q)$ if and only if λ, μ have the same e-core.*

Example: $n = 5$, ℓ divides $q + 1$, $e = 2$. Then χ_λ for $5,\ 32,\ 31^2,\ 2^21,\ 1^5$ are in a block. Same for S_5, $p = 2$.

Example: $n = 4$, ℓ divides $q^2 + 1$, $e = 4$. Then χ_λ for $4,\ 31,\ 21^2,\ 1^4$ are in a block.

$\begin{pmatrix} * & * \\ * & * \end{pmatrix}$ has no 4-hooks.

The ℓ-blocks were then classified for various special cases, and finally Cabanes and Enguehard [13, 22.9] proved the following theorem for unipotent blocks, which are in some sense building blocks for all ℓ-blocks.

Definition. A unipotent block of G is a block which contains unipotent characters.

Theorem 4.3 (Cabanes–Enguehard). *Let B be a unipotent ℓ-block of G, ℓ odd. Then the characters in B are a union of Lusztig families, i.e., a union of constituents of $R_L^G(\psi)$ for various pairs (L,ψ).*

This theorem leads to the following

Surprise: Brauer Theory and Lusztig Theory are compatible!

We have now described a satisfactory solution to Problem 2 (2), i.e., the classification of ℓ-blocks for finite reductive groups. We now consider Problem 2 (3), the decomposition matrix. In the case of $GL(n,q)$ the decomposition matrix is related to that of a q-Schur algebra, and this will be described in Sect. 7. In the case of classical groups, work has been done by Geck, Gruber, Malle and Hiss where ℓ is a so-called "linear prime," using a modular version of Harish–Chandra theory [17, 26]. The problem of finding the decomposition matrix in general is still open, even for unitary groups. For more recent work here see [27].

5 Weyl Groups, Cyclotomic Hecke Algebras, q-Schur Algebras

References are [3, 18, 41].

Our story continues with the introduction of some more major players.

Weyl groups play an important role in the theory of finite reductive groups. For example, the conjugacy classes of maximal tori in a finite reductive group G are parametrized by the F-conjugacy classes in the Weyl group (see [10, 3.3]). There is a set of class functions on G known as "almost characters" which involves the characters of W and plays an important role in the character theory of G, as mentioned in Sect. 3. The ordinary characters of Weyl groups are understood, but as mentioned above the modular theory has open problems even for the type A case, i.e., S_n.

The Hecke algebras which are deformations of Weyl groups and which arise as endomorphism algebras of induced representations of finite reductive groups were studied by Curtis, Iwahori and Kilmoyer [15, Chap. 8]. These Hecke algebras and their characters are studied from the point of view of symmetric algebras in the book by Geck and Pfeifer [24]. With each Hecke algebra we have certain polynomials called generic degrees, which when specialized give the degrees of the constituents of the induced representations mentioned above. Lusztig also showed that more complicated Hecke algebras occur as endomorphism algebras of Harish–Chandra induced cuspidal representations of Levi subgroups, and used these results to classify the unipotent representations. Variants of these Hecke algebras also occur in the work of Howlett and Lehrer on more general induced representations [10, Chap. 10].

Later, Ariki–Koike and Broué–Malle introduced cyclotomic Hecke algebras which are deformations of complex reflection groups [4]. These algebras also arise in the Deligne–Lusztig theory in a mysterious way, that is, they behave as though their specializations are endomorphism algebras of the virtual representations $R_{\mathrm{L}}^{\mathrm{G}}$ defined above. For example, it has been observed that there is a bijection with signs between the constituents of $R_{\mathrm{L}}^{\mathrm{G}}(\lambda)$ for a suitable pair (L, λ) and characters of a complex reflection group [8, 3.2]. In the paper [8] there is developed a theory of "generic groups" which are defined with respect to root data, and the Weyl group plays a central role.

The cyclotomic Hecke algebras which we are interested in are deformations of the groups denoted $G(m, 1, n)$ by Ariki. The group $G(m, 1, n)$ is isomorphic to $\mathbf{Z}_m \wr S_n$.

Definition ([32, p. 420], [3, 12.1]). The cyclotomic Hecke algebra $\mathcal{H}_n$ over a field F has generators $T_1, T_2, \ldots T_n$ and parameters $q, v_1, v_2, \ldots v_m$ which satisfy the relations [3, 12.1]:

$$(T_1 - v_1)(T_1 - v_2)\ldots(T_1 - v_m) = 0, (T_i - q)(T_i + q) = 0 \ (i \geq 2),$$
$$T_1T_2T_1T_2 = T_2T_1T_2T_1, T_iT_j = T_jT_i \ (i \geq j + 2),$$
$$T_iT_{i-1}T_i = T_{i-1}T_iT_{i-1} \ (3 \leq i \leq n).$$

In particular, we get the Hecke algebra of type B or C if $m = 2$. It is a classical result that the ordinary representations of $G(m,1,n)$ are parametrized by m-tuples of partitions whose sizes add up to n. The analogous result holds for the ordinary representations of $\mathcal{H}_n$ in the semisimple case. Now $\mathcal{H}_n$ is semisimple if and only if $q^i v_j - v_k$ $(|i| < n,\ j \neq k)$ and $1 + q + \ldots q^i$ $(1 \leq i < n)$ are all non-zero (see [4, 2.9]). In the non-semisimple case of interest to us where F has characteristic 0, the v_i are powers of q and q is an eth root of unity, we have to take e-restricted partitions (also called Kleshchev partitions) when $m = 1$, and Kleshchev multipartitions when $m > 1$, to parameterize the irreducible modules (see [3, 12.1, 12.2]). Here, a partition λ is e-restricted if $\lambda_i - \lambda_{i+1} < e$ for all $i \geq 1$, where the λ_i are the parts of λ.

In order to study the decomposition numbers for $GL(n,q)$ Dipper and James [18] introduced the q-Schur algebra. In the classical Schur–Weyl theory, the Lie group $GL(n,\mathbf{C})$ and S_d act as centralizers of each other on a tensor space. The Schur algebra $S_{d,n}$ which has a basis indexed by partitions of n with no more than d parts is the image of the action of $GL(n,\mathbf{C})$ on the tensor space. The q-Schur algebra is a deformation of the Schur algebra, defined by Dipper and James as the endomorphism algebra of a sum of permutation modules for the Hecke algebra H_n of type A corresponding to S_n.

The q-Schur algebra $\mathcal{S}_q(n)$ is defined over any field F and involves an element $q \in F$. We use the same q for H_n.

Definition ([41, p. 55]). $\mathcal{S}_q(n) = \mathrm{End}_{H_n} \oplus_\mu M^\mu$.

Here, μ runs over the partitions of n, and M^μ is a permutation module for H_n.

At this stage, we mention another important object, the affine Hecke algebra. This has been studied for a long time, but is of special interest here. In a paper of Ram and Ramagge [45] it is shown how the representation theory (in characteristic 0) of Hecke algebras of classical type can be derived from the representation theory of the affine Hecke algebra of type A. Indeed, there is also a flourishing "Combinatorial Representation Theory," see [6].

6 Lie Algebras

References are [20, 30, 31].

In the classical Cartan–Killing theory, finite-dimensional semisimple Lie algebras are classified by Cartan matrices and Dynkin diagrams. Let $\mathfrak{g}$ be a complex semisimple Lie algebra. There is a root lattice and a weight lattice associated with $\mathfrak{g}$. The root lattice has a basis of simple roots, and there is a concept of dominant weights with respect to this basis. The weight lattice has a basis of fundamental dominant weights, in duality with the basis of simple roots. The Cartan subalgebra $\mathfrak{h}$ acts on a finite-dimensional irreducible module of $\mathfrak{g}$ by linear functions, also called weights, of which one is a "highest weight". The module is then a direct sum of weight spaces.

Kac and Moody generalized this theory to construct infinite-dimensional semisimple Lie algebras which are now known as Kac-Moody algebras. In this theory a generalized Cartan matrix A is a real matrix of the form $A = (a_{ij})$ where $a_{ii} = 2$, a_{ij} are non-positive integers if $i \neq j$, and $a_{ij} = 0$ if and only if $a_{ji} = 0$. Then A is said to be symmetrizable if $A = DB$ where D is diagonal and B is symmetric. The classical representation theory was also extended to the Kac-Moody case, but now we have to consider infinite-dimensional modules. In order to get a theory resembling the finite-dimensional case one considers modules in the category $\mathcal{O}$. Then again the module is a direct sum of weight spaces for the Cartan subalgebra (see [31, Chap. 9]).

Given a generalized Cartan matrix A there are (1) a semisimple Lie algebra (Kac-Moody algebra) $\mathfrak{g} = \mathfrak{g}(A)$, (2) a universal enveloping algebra $U(\mathfrak{g})$, and (3) a quantum enveloping algebra (a Hopf algebra) $U_q(\mathfrak{g})$. The universal enveloping algebra of a finite-dimensional semisimple Lie algebra is a classical object, but the quantum enveloping algebra $U_q(\mathfrak{g})$ has been studied since the 1980s and is now a part of Lie theory. Its representation theory mirrors the classical theory.

Given a Kac-Moody algebra $\mathfrak{g}$ we have a weight lattice P with a basis $\{\Lambda_i, i \in I\}$ of fundamental dominant weights and the notion of a dominant integral weight in P. If Λ is a dominant integral weight then $\mathfrak{g}$ has an irreducible integrable module $V(\Lambda)$ which is the direct sum of weight spaces for the Cartan subalgebra [31, Chaps. 9 and 10].

The quantum enveloping algebra $U_q(\mathfrak{g})$ is generated by $E_i, F_i, K_i^{\pm 1}, i \in I$, with some relations. We have a triangular decomposition (isomorphism of vector spaces) $U_q(\mathfrak{g}) = {U_q}^+(\mathfrak{g}) \otimes {U_q}^0(\mathfrak{g}) \otimes {U_q}^-(\mathfrak{g})$ (see [30, 4.3 and Theorem 4.21], [3, p. 21]).

In the case of the quantum enveloping algebra $U_q(\mathfrak{g})$, for Λ as above we have again an integrable module $V(\Lambda)$ which is the direct sum of weight spaces for ${U_q}^0$, the subalgebra generated by the $K_i^{\pm 1}$. Then $V(\Lambda)$ is irreducible if q is not a root of unity.

Lusztig proved a remarkable result regarding represenations of quantum enveloping algebras and thus of reductive Lie algebras. He defined a "canonical basis" for such an algebra $\mathfrak{g}$ with some remarkable positivity properties, which gives rise to a distinguished basis again called a canonical basis of any irreducible $\mathfrak{g}$-module (see [3, 7.1]). Kashiwara has a similar basis known as the crystal basis.

7 Modular Representations, New

References are [18, 41].

New modular representation theory connects decomposition numbers for symmetric groups, cyclotomic Hecke algebras and q-Schur algebras with Lie theory.

We will first discuss what is meant by "modular representations" for the various objects we have introduced: (1) Finite groups of Lie type (2) Hecke algebras (3) q-Schur algebras.

For (1) the method of reduction mod p has already been described, and this can be applied to S_n and $GL(n,q)$. In the case of S_n we have the classical Specht modules which can be defined over any field.

(2) As in [41, p. 133], [3, 12.1] consider $\mathcal{H}_n$ over a field F. We have elements $q \neq 0$ and $v_1, v_2, \dots v_m$ in F, and a presentation for $\mathcal{H}_n$. Then a "Specht module" can be defined for $\mathcal{H}_n$ corresponding to a multipartition, which is irreducible over F in the semisimple case (see Sect. 5). In the non-semisimple case, the Specht module is no longer irreducible and we can look at its composition factors and talk of the decomposition matrix. When F has characteristic 0 and q is a root of unity Ariki's Theorem, which will be discussed in Sect. 8, describes the decomposition matrices.

(3) As in (2) a "Weyl module" can be defined for $\mathcal{S}_q(n)$, and its composition factors when q is an e-th root of unity give rise to a decomposition matrix. Then Dipper and James proved a remarkable relationship between the decomposition matrices of $\mathcal{S}_q(n)$, $GL(n,q)$ and S_n [41, 6.47]. Their theorem states that the decomposition matrix of $\mathcal{S}_q(n)$, now defined over a field F of characteristic ℓ, with e the order of q mod ℓ, is a submatrix of the decomposition matrix of $FGL(n,q)$. In fact, it is the same as the part of the decomposition matrix of $FGL(n,q)$ corresponding to the unipotent representations. An interesting fact is that both these matrices are square matrices; in the case of $GL(n,q)$ this fact is known by [23].

However, by the work of Varagnolo–Vasserot [41, 47], we know the decomposition matrix of $\mathcal{S}_q(n)$ only in characteristic 0. One then has the James Conjecture, which is still open, which relates the decomposition matrices in characteristic 0 and characteristic ℓ by means of an "adjustment matrix" [41, p.115].

Example, Source: *GAP* An example of a decomposition matrix D for $\mathcal{S}_q(n)$,

$$n = 4, e = 4: \begin{pmatrix} 4|| & 1\,0\,0\,0 \\ 31|| & 1\,1\,0\,0 \\ 211|| & 0\,1\,1\,0 \\ 1111|| & 0\,0\,1\,1 \end{pmatrix}.$$

8 Introducing Lie Theory

References are [3, 25, 41].

The story now continues with a classic paper of Lascoux, Leclerc and Thibon on the Hecke algebra H_n of type A, defined over a field of characteristic 0. They gave an algorithm to compute a matrix, conjectured to be the decomposition matrix of H_n when the parameter q is an e-th root of unity, and connected it with the affine Kac-Moody algebra $\widehat{sl_e}$ (see [41, p. 95]). Ariki proved the conjecture, in fact for cyclotomic Hecke algebras. We state his theorem as in [3, Theorem 12.5].

Consider the cyclotomic Hecke algebra $\mathcal{H}_n$ over a field F of characteristic 0 associated with the group $G(m,1,n)$ with parameters $q, v_1, v_2, \dots v_m$, where each $v_i = q^{\gamma_i}$, $\gamma_i \in \mathbf{Z}/e\mathbf{Z}, i = 1, 2, \dots m$ and q is an e-th root of unity, not equal to 1. Let $\Lambda = \sum_i n_i \Lambda_i$, where v_i occurs with multiplicity n_i. Let $V(F)$ be the complexified Grothiendieck group $\oplus_{n \geq 0} K_0(\mathcal{H}_n - \text{pmod}) \otimes_{\mathbf{Z}} \mathbf{C}$.

Remark. Here all the $\mathcal{H}_n$ are assumed to have the same parameters. Note that q is a special parameter. Then

Theorem 8.1 (Ariki). *The Kac-Moody Lie algebra $\widehat{sl_e}$ acts on $V(F)$ and $V(F)$ is isomorphic to the irreducible $\widehat{sl_e}$-module of highest weight Λ. The canonical basis (in the sense of Lusztig) of the module, specialized at $q = 1$ corresponds under the isomorphism to the basis of $V(F)$ of indecomposable projective modules of the $\mathcal{H}_n$.*

Thus, we introduce Lie theory into the modular representation theory of the Hecke algebras $\mathcal{H}_n$ taken over all n. Now we also have a standard basis for $V(F)$, corresponding to the irreducible modules of $\mathcal{H}_n$. We then have the important fact that the transition matrix between the two bases gives the decomposition matrices for the e-modular theory of the $\mathcal{H}_n$. Furthermore, the decomposition of $V(F)$ into weight spaces (of the Cartan subalgebra) of $\widehat{sl_e}$ corresponds to the decomposition into blocks for the $\mathcal{H}_n$, which is another amazing fact.

The algebra $\widehat{sl_e}$ has generators e_i, f_i, $i = 1, 2, \ldots e$. Their action on $V(F)$ is given by functors called i-restriction, i-induction which can be described combinatorially.

It is worth noting that in the case of the symmetric groups, these operations were defined by de B. Robinson, a pioneer in the representation theory of S_n (see [29, p. 271]). In that case, i-restriction, i-induction correspond to ordinary restriction and induction followed by cutting to a block. We describe these operations below.

Irreducible KS_n-modules are indexed by partitions of n, and a partition of n is represented by a diagram with n nodes. Induction (resp. restriction) from S_n to S_{n+1} (resp. S_n *to* S_{n-1}) is combinatorially represented by adding (resp. removing) a node from a partition.

Now fix an integer $\ell \geq 2$. The residue r at the (i, j)-node of a diagram is defined as $r \equiv (j - i)(\text{mod } \ell)$. We define induction (resp. restriction) from S_n to S_{n+1} (resp. S_n *to* S_{n-1}) by defining operators e_i and f_i, $0 \leq i \leq (\ell - 1)$, which move only nodes with residue i. Then Ind $= \sum_0^{l-1} f_i$ and Res $= \sum_0^{l-1} e_i$. We have similar operations, with multipartitions, in the case of $\mathcal{H}_n$ (see [25] for a discussion).

We now see that given the Hecke algebra $\mathcal{H}_n$ with certain parameters, we have a dominant weight Λ for $\widehat{sl_e}$ associated with it. Thus the Hecke algebra can be parametrized by Λ. In Ariki's theorem we may denote the Hecke algebra as $\mathcal{H}_n^\Lambda$ and the module $V(F)$ as $V(\Lambda)$. This is the view adopted in [32].

The other main idea arising from the work of Lascoux, Leclerc and Thibon and of Ariki is the decomposition of $V(F)$ into blocks of the $\mathcal{H}_n^\Lambda$ coincides with the decomposition of the $\widehat{sl_e}$-module into weight spaces. This leads to a new notation for the blocks: blocks of $\mathcal{H}_n^\Lambda$ can be parametrized by $\nu = \sum_{i \in I} m_i i$, where the m_i are non-negative integers and $I = \mathbf{Z}/\mathbf{eZ}$ [33, 8.1].

9 Categorification

A reference is [43].

Our story takes an abstract direction in this section. "Categorification" is indeed one of the buzzwords of the last few years. The idea is that set-theoretic notions are replaced by category-theoretic notions (see [43]). We state here a definition given in [35].

Let A be a ring, B a left A-module. Let a_i be a basis of A. One should find an abelian category $\mathcal{B}$ such that $K_0(\mathcal{B})$ is isomorphic to B, and exact endofunctors F_i on $\mathcal{B}$ which lift the action of the a_i on B, i.e., the action of $[F_i]$ on $K_0(\mathcal{B})$ descends to the action of a_i on B so that the following diagram commutes. The map $\phi : K_0(\mathcal{B}) \to B$ is an isomorphism.

$$\begin{array}{ccc} K_0(\mathcal{B}) & \xrightarrow{[F_i]} & K_0(\mathcal{B}) \\ \phi\downarrow & & \phi\downarrow \\ B & \xrightarrow{a_i} & B \end{array}$$

Example. Ariki's Theorem. Here $A = \widehat{sl_e}$, $B = V(F)$ and $\mathcal{B} = \oplus_{n\geq 0} (\mathcal{H}_n\text{–pmod})$.

We return to Broué's Abelian Defect Group Conjecture, stated in Sect. 2. Chuang and Rouquier proved the conjecture for S_n and $GL(n,q)$ by introducing the concept of SL_2-categorification. On the way they prove that if two blocks of symmetric groups (possibly not the same S_n) have isomorphic defect groups, the block algebras are derived equivalent, i.e., the derived categories of the module categories are equivalent [16]. This has shown the power of categorification in a seemingly unrelated area such as finite groups. For a discussion of categorical equivalences in the modular representation theory of finite groups, see [42].

10 KLR-Algebras: The Diagrammatic Approach

References are [34, 38].

We have seen that the algebra $\widehat{sl_e}$ is related to the category $\sum_{n\geq 0}(\mathcal{H}_n - \text{pmod})$. We have the following question: Suppose we want to replace $U(\widehat{sl_e})$ by $U(\mathfrak{g})$, where $\mathfrak{g}$ is an arbitrary semisimple Lie algebra. Then how will we define $\mathcal{B}$ to give a categorification as above?

Now we come to the papers of Khovanov–Lauda, which have connections with Knot Theory. Given a symmetrizable Cartan matrix A, there is a Kac-Moody algebra $\mathfrak{g}$ corresponding to A, the "negative part" of the quantum enveloping algebra $U_q^-(\mathfrak{g})$ over $\mathbf{Q}(q)$ and an integral form $\mathfrak{F}$ of $U_q^-(\mathfrak{g})$, a ring defined over $\mathbf{Z}[q,q^{-1}]$. As before (see Sect. 6) I is an index set for the fundamental weights or the simple roots. Khovanov and Lauda then construct an algebra R corresponding to this data as follows (see [38]).

A *diagram* is a collection of (planar) arcs connecting m points on a horizontal line with m points on another horizontal line. Arcs are labeled by elements of I. Let $\mathbf{N}[I] = \langle \nu | \nu = \sum \nu_i i, \nu_i \in \mathbf{N} \rangle$. Khovanov and Lauda first define rings $R(\nu)$ for each $\nu \in \mathbf{N}[I]$. The ring $R(\nu)$ is generated by diagrams in which ν_i arcs have the same label i. The product is defined by concatenation of diagrams, subject to some relations. Then $R = \oplus_\nu R(\nu)$ and $K_0(R) = \oplus_\nu K_0(R(\nu))$.

Theorem 10.1. *The category $R - pmod$ categorifies $\mathfrak{F}$. In other words, we may take $A = \mathfrak{F}$, $B = \mathfrak{F}$, $\mathcal{B} = R - pmod$ in the commutative diagram describing the categorification. In particular, there is an isomorphism between $K_0(R)$ and $\mathfrak{F}$.*

Rouquier defined these algebras independently and they are now called Khovanov-Lauda–Rouquier (KLR) algebras, or sometimes as quiver Hecke algebras. An important property of these algebras is that they are naturally graded. This has ramifications in "graded representation theory" [32, 2.2], to be described in the next section.

11 Graded Representation Theory

Ariki's theorem was regarding representations of $\widehat{sl_e}$, but it can be generalized to those of $U_q(\widehat{sl_e})$ [3, 10.10]. This leads to the idea of graded representation theory, where multiplication by q represents a shift in grading. Indeed, in [32, p. 431] the classification of graded irreducible modules of cyclotomic Hecke algebras is stated as the Main Problem. One can also talk of graded decomposition numbers.

We now briefly mention some of the recent important work in this theory; references for (1) and (2) are references [38, 40, 42] in [32].

(1) (Brundan and Kleshchev) Blocks of cyclotomic Hecke algebras are isomorphic to blocks of Khovanov–Lauda–Rouquier algebras. Since the latter are graded, there is a $\mathbf{Z}$-grading of blocks of cyclotomic Hecke algebras, including group algebras of symmetric groups in positive characteristic.

(2) Building on the above, Brundan, Kleshchev and Wang have constructed gradings of Specht modules of cyclotomic Hecke algebras. Brundan and Kleshchev then describe graded decomposition numbers of these Specht modules over fields of characteristic 0.

(3) Using the connection between the Hecke algebra H_n and the q-Schur algebra $\mathcal{S}_q(n)$, Ariki [5] has shown that $\mathcal{S}_q(n)$ and its Weyl modules can be graded. Moreover, he shows that certain polynomials in v introduced by Leclerc and Thibon give the graded decomposition numbers for $\mathcal{S}_q(n)$ over a field of characteristic 0 with q an e-th root of unity. Here we consider the algebra $U_v(\widehat{sl_e})$ with a parameter v, as distinct from the parameter q in $\mathcal{S}_q(n)$.

Example. Graded decomposition matrix for $n = 4$, $e = 4$: $\begin{pmatrix} 4|| & 1\,0\,0\,0 \\ 31|| & v\,1\,0\,0 \\ 211|| & 0\,v\,1\,0 \\ 1111|| & 0\,0\,v\,1 \end{pmatrix}$

Remark. If $v = 1$, we get the usual decomposition matrix given in Sect. 7. Other decomposition matrices are found in [41, Appendix]. The above matrix corresponds to a block, and the reader can make the connection with the theorem of [23], stated in Sect. 4; these are the partitions with an empty 4-core.

12 Higher Representation Theory

A reference is [43].

Let $\mathcal{C}$ be a 2-category. Then for any $i, j \in \mathcal{C}$ the morphisms form a category whose objects are called 1-morphisms and morphism are called 2-morphisms of $\mathcal{C}$.

We recall the definition of Categorification in Sect. 9, where the action of the algebra A on B was lifted to an action by endofunctors on a category $\mathcal{B}$. In a good situation the image of this lift would be a nice subcategory of the category of endofunctors on $\mathcal{B}$, and we can compose these endofunctors. Here we have an example of a 2-category with one object, whose 1-morphisms are functors and 2-morphisms are natural transformations (see [43, 2.4]).

The concept of "Higher Representation Theory" was introduced by Chuang and Rouquier [16] and continued by Rouquier and Khovanov–Lauda. Here a 2-category is defined corresponding to a Cartan matrix and is a categorification or a "2-analogue" of the enveloping algebras $U(\mathfrak{g})$ and $U_q(\mathfrak{g})$ where $\mathfrak{g}$ is a semisimple Lie algebra. The idea then is to study "2-representations" of a 2-category.

One of the first explicit examples here was a categorification of $U_q(\widehat{sl_2})$ due to Lauda [36]. B.Webster has constructed 2-categories corresponding to $U_q(\mathfrak{g})$ where $\mathfrak{g}$ is a semisimple Lie algebra and categorifies tensor products of irreducible representations of $U_q(\mathfrak{g})$. A generalization of Ariki's theorem in this context can be found in [48, 5.11].

13 End of Story?

We have come to the end of our story. However, as we all know there is no end to mathematical stories, and future generations will continue them.

Acknowledgment I thank the organizers of the interesting and enjoyable conference "Buildings, Finite Geometries, Groups" in Bangalore, August 2010, for their hospitality.

References

1. J.L. Alperin, Local Representation Theory, Cambridge (1986)
2. J. An, Dade's invariant conjectures for the general linear and unitary groups in non-defining characteristics, Transactions of the American Mathematical Society, 353 (2001), 365–390
3. S. Ariki, Representations of Quantum Algebras and Combinatorics of Young Tableaux, University Lecture Series 26, AMS (2002)
4. S. Ariki, Lectures on cyclotomic Hecke algebras, available at mathQA/9908005
5. S. Ariki, Graded q-Schur algebras, available at arXiv:0903.3453v2.
6. H. Barcelo & A. Ram, Combinatorial representation theory, New perspectives in algebraic combinatorics (Berkeley, CA, 1996–97), 23–90, Math. Sci. Res. Inst. Publ., **38**, Cambridge Univ. Press, Cambridge, 1999
7. A. Borel, et al, Seminar on algebraic groups and related finite groups, Springer Lecture Notes 131 (1970)
8. M. Broué, G. Malle & J. Michel, Représentations Unipotentes génériques et blocs des groupes réductifs finis, Astérisque 212 (1993)
9. M. Broué, Isometriés parfaites, types de blocs, catégories dérivées, Astérisque 181–182 (1990), 61–92
10. R. Carter, Finite Groups of Lie Type, Wiley (1985)
11. R. Carter, Simple Groups of Lie Type, Wiley (1972).
12. C. Chevalley, Sur certaines groupes simples, Tohoku Math. J. 7 (1955), 14–66
13. M. Cabanes & M. Enguehard, Representation Theory of Finite Reductive Groups, Cambridge (2004)
14. C. Curtis and I. Reiner, Representation Theory of Finite Groups and Associative Algebras (Wiley, 1962)
15. C. Curtis and I. Reiner, Methods of Representation Theory with applications to finite groups and orders,(Wiley, 1987)
16. J. Chuang & R. Rouquier, Derived equivalences for symmetric groups and sl_2-categorification, Annals of Math. 167 (2008) 245–298
17. R. Dipper, M. Geck, G. Hiss & G.Malle, Representations of Hecke algebras and finite groups of Lie type, in Algorithmic algebra and number theory. Springer, Berlin (1999), 331–378, available at http://www.mathematik.uni-kl.de/ malle/en/publications.html
18. R. Dipper & G.D. James, The q-Schur algebra, Proc. London Math. Soc. 59 (1989), 23–50
19. F. Digne & J. Michel, Representations of finite groups of Lie type, LMS Student Texts 71, Cambridge (1991)
20. J.E. Humphreys, Introduction to Lie algebras and representation tneory, Springer (1972)
21. J.E. Humphreys, Modular representations of finite groups of Lie type, London Mathematical Society lecture note series 326, Cambridge University Press (2006)
22. M. Isaacs, G. Malle, G. Navarro, A reduction theorem for the McKay conjecture, Invent. Math. 170 (2007), 33–101, (MR 2008h:20016)
23. P. Fong & B. Srinivasan, The blocks of finite general linear and unitary groups, Invent. math. 69 (1982), 109–153
24. M. Geck & G. Pfeiffer, Characters of finite Coxeter groups and Iwahori-Hecke algebras, Oxford (2000)
25. M. Geck, Modular Representations of Hecke Algebras, In: Group representation theory (EPFL, 2005; eds. M. Geck, D. Testerman and J. Thevenaz), pp. 301–353, Presses Polytechniques et Universitaires Romandes, EPFL-Press, Lausanne; available at mathRT.0511548
26. M. Geck, Modular Harish-Chandra series, Hecke algebras and (generalized) q-Schur algebras. In: Modular Representation Theory of Finite Groups (Charlottesville, VA, 1998; eds. M. J. Collins, B. J. Parshall and L. L. Scott), p. 1–66, Walter de Gruyter, Berlin 2001
27. M. Geck, Modular principal series representations, Int. Math. Res. Notices, (2006), 1–20, available at mathRT.0603046

28. G.D. James, The Representation Theory of the Symmetric Groups, Springer Laecture Notes 682 (1978)
29. G.D. James, A. Kerber, The Representation Theory of the Symmetric Group, Encyclopedia of Mathematics 16, Addison-Wesley (1981)
30. J.C. Jantzen, Lectures on Quantum Groups, Graduate Studies in Math. 6, AMS (1996)
31. V. Kac, Infinite dimensional Lie algebras, Cambridge University Press (1985)
32. A. Kleshchev, Representation theory of symmetric groups and related Hecke algebras, Bulletin of the AMS 47 (2010), 419–481
33. A. Kleshchev, Linear and Projective Representations of Symmetric Groups, Cambridge University Press (2005).
34. M. Khovanov and A. Lauda, A diagrammatic approach to categorification of quantum groups I. Represent. Theory, 13 (2009), 309–347
35. M. Khovanov, V. Mazorchuk and C. Stroppel, A brief review of abelian categorifications, available at ArXiv.math.RT/0702746v1
36. A. Lauda, A ctegorification of quantum sl_2, available at arXiv:0803.3652
37. G. Lusztig, Characters of reductive groups over a finite field, Ann. of Math. Stud., 107, Princeton Univ. Press, Princeton, NJ (1984)
38. A. Lauda and M. Vazirani, Crystals from categorified quantum groups, available at mathRT.0909.1810
39. G. Malle, The inductive McKay condition for simple groups not of Lie type, Comm. Algebra 36 (2008), 455–463, (MR 2009a:20012)
40. G. Malle, Height 0 characters of finite groups of Lie type. Represent. Theory 11 (2007), 192–220, (MR 2008m:20025)
41. A. Mathas, Iwahori Hecke Algebras and Schur algebras of the symmetric group, University Lecture Series 15, AMS (1999)
42. A. Marcus, Derived equivalences and the abelian defect group conjecture, in Proceedings of the International Conference on Modules and Representation Theory, "Babes-Bolyai" University, Cluj-Napoca (2008), 111–131, available at http : //math.ubbcluj.ro/~marcus/papers/marcus_proc − Cluj08.pdf
43. V. Mazorchuk, Lectures on Algebraic Categorification, available at arXiv:1011.0144v2
44. J. Rickard, The abelian defect group conjecture, in Proceedings of the International Congress of Mathematicians, Berlin, Documenta Mathematica Extra Vol. II, pages 121–128 (electronic), 1998
45. A. Ram & J. Ramagge, Affine Hecke algebras, cyclotomic Hecke algebras and Clifford theory, Trends in Mathematics, (A tribute to C.S. Seshadri (Chennai 2002)), 428–466, available at arXiv:0401.322v2
46. T. Shoji, Lusztig's conjecture for finite classical groups with even characteristic, Contemporary Math. 478, AMS (2009), 207–236
47. M. Varagnolo & E. Vasserot, On the decomposition numbers of the quantized Schur algebra, Duke Math. J. 100 (1999), 267–297
48. B. Webster, Knot invariants and higher representation theory I: diagrammatic and geometric categorification of tensor products, available at arXiv.math.GT/1001.2020v4

The Use of Blocking Sets in Galois Geometries and in Related Research Areas

V. Pepe and L. Storme

Abstract Blocking sets play a central role in Galois geometries. Besides their intrinsic geometrical importance, the importance of blocking sets also arises from the use of blocking sets for the solution of many other geometrical problems, and problems in related research areas. This article focusses on these applications to motivate researchers to investigate blocking sets, and to motivate researchers to investigate the problems that can be solved by using blocking sets. By showing the many applications on blocking sets, we also wish to prove that researchers who improve results on blocking sets in fact open the door to improvements on the solution of many other problems.

Subject Classifications: AMS classification: 05B25, 51E20, 51E21, 94B05

Keywords Blocking sets • Substructures in projective spaces • Linear codes and projective geometry

1 Definitions and Introductory Results

A set B of points of $\mathrm{PG}(n,q)$ is called a *k-blocking set* if every $(n-k)$-dimensional subspace of $\mathrm{PG}(n,q)$ has a non-empty intersection with B, and a set B of points of $\mathrm{PG}(n,q)$ is called a *t-fold* k-blocking set if every $(n-k)$-dimensional subspace contains at least t points of B. In $\mathrm{PG}(2,q)$, the 1-blocking sets are simply called *blocking sets*. Two subspaces Σ_1 and Σ_2 of $\mathrm{PG}(n,q)$ of dimension k and $n-k$ always have a non-empty intersection, hence a set B containing a k-dimensional

V. Pepe (✉) • L. Storme
Department of Mathematics, Ghent University, Krijgslaan 281-S22,
9000 Ghent, Belgium
e-mail: valepepe@cage.ugent.be; ls@cage.ugent.be

N.S.N. Sastry (ed.), *Buildings, Finite Geometries and Groups*, Springer Proceedings in Mathematics 10, DOI 10.1007/978-1-4614-0709-6_15,

subspace is called a *trivial* k-blocking set. Moreover, a k-blocking set B is called *minimal* if it is minimal with respect to the containment relation, i.e., $B \setminus \{P\}$ is not a k-blocking set for every $P \in B$, and it is called *small* if $|B| < \frac{3(q^k+1)}{2}$.

The blocking sets of PG$(2,q)$ have been extensively studied in the last years and there are plenty of results about characterizations (of small ones) and about the bounds on the size. A trivial blocking set of PG$(2,q)$ is a set of points containing a line. The first examples of non-trivial blocking sets of small size of PG$(2,q)$ have been given in the following:

Theorem 1.1. *In PG$(2,q)$, q odd, there exists a projective triangle of side $\frac{q+3}{2}$ that is a minimal blocking set of size $\frac{3(q+1)}{2}$ [12].*

In PG$(2,q)$, q even, there exists a projective triad of side $\frac{q+2}{2}$ that is a minimal blocking set of size $\frac{3q+2}{2}$ [35].

This motivates the choice of the bound $\frac{3(q+1)}{2}$ for the size of a small blocking set in PG$(2,q)$ and then generalized as $\frac{3(q^k+1)}{2}$ for the k-blocking sets of PG(n,q).

In PG$(2,q^2)$, there are other well known examples of minimal blocking sets: the *Baer subplanes* and the *unitals*. A Baer subplane $\mathcal{B}$ of PG$(2,q^2)$ is a plane isomorphic to PG$(2,q)$ contained in PG$(2,q^2)$, hence it has size q^2+q+1 and it has the property that every line of PG$(2,q^2)$ intersects $\mathcal{B}$ in 1 or $q+1$ points. A unital $\mathcal{U}$ is a set of q^3+1 points of PG$(2,q^2)$ such that every line of PG$(2,q^2)$ intersects it in 1 or $q+1$ points and through every point P of $\mathcal{U}$, there exists a unique line ℓ such that $\ell \cap \mathcal{U} = \{P\}$. These two examples are remarkable for the following results:

Theorem 1.2. *[11] Let B be a non-trivial minimal blocking set of PG$(2,q)$, then $|B| \geq q + \sqrt{q} + 1$ and equality holds if and only if q is a square and B is a Baer subplane.*

Theorem 1.3. *Let B be a minimal blocking set of PG$(2,q)$, then*

- *[13] $|B| \leq q\sqrt{q} + 1$,*
- *[14] $|B| = q\sqrt{q} + 1$ if and only if q is a square and B is a unital.*

In [8], the following lower bounds for planes of non-square order are proven:

Theorem 1.4. *If B is a minimal blocking set in PG$(2,q)$ and $q = p^{2h+1}$, p prime, then $|B| \geq q + q^{2/3} + 1$ for $p > 3$ and $|B| \geq q + q^{2/3}/2^{1/3} + 1$ for $p = 2,3$.*

On the other hand, there are results about large minimal blocking sets, that are also called *spectrum results*, like the following:

Theorem 1.5. *[37] There exists a minimal blocking set in PG$(2,q)$, $q > 4$, for every size in $[2q-1, 3q-3]$.*

Theorem 1.6. *[16] For every value k in the interval $[4q\log q, q\sqrt{q} - q + 2\sqrt{q}]$, there exists a minimal blocking set of cardinality k in PG$(2,q)$, q square.*

We now list some results about blocking sets in PG(n,q). We note that most of them are derived from the corresponding planar results. From now on, let θ_k be the number of points of PG(k,q), that is, $\theta_k = (q^{k+1}-1)/(q-1)$.

Theorem 1.7 (Bose and Burton [10]). *Let B be a k-blocking set in* PG(n,q) *that has the smallest possible cardinality. Then B is a k-dimensional subspace of* PG(n,q).

Theorem 1.8. *[5, 31] Let B be a non-trivial k-blocking set in $PG(n,q)$, $q > 2$. Then $|B| \geq \theta_k + r(q)q^{k-1}$, where $q + r(q) + 1$ is the size of the smallest non-trivial blocking set in $PG(2,q)$, and the equality only holds if B is a cone with base a minimal non-trivial blocking set of size $q + r(q) + 1$ of a plane π and vertex a $(k-2)$-dimensional subspace skew to π.*

Theorem 1.9. *[13, 14] Let B be a minimal* 1*-blocking set in $PG(n,q)$, $n \geq 3$, then we have the following:*

- *If $n = 3$, then $|B| \leq q^2 + 1$ and the equality holds if and only if B is an ovoid.*
- *If $n \geq 4$, then $|B| < \sqrt{q^{n+1}} + 1$.*

Another important result about blocking sets that has been widely used to prove many other theorems is the following:

Theorem 1.10. *[61] Let B be a small minimal k-blocking set of $PG(n,q)$, $q = p^h$, $p > 2$ prime. Then every subspace that intersects B intersects it in* $1 \pmod{p}$ *points.*

In light of this, we define the *exponent* of a small minimal k-blocking set B as the largest integer e for which an $(n-k)$-space intersects B in $1 \pmod{p^e}$ points. Then Sziklai proved:

Theorem 1.11. *[58] Let B be a small minimal k-blocking set of $PG(n,q)$, $q = p^h$, $p > 2$ prime, with exponent e. Then $e|h$ and every subspace that intersects B in $1 + p^e$ points intersects it in a subline $PG(1, p^e)$.*

Finally, a *Rédei-type* k-blocking set in PG(n,q) is a blocking set B such that there exists a hyperplane with $|B| - q^k$ points of B. If $|B| < (3q^k+3)/2$, then these k-blocking sets are, in fact, *linear k-blocking sets* (see [3, 57, 62]). A *$\mathbb{F}_q$-linear set* in PG(n,q^h) is a set of points whose defining vectors form a $\mathbb{F}_q$-vector space (for more details about these sets, see [52]). A k-blocking set B that is also a linear set is called a *linear* k-blocking set. If B, $|B| < (3q^k + 3)/2$, is of Rédei-type, then B is linear, but the converse is not true in general (see [50]).

For more results on blocking sets, we refer to the survey articles [6, 60] and also to the chapter on blocking sets [9] in the collected work [20].

The objective of this article is to illustrate the many geometrical problems and problems in related research areas that use blocking sets. The use of blocking sets for solving a great variety of problems gives blocking sets a central place within Galois geometries. By showing the many applications, we wish to motivate researchers to investigate blocking sets, and/or to investigate problems that use blocking sets. We also wish to show to the readers that improved results on blocking sets will imply a large number of improvements to related problems. In particular, in the last section, we state some open problems. One of these open problems is the central problem on blocking sets: the investigation of the linearity conjecture on small (multiple) blocking sets.

To conclude this introduction, we wish to note that p always will denote a prime number and that $\theta_k = (q^{k+1} - 1)/(q - 1)$ is equal to the number of points in the k-dimensional projective space $\mathrm{PG}(k, q)$ over the finite field $\mathbb{F}_q$ of order q.

We now present a classical example of a problem in Galois geometries that uses blocking sets; the investigation of maximal partial spreads in $\mathrm{PG}(3, q)$ of small deficiency δ.

2 Maximal Partial Spreads

A *k-spread* $\mathcal{S}$ of the projective space $\mathrm{PG}(n, q)$ is a partition of the point set of $\mathrm{PG}(n, q)$ in k-dimensional subspaces, that is every point of $\mathrm{PG}(n, q)$ is contained in exactly one element of $\mathcal{S}$. It is well known (see e.g., [35]) that such a k-spread exists if and only if $k + 1$ divides $n + 1$ and then $|\mathcal{S}| = \frac{q^{n+1}-1}{q^{k+1}-1}$. A *partial* k-spread $\mathcal{S}$ of $\mathrm{PG}(n, q)$ is a collection of pairwise disjoint k-subspaces, that is every point of $\mathrm{PG}(n, q)$ is contained in at most one element of $\mathcal{S}$. A *maximal* partial k-spread is a partial k-spread, maximal with respect to the containment relation. The size of the largest maximal partial k-spread, different from a k-spread, has been investigated in great detail, and this is the classical example of a problem that is investigated, using results on blocking sets.

Let $\mathcal{S}$ be a maximal partial line spread of $\mathrm{PG}(3, q)$ and let $|\mathcal{S}| = q^2 + 1 - \delta$ (a line spread of $\mathrm{PG}(3, q)$ has size $q^2 + 1$). We call δ the *deficiency* of $\mathcal{S}$. We will call *holes* the points of $\mathrm{PG}(3, q)$ not contained in any line of $\mathcal{S}$. Since the partial spread is maximal, the set of holes cannot contain lines. Since the lines of $\mathcal{S}$ are pairwise disjoint, every plane contains at most one line of $\mathcal{S}$. The planes containing a line of $\mathcal{S}$ are called *rich*, otherwise they are called *poor*. Let ℓ be a line not in $\mathcal{S}$: the lines of $\mathcal{S}$ meeting ℓ in a point must be contained in distinct planes, hence the number of holes of ℓ is the same as the number of poor planes through ℓ. Let π be a poor plane and let B be the set of holes of π. Every line ℓ of π must contain at least a hole since ℓ is contained in a poor plane. Hence, we have the following:

Lemma 2.1. *[49] The set of holes of a poor plane of a maximal partial spread of size $q^2 + 1 - \delta$ is a non-trivial blocking set of size $q + \delta$ in this plane.*

Hence, results on non-trivial blocking sets in $\mathrm{PG}(2, q)$ give information on maximal partial spreads in $\mathrm{PG}(3, q)$ of small positive deficiency δ.

Let us first consider q to be a square. The largest value for the size of a blocking set B of $\mathrm{PG}(2,q)$ such that B definitely must contain a Baer subplane has been studied in detail; we list here the best known results. As mentioned in the preceding section, p will always denote a prime number.

Theorem 2.2. *Let B be a non-trivial blocking set of $PG(2,q)$, q square.*

1. *[8] Let $q = p^h, h > 2$, $c_2 = c_3 = 2^{-\frac{1}{3}}$ and $c_p = 1$ for $p > 3$. If $|B| < q + c_p q^{\frac{2}{3}} + 1$, then B contains a Baer subplane.*
2. *[59] Let $q = p^2$. If $|B| < 3(q+1)/2$, then B contains a Baer subplane.*

In general, let $\delta(q)$ be the maximum integer such that a non-trivial blocking set of $\mathrm{PG}(2,q)$ of size $q+\delta(q)$ must contain a Baer subplane. Then, for every maximal partial spread $\mathcal{S}$ of positive deficiency $\delta \leq \delta(q)$, the set of holes in a poor plane of $\mathcal{S}$ contains a Baer subplane of holes. One might wonder where all these Baer subplanes of holes in these poor planes arise from. A logical guess would be that the set of holes is the union of Baer subgeometries $\mathrm{PG}(3,\sqrt{q})$, since $|\mathrm{PG}(3,\sqrt{q})| = (\sqrt{q}+1)(q+1) \equiv 0 \pmod{q+1}$. This indeed was proven. Using the results about blocking sets, in [49], the following theorem was proven.

Theorem 2.3. *Let $q > 4$ be a square, let $\delta(q)$ be defined as before, $0 < \delta \leq \min\{(q+1)/2, \delta(q)\}$. If $\mathcal{S}$ is a maximal partial spread of size $q^2+1-\delta$, then*

(a) $\delta = s(\sqrt{q}+1)$ for some integer $s \geq 2$,
(b) The set of holes of $PG(3,q)$ is the union of s pairwise disjoint Baer subgeometries $PG(3,\sqrt{q})$.

However, it is important to mention that no examples of maximal partial spreads in $\mathrm{PG}(3,q)$ having a set of holes equal to the union of $\sqrt{q}-1 > s > 1$ pairwise disjoint Baer subgeometries $\mathrm{PG}(3,\sqrt{q})$ are known. Most likely, such maximal partial spreads do not exist. The proof of the (non-)existence of such maximal partial spreads is one of the open problems we mention in Sect. 7.5.

Let us now consider maximal partial spreads in $\mathrm{PG}(3,q^3)$, where $q = p^h$, h odd, $p \geq 7$.

Under these hypotheses for q, the non-trivial minimal blocking sets of $\mathrm{PG}(2,q^3)$ of the smallest and second smallest size have been completely classified in [7,51, 53,59].

Theorem 2.4. *In $PG(2,q^3)$, $q = p^h$, $p \geq 7$, h odd, p prime, the smallest and the second smallest sizes for a non-trivial minimal blocking set are q^3+q^2+1 and q^3+q^2+q+1, respectively. The minimal blocking sets of these sizes are unique up to projective equivalence and they are of Rédei type. More precisely, they are projected subgeometries $PG(3,q)$ in a plane $PG(2,q^3)$.*

Similarly, as for q square, the set of holes in a poor plane $\mathrm{PG}(2,q^3)$, $q = p^h$, $p \geq 7$, h odd, p prime, of a maximal partial spread, of deficiency $\delta \leq q^2+q+1$, contains a projected subgeometry $\mathrm{PG}(3,q)$. Again, where can all these projected subgeometries $\mathrm{PG}(3,q)$ of holes arise from? A logical guess is a projected subgeometry PG$(5,q)$ in PG$(3,q^3)$ of size $q^5+q^4+q^3+q^2+q+1=(q^2+q+1)(q^3+1)\equiv 0 \pmod{q^3+1}$. This was indeed proven to be the case.

Theorem 2.5. *[49] Let $\mathcal{S}$ be a maximal partial spread of size $q^6+1-\delta$ of $PG(3,q^3)$, where $q=p^h$, h odd, $p\geq 7$, p prime, $0<\delta\leq q^2+q+1$. Then $\delta=q^2+q+1$ and the set of holes forms a projected subgeometry $PG(5,q)$ in $PG(3,q^3)$.*

Finally, let $\mathcal{S}$ be a maximal partial spread of PG$(3,q^3)$, where $q=p^h$, $h\geq 2$ even, $p\geq 7$. Combining the aforementioned results, it is possible to describe the set of holes of $\mathcal{S}$ when its deficiency is small. As before, we first state the results on the small minimal non-trivial blocking sets of the plane.

Theorem 2.6. *[53] In $PG(2,q^3)$, where $q=p^h$, $h\geq 2$ even, $p\geq 7$, the smallest non-trivial blocking sets are:*

(a) A Baer subplane $PG(2,q^{\frac{3}{2}})$ (hence of order $q^3+q^{\frac{3}{2}}+1$);
(b) A minimal blocking set of Rédei type of size q^3+q^2+1 which is a projected subgeometry $PG(3,q)$;
(c) A minimal blocking set of Rédei type of size q^3+q^2+q+1 which is a projected subgeometry $PG(3,q)$.

Theorem 2.7. *Let $\mathcal{S}$ be a maximal partial spread of size $q^6+1-\delta$ of $PG(3,q^3)$, where $q=p^h$, h even, $p\geq 7$, $0<\delta\leq q^2+q+1$. If there is a Baer subplane of holes, then all poor planes contain a Baer subplane of holes, $\delta=s(q^{\frac{3}{2}}+1)$, $s\geq 2$, and the set of holes is the union of s pairwise disjoint subgeometries $PG(3,q^{\frac{3}{2}})$.*

If no poor plane contains a Baer subplane of holes, then $\delta=q^2+q+1$ and the set of holes forms a projected subgeometry $PG(5,q)$ in $PG(3,q^3)$.

It is clear from the preceding results that improved characterization results on small minimal non-trivial blocking sets in PG$(2,q)$ will imply improved characterization results on maximal partial spreads of small positive deficiency δ.

We refer to [5, 24] for a more general application of the results about blocking sets in order to find a lower bound on the size of maximal partial t-spreads in PG(n,q). Precisely, let $\mathcal{S}$ be a maximal partial t-spread of PG(n,q): the maximality of $\mathcal{S}$ implies that the set of points contained in an element of $\mathcal{S}$ is an $(n-t)$-blocking set of PG(n,q). In [24], the author constructs as follows small maximal partial t-spreads: let B be a minimal $(n-t)$-blocking set of PG(n,q), then try to find a partial t-spread $\mathcal{S}_1$ that contains as many points of B as possible; let A be $B\setminus\mathcal{S}_1$, then try to find the smallest set $\mathcal{S}_2$ of mutually disjoint t-spaces that covers all the points of A. In the same paper, the author finds a construction for maximal partial t-spreads:

Theorem 2.8. *In $PG(n,q)$, $n=k(t+1)+t-1+r$, with $k\geq 2$, there exist maximal partial t-spreads of size*

$$q^r\frac{q^{k(t+1)}-1}{q^{t+1}-1}+q^\beta-q^r+1,\ \text{where}\ \begin{cases}\beta=-\infty & \text{if } r=0,\\ \beta=\lceil (t+r-1)/2\rceil+1 & \text{otherwise.}\end{cases} \quad (1)$$

Hence, by *small* maximal partial t-spreads, we mean maximal partial t-spreads of size less than (1). As stated in Sect. 1, the smallest minimal $(n-t)$-blocking set of $PG(n,q)$ is a trivial one, i.e., an $(n-t)$-space, and the second smallest one is given by Theorem 1.8. By counting arguments, it is easy to see that if B is a blocking set as described in Theorem 1.8, a set of t-spaces covering $|B|$ points has size larger than (1); hence, to find lower bounds on small maximal partial t-spreads, it is necessary to start from an $(n-t)$-space; in this way it is possible to find the following bounds:

Theorem 2.9. *[24] In $PG(n,q)$, $n = k(t+1)+t-1+r$, with $k \geq 2$, let $\mathcal{S}$ be a smallest maximal partial t-spread. Then the following hold:*

- *If $r = 0$, then $|\mathcal{S}| = \frac{q^{k(t+1)}-1}{q^{t+1}-1}$ (this is actually proven in [5]);*
- *If $r = 1$, then $|\mathcal{S}| \geq q\frac{q^{k(t+1)}-1}{q^{t+1}-1} + q^2 - q + 1$;*
- *If $r > 1$ and $t+1 \geq 2r$, then $|\mathcal{S}| \geq q^r\frac{q^{k(t+1)}-1}{q^{t+1}-1} + (q^{r+1}-q^r)/2 + 1$;*
- *If $r > 1$ and $t+1 < 2r$, then $|\mathcal{S}| \geq q^r\frac{q^{k(t+1)}-1}{q^{t+1}-1} + (q^{r+1}-q^r+q^{2r-t-1}+3q+1)/2$.*

In some cases, these lower bounds and the upper bound given by (1) coincide and so we have the exact size of a smallest maximal partial t-spread:

Corollary 2.10. *(a) In $PG(2k+1,q)$, $k \geq 2$, the smallest maximal partial line spreads have size $q\frac{q^{2k}-1}{q^2-1} + q^2 - q + 1$.*
(b) In $PG(3k+2,q)$, $k \geq 2$, the smallest maximal partial plane spreads have size $q\frac{q^{3k}-1}{q^3-1} + q^2 - q + 1$.

3 Blocking Sets and Coding Theory

The most natural application of finite geometry to coding theory occurs when the codes are *linear*. Let $\mathbb{F}_q$ be the finite field of order q and let $V := \mathbb{F}_q^n$ be the n-dimensional vector space of the n-tuples over $\mathbb{F}_q$. A linear code C is a k-dimensional subspace of V and it is called an $[n,k]$ *code* over $\mathbb{F}_q$ or an $[n,k]_q$ code. The *Hamming distance* $d(c,c')$ of two codewords c and c' is the number of the components in which they differ and the *minimum distance* d of C is the minimum of the set $\{d(c,c')|c,c' \in C, c \neq c'\}$. A linear $[n,k]$ code or $[n,k]_q$ code having minimum distance d is denoted by $[n,k,d]$ *code* or $[n,k,d]_q$ *code*. The *weight* $w(c)$ of a codeword c is the number of the non-zero components of c. It is well known (see for example [32]) that for a linear code C, $\min\{w(c)|c \in C\setminus\{\mathbf{0}\}\} = \min\{d(c,c')|c,c' \in C, c \neq c'\}$. One of the main reasons why the weight of the codewords of a linear code is investigated is that a code C with minimum distance d can correct up to t errors, where $2t+1 \leq d$.

One of the ways to define a linear code C is by means of its generator matrix H. If H is the incidence matrix of a finite geometry, the properties of the geometries can be translated to properties of C and we have in this case powerful tools to investigate the minimum distance and the weight distribution of C. If H is the incidence matrix of the points and k-spaces of PG(n,q), then we will denote the code generated by H by $C_k(n,q)$. When we say that a vector $v \in C_k(n,q)$ is a subset S of PG(n,q), we mean that v is the incidence vector of S. We let (c_1,c_2) denote the scalar product in $\mathbb{F}_q$ of two codewords c_1 and c_2 of a q-ary linear code C. The *dual code* $C^{\perp}$ of a q-ary linear code C of length m is the set of all vectors orthogonal to all the codewords of C, hence $C^{\perp} = \{v \in V(m,q) | (v,c) = 0, \forall c \in C\}$. If C is a linear $[m,k]$-code, then $C^{\perp}$ is an $[m,m-k]$-code and if H is a generator matrix for C, then H is a parity check matrix for $C^{\perp}$. Hence, $C_k(n,q)^{\perp}$ is the code that has as parity check matrix H the incidence matrix of points and k-dimensional spaces of PG(n,q). The minimum distance d of $C_k(n,q)^{\perp}$ satisfies the following inequality [15]: $(q+p)q^{n-k-1} \leq d \leq 2q^{n-k}$, where $q = p^h$, and for $p = 2$, the lower bound is sharp. In [42], the authors use some properties of blocking sets to prove upper bounds on $d(C_k(n,q)^{\perp})$.

Theorem 3.1. *[42] Let B be a minimal $(n-k)$-blocking set in $PG(n,q)$ of size $q^{n-k} + x$, with $x < (q^{n-k}+3)/2$, such that there exists an $(n-k)$-space μ intersecting B in x points. The difference of the incidence vectors of B and μ is a codeword of $C_k(n,q)^{\perp}$ with weight $2q^{n-k} + \theta_{n-k-1} - x$.*

Theorem 3.2. *[42] There exists a small minimal $(n-k)$-blocking set B of size $q^{n-k}+x$ in $PG(n,q)$, $q = p^h$, such that there is an $(n-k)$-space μ with $|B \cap \mu| = x$ and with $x = q^{n-k-1}(q-1)/(p-1) + \theta_{n-k-2}$.*

Corollary 3.3. *[42] The minimum weight d of $C_k(n,q)^{\perp}$ satisfies the following inequality: $d \leq 2q^{n-k} - q^{n-k-1}(q-p)/(p-1)$.*

There are also applications of the theory of the blocking sets to the study of the weights of the codewords of $C_k(n,q)$. To define $C_k(n,q)$, we have considered the incidence matrix H, that is we have labeled the columns of H by the points of PG(n,q), hence we have ordered the points. Let $c \in C_k(n,q)$ and let $supp(c)$ be the set of indices i for which the ith-component of c is non-zero: this set defines a set S of points of PG(n,q) and we will say that S is defined by c. As a first application, we mention the following theorem:

Theorem 3.4. *[1] The minimum weight of $C_k(n,q)$ is θ_k and a codeword of such a weight is a scalar multiple of the incidence vector of a k-space.*

This can be proven in an easy way (see [62, Theorem 6.3.3]) using the fact that a small weight codeword defines a k-blocking set and using the well known result that a k-blocking set of PG(n,q) of size θ_k is a k-space (Theorem 1.7). Moreover, using the link between k-blocking sets and the codewords of $C_k(n,q)$, it is possible to compute a gap for the weight of the codewords. The first useful result is the following:

Lemma 3.5. *[43] Let $c \in C_k(n,q)$, then there exists a value $a \in \mathbb{F}_p$ such that $(c,\mu) = a$ for all the subspaces μ of $PG(n,q)$ of dimension at least $n-k$.*

Then it is easy to prove the following.

Theorem 3.6. *Let $c \in C_k(n,q)$, $q = p^h$, $p > 3$, with $w(c) < 2q^k$, such that $(c,\mu) \neq 0$ for some $(n-k)$-space μ, then c is a scalar multiple of the incidence vector of a small minimal k-blocking set in $PG(n,q)$.*

Moreover, it is useful to see how the set defined by a codeword intersects a given blocking set.

Lemma 3.7. *[43] Let $c \in C_k(n,q)$, $q = p^h$, $p > 3$, with $w(c) < 2q^k$, such that $(c,\mu) \neq 0$ for some $(n-k)$-space μ, and let B be a small minimal $(n-k)$-blocking set. Then the set S of points defined by c intersects B in 1 mod p points.*

This very strong condition led to the following result about blocking sets.

Theorem 3.8. *[43] If B is a minimal k-blocking set in $PG(n,p^h)$, $p > 2$, $|B| \leq 3(p^{hk} - p^{hk-1})/2$, intersecting every $\mathbb{F}_p$-linear $(n-k)$-blocking set in 1 mod p points, then B is trivial.*

By Lemma 3.5, we have that either $(c,\mu) \neq 0$ for all $(n-k)$-spaces μ, or $(c,\mu) = 0$ for all $(n-k)$-spaces μ, but in the latter case we get $c \in C_k(n,q)^{\perp}$. If $c \notin C_k(n,q)^{\perp}$ and $w(c) < 2q^k$, then by Theorem 3.6 we have that $supp(c)$ defines a small minimal k-blocking set B that by Lemma 3.7 and Theorem 3.8 turns out to be trivial, that is B is a k-subspace and $w(c) = \theta_k$. Hence, we have the following theorem.

Theorem 3.9. *[43] There are no codewords in $C_k(n,q) \setminus C_k(n,q)^{\perp}$, $p > 5$, with weight in the interval $]\theta_k, 2q^k[$.*

Finally, since we have:

Theorem 3.10. *[2] The minimum weight for $C_k(n,q)^{\perp}$ is at least $2(\frac{q^n-1}{q^{n-k}-1}(1-\frac{1}{p})+\frac{1}{p})$.*

We get the following gap for the weight of the codewords of $C_k(n,q)$.

Theorem 3.11. *[43] There are no codewords in $C_k(n,q)$, $p > 5$, with weight in the interval $]\theta_k, 2(\frac{q^n-1}{q^{n-k}-1}(1-\frac{1}{p})+\frac{1}{p})[$.*

4 Minihypers and the Griesmer Bound

4.1 The Griesmer Bound

The Griesmer bound in coding theory gives a lower bound on the length n of a k-dimensional linear code over the finite field $\mathbb{F}_q$ of order q, having minimum distance d.

Theorem 4.1. [27,54] *For every linear* $[n,k,d]_q$ *code,*

$$n \geq \sum_{i=0}^{k-1} \left\lceil \frac{d}{q^i} \right\rceil = g_q(k,d),$$

where $\lceil x \rceil$ *denotes the smallest integer larger than or equal to* x.

Linear codes attaining the Griesmer bound, i.e., with parameters $[g_q(k,d), k,d]_q$, are called *Griesmer codes.*

The problem of whether the lower bound $g_q(k,d)$ is sharp for given k, d, and q is a difficult problem. This problem is investigated by many different techniques, one of which uses the geometrical concept of *minihypers*.

We now give the definition of a minihyper, as stated in the literature, but as immediately can be seen from the definition, a minihyper is nothing else than a blocking set. In the following definition, a multiset $(\mathcal{H}, w)$ in $\mathrm{PG}(N,q)$ is a set of points $\mathcal{H}$ of $\mathrm{PG}(N,q)$ with a weight function w associating a non-negative integer to every point P of $\mathrm{PG}(N,q)$.

Definition 4.2. A multiset (F,w) in $\mathrm{PG}(N,q)$ is called an *$(f,m;N,q)$-minihyper* if

(a) $P \in F \iff w(P) > 0$;
(b) $\sum_{P \in \mathrm{PG}(N,q)} w(P) = f$;
(b) $|F \cap H| = \sum_{P \in F \cap H} w(P) \geq m$ for any hyperplane H, and there exists a hyperplane H_0 with $|F \cap H_0| = m$.

In case $w(P) \in \{0,1\}$, for every point P of $PG(N,q)$, we can omit the weight function w in the definition of the minihyper and simply denote the minihyper by F, and refer to it as *projective minihyper*. We will also speak of *(f,m)-minihypers* if the geometry $\mathrm{PG}(N,q)$ we consider is clear from the context.

It is immediately clear from the definition that an $(f,m;N,q)$-minihyper is in fact an m-fold 1-blocking set of size f. So, in the definition of an $(f,m;N,q)$-minihyper, not only the minimal number m of points the set has in common with every hyperplane is of importance, also the exact size f of the $(f,m;N,q)$-minihyper is of importance.

The reason why both the size f and the minimal number m of points the minihyper has in common with every hyperplane are important follows from the close link between linear codes meeting the Griesmer bound and their geometrical counterpart of the minihypers.

We now describe this link and a general class of minihypers.

4.2 Minihypers and the Belov–Logachev–Sandimirov Construction

The link between minihypers in $\mathrm{PG}(k-1,q)$ and linear $[n,k,d]_q$ codes meeting the Griesmer bound is described in the following way.

For $(s-1)q^{k-1} < d \leq sq^{k-1}$, d can be written uniquely as $d = sq^{k-1} - \sum_{i=1}^{h} q^{\lambda_i}$ such that:

(a) $0 \leq \lambda_1 \leq \cdots \leq \lambda_h < k-1$,
(b) At most $q-1$ of the values λ_i are equal to a given value.

Using this expression for d, the Griesmer bound for a linear $[n,k,d]_q$ code becomes:

$$n \geq s\theta_{k-1} - \sum_{i=1}^{h} \theta_{\lambda_i}.$$

Hamada and Helleseth showed that in the case $d = sq^{k-1} - \sum_{i=1}^{h} q^{\lambda_i}$, there is a one-to-one correspondence between the set of all non-equivalent $[n,k,d]_q$ codes meeting the Griesmer bound and the set of all projectively distinct $(\sum_{i=1}^{h} \theta_{\lambda_i}, \sum_{i=1}^{h} \theta_{\lambda_i - 1}; k-1, q)$-minihypers $\mathcal{F}$ [29]. This correspondence is as follows:

Consider a generator matrix $G = (P_1 \cdots P_n)$ of the $[n,k,d]_q$ code C meeting the Griesmer bound, and let $(s-1)q^{k-1} < d \leq sq^{k-1}$, where d is written as $d = sq^{k-1} - \sum_{i=1}^{h} q^{\lambda_i}$, described above. Then the n columns of G define points of the projective space $\mathrm{PG}(k-1,q)$. *Consider the multiset $s \cdot \mathrm{PG}(k-1,q)$, which is equal to s copies of the projective space* $\mathrm{PG}(k-1,q)$. *Then the multiset $s \cdot \mathrm{PG}(k-1,q) \setminus \{P_1, \ldots, P_n\}$ is equal to a $(\sum_{i=1}^{h} \theta_{\lambda_i}, \sum_{i=1}^{h} \theta_{\lambda_i - 1}; k-1, q)$-minihyper $(\mathcal{F}, w)$. In other words, the weight w of a point P of* $\mathrm{PG}(k-1,q)$ *is equal to s minus the number of columns of G defining the point P of* $\mathrm{PG}(k-1,q)$. *Alternatively, the minihyper corresponding to a Griesmer code is the multiset in $s \cdot \mathrm{PG}(k-1,q)$ obtained by taking the complement of the multiset $\{P_1, \ldots, P_n\}$ in the multiset $s \cdot \mathrm{PG}(k-1,q)$.*

We note that this construction of linking minihypers to Griesmer codes is also known in the literature under the name of *anticodes* [48].

Belov et al. [4] gave a construction method for Griesmer codes, which is easily described by using the corresponding minihypers.

Consider in PG$(k-1,q)$ *a sum of* ϵ_0 *points* $P_1, P_2, \ldots, P_{\epsilon_0}$, ϵ_1 *lines* $\ell_1, \ell_2, \ldots$, ℓ_{ϵ_1}, $\ldots$, ϵ_{k-2} $(k-2)$*-dimensional subspaces* $\pi_1^{(k-2)}, \ldots, \pi_{\epsilon_{k-2}}^{(k-2)}$, *with* $0 \leq \epsilon_i \leq q-1, i = 0, \ldots, k-2$, *then such a sum defines a* $(\sum_{i=0}^{k-2} \epsilon_i \theta_i, \sum_{i=0}^{k-2} \epsilon_i \theta_{i-1}; k-1, q)$*-minihyper* $\mathcal{F}$, *where the weight of a point* R *of* PG$(k-1,q)$ *equals the number of objects, in the description above, in which it is contained.*

Now that the standard examples of minihypers are known, different problems on minihypers arise. One of them is the construction of other examples of minihypers. The Belov–Logachev–Sandimirov construction can be extended by allowing subgeometries and projected subgeometries, but there must be definitely many other types of minihypers. There is also the problem of determining the minihypers which cannot be written as the sum of two other smaller minihypers. One of the problems investigated greatly in recent years is the characterization problem on minihypers, and equivalently on linear codes meeting the Griesmer bound:

Characterize $(f, m; k-1, q)$*-minihypers* $\mathcal{F}$ *for given parameters* $f = \sum_{i=0}^{k-2} \epsilon_i \theta_i$, $m = \sum_{i=0}^{k-2} \epsilon_i \theta_{i-1}$, k, *and* q.

Fundamental research on this problem was performed by Hamada *et al* who, in many articles, obtained a lot of results on minihypers and who developed a great amount of techniques useful in the study of minihypers. Their main results are in [28, 30].

Since the research they performed, the techniques they developed have been extended, including the use of the recent results on blocking sets. This has made it possible to obtain great improvements. Since minihypers are in fact particular blocking sets, the complete characterization of minihypers is greatly based on corresponding characterization results on blocking sets. However, differences occur between the geometrical study of blocking sets, and the corresponding coding-theoretical characterizations of minihypers.

The geometrical interest is mainly the characterization of minimal blocking sets. This characterization includes minimal blocking sets which are cones having a vertex and a base which is a non-trivial minimal blocking set in a smaller space. But such cones are not always minihypers with the correct parameters $(\sum_{i=1}^{h} \theta_{\lambda_i}, \sum_{i=1}^{h} \theta_{\lambda_i - 1}; k-1, q)$. For instance, consider a 2-blocking set B in PG$(3,q)$, q square, which is a cone with vertex the point P and base the Baer subplane PG$(2,\sqrt{q})$ in a plane skew to P. This cone has size $f = q^2 + q\sqrt{q} + q + 1$, and it intersects every plane in at least $m = \sqrt{q} + 1$ points. But it is not possible to write (f, m) in the form $(\sum_{i=1}^{h} \theta_{\lambda_i}, \sum_{i=1}^{h} \theta_{\lambda_i - 1})$. For the main known results on minihypers, we refer to

Table 1 Upper bounds on δ_0

p	s	δ_0
p	1	$\leq (p+1)/2$
p	3	$\leq p^2$
p	Even	$\leq \sqrt{q}$
2	$6m+1, m \geq 1$	$\leq 2^{4m+1} - 2^{4m} - 2^{2m+1}/2$
>2	$6m+1, m \geq 1$	$\leq p^{4m+1} - p^{4m} - p^{2m+1}/2 + 1/2$
2	$6m+3, m \geq 1$	$< 2^{4m+5/2} - 2^{4m+1} - 2^{2m+1} + 1$
>2	$6m+3, m \geq 1$	$\leq p^{4m+2} - p^{2m+2} + 2$
≥5	$6m+5, m \geq 0$	$< p^{4m+7/2} - p^{4m+3} - p^{2m+2}/2 + 1$

the survey articles [55, 56] and also to the chapter on *Galois geometries and coding theory* [40] that appeared in the collected work *Recent research topics in Galois geometry* [20].

We mention here the most recent results on projective minihypers, and on weighted minihypers.

Theorem 4.3 (De Beule et al [19]). *A projective* $(\sum_{i=0}^{k-2} \epsilon_i \theta_i, \sum_{i=0}^{k-2} \epsilon_i \theta_{i-1}; k-1, q)$*-minihyper, where* $\sum_{i=0}^{k-2} \epsilon_i \leq \delta_0$ *with* δ_0 *equal to one of the values in Table 1, is a union of* ϵ_{k-2} *hyperplanes,* ϵ_{k-3} $(k-3)$*-dimensional spaces,* $\ldots, \epsilon_1$ *lines, and* ϵ_0 *points, which all are pairwise disjoint, so is of Belov-Logachev-Sandimirov type.*

In the following table, $q = p^s$, p prime, $s \geq 1$.

Theorem 4.4 (De Beule et al. [18]). *A* $(\sum_{i=0}^{k-2} \epsilon_i \theta_i, \sum_{i=0}^{k-2} \epsilon_i \theta_{i-1}; k-1, q)$*-minihyper, where* $\sum_{i=0}^{k-2} \epsilon_i < \sqrt{q} + 1$*, is a sum of* ϵ_{k-2} *hyperplanes,* ϵ_{k-3} $(k-3)$*-dimensional spaces,* $\ldots, \epsilon_1$ *lines, and* ϵ_0 *points, so it is of Belov–Logachev–Sandimirov type.*

The results on the minihypers are obtained by using results on blocking sets via different kinds of methods.

For instance, consider a $(\sum_{i=0}^{s} \epsilon_i \theta_i, \sum_{i=0}^{s} \epsilon_i \theta_{i-1}; k-1, q)$-minihyper F, $\sum_{i=0}^{s} \epsilon_i = h$ small and with $\epsilon_s \neq 0$. Then there exist $(k-s)$-dimensional subspaces Π_{k-s} intersecting F in $(\epsilon_s(q+1) + m_0^{(s-1)}, \epsilon_s; k-s, q)$-minihypers, with $\epsilon_s + m_0^{(s-1)} \leq h$. This is in fact an ϵ_s-fold 1-blocking set in Π_{k-s}. If these ϵ_s-fold 1-blocking sets in $\mathrm{PG}(k-s, q)$ are characterized, then the $(k+1-s)$-dimensional spaces Π_{k+1-s} intersecting F in $(\epsilon_s(q^2+q+1) + \epsilon_{s-1}(q+1) + m_0^{(s-2)}, \epsilon_s(q+1) + \epsilon_{s-1}; k+1-s, q)$-

minihypers F_{k+1-s}, with $\epsilon_s + \epsilon_{s-1} + m_0^{(s-2)} \le h$, can sometimes be characterized via the many hyperplanes of Π_{k+1-s} that intersect F in $(\epsilon_s(q+1)+m_0^{(s-1)}, \epsilon_s; k-s, q)$-minihypers, with $\epsilon_s + m_0^{(s-1)} \le h$.

Inductively, if all the $(\sum_{i=j+1}^{s} \epsilon_i \theta_{i-j} + m_0^{(j)}, \sum_{i=j+1}^{s} \epsilon_i \theta_{i-j-1}; k-1-j, q)$-minihypers, with $\sum_{i=j+1}^{s} \epsilon_i + m_0^{(j)} \le h$, are characterized, then sometimes all the $(\sum_{i=j}^{s} \epsilon_i \theta_{i+1-j} + m_0^{(j-1)}, \sum_{i=j}^{s} \epsilon_i \theta_{i-j}; k-j, q)$-minihypers, with $\sum_{i=j}^{s} \epsilon_i + m_0^{(j-1)} \le h$, can be characterized. Finally, for $j = 0$, this might lead to the complete characterization of the $(\sum_{i=0}^{s} \epsilon_i \theta_i, \sum_{i=0}^{s} \epsilon_i \theta_{i-1}; k-1, q)$-minihypers F, with $\sum_{i=0}^{s} \epsilon_i = h$.

For instance, this method was used in [18, 19] and in [28, 30].

Since the characterization of minihypers heavily relies on the characterization results on (multiple) blocking sets in the plane PG$(2, q)$, also here we have the phenomenon that improved characterization results on (multiple) blocking sets in PG$(2, q)$ imply improved characterization results on minihypers.

5 Extension Results

Recently, a new link of blocking sets to a coding-theoretical problem has been found. Namely, a link between blocking sets and the extendability of linear codes has been determined. This link has already been used in combination with the two classical results on blocking sets of Bose–Burton (Theorem 1.7) and of Beutelspacher–Heim (Theorem 1.8).

Regarding the extendability problem for linear codes, it is already known for a long time that a binary $[n, k, d]$ code of odd minimum distance d can be extended to an $[n+1, k, d+1]$ code by adding a parity check. This result has been generalized by Hill and Lizak in [33, 34].

Theorem 5.1 (Hill and Lizak [33, 34]). *Let C be an $[n, k, d]_q$ code with* $\gcd(d, q) = 1$ *and with all non-zero weights congruent to* 0 *or* $d \pmod q$. *Then C can be extended to an $[n+1, k, d+1]_q$ code.*

The geometrical version of this result is obtained in the following way. For the details, we refer to the chapter on Galois geometries and coding theory [40] in the collected work [20].

Consider a generator matrix $G = (P_1 \cdots P_n)$ of the linear code C of Theorem 5.1. Then the n columns of G define an $(n, w; k-1, q)$-arc, i.e., a multiset of n points in PG$(k-1, q)$ intersecting every hyperplane in at most $w = n - d$ points, and with moreover $\gcd(n-w,q) = 1$. From the assumptions of Theorem 5.1, the intersection sizes of all hyperplanes with this multiset are congruent to n or $w \pmod q$.

A geometrical proof shows that there are $(q^{k-1} - 1)/(q - 1)$ hyperplanes with intersection size congruent to $n \pmod q$, which form a dual blocking set with respect to the $(k - 3)$-dimensional subspaces of $\mathrm{PG}(k - 1, q)$, i.e., through every $(k - 3)$-dimensional subspace of $\mathrm{PG}(k - 1, q)$, there passes at least one hyperplane of $\mathrm{PG}(k - 1, q)$ with intersection size congruent to $n \pmod q$. By the dual of the Bose-Burton theorem (Theorem 1.7), these hyperplanes pass through a fixed point P. Hence, we can construct an $(n+1, w; k-1, q)$-arc by increasing the multiplicity of P by 1.

Then $\hat{G} = (P_1 \cdots P_n P)$ is the generator matrix of an $[n + 1, k, d + 1]_q$ code $\hat{C}$ which is an extension of C.

Using the result of Beutelspacher and Heim (Theorem 1.8) and the geometrical ideas described above, the preceding result was improved.

Theorem 5.2. (Landjev and Rousseva [39]) *Let $\mathcal{K}$ be an $(n, w; k - 1, q)$-arc, $q = p^s$, with spectrum $(a_i)_{i \geq 0}$. Let $w \not\equiv n \pmod q$ and*

$$\sum_{i \not\equiv w \pmod q} a_i < q^{k-2} + q^{k-3} + \cdots + q + 1 + q^{k-3} \cdot r(q), \qquad (2)$$

where $q + r(q) + 1$ is the minimal size of a non-trivial blocking set of $\mathrm{PG}(2, q)$. *Then $\mathcal{K}$ is extendable to an $(n + 1, w; k - 1, q)$-arc.*

This result is as follows restated into a non-extendability result for linear codes.

Theorem 5.3. *Let C be a non-extendable $[n, k, d]_q$ code, $q = p^s$, with* $\gcd(d, q) =$ 1. *If $(A_i)_{i \geq 0}$ is the spectrum of C, then $\sum_{i \not\equiv 0,d \pmod q} A_i \geq q^{k-3} \cdot r(q)$, where $r(q)$ is the same as in Theorem 5.2.*

6 Blocking Sets and Cryptography

We now mention an application of blocking sets in cryptography.

In [38], the authors designed a new key distribution scheme for TV-nets. A major problem in the design of such a key distribution scheme is the fact that sometimes subscribers to such a TV-net become compromised; this means that they no longer wish to pay for receiving the programs of this latter TV-net. If this happens, the codes (called *keys*) of these compromised subscribers become themselves compromised, and must be no longer valid. The system of [38], based on points and lines of projective planes, enables a TV-net to distribute keys to their subscribers in such a way that keys still remain valid, even if subscribers become compromised. Only if certain subsets of compromised subscribers are reached, the TV-net has to distribute a new set of keys to all of their subscribers. We now describe this key distribution scheme.

We identify the subscribers with the points of a projective plane PG$(2,q)$. The keys are identified with the lines of the projective plane PG$(2,q)$.

If a person subscribes to the TV-net, he is identified with a point P of PG$(2,q)$ and he receives the keys of all the lines of PG$(2,q)$ passing through P. This enables him to receive the programs of the TV-net.

If however a subscriber P decides not to pay any longer for receiving the programs of the TV-net, the keys of the lines of PG$(2,q)$ passing through the point P become invalid. This is no problem for the other subscribers since they still lie on q lines not passing through P, so they still have q valid keys for receiving the programs of the TV-net.

It is clear that if at most q subscribers become compromised, then every non-compromised subscriber still has at least one valid key to receive the programs of the TV-net. More precisely, when does a non-compromised subscriber Q loose all his keys to view the programs? He looses all keys when all the keys of the lines of PG$(2,q)$ through Q become invalid. In other words, when the set of compromised subscribers forms a trivial dual blocking set in PG$(2,q)$, containing all lines through Q.

Only when the compromised subscribers form a trivial dual blocking set in PG$(2,q)$, the TV-net has to distribute a new set of keys to all of their subscribers.

7 Other Applications and Open Problems

In this section, we shall give a brief overview on other applications of blocking sets. The first to mention is on the theory of *t-covers*.

7.1 Covers in Galois Geometries

A *t-cover* $\mathcal{C}$ of PG(n,q) is a set of t-subspaces of PG(n,q) that cover the point set of PG(n,q), i.e., every point of PG(n,q) is contained in at least one element of $\mathcal{C}$. Obviously, $|\mathcal{C}| \geq \lceil \frac{\theta_n}{\theta_t} \rceil$ (we remind that a t-spread, which is a particular case of a t-cover, has size $\frac{\theta_n}{\theta_t}$) and we call the *excess* of $\mathcal{C}$ the integer $\epsilon = |\mathcal{C}| - \lceil \frac{\theta_n}{\theta_t} \rceil$. The *multiple points* of $\mathcal{C}$ are the points contained in more than one element of $\mathcal{C}$ and if P is a multiple point, then the *surplus* of P is the number of elements of $\mathcal{C}$ containing P minus one. One can consider the surplus as a weight function mapping a point $P \in$ PG(n,q) to a non-negative integer $surplus(P)$. So we have $surplus(P) > 0$ if and only if P is a multiple point and, if $t+1$ divides $n+1$, then $\sum_{P\in PG(n,q)} surplus(P) = \epsilon\theta_t$. The following theorem shows the link between blocking sets, more precisely minihypers, and the multiple points of a t-cover.

Theorem 7.1. *[26] Let $\mathcal{C}$ be a t-cover of $PG(n,q)$, $(t+1)|(n+1)$, with excess $\epsilon < q$. Let F be the set of multiple points of $\mathcal{C}$ and let $w(P) = surplus(P)$, $\forall P \in PG(n,q)$. Then (F,w) is an $(\epsilon\theta_t, \epsilon\theta_{t-1}; n, q)$-minihyper.*

By a classification result about minihypers of [26], it is possible to describe the set of multiple points:

Corollary 7.2. *[26] Let $\mathcal{C}$ be a t-cover of $PG(n,q)$, $(t+1)|(n+1)$, with excess $\epsilon < \epsilon_q$ where ϵ_q is such that $q + \epsilon_q$ is the size of the smallest non-trivial blocking sets of $PG(2,q)$. Then the multiple points form a sum of ϵ t-subspaces.*

7.2 Minihypers and *i*-Tight Sets

By a *polar space*, we will mean the lattice of subspaces of a projective space contained in a non-singular quadric or Hermitian variety, or the lattice of subspaces totally isotropic with respect to a symplectic polarity (see for example [36]). A subset $\mathcal{T}$ of a polar space of rank $r \geq 2$ over $\mathbb{F}_q$ is *i-tight* if

$$|P^{\perp} \cap \mathcal{T}| = \begin{cases} i\frac{q^{r-1}-1}{q-1} + q^{r-1} & \text{if } P \in \mathcal{T}, \\ i\frac{q^{r-1}-1}{q-1} & \text{if } P \notin \mathcal{T}. \end{cases}$$

The easiest example of an i-tight set of a polar space is the union of i pairwise disjoint generators.

In [17], it has been shown that an i-tight set $\mathcal{T}$ of $W(2r+1,q)$, $Q^+(2r+1,q)$, or $H(2r+1,q)$ is a set of $i\theta_r$ points intersecting every hyperplane in at least $i\theta_{r-1}$ points, hence $\mathcal{T}$ is an $(i\theta_r, i\theta_{r-1}; 2r+1, q)$-minihyper. Using characterization results about minihypers of [17, 25], it is possible to characterize i-tight sets in the aforementioned polar spaces.

Theorem 7.3. *[17] An i-tight set on $Q^+(2r+1,q)$, with $2 < i < q/2-1$, can only exist for r odd; then such an i-tight set is the union of i pairwise disjoint generators of $Q^+(2r+1,q)$. For every $r \geq 1$, a 1- or 2-tight set on $Q^+(2r+1,q)$ consists of one or two disjoint generators.*

Theorem 7.4. *[17] Let $\mathcal{T}$ be an i-tight set of $H(2r+1,q)$, with $q > 16$ and $i < q^{\frac{5}{8}}/\sqrt{2}+1$, then $\mathcal{T}$ is the union of pairwise disjoint generators and Baer subgeometries $PG(2r+1,\sqrt{q})$, such that the Hermitian polarity induces a symplectic polarity on these Baer subgeometries.*

Theorem 7.5. *[17] Let $\mathcal{T}$ be an i-tight set of $W(2r+1,q)$, with q square and $i < q^{\frac{5}{8}}/\sqrt{2}+1$, then $\mathcal{T}$ is the union of pairwise disjoint r-dimensional* *subspaces and Baer subgeometries $PG(2r+1, \sqrt{q})$. The r-dimensional subspaces are either generators or pairs $\{U, U^{\perp}\}$, with $U \cap U^{\perp} = \emptyset$; the subgeometries are either Baer subgeometries invariant under the symplectic polarity or pairs of Baer subgeometries $\{\mathcal{B}_1, \mathcal{B}_2\}$, where $P^{\perp} \cap \mathcal{B}_2 = PG(2r, \sqrt{q}), \forall P \in \mathcal{B}_1$.*

Since an i-tight set $\mathcal{T}$ of $W(2r+1, q)$, $Q^+(2r+1, q)$, or $H(2r+1, q)$ is an $(i\theta_r, i\theta_{r-1}; 2r+1, q)$-minihyper, improved results on $(i\theta_r, i\theta_{r-1}; 2r+1, q)$-minihypers will imply improved results on these i-tight sets $\mathcal{T}$. As minihypers are blocking sets, we can again state that improved results on blocking sets imply improved results on another topic, i.e., that of i-tight sets of polar spaces.

7.3 t-Fold k-Blocking Sets in $PG(n, q)$

Recently, results on onefold blocking sets have been extended to t-fold blocking sets. This includes in particular the $1 \pmod p$ result (Theorem 1.10), which was extended to a $t \pmod p$ result. Since such $t \pmod p$ results are very strong results, implying strong characterization results, we mention this $t \pmod p$ result on t-fold blocking sets, and give also a characterization result on t-fold blocking sets heavily relying on this $t \pmod p$ result.

Theorem 7.6. *[22] Let B be a minimal weighted t-fold k-blocking set of* $\mathrm{PG}(n, q)$, *$q = p^h$, p prime, $h \geq 1$, of size $|B| = tq^k + t + k'$, with $t + k' \leq (q^k - 1)/2$.*

Then B intersects every $(n-k)$-dimensional subspace in $t \pmod p$ points.

Theorem 7.7. *[22] Let B be a minimal weighted t-fold k-blocking set of* $\mathrm{PG}(n, q)$, *$q = p^h$, p prime, $h \geq 1$, of size $|B| = tq^k + t + k'$, with $t + k' \leq (q^k - 1)/2$.*

Let $e \geq 1$ be the largest integer such that each $(n-k)$-dimensional subspace intersects B in $t \pmod{p^e}$ points. Then, for $0 \leq s \leq n-k$ and every s-dimensional subspace Π_s, $|B \cap \Pi_s| \in \{0, 1, ..., t\} \pmod{p^e}$.

Theorem 7.8. *[23] Let B be a minimal t-fold k-blocking set in* $\mathrm{PG}(n, q)$, *q square, $q \geq 661$, $t < c_p q^{1/6}/2$, of size at most $|B| \leq tq^k + 2tq^{k-1}\sqrt{q} < tq^k + c_p q^{k-1/3}$, with $c_2 = c_3 = 2^{-\frac{1}{3}}$ and $c_p = 1$ for $p > 3$.*

Then B is a union of t pairwise disjoint cones $\langle \pi_{m_i}, \mathrm{PG}(2(k - m_i - 1), \sqrt{q}) \rangle$, $-1 \leq m_i \leq k-1$, $i = 1, \ldots, t$.

Theorem 7.9. *[23] Let B be a minimal t-fold k-blocking set in* $\mathrm{PG}(n, q)$, *q square, $t \geq 2$, which is a union of t pairwise disjoint cones $\langle \pi_{m_i}, \mathrm{PG}(2(k - m_i - 1), \sqrt{q}) \rangle$,* $\max\{-1, 2k-n-1\} \leq m_i \leq k-1$. *Then $k < n/2$ if B contains at least one k-dimensional space* $\mathrm{PG}(k, q)$ *and $k \leq n/2$ in the other cases.*

7.4 Blocking Sets and Semifields

A finite *semifield* $\mathbb{S}$ is a finite algebraic structure satisfying all the axioms of a skew field except (possibly) associativity. These algebraic structures are of considerable interest because they coordinatize certain *translation planes*, called *semifield planes* and two semifields are *isotopic* if and only if the corresponding translation planes are isomorphic. A semifield coordinatizing a Desarguesian plane is isotopic to a field. A geometric construction to get a translation plane is the following: let $\mathcal{S}$ be a $(t-1)$-spread of PG$(2t-1,q)$, embed PG$(2t-1,q)$ as a hyperplane into PG$(2t,q)$, and let $A(\mathcal{S})$ be a point-line geometry such that the points are the points of PG$(2t,q)\backslash$PG$(2t-1,q)$ and the lines are the t-dimensional subspaces of PG$(2t,q)$ intersecting PG$(2t-1,q)$ in an element of $\mathcal{S}$ and the incidence is the containment. Then $A(\mathcal{S})$ is a translation plane of order q^t (see e.g., [21]). A spread $\mathcal{S}$ is called a *semifield spread* if $A(\mathcal{S})$ is coordinatized by a semifield. For PG$(2t-1,q)$, it is always possible to choose as homogeneous coordinates $(x_1,\ldots,x_t;y_1,\ldots,y_t)=(\mathbf{x};\mathbf{y})$ in such a way that the space A of equation $\mathbf{x}=\mathbf{0}$, B of equation $\mathbf{y}=\mathbf{0}$ and C of equation $\mathbf{x}=\mathbf{y}$ are elements of the spread $\mathcal{S}$. Then for every element D of $\mathcal{S}$ distinct from A, there is a unique $t\times t$ matrix J_D over $\mathbb{F}_q$ such that the point $(\mathbf{a};\mathbf{b})$ belongs to D if and only if $\mathbf{b}=\mathbf{a}J_D$. The set $\mathbb{C}=\{J_D|D\in\mathcal{S}, D\neq A\}$ is called the *spread set* associated to $\mathcal{S}$ with respect to A, B and C. The spread $\mathcal{S}$ is a semifield spread (with respect to A) if and only if $\mathbb{C}$ is closed under the sum (for more details, see [21, Section 5.1]). Since $\mathbb{C}$ is closed under the sum, there exists a subfield $\mathbb{F}_s$ of $\mathbb{F}_q$ such that $\mathbb{C}$ is a vector space of rank tn over $\mathbb{F}_s$, where $q=s^n$. Embed PG$(2t-1,q)$ as a canonical subgeometry into PG$(2t-1,q^t)$ in such a way that $(\mathbf{x};\mathbf{y})\in$PG$(2t-1,q)$ if and only if $\mathbf{x}^q=\mathbf{x}$ and $\mathbf{y}^q=\mathbf{y}$, where $\mathbf{x}^q=(x_1^q,\ldots,x_t^q)$. Let A^* and B^* be the spaces of equation $\mathbf{x}=0$ and $\mathbf{y}=0$ respectively. If $P=(\mathbf{b};\mathbf{0})$ is a fixed imaginary point of B^*, then $I=\{(\mathbf{b};\mathbf{b}X)|X\in\mathbb{C}\}$ is the *indicator set* of $\mathcal{S}$ (actually the definition of indicator set appears in a different, more involved way in [44], but here, for the sake of brevity, we give this equivalent way to introduce them) and let $I^*=\{(\lambda\mathbf{b};\mathbf{b}X)|\lambda\in\mathbb{F}_s, X\in\mathbb{C}\}$. Let us now focus on the case $t=2$. Embed PG$(3,q)$ as a canonical Baer subgeometry into PG$(3,q^2)$ and let $A^*=\{(x_1,x_2;0,0)|x_i\in\mathbb{F}_{q^2}\}$, then $T=\langle A^*,P\rangle$ is a plane isomorphic to PG$(2,q^2)$. Then, with the notations as above, we have the following theorem.

Theorem 7.10. *[44] The spread $\mathcal{S}$ is a semifield spread for PG$(3,q)$ if and only if I^* is a Rédei-type blocking set of $T=$PG$(2,q^2)$ disjoint from the Baer subline A of A^* which is a Rédei line of I^*.*

The semifields are not only linked to the spreads of projective spaces, but also to *flocks*, and, in [44], it is also shown when a Rédei-type blocking set gives rise to a semifield flock (for the definitions and details, we refer to [44]).

For a survey on the most important results on finite semifields, we refer to the article [41] in the collected work [20].

7.5 Open Problems

1. The linearity conjecture.

   All the known examples of small minimal k-blocking sets in $\mathrm{PG}(n,q)$ are linear k-blocking sets. If t such pairwise disjoint small linear k-blocking sets in $\mathrm{PG}(n,q)$ exist, their union is a t-fold k-blocking set in $\mathrm{PG}(n,q)$.

   It is generally believed that all small minimal k-blocking sets in $\mathrm{PG}(n,q)$ are linear k-blocking sets, and that all small minimal t-fold k-blocking sets in $\mathrm{PG}(n,q)$ are the union of linear sets, under suitable conditions on the parameters t,k,n,q, and $|B|$. This had led to the following linearity conjecture on blocking sets.

   Conjecture 7.11 (Linearity conjecture for multiple blocking sets [58]). In $\mathrm{PG}(n,q)$, any t-fold minimal k-blocking set B is the union of some (not necessarily disjoint) linear point sets $B_1,\ldots,B_s$, where B_i is a t_i-fold k-blocking set, and $t_1+\cdots+t_s=t$; provided that t and $|B|$ are small enough ($t \leq T(n,q,k)$ and $|B| \leq S(n,q,k)$ for two suitable functions T and S).

   The proof of this linearity conjecture or parts of this conjecture will imply many new results. As we have mentioned in the preceding sections, improvements to characterization results on blocking sets will open the door to new results on many other problems. That is why we present the problem of proving the linearity conjecture as the central problem on blocking sets.

2. Maximal partial spreads.

   In Theorem 2.3, a theorem on maximal partial spreads in $\mathrm{PG}(3,q)$, q square, is stated where it is mentioned that the set of holes is the union of s, with $\sqrt{q}-1 > s > 1$, Baer subgeometries $\mathrm{PG}(3,\sqrt{q})$. Presently, no such examples of maximal partial spreads are known.

   Therefore, as a second open problem, we present the problem of proving or disproving the existence of maximal partial spreads in $\mathrm{PG}(3,q)$, q square, satisfying the conditions of Theorem 2.3.

3. Extendability of linear codes.

   Improvements to the results on the extendability of linear codes are stated in the chapter on Galois geometries and coding theory [40] of [20]. For instance, Maruta proved other extension results [45–47], including a doubly-extendability result. The conditions that need to be satisfied however are very technical. An open problem is to simplify these conditions. Here, a link with blocking sets or with (weighted) 2-fold blocking sets might occur. In general, is it possible to prove a t-fold extendability result on linear codes involving (weighted) t-fold blocking sets?

References

1. E.F. Assmus, Jr. and J.D. Key, Designs and their codes, *Cambridge University Press*, 1992
2. B. Bagchi and S.P. Inamdar, Projective Geometric Codes, *J. Combin. Theory, Ser. A* **99 (1)** (2002), 128–142
3. S. Ball, The number of directions determined by a function over a finite field, *J. Combin. Theory, Ser. A* **104** (2003), 341–350
4. B.I. Belov, V.N. Logachev and V.P. Sandimirov, Construction of a class of linear binary codes achieving the Varshamov-Griesmer bound, *Problems of Info. Transmission* **10** (1974), 211–217
5. A. Beutelspacher, Blocking sets and partial spreads in finite projective spaces, *Geom. Dedicata* **9 (4)** (1980), 425–449
6. A. Blokhuis, Blocking sets in Desarguesian planes. Combinatorics, Paul Erdős is eighty, Vol. 2 (Keszthely, 1993), 133–155, Bolyai Soc. Math. Stud., 2, János Bolyai Math. Soc., Budapest, 1996
7. A. Blokhuis, S. Ball, A.E. Brouwer, L. Storme and T. Szőnyi, On the number of slopes of the graph of a function defined on a finite field, *J. Combin. Theory, Ser. A* **86 (1)** (1999), 187–196
8. A. Blokhuis, L. Storme and T. Szőnyi, Lacunary polynomials, multiple blocking sets and Baer subplanes, *J. London Math. Soc.* **60 (2)** (1999), 321–332
9. A. Blokhuis, P. Sziklai and T. Szőnyi, Blocking sets in projective spaces. *Current research topics in Galois geometry* (J. De Beule and L. Storme, Eds.), NOVA Academic Publishers, (2011), 61–82
10. R.C. Bose and R.C. Burton, A characterization of flat spaces in a finite geometry and the uniqueness of the Hamming and the MacDonald codes, *J. Combin. Theory* **1** (1966), 96–104
11. A.A. Bruen, Baer subplanes and blocking sets, *Bull. Amer. Math. Soc.* **76** (1970), 342–344
12. A.A. Bruen, Blocking sets in finite projective planes, *SIAM J. Applied Math.* **21** (1971), 380–392
13. A.A. Bruen and J.A. Thas, Blocking sets, *Geom. Dedicata* **6 (2)** (1977), 193–203
14. A.A. Bruen and J.A. Thas, Hyperplane coverings and blocking sets, *Math. Z.* **81 (3)** (1982), 407–409
15. N.J. Calkin, J.D. Key and M.J. de Resmini, Minimum weight and dimension formulas for some geometric codes, *Des. Codes Cryptogr.* **17 (1)** (1999), 105–120
16. A. Cossidente, A. Gács, C. Mengyán, A. Siciliano, T. Szőnyi and Zs. Weiner, On large minimal blocking sets in $PG(2,q)$, *J. Combin. Des.* **13 (1)** (2005), 25–41
17. J. De Beule, P. Govaerts, A. Hallez and L. Storme, Tight sets, weighted m–covers, weighted m-ovoids and minihypers, *Des. Codes Cryptogr.* **50** (2009), 187–201
18. J. De Beule, K. Metsch and L. Storme, Characterization results on arbitrary weighted minihypers and on linear codes meeting the Griesmer bound, *Adv. Math. Commun.* **2** (2008), 261–272
19. J. De Beule, K. Metsch and L. Storme, Characterization results on arbitrary non-weighted minihypers and on linear codes meeting the Griesmer bound, *Des. Codes Cryptogr.* **49** (2008), 187–197
20. J. De Beule and L. Storme (Editors), Current research topics in Galois Geometry, NOVA Academic Publishers, (2011), ii+271pp
21. P. Dembowski, Finite geometries, *Springer, Berlin-New York*, 1968
22. S. Ferret, L. Storme, P. Sziklai and Zs. Weiner, A t (mod p) result on multiple $(n-k)$-blocking sets in $PG(n,q)$, *Innov. Incidence Geom.* **6–7** (2007-2008), 169–188
23. S. Ferret, L. Storme, P. Sziklai and Zs. Weiner, A characterization of multiple $(n-k)$-blocking sets in projective spaces of square order, *Adv. Geom.*, submitted
24. P. Govaerts, Small maximal partial t–spreads, *Bull. Belg. Math. Soc. Simon Stevin* **12 (4)** (2005), 607–615
25. P. Govaerts and L. Storme, On a particular class of minihypers and its applications, II: Improvements for q square, *J. Combin. Theory, Ser. A* **97 (2)** (2002), 369–393

26. P. Govaerts and L. Storme, On a particular class of minihypers and its applications, I: The result for general q, *Des. Codes Cryptogr.* **28 (1)** (2003), 51–63.
27. J.H. Griesmer, A bound for error-correcting codes, *IBM J. Res. Develop.* **4** (1960), 532–542
28. N. Hamada and T. Helleseth, A characterization of some q-ary codes ($q > (h-1)^2$, $h \geq 3$) meeting the Griesmer bound, *Math. Japonica* **38** (1993), 925–940
29. N. Hamada and T. Helleseth, Codes and minihypers. *Optimal codes and related topics.* Proceedings of the EuroWorkshop on Optimal codes and related topics (Sunny Beach, Bulgaria, June 10-16, 2001), pp. 79–84
30. N. Hamada and T. Maekawa, A characterization of some q-ary codes ($q > (h-1)^2, h \geq 3$) meeting the Griesmer bound: Part 2, *Math. Japonica* **46** (1997), 241–252
31. U. Heim, Blockierende Mengen in endlichen projektiven Raumen, *Mitt. Math. Semin. Giessen* **226** (1996), 4–82
32. R. Hill, *A first course in coding theory*, Oxford Applied Mathematics and Computing Science Series (1986)
33. R. Hill, An extension theorem for linear codes, *Des. Codes Cryptogr.* **17** (1999), 151–157
34. R. Hill and P. Lizak, Extensions of linear codes, Proc. Intern. Symposium on Inform. Theory, Canada, 1995, p. 345
35. J.W.P. Hirschfeld, Projective Geometries over Finite Fields, *Oxford University Press*, Oxford, 1979
36. J.W.P. Hirschfeld and J.A. Thas, General Galois Geometries, *Oxford University Press*, Oxford, 1991
37. S. Innamorati and A. Maturo, The spectrum of minimal blocking sets, *Discrete Math.* **208/209** (1999), 339–347
38. V. Korjik, M. Ivkov, Y. Merinovich, A. Barg and H.C.A. van Tilborg, A broadcast key distribution scheme based on block designs. Cryptography and coding, Lecture Notes of Comput. Sci. (Ed., C. Boyd) 1025, Springer (1995), pp. 3–12
39. I. Landjev and A. Rousseva, An extension theorem for arcs and linear codes, *Problems of Info. Transmission* **42**(4) (2006), 65–76
40. I. Landjev and L. Storme, Galois geometries and coding theory. *Current research topics in Galois geometry* (J. De Beule and L. Storme, Eds.), NOVA Academic Publishers, (2011), 183–210
41. M. Lavrauw and O. Polverino, Finite semifields. *Current research topics in Galois geometry* (J. De Beule and L. Storme, Eds.), NOVA Academic Publishers, (2011), 127–155
42. M. Lavrauw, L. Storme, and G. Van de Voorde, On the code generated by the incidence matrix of points and k-spaces in PG(n, q) and its dual, *Finite Fields Appl.* **14 (4)** (2008), 1020–1038
43. M. Lavrauw, L. Storme, P. Sziklai, and G. Van de Voorde, An empty interval in the spectrum of small weight codewords in the code from points and k-spaces of PG(n, q), *J. Combin. Theory, Ser. A* **116 (4)** (2009), 996–1001
44. G. Lunardon, Blocking sets and semifields, *J. Combin. Theory, Ser. A* **113** (2006), 1172–1188
45. T. Maruta, On the extendability of linear codes, *Finite Fields Appl.* **7** (2001), 350–354
46. T. Maruta, Extendability of linear codes with minimum distance d, $\gcd(d, q) = 1$, *Discrete Math.* **266** (2003), 377–385.
47. T. Maruta, A new extension theorem for linear codes, *Finite Fields Appl.* **10** (2004), 674–685
48. F.J. MacWilliams and N.J.A. Sloane, The theory of error-correcting codes, *North-Holland Mathematical Library*, Amsterdam-New York-Oxford (1977)
49. K. Metsch and L. Storme, Partial t-spreads in PG$(2t+1, q)$, *Des. Codes Cryptogr.* **18 (1–3)** (1999), 199–216
50. P. Polito and O. Polverino, On small blocking sets, *Combinatorica* **18 (1)** (1998), 133–137
51. O. Polverino, Small blocking sets in PG$(2, p^3)$, *Des. Codes Cryptogr.* **20 (3)** (2000), 319–324
52. O. Polverino, Linear sets in finite projective spaces, *Discrete Math.* **310** (2010), 3096–3107
53. O. Polverino and L. Storme, Small minimal blocking sets in PG$(2, q^3)$, *European J. Combin.* **23 (1)** (2002), 83–92
54. G. Solomon and J.J. Stiffler, Algebraically punctured cyclic codes, *Inform. and Control* **8** (1965), 170–179

55. L. Storme, Linear codes meeting the Griesmer bound, minihypers, and geometric applications, *Le Matematiche* **LIX** (2004), 367–392.
56. L. Storme, Linear codes meeting the Griesmer bound. Proceedings of the Contact Forum *Coding Theory and Cryptography*, October 7, 2005, at The Royal Flemish Academy of Belgium for Science and the Arts, Brussels, Belgium, (2006), 85–112
57. L. Storme and P. Sziklai, Linear pointsets and Rédei type k–blocking sets in $\mathrm{PG}(n,q)$, *J. Algebraic Combin.* **14 (3)** (2001), 221–228
58. P. Sziklai, On small blocking sets and their linearity, *J. Combin. Theory, Ser. A* **115 (7)** (2008), 1167–1182
59. T. Szőnyi, Blocking sets in Desarguesian affine and projective planes, *Finite Fields Appl.* **3 (3)** (1997), 187–202
60. T. Szőnyi, A. Gács and Zs. Weiner, On the spectrum of minimal blocking sets in $\mathrm{PG}(2,q)$, Combinatorics, 2002 (Maratea), *J. Geom.* **76 (no. 1-2)** (2003), 256–281
61. T. Szőnyi and Zs. Weiner, Small blocking sets in higher dimensions, *J. Combin. Theory, Ser. A* **95 (1)** (2001), 88–101
62. G. Van de Voorde, Blocking Sets in Finite Projective Spaces and Coding Theory, PhD Thesis, 2010

Quadratic Action

F.G. Timmesfeld

Abstract Let V be a vectorspace over a field of char. different from two and x be an element of $GL(V)$ different from id. Then x acts quadratically on V, if $(x - id)^2 = 0$. In this survey we study and classify subgroups of $GL(V)$ generated by quadratically acting elements.

Subject Classifications: AMS classifications: 20E32, E34 and E42

Keywords Quadratic action

1 History

Let G be a group and V a $\mathbb{Z}G$-module, i.e., an abelian group on which G acts. Then we say that the element $x \in G$ acts *quadratically* on V if $V(x - \mathrm{id})^2 = 0 \neq V(x - \mathrm{id})$. In multiplicative notation, which we prefer, the condition is

$$[V, x, x] = 1 \neq [V, x],$$

where the commutator is computed in the semidirect product of V and G. A subgroup A of G acts quadratically on V, if $C_A(V) = 1$ and $[V, A, A] = 1$.

For finite groups the concept of quadratic action is as old as modern group theory and goes back to J. Thompson. It underlies Glauberman's theorem, see [7], and also the Glauberman–Thompson normal p-complement theorem. The major classification result was announced by Thompson [12] at the ICM in Nice 1970:

F.G. Timmesfeld (✉)
Mathematisches Institut der Justus-Liebig Universitaet, 35390 Giessen, Germany
e-mail: Franz.Timmesfeld@math.uni-giessen.de

N.S.N. Sastry (ed.), *Buildings, Finite Geometries and Groups*, Springer Proceedings in Mathematics 10, DOI 10.1007/978-1-4614-0709-6_16,

Theorem 1.1. *Suppose the finite group G acts faithfully and irreducibly on the $\mathbb{Z}_p G$-module V, where $p \geq 5$ is a prime, and G is generated by quadratically acting elements. Then G is a central product of finite Lie-type groups in char. p different from $E_8(q), q = p^n$.*

The proof of this theorem still remains unpublished, although there exists a preprint containing many important features. It is obvious that, to prove this theorem, Thompson had to develop a new theory. I will mention some of its highlights.

Lemma 1.2 (SL_2-Lemma). *Let G and V be as in Theorem 1.1. Among all quadratically acting elements of G choose x such that* $\dim[V, x]$ *is minimal. Choose x^g such that $\langle x, x^g \rangle$ is not unipotent on V. Then $\langle x, x^g \rangle \simeq SL_2(q)$, q a power of p.*

The existence of x^g is guaranteed by the so called Baer–Suzuki Theorem, which says that either $x \in O_p(G)$ or there exists an x^g such that $\langle x, x^g \rangle$ is not a p-group. Now $O_p(G) = 1$, since G acts irreducibly on V.

The proof of Lemma 1.2 depends on the well known Dixon-lemma determining subgroups of $SL_2(q)$, $q = p^n$ generated by non commuting p-elements. If $p = 3$ there is an additional case in Dixon's lemma, namely $SL_2(5)$ as subgroup of $SL_2(q)$. This (and the fact that $SL_2(3)$ is solvable) prevent, that the theory of quadratic action works as well in case $p = 3$. (If $p = 2$, then all involutions of G act quadratically on V. So there cannot be such a classification.)

From Lemma 1.2 one obtains

Corollary 1.3. *Let x and x^g be as in (1) and set $E_x = \{y \in G \mid [V, y] = [V, x]$ and $C_V(y) = C_V(x)\}$ and $E(x) = \langle E_x \rangle$. Then the following hold:*

(1) $E(x) = E_x \cup \{id\}$ is an elementary abelian p-group.
(2) $\langle E(x), E(x)^g \rangle \simeq SL_2(|E(x)|)$.

It can be shown that the central involution in this $SL_2(|E(x)|)$ is a classical involution of G in the sense of Aschbachers' classical involution theorem, [1], for which he won the Cole-price in Algebra. That is, if i is this central involution, then $SL_2(|E(x)|)$ is a subnormal subgroup of $C(i)$.

If $p = 3 = |E(x)|$ some work on quadratic action has been done by Ho [8] and Chermak [4]. In particular Ho determined the groups $\langle x, x^g \rangle$ when $\langle x, x^g \rangle$ is not a 3-group.

2 Connections with Abstract Root Subgroups

If one thinks about generalizing Theorem 1.1 to arbitrary fields k with char. $k \neq 2$ and $k \neq GF(3)$ then one realizes that one needs a condition, which prevents the groups from being defined over a ring. (For example $SL_n(\mathbb{Z})$ is generated by quadratically acting elements on $\mathbb{Q}^n$.) This condition should be automatically satisfied if $k = \mathbb{Z}_p$.

Definition 2.1. Suppose V is a kG-module and $x \in G$ acts quadratically on V. Then $x = \mathrm{id} + \alpha$, $\alpha^2 = 0 \neq \alpha$. For $c \in k$ let

$$c \circ x = \mathrm{id} + c\alpha.$$

Then $k \circ x = \{c \circ x \mid c \in k\} \simeq (k, +)$ with multiplication $(c \circ x)(d \circ x) = (c+d)x$. Further, if $k = \mathbb{Z}_p$, then $k \circ x = \langle x \rangle$.

Now the appropriate condition is that G should be generated by quadratic elements x with $k \circ x \subseteq G$. With this condition we obtain:

Theorem 2.2 (Root-structure theorem). *Suppose the group G acts faithfully on the finite dimensional k-vector space V, char. $k \neq 2$ and $k \neq GF(3)$, such that the maximal unipotent normal sugbroup $R_u(G) = 1$ and G contains a quadratic element x with $k \circ x \subseteq G$. Under all such quadratic elements pick x such that* $\dim_k[V, x]$ *is minimal. Let*

$$E_x = \{y \mid k \circ y \subseteq G, [V, y] = [V, x] \text{ and } C_V(y) = C_V(x)\}$$

and $E(x) = \langle E_x \rangle$. ($y \in E_x$ automatically acts quadratically on V.) Then the following hold:

(1) $E(x) = E_x \cup \{id\}$ is either an elementary abelian p-group (if char. $k = p$) or torsionfree and divisible (if char. $k = 0$).
(2) Let $\Sigma = E(x)^G = \{E(x)^g \mid g \in G\}$. Then for $A, B \in \Sigma$ one of the following holds:

  (i) $[A, B] = 1$.
  (ii) $\langle A, B\rangle$ is nilpotent of class two with $[A, B] \in \Sigma$.
  (iii) $\langle A, B\rangle$ is a "special rank one group," i.e., for each $1 \neq a \in A$ there exists a $1 \neq b \in B$ with

$$a^b = b^{-a} \text{ and } A^b = B^a \text{ and vice versa.}$$

For proof see [14, Th.2 and Th.1] or [13, V(1.28)]. In fact, certain technical matrix calculations are not contained in [13]. Under a slightly different condition it has been shown in [6] that (2)(i) or (ii) holds, when $\langle A, B\rangle$ is unipotent on V. Unfortunately, I assumed in the statement of Th.2 in [14] that G acts irreducibly on V, which is only needed to obtain $R_u(G) = 1$.

Definition 2.3. A conjugacy class Σ of abelian subgroups generating a group R, such that for each pair $A, B \in \Sigma$ one of the cases (2)(i)–(iii) is satisfied, is called a class of *abstract root subgroups* of R.

With this notation we obtain:

Theorem 2.4 (Central-product-theorem). *Suppose G acts faithfully on the k-vector space V, k as in Theorem 2.2, such that $R_u(G) = 1$ and G is generated by quadratic elements x with $k \circ x \subseteq G$. Then $G = G_1 * \cdots * G_\ell$ ($*$ means central product) such that each G_i is generated by a class Σ_i of abstract root subgroups.*

The central-product theorem is a consequence of Theorem 2.2 and Th.1 of [14]. Again this theorem is stated in Th.1 of [14] resp. in V (1.6) of [13] only under the hypothesis that G acts irreducibly on V.

Now the problem of determining groups generated by quadratically acting elements is reduced to the determination of groups G which are generated by a class Σ of abstract root groups, and which satisfy $R_u(G) = 1$. In fact the latter problem also makes sense in char. 2, i.e., when the elements of Σ are elementary abelian 2-groups. This problem has been solved completely in [Ti1] under the additional hypothesis that Σ contains a pair of commuting root groups. The classification depends on the classification of spherical buildings by Tits resp. the classification of Moufang polygons by Tits and Weiss [20]. The condition that there exist commuting root subgroups is needed to obtain lines of certain geometries related to the building of G.

Remark 2.5. (1) Similarly as in Sect. 1, the proof of Theorem 2.2 resp. 2.4 depends on an extension of the Baer–Suzuki theorem to arbitrary linear groups and on an extension of the Dickson lemma to infinite fields. The first was given by Aschbacher in [2], see also [17, (2.1)]. The second has been obtained by Bashkirov [3] resp. unpublished by Chernousov, see also the Appendix of [13]. The result is:

Let k be an infinite field of char. $\neq 2$ and λ algebraic over k. Then

$$\left\langle \begin{pmatrix} 1 & \\ c & 1 \end{pmatrix}, \begin{pmatrix} 1 & c\lambda \\ & 1 \end{pmatrix} \middle| c \in k \right\rangle = \mathrm{SL}_2(k(\lambda)).$$

(2) Groups X generated by two different (nilpotent) subgroups A and B satisfying:

For each $1 \neq a \in A$ there exists a $1 \neq b \in B$ with $A^b = B^a$ and vice versa are called (abstract) *rank one groups*.

These rank one groups are nearly equivalent to the Moufang sets introduced by Tits [19]. This is a set Ω together with a family $(U_\alpha)_{\alpha \in \Omega}$ of subgroups of $\mathrm{Sym}(\Omega)$ satisfying:

(i) U_α fixes α and acts regularly on $\Omega - \{\alpha\}$.
(ii) U_α normalizes $\{U_\beta \mid \beta \in \Omega\}$ (as subgroup of $\mathrm{Sym}(\Omega)$.)

In fact, given a Moufang set, then the subgroup $\langle U_\alpha, U_\beta \mid \alpha \neq \beta \in \Omega \rangle$ of $\mathrm{Sym}(\Omega)$ is a rank one group with $A = U_\alpha$ and $B = U_\beta$. Conversely, if $X = \langle A, B \rangle$ is a rank one group and $\overline{X} = X/Z(X)$, then setting $\Omega = \{A^x \mid x \in X\}$ and $U_\alpha = \overline{A^x}$ for $\alpha = A^x$, Ω together with $(U_\alpha)_{\alpha \in \Omega}$ is a Moufang set. For details see [13, I(1.3)].

There is recently a large amount of literature on Moufang sets. I only mention two papers, which are of particular interest for us:

[5] Moufang sets and Jordan division algebras, where the connection between special rank one groups and Jordan Division algebras is discovered. This will be exploited in the next section.

[10] Proper Moufang sets with abelian root groups are special. In this paper it is shown, that if $X = \langle A, B\rangle$ is a rank one group with A abelian, then either $X/Z(X)$ is sharply doubly transitive on A^X or the special equation $a^b = b^{-a}$ holds, for all $a \in A$, $b \in B$ satisfying $A^b = B^a$.

3 Quadratic Action Without Commuting Root Groups

One is naturally led to ask what can be said in Theorem 2.2 about the normal subgroup $Y = \langle \Sigma \rangle$ of G, if Σ does not contain a pair of commuting root groups. If this is the case, then there does not exist a pair of elements of Σ satisfying (2)(ii) too. This means that $\langle A, B\rangle$ is a rank one group for all $A \neq B \in \Sigma$. In this case the geometric arguments of [13, 14] to classify Y don't work, since there is no geometry around.

In case G is finite, the only possibilities for the group Y are $\mathrm{SL}_2(q)$ and $SU_3(q)$. This has been shown by Thompson in his preprint, but also follows from papers of Aschbacher and Hall. In the infinite case the problem is much more substantial, since one does not only obtain unitary groups of Witt-index one of arbitrary dimension, but also algebraic groups of relative rank one like ${}^2E_{6,1}$, the root groups of which act quadratically on the 27-dimensional module for E_6.

There are the following results:

Theorem 3.1. *[15] Suppose G satisfies the hypothesis of Theorem 2.2 and $\langle A, B\rangle$ satisfies (2)(iii) for $A, B \in \Sigma$. Then $\langle A, B\rangle = SL_2(J, R)$, where $R = End_k V$ and J is a Jordan division subalgebra of R. (Here $SL_2(J, R) = \langle \left(\begin{smallmatrix} 1 & \lambda \\ & 1 \end{smallmatrix}\right), \left(\begin{smallmatrix} 1 & \\ \mu & 1 \end{smallmatrix}\right) \mid \lambda, \mu \in J\rangle$ considered as subgroup of $GL_2(R)$.)*

Actually, the theorem also holds in char. 2, see [15]. Now it is not clear how useful Theorem 3.1 is, since the group $\mathrm{SL}_2(J, R)$ might be unknown. In particular the latter group does not only depend on J, but also on R. But using Zelmanov's famous classification of Jordan division algebras one obtains the following:

Corollary 3.2. *[15] Let $Y = \langle A, B\rangle$ be as in Theorem 3.1. Then one of the following holds for $\overline{Y} = Y/Z(Y)$:*

(1) $J = K$ is a division ring and $\overline{Y} \simeq PSL_2(K)$.
(2) $J = K_\sigma$, where K is a division ring with involution σ and $\overline{Y} \simeq PSU_2(K, f)$, where $f : K^2 \times K^2 \to K$ is a σ-hermitian form of Witt-index one.
(3) J is the Jordan Clifford algebra of some anisotropic quadratic form $q : W \to \ell$, ℓ an extension field of k and $\overline{Y} \simeq P\Omega(W', q')$; where $W' = W \perp H$, H a hyperbolic plane and q' is the extension of q to W'.

I actually think that the outcome of the classification of the "quadratic rank one groups" in Corollary 3.2 is too complicated to be of much use for the classification of the higher rank groups in [13, 14]. It is the definition of "special rank one groups" which somehow picks the right properties for a classification of groups generated by abstract root subgroups.

Next, a Proposition, which connects the case where Σ does not contain different commuting root subgroups with rank one groups.

Proposition 3.3. *[16] Suppose G satisfies the hypothesis of Theorem 2.2, but there are no commuting root subgroups $A \neq B$ in Σ. Then*

$$Y = \langle \Sigma \rangle = \langle U(A), U(B) \rangle \text{ is a rank one group with respect to } U(A), U(B) \text{ for } A \neq B \in \Sigma, \text{ where}$$

$$U(A) = \{\sigma \in Y \mid [V,\sigma] \subseteq C_V(A), [C_V(A),\sigma] \subseteq [V,A] \text{ and } [V,A,\sigma] = 0\} \text{ is nilpotent of class two.}$$

In particular, Y acts doubly transitively on Σ.

In the same paper, I determined Y in case Σ does not contain commuting root subgroups and Y is generated by 3 root subgroups.

Theorem 3.4. *[16] Let G, Σ and Y be as in Theorem 2.2 and assume that Σ does not contain a pair of commuting root subgroups and Y is generated by three elements of Σ. Let J be the Jordan division algebra of Theorem 3.1. Then one of the following holds:*

(1) J is a field and Y is a 3-dimensional unitary group of Witt-index one over a quadratic extension field K of J.
(2) $J = K_\sigma$, where K is a division ring with involution σ and $Y = SU_3(K, f)$, where $f : K^3 \times K^3 \to K$ is a σ-hermitian form of Witt-index one. In particular K is the associative envelope of J and whence uniquely determined by J.

The main problem in this area is, in my view, the determination of the algebraic groups of relative rank one by quadratic action. (All such groups, except of type E_8, admit a "quadratic module.") This is a difficult problem, since there does not exist any group theoretic classification of these groups so far. By Proposition 3.3 this problem is closely related to the determination of non special Moufang sets.

I spent some time on the problem of determining Y, if it is generated by 4 root subgroups. But finally I decided that this is not the right way to proceed, since it gets more and more complicated, but one still does not obtain the more interesting groups. (For example, ${}^2E_{6,1}$ needs at least 5 root groups to generate it, since the irreducible "quadratic" module is the 27-dimensional module for E_6 and the commutator space of a root group with this module is of dimension 6.)

I recently discovered an argument which is somewhat more in the spirit of algebraic groups. Namely, I showed that under the hypothesis of Theorem 3.4 you can embed Y into some $\mathrm{SL}_3(K)$ which acts on the original quadratic module for Y. Now suppose that $Y = \langle \Sigma \rangle$ is generated by an arbitrary number of root subgroups and Σ does not contain a pair of commuting root subgroups. Pick a 3-generator subgroup Y_1 which is known by Theorem 3.4 and whence can be embedded into $\overline{Y}_1 \simeq \mathrm{SL}_3(K)$ acting on the original module V for Y. Hence, $\overline{Y} = \langle Y, \overline{Y}_1 \rangle$ acts on V, satisfies the hypothesis of Theorem 2.2 and contains commuting root subgroups (obtained from $\overline{Y}_1$). Thus, by the classification in Sect. 2 $\overline{Y}$ is known. Hence, the problem of determining Y is reduced to the problem of determining subgroups of a known group $\overline{Y}$ generated by a class $\overline{\Sigma}$ of abstract root groups, which act quadratically on some module V, which is irreducible for $\overline{Y}$ and for Y. Moreover $Y = \langle A \cap Y \mid A \in \overline{\Sigma}, A \cap Y \neq 1 \rangle$ and the set of such $A \cap Y$ is a class of K root subgroups, K a proper subfield of the field of definition of $\overline{Y}$. This could be split into a number of special problems like: Suppose $\overline{Y} = E_6(k)$ and V is one of the 27-dimensional kY-modules on which the long root groups of $\overline{Y}$ act quadratically. Let Σ be the class of long root groups of $\overline{Y}$. Suppose Y is a proper subgroup of $\overline{Y}$ acting irreducibly on V and satisfying:

> Let $\Delta = \{A \cap Y \mid A \in \Sigma \text{ and } A \cap Y \neq 1\}$. Then $Y = \langle \Delta \rangle$ and Δ is a class of ℓ root subgroups of Y, ℓ a proper subfield of k, containing no commuting root groups. Show that then k is a quadratic extension field of ℓ and $Y \simeq {}^2E_{6,1}(k)$.

4 Quadratic Modules

Let G be a finite Lie-type group in char. $p \neq 2$. Then Premet and Suprunenko [9] have determined the irreducible "quadratic" kG-modules, i.e., when an element of G acts quadratically, where k is the algebraic closure of $\mathbb{Z}_p$. It follows from the theory described in Sect. 2, that if one element of G acts quadratically, then the long root subgroups must act quadratically. (This can also be shown directly.) Now for group theoretic purposes I recently have shown:

Theorem 4.1. *[18] Let G be a finite Lie type group in char. $p \neq 2$ different from $SL_2(3)$ and V a $\mathbb{Z}_pG$-module on which the root groups of G act quadratically. Then*

$$(*) \qquad V = C_V(G) \oplus [V, G]$$

and $[V, G]$ is a direct sum of nontrivial irreducible modules in some list.

Of course, if one knows that $(*)$ holds, the list of irreducible summands of $[V, G]$ can be obtained from [9]. But the proof of $(*)$ is constructive, i.e., we have to construct the irreducible direct summands of $[V, G]$ by induction on the rank of G, knowing that by [13, I(3.4) and (3.8)(1)] the result is true for $\mathrm{SL}_2(k)$, when char. $k \neq 2$ and $k \neq GF(3)$. In fact the proof of Theorem 4.1 works, except for E_6 and E_7, for arbitrary fields k.

Unfortunately, Theorem 4.1 is not the theorem one would really like to have. As the remarks at the beginning of this section show, one would like to have the same theorem just under the hypothesis that the long root subgroups act quadratically. Unfortunately, this more general theorem is false. Counterexamples are given by the Weyl-modules with fundamental weight ω_i, $2 \leq i \leq \ell$ for $\mathrm{Sp}(2\ell, q), q = p^n$. In fact the Weyl modules are always indecomposable, see [11, p. 211]. Moreover, since they are submodules of the ith exterior power of the natural module for $\mathrm{Sp}(2\ell, q)$, the long root groups, which are transvection groups, act quadratically. But they are not always irreducible. (The Weyl-module corresponding to ω_2 is not irreducible if $p \mid \ell$.)

References

1. Aschbacher, M., A characterization of Chevalley groups over fields of odd order I,II. Ann. of Math., 106 (1977), 353–468
2. Aschbacher, M., The 27-dimensional module for E_6 IV. J. of Algebra 131 (1990), 23–39
3. Bashkirov, E. L., On subgroups of the group SL_2 over an infinite field. Manuscript is kept in All Union Institute of scientific and technical information. (VINITI) Moscow, No 7369–7384 Dep
4. Chermak, A., Quadratic pairs. J. of Algebra 277 (2004), 36 – 72.
5. DeMedts, T., Weiss, R. M., Moufang sets and Jordan division algebras. Math. Ann. 335 (2006) 415–432
6. Glauberman, G., Quadratic elements in unipotent linear groups. J. of Algebra 20 (1972), 637–654
7. Gorenstein, D., Finite groups. Harper Row, 1968
8. Ho, C., On the quadratic pairs. J. of Algebra 43 (1976), 338–353
9. Premet, A., Suprunenko, I., Quadratic modules for Chevalley groups over fields of odd characteristic. Math. Nachr. 110 (1983), 65–96
10. Segev, Y., Proper Moufang sets with abelian root groups are special. J. Amer. Math. Soc., Vol. 22 (2009), 889–908
11. Steinberg, R., Lectures on Chevalley groups. Yale Univ. (1967)
12. Thompson, J., Quadratic Pairs. Proc. of the ICM, Nice (1970)
13. Timmesfeld, F. G., Abstract root subgroups and simple groups of Lie type. Monographs in Math., Vol. 95, Birkhäuser, 2001
14. Timmesfeld, F. G., Abstract root subgroups and quadratic action. Advances in Math. 142, (1999), 1 – 150.
15. Timmesfeld, F.G., Quadratic rank one groups and quadratic Jordan division algebras. Proc. Lond. Math. Soc. 95 (2007) 156–178
16. Timmesfeld, F.G., Quadratic pairs without commuting root subgroups. J. of Alg. 318 (2007), 111–134
17. Timmesfeld, F. G., Subgroups of Lie type groups containing a unipotent radical. J. of Alg. 323 (2010), 1408–1431
18. Timmesfeld, F. G., Complete reducibility of quadratic modules for finite Lie type groups. To appear
19. Tits, J., Twin buildings and groups of Kac-Moody type. Proc. of the LMS Durham symposium, LMS Lecture Note Series 165 (1992), 249–286
20. Tits, J., Weiss, R., Moufang Polygons, Monographs in Math., Springer, 2002

Problems Set

Two Problems on Finite Sets

Bhaskar Bagchi

1. An MIF(k) is a maximal intersecting family of k-sets; i.e., it is a pairwise intersecting family of sets of size k such that at least one member of the family is disjoint from any k-set from outside the family. The elements of the union of a MIF(k) will be called its points. The number n of points will be called its size. Though it may look surprising, any MIF(k) is finite: indeed one has the bound $n \leq k^{k+1}$ on its size. (One may prove by backward induction on ℓ that for $0 \leq \ell \leq k$, any ℓ-set is contained in at most $k^{k-\ell}$ members of the MIF(k). The case $\ell = 0$ gives the bound.) An example of a MIF(k): all k-subsets of a given $(2k-1)$-set. This is clearly the smallest MIF(k).
   We shall say that a MIF(k) is *special* if each set of $k+1$ points contains either 0 or 2 members of the MIF(k).

   Problem. Prove (or disprove!) that any special MIF(k) has size $3k-3$.

   (Very few special MIF(k) are known: one each with $k = 3, 4$ and six with $k = 6$.)
2. A linear space is a pair (P, L) where P is a finite set (whose elements are called points) and L is a collection of subsets of P, each of size 2 or more (the elements of L are called the lines) such that any two distinct points are together in a unique line. For a positive integer n, we say that a linear space is of order n if each point is in exactly $n+1$ lines. For a prime p, we say that the linear space (P, L) is *non-trivial at p* if there is a function $\phi : P \to Z/pZ$, not identically zero, such that $\sum_{x \in \ell} \phi(x) = 0$ for each $\ell \in L$.

   Problem. (a) If a linear space (P, L) of order p is non-trivial at the prime p, then show that $\#(P) \geq 2p$ and equality holds iff P is the disjoint union of two lines of size p.

N.S.N. Sastry (ed.), *Buildings, Finite Geometries and Groups*, Springer Proceedings in Mathematics 10, DOI 10.1007/978-1-4614-0709-6,

(This is a theorem due to Inamdar. But we are looking for an elementary proof.)

(b) If a linear space (P, L) of odd prime order p is non-trivial at p and if $\#(P) > 2p$, then prove (or disprove!) that $\#(P) \geq 3p - 3$ and equality holds iff P is the disjoint union of three lines of size $p - 1$.
(This is known to be true for $p \leq 7$.)

Remark. (1) A pure simplicial complex of dimension d is said to be a (weak) pseudo-manifold if each of its $(d-1)$-dimensional faces is contained in exactly two d-dimensional faces (facets). It is said to satisfy *complementarity* if, for each partition $A \sqcup B$ of its vertex set into two subsets, exactly one of A, B is a face. It is known that any triangulation of a projective plane over a real division algebra of real dimension d requires at least $3d + 3$ vertices, and any triangulation attaining this lower bound must satisfy complementarity. In [1], we conjectured that any d-dimensional pseudomanifold with complementarity must have exactly $3\frac{d}{2} + 3$ vertices (in particular d must be even). It is easy to see that the complements of the facets in any pseudomanifold with complementarity form a special MIF, as defined in Problem 1 above. Thus the known examples of special MIFs mentioned above arise from the well known minimal triangulations of the projective planes over reals, complex numbers and quaternions (using 6, 9, 15 vertices respectively). A positive resolution of Problem 1 will lead to a proof of the conjecture in [1].

(2) Problem 2(b) is an easy to understand reformulation of the $3p-3$ conjecture in [2] and [3]. Problem 2(a) is a theorem from [4].

References

1. B. Bagchi and B. Datta, *Non-existence of six-dimensional pseudomanifolds with complementarity*, Advances in Geometry **4** (2004), 537–550
2. B. Bagchi, *A conjecture on linear spaces of prime order*, J. Combin. Inform. & System Sciences **34** (2009), 23–31
3. B. Bagchi, *On characterizing designs by their codes*, This volume (2011)
4. S.P. Inamdar, *Rigidity Theorems for partial linear spaces*, J. Combin. Theory (A) **96** (2001), 388–395

Embedding the Dual of a Non-Embeddable Polar Space

A. Pasini

It is well known that only one family of non-classical (i.e. non-embeddable) thick polar spaces exist of rank $n \geq 3$. They have rank 3 and their planes are moufang but not Desarguesian (see Tits [[1], Chap. 9]; also Mühlherr [2]). Each of these polar spaces can be obtained as a subgeometry of an $E_7(k)$-building for a suitable field k.

Let Δ be such a building, $\mathcal{S}$ a polar space as above realized as a subgeometry of Δ and $\mathcal{D}$ the dual of $\mathcal{S}$.

It is not difficult to see that the 56-dimensional embedding $e : \Delta \to V = V(56, K)$ induces an embedding $e_{\mathcal{D}}$ of $\mathcal{D}$ in the subspace $\langle e(\mathcal{D})\rangle$ of V spanned by the e-image $e(\mathcal{D})$ of $\mathcal{D}$. It is known that actually $e(\mathcal{D})$ spans V. I have been told by A. Cohen that this fact has been proved by Freudenthal many years ago in the context of his investigations of buildings of exceptional Lie type, although he has never published that proof. Anyway, $\langle e(\mathcal{D})\rangle = V$. Namely, $e_{\mathcal{D}}$ is an embedding of $\mathcal{D}$ in V. However, a number of questions remain to answer. Here are two of them.

1. Is $e_{\mathcal{D}}$ universal?
2. Compute the generating rank and the embedding rank of $\mathcal{D}$.

References

1. J. Tis, *Buildings of Spherical Type and Finite BN-pairs*. Lecture Notes in Mathematics 386, Springer, Berlin 1974
2. B. Mühlherr, A Geometric approach to non-embeddable polar spaces of rank 3. *Bull. Soc. Mathematique Belgique*, XLII (1990), 577–594

Linearity Conjecture on Blocking Sets in *PG*(*n*, *q*)

Leo Storme

In $PG(2, q)$:

Definition. (1) A *blocking set* B in $PG(2, q)$ is a set of points intersecting every line in at least one point.

(2) A blocking set B in $PG(2, q)$ is *minimal* when no proper subset of B still is a blocking set.

(3) A blocking set B in $PG(2, q)$ is *small* when $|B| < \frac{3(q+1)}{2}$.

Known Examples of Small Minimal Blocking Sets

Let $q = p^h$, p prime, and let $e|h$, then the projection of a subgeometry $PG(\frac{h}{e}, p^e)$ in $PG(2, q)$ is a minimal blocking set of $PG(2, q)$, called a *linear* blocking set.

(It has the properties that it intersects every line of $PG(2, q)$ in 1 (mod p^e) points and every $(1 + p^e)$-secant is a subline $PG(1, p^e)$.)

Known Results on Small Minimal Blocking Sets

Let B be a small minimal blocking set in $PG(2,q)$, $q = p^h$, p prime, $h \geq 1$, then

(1) B intersects every line in 1 (mod p) points.
(2) Let e be the largest integer such that B intersects every line in 1 (mod p^e) points. (This number is called the *exponent* of B.) Then $e|h$ and every $(1+p^e)$-secant to B is a subline $PG(1, p^e)$.

Conjucture. All small minimal blocking sets in $PG(2,q)$, $q = p^h$, p prime, $h \geq 1$, are linear blocking sets.

In $PG(n,q)$:

k-Blocking Sets in $PG(n,q)$

Definition. (1) A *k-blocking set* B in $PG(n,q)$ is a set of points intersecting every $(n-k)$-dimensional subspace of $PG(n,q)$ in at least one point.
(2) A k-blocking set B in $PG(n,q)$ is called *small* when $|B| < 3(q^k+1)/2$.
(3) The known examples of small minimal k-blocking sets in $PG(n,q)$ are again *linear*, i.e. they are (possibly) projected subgeometries of appropriate dimensions and defined over appropriate subfields $\mathbb{F}_{p^e}$ of $\mathbb{F}_q$, $q = p^h$, p prime, $h \geq 1$, $e|h$, in $PG(n,q)$.

Conjucture. Every small minimal k-blocking set in $PG(n,q)$ is a linear k-blocking set.

"Which (Families of) Finite Groups Admit Neat/Natural Gelfand Model?"

D.-N. Verma, dnverma@math.tifr.res.in

Given a group G, a *locally finite*(-dimensional) G-module M is to be called a *Gelfand model* of G iff M is semisimple ($\equiv$ direct sum of simple modules), *and every* simple G-module occurs once and exactly once in M. But, we ought to rule out the "unnatural" construction for M if it were to be *defined* as the direct sum of one copy of each G-irreducible; indeed, the quest of a Gelfand model not only presumes a lack of prior knowledge of any description of the G-irreducibles, but needs also to be supplemented with the TASK of describing how to find each of the minimal G-stable subspaces.

This definition is motivated from the (semisimple) "Lie Representation Theory", as all reductive algebraic groups in characterization 0 possess one (or more) natural

Gelfand models, which are obtained intrinsically from the given group structure and the underlying Geometry. A priori, no one seems to have ever suggested of a similar possibility for the class of all finite groups (or even finite simple groups).

Yet, there are reasons to suspect of some, or many, large classes of finite groups that may admit some (or even many) Gelfand models. At the root of this lies the relatively new awakening (over recent decades) that the vectorspace spanned by (or in some way *related to*) the set of all involutions in G is suspected to be a candidate for M; to be precise, if F is "totally real" (in that $\mathbb{R}G$ is a direct sum of matrix algebras over $\mathbb{R}$, i.e. matrix summands over complex numbers and quaternions do not occur), it is known that dim M = expect sum of dimensions of G-irreducibles, which makes this suspicion natural. While many families of finite groups are known to be totally Real, such as the class of "Weyl groups" (= crystallographic finite coxeter groups), the said suspicion is still conjectural in that generality. For the case of Symmetric Groups S_n (= Type-A_{n-1} Weyl group = the group of all $n!$ permutations of n given objects), however, this has been established during the past decade, through many independent efforts (indicated below).

The most notable reference is to the continuing work of Jorge Soto-Andriade (from Chile); full references to his early works, as also to previous relevant works of Klyachko, Inglis and Richardson's 2 page paper, and to J.L. Aguado and J. Araujo (from Argentina) can be found in the unpublished manuscript of D.-N. Verma and Vijay Kodiyalam at 2003 ArXiv. These were followed by some more general results of D. Bump and D. Ginzburg (J. Alg. 04), referred in later papers of Soto-Andrade on ArXiv, – who also gave a related talk at the ICM 2010 very recently.